D1710379

Selected Titles in This Series

94 **Mara D. Neusel and Larry Smith,** Invariant theory of finite groups, 2002
93 **Nikolai K. Nikolski,** Operators, functions, and systems: An easy reading. Volume 2: Model operators and systems, 2002
92 **Nikolai K. Nikolski,** Operators, functions, and systems: An easy reading. Volume 1: Hardy, Hankel, and Toeplitz, 2002
91 **Richard Montgomery,** A tour of subriemannian geometries, their geodesics and applications, 2002
90 **Christian Gérard and Izabella Łaba,** Multiparticle quantum scattering in constant magnetic fields, 2002
89 **Michel Ledoux,** The concentration of measure phenomenon, 2001
88 **Edward Frenkel and David Ben-Zvi,** Vertex algebras and algebraic curves, 2001
87 **Bruno Poizat,** Stable groups, 2001
86 **Stanley N. Burris,** Number theoretic density and logical limit laws, 2001
85 **V. A. Kozlov, V. G. Maz′ya, and J. Rossmann,** Spectral problems associated with corner singularities of solutions to elliptic equations, 2001
84 **László Fuchs and Luigi Salce,** Modules over non-Noetherian domains, 2001
83 **Sigurdur Helgason,** Groups and geometric analysis: Integral geometry, invariant differential operators, and spherical functions, 2000
82 **Goro Shimura,** Arithmeticity in the theory of automorphic forms, 2000
81 **Michael E. Taylor,** Tools for PDE: Pseudodifferential operators, paradifferential operators, and layer potentials, 2000
80 **Lindsay N. Childs,** Taming wild extensions: Hopf algebras and local Galois module theory, 2000
79 **Joseph A. Cima and William T. Ross,** The backward shift on the Hardy space, 2000
78 **Boris A. Kupershmidt,** KP or mKP: Noncommutative mathematics of Lagrangian, Hamiltonian, and integrable systems, 2000
77 **Fumio Hiai and Dénes Petz,** The semicircle law, free random variables and entropy, 2000
76 **Frederick P. Gardiner and Nikola Lakic,** Quasiconformal Teichmüller theory, 2000
75 **Greg Hjorth,** Classification and orbit equivalence relations, 2000
74 **Daniel W. Stroock,** An introduction to the analysis of paths on a Riemannian manifold, 2000
73 **John Locker,** Spectral theory of non-self-adjoint two-point differential operators, 2000
72 **Gerald Teschl,** Jacobi operators and completely integrable nonlinear lattices, 1999
71 **Lajos Pukánszky,** Characters of connected Lie groups, 1999
70 **Carmen Chicone and Yuri Latushkin,** Evolution semigroups in dynamical systems and differential equations, 1999
69 **C. T. C. Wall (A. A. Ranicki, Editor),** Surgery on compact manifolds, second edition, 1999
68 **David A. Cox and Sheldon Katz,** Mirror symmetry and algebraic geometry, 1999
67 **A. Borel and N. Wallach,** Continuous cohomology, discrete subgroups, and representations of reductive groups, second edition, 2000
66 **Yu. Ilyashenko and Weigu Li,** Nonlocal bifurcations, 1999
65 **Carl Faith,** Rings and things and a fine array of twentieth century associative algebra, 1999
64 **Rene A. Carmona and Boris Rozovskii, Editors,** Stochastic partial differential equations: Six perspectives, 1999
63 **Mark Hovey,** Model categories, 1999
62 **Vladimir I. Bogachev,** Gaussian measures, 1998

(Continued in the back of this publication)

Mathematical
Surveys
and
Monographs

Volume 94

Invariant Theory of Finite Groups

Mara D. Neusel
Larry Smith

American Mathematical Society

2000 *Mathematics Subject Classification.* Primary 13A50, 55S10.

ABSTRACT. This book gives a comprehensive overview of the invariant theory of finite groups acting linearly on polynomial algebras. It spans the gamut from the classical methods and results of Emmy Noether, T. Molien, D. Hilbert and L. E. Dickson to the modern methods and insights obtained from using Steenrod algebra technology, as in the proof of the Landweber-Stong conjecture. Numerous examples illustrate the theory and techniques introduced.

Library of Congress Cataloging-in-Publication Data

Neusel, Mara D., 1964–
Invariant theory of finite groups / Mara D. Neusel, Larry Smith.
p. cm. — (Mathematical surveys and monographs, ISSN 0076-5376 ; v. 94)
Includes bibliographical references and index.
ISBN 0-8218-2916-5 (alk. paper)
1. Finite groups. 2. Invariants. I. Smith, L. (Larry), 1942– II. Title. III. Mathematical surveys and monographs ; no. 94.

QA177 .N46 2001
512′.2—dc21 2001053841

Typeset by $\mathcal{AMS}$-TEX.
Printed in the United States of America.

∞ The paper used in this book is acid-free and falls within the guidelines established to ensure permanence and durability.
Visit the AMS home page at URL: http://www.ams.org/

10 9 8 7 6 5 4 3 2 1 07 06 05 04 03 02

Contents

Chapter 1
Invariants, their Relatives, and Problems

INVARIANT theory is primarily concerned with the study of group actions, their fixed points, and their orbits. The actions are usually on algebras of various sorts, the fixed points are subalgebras, and the orbits form a variety. In this book we will concentrate on *linear* actions of *finite* groups on polynomial (and related) algebras over a field.

The purpose of this chapter is to set the stage. We define the basic objects of study, establish some of the notation used throughout this book, and present some examples to illustrate concepts particular to invariant theory. With this in hand we can describe the types of problems that have concerned invariant theorists in the twentieth century and on into the twenty-first. This allows us to survey, without proofs, the topics we will cover in detail in the succeeding chapters.

We emphasize that in this chapter we will only try to indicate in general terms how the problems we present have been solved, or partially solved, without going into details, or proofs. These can be found either in the later chapters of this book or, where the topic is inappropriate for us, in the references to the literature provided. Many interesting questions arise from the solutions to the problems discussed in this chapter, and will be presented later in the book. By no means can we mention all the problems of interest to invariant theorists, or the invariant-theoretic problems of interest to other specialists: That alone would take a book in itself; see, e.g., [297].

Many of the theorems in commutative algebra (e.g., Hilbert's basis theorem, Hilbert's Nullstellensatz, Hilbert's syzygy theorem, the Lasker-Noether theorem, and many, many more) were originally discovered in connection with problems in invariant theory. For the convenience of the reader we have collected in Appendix A the basic definitions and theorems from commutative algebra of which we make extensive use. Depending on what seemed appropriate to us, i.e., did we have something new to say, or not, we have included our own proofs, or referred to the literature. This appendix is not meant to be a course in commutative algebra, so we have assumed that the reader has at least a passing familiarity with most of these results: It also serves to indicate what is needed in the way of prerequisites to read this book, and to establish many of the notational conventions from commutative algebra that we use.

Invariant theory draws upon the representation theory of finite groups, which provides the basis for many problems in invariant theory. In representation theory there is a strong dichotomy between the nonmodular and modular cases. In the **nonmodular case** the characteristic of the ground field $\mathbb{F}$ is either 0 or a prime integer p that does not divide the order $|G|$ of the group G. In this case the representation theory of G looks much as it does over $\mathbb{C}$, the complex field, at least if $\mathbb{F}$ contains sufficiently many roots of unity. This means that character theory can be used to advantage, and in many cases it provides complete information about the finite-dimensional representations of G. In the **modular case**, i.e., where the characteristic of the ground field is a prime integer dividing the order of the group, things are much less clear. Something similar happens in invariant theory, and, as we will see, the dichotomy between the modular and nonmodular cases becomes even more extreme from the invariant theorist's viewpoint. In fact, many of the most interesting problems in the invariant theory of finite groups occur *only* in the modular case, and have been the focus of much recent research.

Classical invariant theory was concerned primarily with the real and complex fields $\mathbb{R}$ and $\mathbb{C}$. The case of finite fields (also referred to as Galois fields) has become important in recent years for applications, particularly to algebraic topology, and will play a prominent role in our exposition. We will use the notation $\mathbb{F}_q$ to denote the finite field with $q = p^{\nu}$ elements, where p is a prime integer. It is in the case of finite groups over finite fields that a number of new ideas emanating from algebraic topology have enriched invariant theory and provided new tools: these tools are new in the sense that they were unknown before ca. 1940 and were not used in invariant theory until ca. 1980. Much work still needs to be done in the modular case to bring order to the current state of knowledge, and we will try to indicate problems that might serve as guides for future research.

1.1 Polynomial Invariants of Linear Groups

To begin, let V be a finite-dimensional vector space over the field $\mathbb{F}$, which will remain the fixed ground field in the discussion. We denote by $\mathbb{F}[V]$ the **graded algebra of polynomial functions**[1] **on** V, which is defined to be the symmetric algebra on V^*, the dual of V. In other words, the

[1] We will have little occasion to work with nonhomogeneous polynomials in $\mathbb{F}[V]$. To

homogeneous component of $\mathbb{F}[V]$ of degree m, denoted by $\mathbb{F}[V]_m$, is $S^m(V^*)$, the mth symmetric power of V^*; see Appendix A, Section A.1, for a discussion of gradings. With this grading convention linear forms have degree one.

It is worthwhile emphasizing here that $S^m(V^*)$ may not be what a representation theorist might think it to be: Namely, it is not the subspace of the tensor product $\bigotimes_m V^* = \underbrace{V^* \otimes \cdots \otimes V^*}_{m}$ spanned by the symmetric tensors.[2] Rather it is the quotient of $\bigotimes_m V^*$ by the subspace spanned by all elements of the form $w_1 \otimes \cdots \otimes w_m - w_{\sigma(1)} \otimes \cdots \otimes w_{\sigma(m)}$ as σ ranges over all permutations of $\{1, \ldots, m\}$.

If $z_1, \ldots, z_n \in V^*$ is a basis, we also denote $\mathbb{F}[V]$ by $\mathbb{F}[z_1, \ldots, z_n]$. The elements of $\mathbb{F}[z_1, \ldots, z_n]$ are just homogeneous polynomials in the linear forms $z_1, \ldots, z_n$ with coefficients in $\mathbb{F}$. A **monomial** is an element of $\mathbb{F}[V] = \mathbb{F}[z_1, \ldots, z_n]$ that may be written as a product of the basis elements for V^*, so after rearranging terms, it has the form $z_1^{k_1} \cdots z_n^{k_n}$. The **degree** of this monomial is $k = k_1 + \cdots + k_n$, and every homogeneous form f of degree k can be written in a unique way as a linear combination of monomials, viz.,

$$f = \sum_{|K|=k} a_K(f) z^K, \qquad a_K(f) \in \mathbb{F}.$$

Here we have introduced the **multi-index** $K = (k_1, \ldots, k_n)$ and the notation $z^K = z_1^{k_1} \cdots z_n^{k_n}$ for monomials, as well as $|K| = k_1 + \cdots + k_n$ for the degree of z^K. A **term** is a monomial with a nonzero coefficient $a \in \mathbb{F}$. Thus any homogeneous form is a unique sum of terms.

Note that the functor

$$\mathbb{F}[-] : \mathcal{V}ect_{\mathbb{F}} \rightsquigarrow \mathcal{A}lg_{\mathbb{F}}$$

$\mathcal{V}ect_{\mathbb{F}}$ denotes the category of finite-dimensional vector spaces over $\mathbb{F}$ and $\mathcal{A}lg_{\mathbb{F}}$ that of graded connected algebras over $\mathbb{F}$.

is contravariant. This historical accident is fortuitous for the applications of invariant theory in algebraic topology and to the cohomology of groups.

It is often convenient to think of forms as being functions. The naive way of doing this would be to define

$$f(v) = \sum_{|K|=k} a_K(f) z^K(v) = \sum_{|K|=k} a_K(f) z_1(v)^{k_1} \cdots z_n(v)^{k_n} \in \mathbb{F},$$

for $v \in V$. This is fine, but unfortunately, the map $\mathbb{F}[V] \longrightarrow \mathrm{Fun}(V, \mathbb{F})$ need not be injective, i.e., V may not contain enough points to separate

---

emphasize the homogeneity we refer to the elements of $\mathbb{F}[V]$ as **forms**, as in the classical literature.

[2] The natural product on the graded module of symmetric tensors is the *shuffle product*, see, e.g., [232] page 243, and leads to a a *divided power algebra* in nonzero characteristic; see, e.g., [66] Exposé 7.

distinct polynomials into distinct functions. For example, if $\mathbb{F}$ is a finite field with p elements, then the polynomials $z, z^p \in \mathbb{F}[z]$ are not separated by their values on the points of $V = \mathbb{F}^1$. However, if $\mathbb{F}$ has p^2 elements, then they are separated,[3] but this time, z cannot be separated from z^{p^2}. It is pretty clear how to remedy this situation: We allow the functions to be evaluated over a larger field. Essentially, this works because a homogeneous polynomial function $\mathbb{F} \longrightarrow \mathbb{F}$ of degree d can have at most d roots. If $\mathbb{F}$ is large enough, then there will be points in $\mathbb{F}$ where any nonzero polynomial f has a nonzero value, so we can separate two distinct polynomials by evaluating them at an appropriate point.

To be more formal, let $\overline{\mathbb{F}}$ be an algebraic closure of $\mathbb{F}$ and L a 1-dimensional vector space over $\mathbb{F}$ with $z \in L^*$ a basis element. Then the preceding discussion shows that the evaluation map, viz., $\varepsilon : \mathbb{F}[L] = \mathbb{F}[z] \longrightarrow \mathrm{Fun}(L \otimes_{\mathbb{F}} \overline{\mathbb{F}}, \overline{\mathbb{F}})$, is a monomorphism. If V is any finite-dimensional vector space over $\mathbb{F}$ and $z_1, \ldots, z_n \in V^*$ is a basis, then $\mathbb{F}[V] = \mathbb{F}[z_1] \otimes \cdots \otimes \mathbb{F}[z_n]$. Since taking tensor products over a field is exact, the composition ε defined by the commutative square (the $=\!=\!=$ is a stretched equality sign)

$$\begin{array}{ccc} \mathbb{F}[V] & =\!=\!=\!=\!=\!= & \mathbb{F}[z_1] \otimes \cdots \otimes \mathbb{F}[z_n] \\ \downarrow{\scriptstyle \varepsilon} & & \downarrow \\ \mathrm{Fun}(V \otimes_{\mathbb{F}} \overline{\mathbb{F}}, \overline{\mathbb{F}}) & \longleftarrow & \underbrace{\mathrm{Fun}(L_1 \otimes_{\mathbb{F}} \overline{\mathbb{F}}, \overline{\mathbb{F}}) \otimes \cdots \otimes \mathrm{Fun}(L_n \otimes_{\mathbb{F}} \overline{\mathbb{F}}, \overline{\mathbb{F}})}_{\longleftarrow n \longrightarrow} \end{array}$$

is a monomorphism, where $L_i = \mathrm{Span}_{\mathbb{F}}(z_i) \subseteq V^*$ for $i = 1, \ldots, n$.

Thus, we may think of the elements of $\mathbb{F}[V]$ as functions; the **polynomial functions** on $\overline{V} = V \otimes_{\mathbb{F}} \overline{\mathbb{F}}$ defined over $\mathbb{F}$. In this way we have identified $\mathbb{F}[V]$ with the subalgebra of the algebra of all functions from $\overline{V}$ to $\overline{\mathbb{F}}$ generated by the linear forms defined over V, i.e., obtained from V^* by extension of scalars from $\mathbb{F}$ to $\overline{\mathbb{F}}$.

If G is a finite group, $\mathbb{F}$ a field, and $\rho : G \longrightarrow \mathrm{GL}(n, \mathbb{F})$ a representation of G over $\mathbb{F}$, then, via ρ, G acts on the left of the vector space $V = \mathbb{F}^n$. A central theme in invariant theory is the study of the induced action on the algebra of polynomial functions $\mathbb{F}[V]$ on V. This action arises from the left action of G on V^* defined by[4] $(g \cdot z)(v) = z(\rho(g)^{-1} \cdot v)$ for $g \in G$, $z \in V^*$, and $v \in V$, and its extensions to $S^m(V^*)$, $m \in \mathbb{N}_0$, which fit together to give a left G-action on $\mathbb{F}[V]$ by algebra automorphisms. By definition, the **ring**, or **algebra, of invariants**, denoted by $\mathbb{F}[V]^G$, is the fixed subalgebra, i.e.,

$$\mathbb{F}[V]^G = \{f \in \mathbb{F}[V] \mid g \cdot f = f \ \forall \, g \in G\}.$$

---

[3] Since $a^p = a$ precisely for those elements in $\mathbb{F}$ that lie in the prime subfield $\mathbb{F}_p \subsetneq \mathbb{F}$.

[4] In the sequel, we will adopt the viewpoint that a pair of parentheses that contribute nothing to the understanding, and might detract from the legibility, will simply be left out. So we would write $g \cdot z(v)$ instead of $(g \cdot z)(v)$. A similar remark applies to excessive $\cdot$s.

Note that we should really indicate the representation ρ in the notation for the ring of invariants $\mathbb{F}[V]^G$, since it depends on ρ and not just G. However, no confusion ought to arise as to which representation is meant in any particular circumstance. If confusion might arise, we write $\mathbb{F}[V]^{\rho(G)}$, because of course $\mathbb{F}[V]^G$ really depends on the image of G under ρ in $GL(n, \mathbb{F})$. Indeed, this being the case, we see that for the sake of invariant theory we may as well suppose that ρ is faithful, for otherwise we could replace G by $\rho(G)$ without changing the ring of invariants. To indicate that ρ is assumed faithful we use the notation $\hookrightarrow$ instead of $\longrightarrow$, i.e., we write $\rho : G \hookrightarrow GL(n, \mathbb{F})$. This will allow us to make statements about the behavior of the invariants with respect to the order $|G|$ of G that otherwise would require unnecessary qualification.

If we think of $\mathbb{F}[V]$ as an algebra of functions as described above, then the action of G on $\mathbb{F}[V]$ is given by the requirement

$$g \cdot f(v \otimes_{\mathbb{F}} \bar{a}) = f(\rho(g)^{-1}v) \cdot \bar{a} \qquad \forall\, v \in V, g \in G, \bar{a} \in \overline{\mathbb{F}}.$$

Composing ρ with the inversion map of G is introduced so that G still acts on the left of $\mathbb{F}[V]$.

EXAMPLE 1 (Symmetric Polynomials): Let the **symmetric group** Σ_n act via its defining representation as a permutation group of $x_1, \ldots, x_n$. As is well known ([372] Theorem 1.2.3),

$$\mathbb{F}[x_1, \ldots, x_n]^{\Sigma_n} = \mathbb{F}[e_1, \ldots, e_n],$$

where $e_1, \ldots, e_n$ are the elementary symmetric polynomials in $x_1, \ldots, x_n$.

EXAMPLE 2 (Alternating Polynomials): The invariants of the **alternating group** A_n in its tautological representation are probably less familiar. $\mathbb{F}[x_1, \ldots, x_n]^{A_n}$ is generated as an algebra by $e_1, \ldots, e_n$, and the polynomial ∇ obtained by summing all the elements of the A_n-orbit of the monomial $x_1^1 \cdot x_2^2 \cdots x_{n-1}^{n-1}$. These polynomials are not algebraically independent: ∇ is a root of a quadratic polynomial with coefficients in $e_1, \ldots, e_n$. If the characteristic of $\mathbb{F}$ is not 2, we could replace ∇ by Δ, the discriminant (see [372] Section 1.3), and Δ^2 is also a polynomial in $e_1, \ldots, e_n$. The homogeneous polynomials in $\mathbb{F}[x_1, \ldots, x_n]^{A_n}$ are sums of symmetric and **sign-symmetric** polynomials, which are polynomials f such that

$$f(x_{\sigma(1)}, \ldots, x_{\sigma(n)}) = (-1)^{\mathrm{sgn}(\sigma)} f(x_1, \ldots, x_n),$$

where $\mathrm{sgn}(\sigma)$ denotes the *signum*, or *sign*, of the permutation $\sigma \in \Sigma_n$.

The sign-symmetric polynomials just introduced are a special case of a more general construction, the relative invariants. These may be defined as follows: Let $\rho : G \hookrightarrow GL(n, \mathbb{F})$ be a representation of a finite group over the field $\mathbb{F}$ and $\chi : G \longrightarrow \mathbb{F}^{\times}$ a homomorphism, i.e., a 1-dimensional representation of G. The χ**-relative invariants** $\mathbb{F}[V]^G_\chi$ are defined by

$$\mathbb{F}[V]^G_\chi = \{f \in \mathbb{F}[V] \mid gf = \chi(g) \cdot f \ \forall\, g \in G\}.$$

The χ-relative invariants are a module over $\mathbb{F}[V]^G$. Using this terminology we could interpret the previous example for fields of characteristic not 2 as saying

$$\mathbb{F}[x_1, \ldots, x_n]^{A_n} \cong \mathbb{F}[x_1, \ldots, x_n]^{\Sigma_n} \oplus \mathbb{F}[x_1, \ldots, x_n]^{\Sigma_n}_{\mathrm{sgn}},$$

Moreover, $\mathbb{F}[x_1, \ldots, x_n]^{\Sigma_n}_{\mathrm{sgn}}$ is a free module over $\mathbb{F}[x_1, \ldots, x_n]^{\Sigma_n}$ on the single generator Δ. Relative invariants have been studied in, e.g., [275], [321], and [399].

More generally, for any representation $\vartheta : G \longrightarrow \mathrm{GL}(k, \mathbb{F})$ we may define an $\mathbb{F}[V]^G$ module depending on ϑ in the following way. Denote by $\mathbb{F}(G)$ the group algebra of G over $\mathbb{F}$. If W denotes the G-vector space corresponding to the representation $\vartheta : G \longrightarrow \mathrm{GL}(k, \mathbb{F})$ and V to $\rho : G \hookrightarrow \mathrm{GL}(n, \mathbb{F})$, then one could define

$$\mathbb{F}[V]^G_\vartheta = \mathrm{Hom}_{\mathbb{F}(G)}(W, \mathbb{F}[V]),$$

which is a module over $\mathbb{F}[V]^G$; see, e.g., [345] Sections 2.6, 2.7 and 8.1, and [372] Section 1.2. In the classical case $\mathbb{F} = \mathbb{C}$, the representation $\vartheta : G \longrightarrow \mathrm{GL}(k, \mathbb{C})$ is an irreducible representation of G. Then Schur's Lemma implies that the degree m component of $\mathbb{C}[V]^G_\vartheta$ is the largest G-stable subspace[5] of $\mathbb{C}[V]_m$ that decomposes into a direct sum of copies of ϑ; see [345] Section 2.7 Proposition 8. This is called the ϑ**-isotypic component** of $\mathbb{C}[V]_m$, and the number of summands, $\mathrm{mult}_{\mathbb{C}[V]_m}(\vartheta)$, is the **multiplicity** of ϑ in $\mathbb{C}[V]_m$.

It is not hard to prove in general that the evaluation map is an isomorphism of G-modules

$$(\mathbb{F}[V]^G_\vartheta)_m \cong \mathrm{Hom}_{\mathbb{F}(G)}(W, \mathbb{F}[V]_m) \otimes W,$$

where $\mathrm{Hom}_{\mathbb{F}(G)}(W, \mathbb{F}[V]_m)$, is called the **module of covariants of type** ϑ **of degree** m. From this it follows that in the classical case

$$\mathrm{mult}_{\mathbb{C}[V]_m}(\vartheta) = \dim_{\mathbb{C}}\left(\mathrm{Hom}_{\mathbb{C}(G)}(W, \mathbb{C}[V]_m)\right),$$

and the multiplicity can be computed from the character of ϑ; see, e.g., [401] Theorem 2.1. We refer to Sections 3.1.4 and 12.1.1 of [141] for a more detailed discussion. See also [69] and [209].

The special case of finding the decomposition of the homogeneous components $\mathbb{F}[V]_k$ as a $\mathrm{GL}(n, \mathbb{F})$-module received much attention in the classical literature, see, e.g., [451], [173], and the references there. In the modular case, particularly over finite fields, this problem becomes two problems: to

[5] If U is a G-module, a subspace is called G**-stable** if it is mapped into itself by all the elements of G.

find the decomposition into indecomposables and to find the composition factors of the homogeneous components of $\mathbb{F}[V]$ as a $GL(n, \mathbb{F})$-module. Both these problems have attracted quite a bit of recent attention, see, e.g., [133], which found unusual applications in [60], as well as [439] and [460] for composition factors, and [192], [420] for information on the decomposition into indecomposable modules.

Even for very small groups a number of surprisingly hard problems can occur. The cyclic groups of orders 2 and 3 fall under the symmetric and alternating groups as the first nontrivial cases, so their invariants for the representation where they act by cyclic permutation are special cases of Examples 1 and 2. The next largest cyclic group, $\mathbb{Z}/4$, already serves to illustrate how easy it is to formulate problems that can be quite hard to solve. It shows also how strongly the situation changes when we leave the nonmodular case and enter the modular world.

EXAMPLE 3 (M.-J. Bertin [36]: The Regular Representation of $\mathbb{Z}/4$): The group $\mathbb{Z}/4$ acts on the set $X = \{x_1, x_2, x_3, x_4\}$ by cyclic permutation. For any field $\mathbb{F}$ we can consider the corresponding linearization of this permutation representation: This is the regular representation of the group $\mathbb{Z}/4$ over the field $\mathbb{F}$.

If $\mathbb{F}$ has characteristic different from 2 it is relatively straightforward to find a set of algebra generators for $\mathbb{F}[x_1, x_2, x_3, x_4]^{\mathbb{Z}/4}$; see, e.g., Lemma 7.3.5. In particular, in all these cases, $\mathbb{F}[x_1, x_2, x_3, x_4]^{\mathbb{Z}/4}$ is generated as an algebra by forms of degree at most 4, the group order. By contrast, if $\mathbb{F}$ has characteristic 2, then the polynomial of degree 5

$$x_1^3x_2^2 + x_2^3x_3^2 + x_3^3x_4^2 + x_4^3x_1^2$$

is not only invariant, it is indecomposable, [292], so must, up to decomposables, be part of any minimal set of algebra generators.

Furthermore, the elementary symmetric polynomials e_1, e_2, e_3, e_4 belong to $\mathbb{F}[x_1, x_2, x_3, x_4]^{\mathbb{Z}/4}$, and hence $\mathbb{F}[e_1, e_2, e_3, e_4] \subset \mathbb{F}[x_1, x_2, x_3, x_4]^{\mathbb{Z}/4}$. If $\mathbb{F}$ has characteristic unequal to 2, then $\mathbb{F}[x_1, x_2, x_3, x_4]^{\mathbb{Z}/4}$ is easy to describe as a module over $\mathbb{F}[e_1, e_2, e_3, e_4]$: It is a free module with 6 generators; see, e.g., Theorem 5.5.2 and Proposition 3.3.2. In other words, the ring of invariants is Cohen-Macaulay. However, if $\mathbb{F}$ has characteristic 2, then this is no longer the case, i.e., $\mathbb{F}[x_1, x_2, x_3, x_4]^{\mathbb{Z}/4}$ is not free as an $\mathbb{F}[e_1, e_2, e_3, e_4]$-module (see Section 5.8, Example 1): It has projective dimension 1, [36]. This example was placed in a wider context in [124] and [105]. In fact, this is an example of minimal dimension of a ring of invariants of a finite group that is not Cohen-Macaulay; see [369]. An example with minimal group order, namely $\mathbb{Z}/2$ in characteristic 2, is discussed in Section 5.5, Example 2. This, however, needs dimension 6 to be realized (cf. Problem 4 in Section 1.4).

We present problems in connection with degree bounds on generators, such as those encountered in the preceding example, in more detail in Section 1.4, in particular Problem 2, and in Section 2.3.

Example 3 also illustrates a bit of the strong dichotomy between the modular and nonmodular cases of invariant-theoretic problems. Indeed, as soon as we consider modular, as opposed to nonmodular, representations, we lose many of the good properties of rings of invariants, and the situation becomes more delicate, and more interesting.

An exception to this rule of thumb is the class of permutation representations: For such representations many statements remain true in arbitrary characteristic, because they depend on counting algorithms. See, for example, the discussion of Molien's theorem (Propositions 3.2.2 and 3.2.5) for permutation representations in Section 3.2, and Göbel's theorem (Theorem 3.4.2 and Corollary 3.4.3) on degree bounds in Section 3.4. This is also an indication that permutation representations are best classified under the *combinatorial* aspects of invariant theory, as opposed to the algebraic ones.

1.2 Coinvariants and Stable Invariants

In addition to the algebra of invariants $\mathbb{F}[V]^G$ and the modules of relative invariants $\mathbb{F}[V]^G_\chi$, another basic object of study in invariant theory is the **algebra of coinvariants**.

DEFINITION: *Let* $\rho : G \hookrightarrow \mathrm{GL}(n, \mathbb{F})$ *be a representation of a finite group* G *over the field* $\mathbb{F}$. *The* **ring,** *or* **algebra, of coinvariants,** *denoted by* $\mathbb{F}[V]_G$, *is the quotient of* $\mathbb{F}[V]$ *by the ideal generated by the invariant polynomials of positive degree.*

In other words, let

$$\overline{\mathbb{F}[V]^G} = \{f \in \mathbb{F}[V]^G \mid \deg(f) > 0\}$$

be the **augmentation ideal** of $\mathbb{F}[V]^G$. This is the unique graded maximal ideal in $\mathbb{F}[V]^G$, see Appendix A. Then the coinvariants are defined by

$$\mathbb{F}[V]_G = \mathbb{F}[V] \Big/ \left(\overline{\mathbb{F}[V]^G}\right)^{\mathrm{e}}.$$

Here I^{e}, or $(I)^{\mathrm{e}}$, denotes the ideal in $\mathbb{F}[V]$ obtained by extending[6] the ideal I from $\mathbb{F}[V]^G$ to $\mathbb{F}[V]$, i.e., I^{e} is the ideal in $\mathbb{F}[V]$ generated by the elements of I. Using tensor products, this may be written as

$$\mathbb{F}[V]_G = \mathbb{F} \otimes_{\mathbb{F}[V]^G} \mathbb{F}[V],$$

where $\mathbb{F}[V]^G$ acts on $\mathbb{F}[V]$ by multiplication of polynomials, and on $\mathbb{F}$

[6] The ^{e} stands for the word *extended*, and is not an exponent.

via the **augmentation homomorphism**

$$\varepsilon : \mathbb{F}[V]^G \longrightarrow \mathbb{F}, \qquad \varepsilon(f) = \begin{cases} f & \text{if } f \text{ has degree } 0, \\ 0 & \text{if } f \text{ has positive degree,} \end{cases}$$

whose kernel is the augmentation ideal. In the definition of ε we identify homogeneous polynomials of degree 0 with elements of $\mathbb{F}$.

It will be convenient to have a name for the ideal $\left(\overline{\mathbb{F}[V]^G}\right)^{\mathrm{e}} \subseteq \mathbb{F}[V]$.

DEFINITION: *Let $\rho : G \hookrightarrow \mathrm{GL}(n, \mathbb{F})$ be a representation of a finite group over the field $\mathbb{F}$. The* **Hilbert ideal** *of ρ is the ideal $\mathfrak{h}(\rho)$ in $\mathbb{F}[V]$ generated by the G-invariant forms of positive degree. If ρ is clear from context, we write $\mathfrak{h}(G)$ instead.*

The Hilbert ideal appears implicitly in D. Hilbert's paper [163] (n.b. Emmy Noether had not yet organized ideal theory [302]). We have that $\mathbb{F}[V]_G = \mathbb{F}[V]/\mathfrak{h}(\rho)$. Note that the action of G on $\mathbb{F}[V]$ stabilizes the ideal $\mathfrak{h}(\rho) \subseteq \mathbb{F}[V]$, but it does not fix it pointwise. The group G also acts on the quotient algebra $\mathbb{F}[V]_G$, and as we shall see, this is a graded finite-dimensional representation of G (Corollary 2.1.6), which is often quite difficult to determine.

Already in the examples of the symmetric and alternating groups the structure of the ring $\mathbb{F}[V]_G$ is not so apparent.

EXAMPLE 1: Consider the tautological representation of Σ_3 on $V = \mathbb{F}^3$. If $\{x, y, z\}$ is the standard basis for V^*, which is permuted by Σ_3, then

$$\mathbb{F}[x, y, z]^{\Sigma_3} = \mathbb{F}[x + y + z, xy + yz + zx, xyz],$$

and the ring of coinvariants is

$$\mathbb{F}[x, y, z]_{\Sigma_3} = \mathbb{F}[x, y, z]/(x + y + z, xy + yz + zx, xyz).$$

The relation $x + y + z = 0 \in \mathbb{F}[x, y, z]_{\Sigma_3}$ allows us to express z in terms of x and y, and doing so, we obtain the isomorphism

$$\mathbb{F}[x, y, z]_{\Sigma_3} \cong \mathbb{F}[x, y]/(x^2 + xy + y^2, x^2y + xy^2).$$

This algebra is finite-dimensional, since $\mathbb{F}[x, y, z]$ is a finitely generated $\mathbb{F}[e_1, e_2, e_3]$-module. One way to visualize this ring of coinvariants is with the aid of Diagram 1.2.1, where the nodes on a horizontal level indicate basis vectors for the homogeneous component of the algebra of coinvariants whose degree is equal to the height of the node above the node labeled 1, which has degree zero. Since

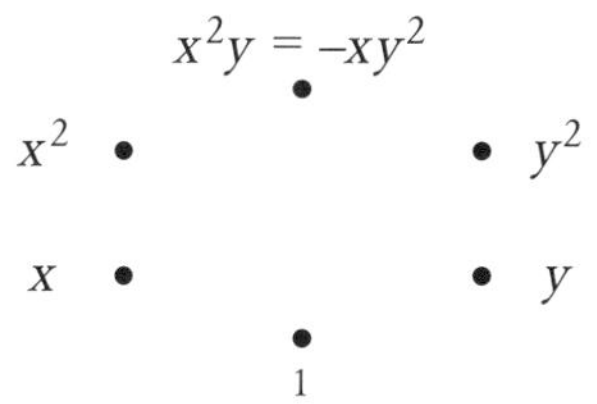

$x^2y = -xy^2$

x^2 • • y^2

x • • y

1

DIAGRAM 1.2.1: $\mathbb{F}[x, y, z]_{\Sigma_3}$

$\mathbb{F}[x, y, z]_{\Sigma_3}$ is a finite-dimensional representation of Σ_3, we might ask for a description of it in representation-theoretic terms.

If $\mathbb{F}$ is a field of characteristic different from 2 or 3, then it is the regular representation[7] of Σ_3: In degree 0 we have the trivial 1-dimensional representation, the irreducible 2-dimensional representation occurs twice, as the homogeneous components of degrees 1 and 2; and the determinant representation appears in degree 3. Note how the grading has nicely separated out the irreducible representations of Σ_3, and that in any case, an irreducible representation must be concentrated in a single homogeneous degree, for if it had more than one nonzero homogeneous piece, it would have a nontrivial direct sum decomposition. If the characteristic of $\mathbb{F}$ is 2 or 3, then this cannot be the regular representation! To see this, note that being graded, the homogeneous component of degree 0, which is just the 1-dimensional trivial representation, is a direct summand. If $\mathbb{F}[x, y, z]_{\Sigma_3}$ were the regular representation, then the trivial representation, being a direct summand, would have to be projective. It is of course not projective[8] in this case, so $\mathbb{F}[x, y, z]_{\Sigma_3}$ is not the regular representation.

Again, this illustrates how the modular and nonmodular cases can differ. Indeed, we can exploit this difference to introduce another algebra of invariants, the algebra of stable coinvariants.

Let $\rho : G \hookrightarrow GL(n, \mathbb{F})$ be a representation of a finite group G over the field $\mathbb{F}$. Then, as we noted above, G acts by algebra automorphisms on $\mathbb{F}[V]_G$, and so we could consider the invariant subalgebra $(\mathbb{F}[V]_G)^G$, and the quotient of $\mathbb{F}[V]_G$ by the extended ideal $\left((\overline{\mathbb{F}[V]_G)^G}\right)^{\mathrm{e}} \subset \mathbb{F}[V]_G$. This is a process that lends itself to repetition.

To formalize this process, consider a positively graded connected algebra A over a field $\mathbb{F}$ and G a group of algebra automorphisms of A. Denote by $A^G \subseteq A$ the subalgebra of G-invariant elements and set $A_G = \mathbb{F} \otimes_{A^G} A$, where A is regarded as an A^G-module via the inclusion, and $\mathbb{F}$ via the augmentation. This is just the algebra of coinvariants in a more general setting. The kernel of the quotient map $A \longrightarrow A_G$ is stable under the action of the group G on A, and hence A_G inherits a G-action from A. We may therefore iterate the construction of the coinvariants and inductively define

---

[7] This is a special case of a theorem of C. Chevalley, Theorem 7.2.1.

[8] This is a consequence of the following lemma:

LEMMA: *Let G be a finite group and $\mathbb{F}$ a field. If $\mathbb{F}$, regarded as a trivial G-module, is projective, then $|G| \in \mathbb{F}^\times$.*

PROOF: Let $\varepsilon : \mathbb{F}(G) \longrightarrow \mathbb{F}$ be the augmentation homomorphism of the group algebra $\mathbb{F}(G)$. If $\mathbb{F}$ is projective, then this splits. Let $\mathbb{F}(G) \underset{\varepsilon}{\overset{\sigma}{\leftrightarrows}} \mathbb{F}$ be such a splitting. Then $x = \sigma(1)$ lies in the fixed-point set $\mathbb{F}(G)^G$, which is the 1-dimensional subspace spanned by $\mathfrak{S}$, the sum of the group elements. Hence $x = \lambda \cdot \mathfrak{S}$ for some $\lambda \in \mathbb{F}$. If we apply ε to this equation, we get $1 = \varepsilon(x) = \lambda \cdot \varepsilon(\mathfrak{S}) = \lambda \cdot |G|$, which shows that $|G|$ is invertible in $\mathbb{F}$. □

A_{G^m} for $m \in \mathbb{N}_0$ by $A_{G^m} = (A_{G^{m-1}})_G$. (Of course, A_{G^0} is just A, and A_{G^1} is A_G.) Denote by $\mathcal{I}_m \subset A$ the kernel of the natural map $A \longrightarrow A_{G^m}$. The ideals $\mathcal{I}_m$, $m \in \mathbb{N}_0$, form an ascending chain

$$(0) = \mathcal{I}_0 \subseteq \mathcal{I}_1 \subseteq \mathcal{I}_2 \subseteq \cdots \subseteq \mathcal{I}_m \subseteq \cdots \subset A,$$

and we set $\mathcal{I}_\infty = \bigcup_{m \in \mathbb{N}_0} \mathcal{I}_m$, which is called the **ideal of stable invariants**. The quotient algebra $A_{G_\infty} = A/\mathcal{I}_\infty$ is the **algebra of stable coinvariants**. Alternatively, the ideals $\mathcal{I}_m$, $m \in \mathbb{N}_0$, may be defined inductively[9] by

$$\mathcal{I}_m = \begin{cases} (0) & \text{for } m = 0, \\ \left(\left\{a \in A \mid ga - a \in \mathcal{I}_{m-1}\right\}\right) & \text{for } m > 0. \end{cases}$$

If B is a graded connected algebra over $\mathbb{F}$ on which G acts, we say that G acts **fixed point freely** on B if $B^G = \mathbb{F}$, i.e., the action of G on the homogeneous component B_m of B of degree m has only $0 \in B_m$ as a fixed point for all $m > 0$ and fixes $B_0 = \mathbb{F}$. The action of G on $A/\mathcal{I}_\infty = A_{G^\infty}$ is fixed point free in this sense. The following lemma from [285] is easily proven and shows how this property characterizes $\mathcal{I}_\infty$.

LEMMA 1.2.1: *Let A and B be positively graded commutative connected algebras over a field $\mathbb{F}$ on which the group G acts by algebra automorphisms. Assume that the action of G on B is fixed-point free. Then the quotient map $q : A \longrightarrow A_{G^\infty}$ has the following universal property: For any homomorphism $\varphi : A \longrightarrow B$ of algebras commuting with the G-action, there exists a unique algebra homomorphism $\tilde{\varphi} : A_{G^\infty} \longrightarrow B$ making the triangle commute.* □

$$\begin{array}{ccc} A & \xrightarrow{q} & A_{G_\infty} \\ {\scriptstyle\varphi}\downarrow & \swarrow {\scriptstyle\tilde{\varphi}} & \\ B & & \end{array}$$

Let $\rho : G \hookrightarrow \mathrm{GL}(n, \mathbb{F})$ be a representation of a finite group. The totalization (see Section A.1) of the algebra of coinvariants $\mathbb{F}[V]_G$ is finite-dimensional, and generated by the homogeneous component of degree 1 $(\mathbb{F}[V]_G)_1$, which is the image of V^* under the quotient map $\mathbb{F}[V] \longrightarrow \mathbb{F}[V]_G$. Since the dimensions of the totalization of $\mathbb{F}[V]_{G^i}$ decrease with i, the algebras $\mathbb{F}[V]_{G^i}$ become isomorphic after at most $\dim_{\mathbb{F}}(\mathrm{Tot}(\mathbb{F}[V]_G))$ steps and the chain of ideals

$$(0) = \mathcal{I}_0 \subseteq \mathcal{I}_1(\mathbb{F}[V]) \subseteq \mathcal{I}_2(\mathbb{F}[V]) \subseteq \cdots \subseteq \mathcal{I}_m(\mathbb{F}[V]) \subseteq \cdots \subset \mathbb{F}[V]$$

stabilizes after at most this many steps. The totalization of any of the quotient algebras $\mathbb{F}[V]/\mathcal{I}_i$ is finite dimensional, so the radical of the ideal $\mathcal{I}_i$ is the maximal ideal of $\mathbb{F}[V]$. Hence, each of the ideals $\mathcal{I}_i \subset \mathbb{F}[V]$, $i \in \mathbb{N}$, is $\overline{\mathbb{F}[V]}$-primary.

[9] If $X \subset A$ then (X), denotes the ideal of A generated by X.

EXAMPLE 2: Let p be an odd prime and $\mathbb{F}$ a field of characteristic p. The matrices

$$\mathbf{S} = \begin{bmatrix} 1 & 1 \\ 0 & 1 \end{bmatrix}, \quad \mathbf{T} = \begin{bmatrix} -1 & 0 \\ 0 & 1 \end{bmatrix} \in \mathrm{GL}(2, \mathbb{F})$$

generate a dihedral group of order $2p$. We recall ([372] §5.6 Example 1, or Example 13 in Section 7.4) that

$$\mathbb{F}[x, y]^{D_{2p}} \cong \mathbb{F}[y, (xy^{p-1} - x^p)^2],$$

where $x, y \in V^* = \mathrm{Hom}_{\mathbb{F}}(V, \mathbb{F})$ is the dual of the canonical basis for $\mathbb{F}^2$. Therefore,

$$\mathbb{F}[x, y]_{D_{2p}} \cong \frac{\mathbb{F}[x, y]}{(y, (xy^{p-1} - x^p)^2)} \cong \frac{\mathbb{F}[\overline{x}]}{(\overline{x}^{2p})},$$

where $\overline{x} \in \mathbb{F}[x, y]_{D_{2p}}$ is the residue class of x. The action of D_{2p} on x, y is given the formulae

$$\begin{aligned} \mathbf{S}(x) &= x + y, & \mathbf{T}(x) &= -x, \\ \mathbf{S}(y) &= y, & \mathbf{T}(y) &= y. \end{aligned}$$

From these formulae it follows that

$$(\mathbb{F}[x, y]_{D_{2p}})^{D_{2p}} \cong \frac{\mathbb{F}[\overline{x}^2]}{(\overline{x}^{2p})},$$

and hence

$$\mathbb{F}[x, y]_{D_{2p} D_{2p}} \cong \frac{\mathbb{F}[\overline{x}]}{(\overline{x}^2)}.$$

The action of D_{2p} on $\mathbb{F}[x, y]_{D_{2p} D_{2p}}$ is fixed-point free, so $\mathcal{I}_2 = \cdots = \mathcal{I}_\infty = (x^2, y) \subset \mathbb{F}[x, y]$ is the ideal of stable invariants.

The stable invariants are a phenomenon of the modular case, and not a great deal is known about them; see, however, [186], [187], [285], and [286].

1.3 Basic Problems in Invariant Theory

Since we have the basic definitions at hand, we can begin to discuss the problems that have animated invariant theory. To recapitulate, the ingredients of an invariant-theoretic setup consist of a group G, a field $\mathbb{F}$, and a representation $\rho : G \hookrightarrow \mathrm{GL}(n, \mathbb{F})$. The group G acts via ρ on the algebra $\mathbb{F}[V]$ of polynomial functions on the representation space V, and a central object of study for us is the ring of invariants $\mathbb{F}[V]^G$. So what would we want to know about this ring? It certainly would be of great significance to know that $\mathbb{F}[V]^G$ has a *finite description*. There are several ways to interpret this. Here is perhaps the most basic.

PROBLEM 1 (Algebraic Finiteness): *Let $\rho : G \hookrightarrow \mathrm{GL}(n, \mathbb{F})$ be a representation of a finite group G over the field $\mathbb{F}$. Is $\mathbb{F}[V]^G$ finitely generated as an algebra over $\mathbb{F}$? Is it finitely related? That is to say, does it have a* *presentation as a quotient of a finitely generated polynomial algebra over $\mathbb{F}$ by an ideal that is finitely generated?*

As was shown by Emmy Noether in two remarkable papers, [301] and [303], the answer for finite groups G is always yes; see Chapter 2. This establishes the basic *algebraic* finiteness of the rings of invariants of finite groups. Previously, D. Hilbert had proven in [163] that the answer is yes for $\mathrm{SL}(n, \mathbb{C})$, and his argument was eventually extended[10] with little change to any reductive algebraic group over $\mathbb{C}$. If G is not finite, nor reductive, then many years later M. Nagata showed in [268] that the answer can be no, and V. L. Popov demonstrated in [318] that the groups whose invariants are always finitely generated are precisely the reductive groups. Emmy Noether's theorems solve the basic problem of algebraic finiteness for finite groups. But, why should we stop with just the first-order relations? We could ask for relations between relations, and so on. This is tantamount to asking whether $\mathbb{F}[V]^G$ admits a finite projective resolution by finitely generated modules over a polynomial algebra.

Specifically, using the algebraic finiteness theorems (Emmy Noether's theorem if G is finite, and D. Hilbert's theorem if G is reductive), choose a minimal generating set $f_1, \ldots, f_m \in \mathbb{F}[V]^G$ for $\mathbb{F}[V]^G$ as an algebra over $\mathbb{F}$. Form the polynomial algebra $\mathbb{F}[F_1, \ldots, F_m]$ whose generators are in bijective correspondence with $f_1, \ldots, f_m$ and define an epimorphism $\varphi : \mathbb{F}[F_1, \ldots, F_m] \longrightarrow \mathbb{F}[f_1, \ldots, f_m]$, the **remembering map**, by requiring that $\varphi(F_i) = f_i$ for $i = 1, \ldots, m$. The kernel of φ is the **first syzygy module** Syz_1. Suppose that $\mathrm{Syz}_1, \ldots, \mathrm{Syz}_k$ have already been defined. Choose a minimal generating set (we do not assume that it is finite) for Syz_{k+1} as an $\mathbb{F}[F_1, \ldots, F_m]$-module and let L_{k+1} be a free $\mathbb{F}[F_1, \ldots, F_m]$-module with basis in bijective correspondence with this generating set. The kernel of the obvious $\mathbb{F}[F_1, \ldots, F_m]$-module epimorphism, $L_{k+1} \longrightarrow \mathrm{Syz}_k$, is the next syzygy module Syz_{k+1}, the $(k+1)$**st syzygy module**. The problem at hand becomes the following:

PROBLEM 2 (Homological Finiteness): *Let $\rho : G \hookrightarrow \mathrm{GL}(n, \mathbb{F})$ be a representation of a finite group G over the field $\mathbb{F}$. Are the syzygy modules of $\mathbb{F}[V]^G$ finitely generated, and is the syzygy chain finite?*

D. Hilbert, in a ground breaking paper [163], answered this question in the affirmative, proving the basic *homological* finiteness of $\mathbb{F}[V]^G$ (see Chapter 5), establishing the field of homological algebra and ushering in a whole new era in algebra by creating completely new paradigms.

We could, of course, understand the discussion leading to Problems 1 and 2 in another way: Namely, does $\mathbb{F}[V]^G$ contain a well-understood subalgebra of which it is a *finite* extension? For example, it might happen that $\rho : G \hookrightarrow \mathrm{GL}(n, \mathbb{F})$ is a permutation representation, in which case the

[10] The history here is a bit complicated. See the discussion in [267] Section 2.

elementary symmetric polynomials $e_1, \ldots, e_n$, and therefore also the subalgebra they generate, $\mathbb{F}[e_1, \ldots, e_n]$, would belong to $\mathbb{F}[V]^G$. Moreover, since $\mathbb{F}[V]$ is a finitely generated $\mathbb{F}[e_1, \ldots, e_n]$-module, $\mathbb{F}[e_1, \ldots, e_n]$ a Noetherian ring, and $\mathbb{F}[V]^G \subseteq \mathbb{F}[V]$ an $\mathbb{F}[e_1, \ldots, e_n]$-submodule, $\mathbb{F}[V]^G$ is a also a finitely generated $\mathbb{F}[e_1, \ldots, e_n]$-module. Which means, of course, that $\mathbb{F}[e_1, \ldots, e_n] \subseteq \mathbb{F}[V]^G$ is a finite ring extension. Is this an accident? In other words we have the following question:

PROBLEM 3 (Noetherian Finiteness): *Let $\rho : G \hookrightarrow \mathrm{GL}(n, \mathbb{F})$ be a representation of a group G over the field $\mathbb{F}$. Does there exist a polynomial subalgebra $\mathbb{F}[h_1, \ldots, h_n] \subseteq \mathbb{F}[V]^G$ over which $\mathbb{F}[V]^G$ is finite?*

Combined results of Emmy Noether, [302], and D. Hilbert, [163], show that the answer for a finite group G is affirmative; see Chapter 4. This follows from what is usually referred to as the **Noether normalization theorem**; see Theorem A.3.1 for the statement, and any of the standard references, such as [25] or [20], for a proof. The forms $h_1, \ldots, h_n \in \mathbb{F}[V]^G$ whose existence is guaranteed by this theorem are called a **system of parameters**.

For an algebra of invariants $\mathbb{F}[z_1, \ldots, z_n]^G$ it is easy to *count* in an appropriate sense the number of invariants in a given homogeneous degree k: One simply computes the dimension over $\mathbb{F}$ of $\mathbb{F}[z_1, \ldots, z_n]^G_k$. It has proven efficacious to pack all these dimensions into a formal series,

$$P(\mathbb{F}[z_1, \ldots, z_n]^G, t) = \sum_{k=0}^{\infty} \dim_{\mathbb{F}}(\mathbb{F}[z_1, \ldots, z_n]^G_k) \cdot t^k,$$

which we call the **Poincaré series**[11] of $\mathbb{F}[h_1, \ldots, h_n]$.

PROBLEM 4 (Combinatorial Finiteness): *Let $\rho : G \hookrightarrow \mathrm{GL}(n, \mathbb{F})$ be a representation of the group G over the field $\mathbb{F}$. Is there a simple formula for the Poincaré series*

$$P(\mathbb{F}[V]^G, \mathbb{F}) = \sum_{k=0}^{\infty} \dim_{\mathbb{F}}(\mathbb{F}[V]^G_k) \cdot t^k$$

of the ring of invariants? What properties does it have as a function of t?

A straightforward induction argument shows that

$$P(\mathbb{F}[z_1, \ldots, z_n], t) = \left(\frac{1}{1-t}\right)^n,$$

[11] Alternatively, one could study the **Hilbert function** $H(\mathbb{F}[z_1, \ldots, z_n]^G, k)$, which is defined by

$$H(\mathbb{F}[z_1, \ldots, z_n]^G, -) : \mathbb{N}_0 \longrightarrow \mathbb{N}_0, \quad H(\mathbb{F}[z_1, \ldots, z_n]^G, k) = \dim_{\mathbb{F}}(\mathbb{F}[z_1, \ldots, z_n]^G_k).$$

The study of the Hilbert function and the Poincaré series are more or less equivalent; see, e.g., [400].

but what happens after we take the invariants is not so clear. In a marvelous[12] paper that still seems as modern as if it were written yesterday,[13] T. Molien shows how to derive such a formula for Poincaré series of the ring of invariants of any finite group if the base field is $\mathbb{C}$. With a bit of good will, and hindsight, we see that his proof works equally well in the nonmodular case; see Section 3.1. Even in the simplest modular cases (see, e.g., [8], [10] and the references there, as well as [177]) no analogue is known.

1.4 Problems for Finite Groups

The results mentioned in the preceding section are basically of a theoretical nature: They asserted the existence of something without giving a means of constructing it. They also make sense in some cases for nonfinite groups, e.g., algebraic groups. To find invariants we need some tools to construct G-invariant polynomials. It is here that the cases of finite and nonfinite groups seemingly diverge. We will consider only the case of finite groups; see, e.g., [83], [164], [319], and [447] for the reductive case.

A basic tool for the construction of invariants of finite groups is the **transfer homomorphism** $\mathrm{Tr}^G : \mathbb{F}[V] \longrightarrow \mathbb{F}[V]^G$. It was probably introduced by H. Maschke [238], but see [95] for another possible source, and is defined by the formula

$$\mathrm{Tr}^G(f) = \sum_{g \in G} g \cdot f \qquad f \in \mathbb{F}[V].$$

On the face of it Tr^G is only a map from $\mathbb{F}[V]$ to itself. However, for any $h \in G$ the sum $h \cdot \mathrm{Tr}^G(f) = \sum_{g \in G} hg \cdot f$ is the same as the sum defining $\mathrm{Tr}^G(f)$ except for the order of the summands. It follows that $\mathrm{Tr}^G(f) \in \mathbb{F}[V]^G$ for any $f \in \mathbb{F}[V]$, and Tr^G is usually regarded as a map $\mathrm{Tr}^G : \mathbb{F}[V] \longrightarrow \mathbb{F}[V]^G$ as indicated. A detailed discussion of the transfer appears in Chapters 2 and 9.

Elements in the image of the transfer are G-invariant, so the transfer provides a tool to construct invariant forms of arbitrarily large degree. Notice that Tr^G is an $\mathbb{F}[V]^G$-module homomorphism, and therefore $\mathrm{Im}(\mathrm{Tr}^G) \subseteq \mathbb{F}[V]^G$ is an ideal. Moreover, for $f \in \mathbb{F}[V]^G$ one has $\mathrm{Tr}^G(f) = |G| \cdot f$, and hence in the nonmodular case, $\frac{1}{|G|} \cdot \mathrm{Tr}^G$ is a well-defined projection from $\mathbb{F}[V]$ onto $\mathbb{F}[V]^G$, and $\mathrm{Tr}^G : \mathbb{F}[V] \longrightarrow \mathbb{F}[V]^G$ is surjective.

In the modular case, the transfer is never zero, but also is never onto in positive degrees, [114], [373] Section 2, and [216]. The size of the image has been the subject of much recent research; see, e.g., [119], [215], [216],

[12] One has to realize that F. G. Frobenius had invented character theory only a few years earlier.

[13] Only the reference to a computer algebra program for the results of the computation are missing. Indeed, T. Molien does not tell us how he made his computations.

[292], [295], [349], [373], [378], and the references there. By contrast, in the nonmodular case, the transfer provides us with a means of constructing[14] all invariants, and in particular, algebra generators for $\mathbb{F}[V]^G$; if only we knew when to stop!

PROBLEM 1 (Upper Bounds): *Let* $\varrho : G \hookrightarrow \mathrm{GL}(n, \mathbb{F})$ *be a representation of a finite group over the field* $\mathbb{F}$ *and denote by* $\beta(\varrho)$ *the maximal degree of a generator for* $\mathbb{F}[V]^G$ *in a minimal algebra generating set. Find an a priori upper bound for* $\beta(\varrho)$*, and for* $\beta_{\mathbb{F}}(G) = \max\{\beta(\varrho) \mid \varrho$ *is defined over* $\mathbb{F}\}$.

Algebraic finiteness for finite groups tells us that rings of invariants of finite groups are finitely generated. If the ground field is $\mathbb{C}$, the complex numbers, then in [301] Emmy Noether provides

(1) a means of constructing sufficiently many forms to generate the ring of invariants, and
(2) an a priori upper bound, namely the group order $|G|$, on their degrees, and hence also on the number of such forms required.

This supplies us with a criterion[15] over $\mathbb{C}$ to determine when we are finished with a computation; namely, if we know all the invariants up to degree $|G|$, then we certainly have found a set of algebra generators. This upper bound for $\beta_{\mathbb{C}}(G)$ is known as **Noether's bound** for $\mathbb{C}$. What about other ground fields? In the modular case there is no upper bound for $\beta_{\mathbb{F}}(G)$; see Chapter 6. Trying to extend Noether's bound to $\beta_{\mathbb{F}}(G)$ in all nonmodular cases, i.e., to all cases where the characteristic of $\mathbb{F}$ does not divide the order of the group G, has led to a number of new results and tools.

By distilling the essence of one[16] of Emmy Noether's arguments in [301] one can show that $\beta_{\mathbb{F}}(G) \leq |G|$, provided that G contains a chain of subgroups for which this property holds. The class of groups to which this applies includes the solvable groups; see Section 7.3, [211] and [376]. By reworking the other argument, one arrives first at the notion of the orbit Chern classes, [387] (and Euler classes), and then at various notions that refine them which are referred to collectively as fine orbit Chern classes; see Chapter 4.

The **orbit Chern classes** (see Section 4.1 for a precise definition) are the elementary symmetric polynomials in the elements of an orbit of G acting on the space of linear forms V^* regarded as elements of $\mathbb{F}[V]^G$. They turn out to be sufficient to generate $\mathbb{F}[V]^G$ in many cases:

- if the characteristic of $\mathbb{F}$ is strictly larger than $|G|$, [387];

[14] However, be warned: The computation of the image of the transfer in degree k a priori requires that we compute the matrix of $\mathrm{Tr}^G : \mathbb{F}[V]_k \longrightarrow \mathbb{F}[V]_k$. The dimension of this vector space is $\binom{n+k-1}{k}$, so we are talking about a square matrix of this size. The number of entries is of order of magnitude $k^{2(n-1)}$ and leads to prohibitive storage problems when implemented on a computer.

[15] Although not an efficient one.

[16] The paper [301] contains not one, *but two*, proofs of the algebraic finiteness for finite groups.

- if G is solvable and the characteristic of $\mathbb{F}$ does not divide the order of G, [370];
- if $G = A_n$ is the alternating group and the characteristic of $\mathbb{F}$ is prime to the order of A_n, [379] and [376];
- if G is a Coxeter group containing no factors of the form E_6, E_7, or E_8, [211].

Since by construction the orbit Chern classes have degree at most equal to the order of the group, all this served as evidence indicating that the following problem should have a positive solution.

PROBLEM 2 (Emmy Noether's Bound): *Let G be a finite group and $\mathbb{F}$ a field. If $|G| \in \mathbb{F}^{\times}$, is $\beta_{\mathbb{F}}(G) \leq |G|$?*

As this book was being written P. Fleischmann, [120], and J. Fogarty, [122], independently succeeded in extracting from the circle of ideas emanating from Emmy Noether's paper [301] a positive solution for this problem. We will present an elegant proof of this due to D. J. Benson that Noether's bound holds in the nonmodular case in Section 2.3.

In the nonmodular case, if the ground field contains enough roots of unity, then every finite-dimensional representation is a sum of irreducible ones. One could therefore reasonably ask whether there is a worst case for $\beta_{\mathbb{F}}(\varrho)$. If the ground field $\mathbb{F}$ is $\mathbb{C}$, then it was shown by B. Schmid that the answer is yes: The regular representation is the worst case. She also showed that $\beta_{\mathbb{C}}(G)$ is strictly less than $|G|$ unless G is a cyclic group. For other ground fields, if the characteristic of $\mathbb{F}$ is strictly greater than $|G|$, then the regular representation is again the worst case, [381]. This was generalized still further by D. Krause in her diploma thesis [211] (see Section 7.3 Example 2 and Theorem 4.2.4) by reducing the requirement that $p > |G|$. For example, if G is solvable with lower central series $\{1\} = G_0 < G_1 < \cdots < G_s = G$ and $\mathbb{F}$ has characteristic $p > \max_i\{|G_i : G_{i-1}|\}$ then $\beta_{\mathbb{F}}(G) = \beta_{\mathbb{F}}(\mathrm{reg})$, where reg is the regular representation of G over $\mathbb{F}$. Other results along these lines appear in [92]. There is still a gray area here, where the characteristic of $\mathbb{F}$ does not divide $|G|$ but is smaller than $|G|$.

In the modular case there is no such worst case, and precious little is known, but see, e.g., [177], [211], [288], [292], [297], and [386]. Examples show that the invariants of $\mathbb{Z}/2$ over a field of characteristic 2 can require a generator of arbitrarily large degree; see [325], or [205]. For example, if $\mathbb{Z}/2$ axts on the set $XY = \{x_1, \ldots, x_n, y_1, \ldots, y_n\}$ by simultaneously exchanging x_i and y_i for $i = 1, \ldots, n$, then $x_1 \cdots x_n + y_1 \cdots y_n$ is an indecomposable invariant of degree n. This is not an isolated phenomenon, as many other examples of this type exist; see, e.g., [327], [54], and the last section of [115]. The evidence seems to indicate there is a bound for $\beta_{\mathbb{F}}(\varrho)$ that depends only on the order of the group and the dimension of the representation, but the jury is still out on this.

As mentioned previously, invariant rings of permutation representations often behave similarly regardless of the characteristic, and this is the case for Problem 1 restricted to this class of representations. In his diploma thesis M. Göbel; see [134] showed that the polynomial invariants of permutation representations of degree n are always generated by invariant forms of degree at most $\max\{n, \binom{n}{2}\}$ independent of the characteristic of the ground field. He also gave an algorithm, again characteristic free, that constructs a set of algebra generators. For a discussion of these results see Section 3.4 and for further developments see [136], [137], and [138].

What about Noetherian finiteness? Can we construct a system of parameters whose existence is asserted by Noetherian finiteness? There is a very general construction due to E. Dade, [323], [373], and [401], of such a system of parameters that we will examine in Section 4.3, but it lacks a certain optimal property, which can best be captured in the following:

PROBLEM 3 (Optimal System of Parameters): *Let $\rho : G \hookrightarrow \mathrm{GL}(n, \mathbb{F})$ be a representation of a finite group over the field $\mathbb{F}$. Find a system of parameters $h_1, \ldots, h_n \in \mathbb{F}[V]^G$ for $\mathbb{F}[V]^G$ such that as a module over $\mathbb{F}[h_1, \ldots, h_n]$ a minimal number of generators is required. Give an a priori estimate of this minimum.*

A system of parameters $h_1, \ldots, h_n \in \mathbb{F}[V]^G$ such that the number of generators of $\mathbb{F}[V]^G$ as a module over $\mathbb{F}[h_1, \ldots, h_n]$ is minimal is called an **optimal system of parameters**. It is not hard to see by computing with Poincaré series that in the nonmodular case $|G|$ will divide the product $\deg(h_1) \cdots \deg(h_n)$ and that the quotient is the number of module generators needed. There are other ways to demand optimality of a system of parameters. For example, a system of parameters $h_1, \ldots, h_n$ is called **pseudo-optimal** if it minimizes the maximal degree of a generator of $\mathbb{F}[V]^G$ over $\mathbb{F}[h_1, \ldots, h_n]$. These notions were introduced by D. Engelmann in [108] in connection with finding the invariants of the symmetric group Σ_n acting on the vector space of $n \times n$ matrices by simultaneous permutation of the rows and columns. His computations showed the effectiveness of this idea when attempting machine computations.

Given a system of parameters for $\mathbb{F}[V]^G$ we could also ask: How complicated is $\mathbb{F}[V]^G$ as a module over the subalgebra they generate? One way of formulating this appears in the following problem:

PROBLEM 4 (Cohomological Finiteness): *Let $\rho : G \hookrightarrow \mathrm{GL}(n, \mathbb{F})$ be a representation of a finite group over the field $\mathbb{F}$ and $h_1, \ldots, h_n \in \mathbb{F}[V]^G$ a system of parameters. What kind of a module is $\mathbb{F}[V]^G$ over $\mathbb{F}[h_1, \ldots, h_n]$? In particular, is it a free module? If not, what is its projective dimension?*

Indeed, we need to be a bit careful here with the types of properties we consider. They should be independent of the choice of the system of parameters, unless, of course, we can pin down a natural, particularly good, system of parameters. The projective dimension is one such property, and the theorem of J. E. Eagon and M. Hochster in [172] tells us that $\mathbb{F}[V]^G$ is a free module over $\mathbb{F}[h_1, \ldots, h_n]$ in the nonmodular case. In other words, $\mathbb{F}[V]^G$ is a Cohen-Macaulay algebra in the nonmodular case. It need not be so in the modular case, [36] (this is Example 3 in Section 1.1), and this is the source of many of the most interesting problems that still need clarification. See, e.g., Section 5.5 and the references there.

If $\rho : G \hookrightarrow \mathrm{GL}(n, \mathbb{F})$ is a representation of a finite group over the field $\mathbb{F}$ and $h_1, \ldots, h_n \in \mathbb{F}[V]^G$ a system of parameters, and d is the projective dimension of $\mathbb{F}[V]^G$ over $\mathbb{F}[h_1, \ldots, h_n]$, then d is independent of the choice of the system of parameters. The difference $n - d$ is called the **depth** or **homological codimension** of $\mathbb{F}[V]^G$. There are other ways to define the depth (see Section 5.5), and the terminology codimension is justified by the Auslander-Buchsbaum equality, see [21] or [47] Theorem 1.3.3. As we will discuss in the next section, for finite fields there is a universal test set of polynomials, the Dickson polynomials (see Chapter 6), for checking the depth of $\mathbb{F}[V]^G$.

What else can be said about homological finiteness? Of course, the simplest possible situation would be for there to be no syzygies, i.e., the ring of invariants would itself be a polynomial algebra.

PROBLEM 5 (The Polynomial Algebra Problem): *For which groups G, and which representations $\rho : G \hookrightarrow \mathrm{GL}(n, \mathbb{F})$ over which fields $\mathbb{F}$, is $\mathbb{F}[V]^G$ a polynomial algebra?*

This is a problem with a long and rich history in the classical case. It starts with the finite real reflection groups and ends with their complex analogues, the finite complex pseudoreflection groups, and reaches a high point for the complex case in the paper of G. C. Shephard and J. C. Todd, [352]. Here one finds a classification of the groups, and the representations thereof, whose rings of invariants are polynomial. The representations that occur are those where $\rho(G)$ is generated by **pseudoreflections**, i.e., by elements of finite order that leave a hyperplane pointwise fixed. We will touch on these matters in Section 7.1.

As usual, the modular case is more obscure. For p-groups over the prime field $\mathbb{F}_p$ there is a satisfactory solution to Problem 5 due to H. Nakajima, [277]. However, the result as stated does not extend to other fields of characteristic p: There is a counterexample due to R. E. Stong, [413]. We will discuss this in more detail in Section 6.3.

In the general case there is a theorem due to J.-P. Serre that says: *If $\mathbb{F}[V]^G$ is a polynomial algebra, then G (or better put $\rho(G)$) must be generated by pseudoreflections* ([346], [39] Chapitre V, EXERCISES §5, Exercise 8 and 9, or [30] Theorem 7.2.1). If *all* the elements of G except the identity are pseudoreflections, then the invariants are a polynomial algebra and there is a complete classification of the groups and their invariants ([224], [273], and [372] Chapter 8 Section 2, as well as Section 7.1 and the references there).

If $\mathbb{F}[V]^G$ is a polynomial algebra, then the same is true of $\mathbb{F}[V]^{G_U}$, where $U \subseteq V$ is a subspace, and G_U the **pointwise stabilizer** of U. This result is due to R. Steinberg, [407], in the case that the ground field is the complex numbers $\mathbb{C}$. A generalization to arbitrary characteristic due to J.-P. Serre is indicated in [39], Chapitre V, §6, Exercise 8, and appears with a full proof in H. Nakajima's [271]. C. W. Wilkerson and W. G. Dwyer gave a proof in the special case where the ground field is the prime field $\mathbb{F}_p$ using Steenrod operations, [101], and this was generalized in [380] to arbitrary Galois fields. We will discuss these and related matters in Chapter 10.

The significance of pointwise stabilizers is a central theme of the paper [199] of G. Kemper and G. Malle who proved that for irreducible representations, the ring of invariants $\mathbb{F}[V]^G$ is polynomial if and only if $\rho(G)$ is generated by pseudoreflections and the ring of invariants $\mathbb{F}[V]^{G_U}$ for any nonzero subspace $U \leq V$ is a polynomial algebra. Their proof is based on the classification of the irreducible linear groups generated by pseudoreflections ([191], [436], and [463]). This brings the situation for such representations up to the level of [352].

The next simplest case of homological finiteness is where there is only one syzygy module, Syz_1, the first one. This means that $\mathbb{F}[V]^G$ is a complete intersection. In this case one has the results of K. Watanabe [445], V. Kac and K. Watanabe [188], as well as the result of [380] that the complete intersection property is inherited in passing from a group to the pointwise stabilizer of a subspace; see Chapter 10. The classification of subgroups of $\mathrm{GL}(n, \mathbb{C})$ with invariant rings that are complete intersections is, however, rather long and complicated. We therefore refer the reader to the original sources [147], [279], [280], [282], and [330], as well as Section 5.6.

1.5 Problems for Finite Groups over Finite Fields

Not only is the case of finite groups special; the case where in addition the ground field $\mathbb{F}$ is also finite offers a number of very interesting aspects, results, and open problems. The most obvious additional feature in this case is that the full group of linear transformations $\mathrm{GL}(n, \mathbb{F}_q)$ is itself a finite group. If $\mathbb{F}_q$ has $q = p^\nu$ elements, then $|\mathrm{GL}(n, \mathbb{F}_q)| = \prod_{i=0}^{n-1}(q^n - q^i)$.

NOTATION: *Throughout this section $q = p^\nu$, where p is a prime integer.*

The invariants of the full general linear group $\mathrm{GL}(n, \mathbb{F}_q)$ were computed by L. E. Dickson, [85], and the result is decidedly pleasant; namely, it is a polynomial algebra, viz.,

$$\mathbb{F}_q[V]^{\mathrm{GL}(n,\mathbb{F}_q)} = \mathbb{F}_q[\mathbf{d}_{n,0}, \mathbf{d}_{n,1}, \ldots, \mathbf{d}_{n,n-1}].$$

The degree of the polynomial $\mathbf{d}_{n,i}$ is $q^n - q^i$, and there are a number of implicit and explicit formulas for them. For example, they are the orbit Chern classes of the orbit $V^* \setminus \{0\}$. If the ground field is clear from context, we will denote $\mathbb{F}_q[V]^{\mathrm{GL}(n,\mathbb{F}_q)}$ by $\mathbf{D}(n)$, and call it the **Dickson algebra of $\mathbb{F}_q$ of degree n**, or just the Dickson algebra if $\mathbb{F}_q$ and n are clear from the context. Notice that the **Dickson polynomials, $\mathbf{d}_{n,0}, \ldots, \mathbf{d}_{n,n-1}$**, being invariants of the full linear group $\mathrm{GL}(n, \mathbb{F}_q)$, are present in the ring of invariants of any n-dimensional representation of any group G over $\mathbb{F}_q$. In other words, for any representation $\rho : G \hookrightarrow \mathrm{GL}(n, \mathbb{F}_q)$ of a finite group G over the field $\mathbb{F}_q$ we have $\mathbf{D}(n) \subseteq \mathbb{F}_q[V]^G \subseteq \mathbb{F}_q[V]$. Since $\mathbb{F}_q[V]$ is a finitely generated $\mathbf{D}(n)$-module (see Theorem 2.1.4) and $\mathbf{D}(n)$ is a Noetherian ring, it follows that $\mathbb{F}_q[V]^G$ is also a finitely generated $\mathbf{D}(n)$-module. The Dickson polynomials $\mathbf{d}_{n,0}, \ldots, \mathbf{d}_{n,n-1} \in \mathbb{F}_q[V]^G$ are therefore a *universal* system of parameters for $\mathbb{F}_q[V]^G$.

The existence of such a universal system of parameters has many useful consequences. In particular, $\mathbb{F}_q[V]^G$ is Cohen-Macaulay if and only if the Dickson polynomials $\mathbf{d}_{n,0}, \ldots, \mathbf{d}_{n,n-1} \in \mathbb{F}_q[V]^G$ are a regular sequence; see Section 5.5. Much more remarkable is the following explicit criterion due to D. Bourguiba and S. Zarati, [41], for $\mathbb{F}_q[V]^G$ to have depth k: The depth of $\mathbb{F}_q[V]^G$ is at least k if and only if $\mathbf{d}_{n,n-1}, \ldots, \mathbf{d}_{n,n-k} \in \mathbb{F}_q[V]^G$ is a regular sequence; see Section 8.5. Indeed, it is not possible to replace this sequence of the k *lower* Dickson polynomials with another sequence of k Dickson polynomials; see Section 8.5 or Section 8.4 in [290].

Another special feature of algebras over finite fields is the Frobenius homomorphism. For the purposes of invariant theory over a Galois field $\mathbb{F}_q$ with $q = p^\nu$ elements we take the **Frobenius homomorphism** to be the $\mathbb{F}_q$-linear map $\mathbb{F}_q[V] \longrightarrow \mathbb{F}_q[V]$ induced by the map on linear forms $\Phi : \ell \longrightarrow \ell^q$.[17]

Note that for a representation $\rho : G \hookrightarrow \mathrm{GL}(n, \mathbb{F}_q)$, the G-action on $\mathbb{F}_q[V]$ commutes with the Frobenius homomorphism Φ acting on $\mathbb{F}_q[V]$, and therefore the algebra of invariants $\mathbb{F}_q[V]^G$ must be closed (i.e., mapped into itself) under the Frobenius homomorphism. Certainly, not every subalgebra of $\mathbb{F}_q[V]$ has this property, and so this imposes an additional structure on algebras of invariants not shared by all subalgebras, not even the finitely generated ones contained in $\mathbb{F}_q[V]$. (See, e.g., [343] for a nice example that illustrates this viewpoint.)

One way to organize the information hidden in the Frobenius homomorphism is by means of the Steenrod operations. These originated in algebraic

[17] This is neither the usual, nor the only, definition of a Frobenius homomorphism.

topology ([65], [252], [341], [405] and [461] to name but a few sources) and have proven extremely useful in modular invariant theory. For example, the criterion of D. Bourguiba and S. Zarati just cited has no proof without them.

Here is a quick definition of the **Steenrod operations**: Define a map[18]

$$\boldsymbol{P}(\xi) : \mathbb{F}_q[V] \longrightarrow \mathbb{F}_q[V][[\xi]],$$

by the rules

(i) $\boldsymbol{P}(\xi)$ is $\mathbb{F}_q$-linear,
(ii) $\boldsymbol{P}(\xi)(v) = v + v^q\xi$ for $v \in V^*$,
(iii) $\boldsymbol{P}(\xi)(u \cdot w) = \boldsymbol{P}(\xi)(u) \cdot \boldsymbol{P}(\xi)(w)$ for $u, w \in \mathbb{F}_q[V]$,
(iv) $\boldsymbol{P}(\xi)(1) = 1$.

We consider $\boldsymbol{P}(\xi)$ as a ring homomorphism of degree 0 by giving ξ the degree $(1-q)$. If we do this, then (i), (iii), and (iv) amount to saying that $\boldsymbol{P}(\xi)$ is a homomorphism of $\mathbb{F}_q$-algebras, and (ii) normalizes what happens on linear forms. By separating out homogeneous components we obtain for all $i \in \mathbb{N}_0$ $\mathbb{F}_q$-linear maps

$$\mathcal{P}^i : \mathbb{F}_q[V] \longrightarrow \mathbb{F}_q[V]$$

by the requirement

$$\boldsymbol{P}(\xi)(f) = \sum_{i=0}^{\infty} \mathcal{P}^i(f)\xi^i,$$

called the **Steenrod reduced power operations**, or simply the **Steenrod operations**. See Section 8.1 for a more detailed account. Other algebraic introductions can be found in [372] Chapters 10 and 11, [383], [458], [213], and for $q=2$, [37]. For $q=2$ these operations are usually denoted by $\mathrm{Sq}^i(f)$, and called **Steenrod squares**.

If we let G act on $\mathbb{F}_q[V][[\xi]]$ by demanding that G acts via ρ on $\mathbb{F}_q[V]$, and trivially on ξ, then $\boldsymbol{P}(\xi)$ is a G-equivariant map, so $\boldsymbol{P}(\xi)(\mathbb{F}_q[V]^G) \subseteq (\mathbb{F}_q[V][[\xi]])^G = \mathbb{F}_q[V]^G[[\xi]]$. Hence, for all $k \in \mathbb{N}_0$, $\mathcal{P}^k(f) \in \mathbb{F}_q[V]^G$ whenever $f \in \mathbb{F}_q[V]^G$, so not only do the Steenrod operations provide a theoretical tool, they also allow us to construct new invariants from old ones. This can often be used to good advantage. Furthermore, the action of the Steenrod algebra on the Dickson algebra is very rigid, see Section 8.6, and is central to the solution of the inverse invariant theory problem, [290], which is addressed in Section 8.4.

PROBLEM 1 (Inverse Invariant Theory): *Which subalgebras of $\mathbb{F}_q[V]$ occur as rings of invariants?*

Another way in which the Steenrod operations enter into invariant theory

---

[18] If A is a ring, then $A[[\xi]]$ denotes the ring of formal power series over A in the variable ξ.

is through the simple observation that $\mathrm{Im}(\mathrm{Tr}^G) \subseteq \mathbb{F}_q[V]^G$ is invariant under the Steenrod operations. This places a considerable restraint on the possible ideals that can occur as the image of the transfer. In particular, since $\mathbb{F}_q[V]^G$ contains only finitely many prime ideals invariant under the action of the Steenrod operations ([343], [371] Theorem 4.8, [372] Theorem 11.4.8, and Section 9.2) and the radical of $\mathrm{Im}(\mathrm{Tr}^G)$ is also invariant under Steenrod operations, there are only a small number, among all possibilities, for $\sqrt{\mathrm{Im}(\mathrm{Tr}^G)}$. The interaction between the Dickson algebra and the Steenrod operations also shows that some power of the top-degree Dickson polynomial $\mathbf{d}_{n,0}$ always belongs to the image of the transfer, whereas if the characteristic of $\mathbb{F}_q$ divides the order of G, no power of the bottom Dickson class $\mathbf{d}_{n,n-1}$ does. We will examine these results about invariant ideals in Chapter 9.

1.6 Problems for Special Representations

In this section we consider some of the problems connected with the study of the rings of invariants of special classes of groups or representations. We have already mentioned the family of solvable groups in connection with Emmy Noether's bound in the nonmodular case. For the more restricted classes of nilpotent and p-groups D. Krause [211] has a number of further results in this direction; see Section 7.3.

Permutation representations are a logical place to look for interesting examples and problems. After all, the ur-theorem of invariant theory might well be regarded as the theorem on elementary symmetric functions: *Every symmetric polynomial can be written in a unique way as a polynomial in the elementary symmetric functions.*

As an example of what we mean, the Poincaré series of the ring of invariants of a permutation representation turns out to be independent of the ground field (Proposition 3.2.5); as an algebra the ring of invariants is generated by forms of degree at most $\max\{n, \binom{n}{2}\}$ (Corollary 3.4.3); and M. Göbel has also provided a number of algorithms, [135], for computing an additive basis for the homogeneous components of such a ring of invariants, and hence implicitly to construct algebra generators. This extends the corresponding result in characteristic zero due to A. M. Garsia and D. Stanton, [129], to any ground field. We will examine some of these results in Section 3.4.

This more or less ends the good news for permutation representations, because already the permutation representations of $\mathbb{Z}/2$ in characteristic 2 provide a rich source of nasty examples.

EXAMPLE 1 (Vector Invariants of $\mathbb{Z}/2$): Consider the action of $\mathbb{Z}/2$ on the set $\{x_1, \ldots, x_k, y_1, \ldots, y_k\}$ on which $\mathbb{Z}/2$ acts by simultaneous permutation of $x_1, \ldots, x_k$ with $y_1, \ldots, y_k$. Then:

① $\mathbb{F}_2[x_1, \ldots, x_k, y_1, \ldots, y_k]^{\mathbb{Z}/2}$ contains an indecomposable algebra generator of degree k, namely $x_1 \cdots x_k + y_1 \cdots y_k$ ([325], and [115] Remark 6.2), so no analogue of Emmy Noether's bound (depending only on the order of the group) can exist in the modular case.

② $\mathbb{F}[x_1, \ldots, x_k, y_1, \ldots, y_k]^{\mathbb{Z}/2}$ is Cohen-Macaulay if and only if k is 1 or 2. In fact, the depth, which we defined to be $2k - h$, where h is the projective dimension of $\mathbb{F}[x_1, \ldots, x_k, y_1, \ldots, y_k]$ over the subalgebra $\mathbb{F}[x_1 + y_1, x_1y_1, \ldots, x_k + y_k, x_ky_k]$, is $k + 2$ if $k \geq 3$, not $2k$, [105], [375].

③ There can be at most two linear forms in a regular sequence of maximal length, which if $k \geq 3$ is $k + 2$ and not $2k$, [375].

Similar results hold for the multiples of the permutation representation of $\mathbb{Z}/p$ on the set $z_1, \ldots, z_p$ by cyclic permutation over the ground field $\mathbb{F}_p$, [105], [326], [327], and [375].

The regular representation is a permutation representation. One certainly has a right to wonder whether its invariants are particularly nice, or decisive: They are in the theorem of B. Schmid (see [333], [381], or Theorem 4.1.4) on degree bounds for algebra generators. For example, even in the modular case, are they always Cohen-Macaulay? This turns out to be false: The only modular cases of a regular representation having Cohen-Macaulay invariants are $\mathbb{Z}/2$, $\mathbb{Z}/3$, and $\mathbb{Z}/2 \times \mathbb{Z}/2$, [196].

The discussion of permutation representations, particularly the action of $\mathbb{Z}/2$ on the finite set $\{x_1, \ldots, x_k, y_1, \ldots, y_k\}$, suggests considering the problem of computing the invariants of the multiples of some fixed representation. Such rings of invariants are usually referred to as **vector invariants**. The following is perhaps the basic problem concerning them.

PROBLEM 1 (Fundamental Theorems): *Let* $\rho : G \hookrightarrow \mathrm{GL}(n, \mathbb{F})$ *be a representation of a finite group* G *over the field* $\mathbb{F}$*. Introduce the sequence of representations* $k \cdot \rho : G \hookrightarrow \mathrm{GL}(kn, \mathbb{F})$ *defined by the block matrices*

$$\begin{bmatrix} \rho & \cdots & 0 & 0 \\ 0 & \rho & \cdots & 0 \\ & & & \\ 0 & 0 & \ddots & 0 \\ 0 & 0 & \cdots & \rho \end{bmatrix} \in \mathrm{GL}(kn, \mathbb{F}).$$

The corresponding representation spaces are $\oplus_k V = V \oplus \cdots \oplus V$*. Find a description of* $\mathbb{F}[\oplus_k V]^G$ *in terms of* $\mathbb{F}[V]^G$*, or perhaps* $\mathbb{F}[\oplus_j V]^G$ *for* $j = 1, \ldots, k-1$*.*

As was noted above, already starting with the smallest possible nontrivial group, namely $\mathbb{Z}/2$, and its smallest possible nontrivial representation, the permutation representation σ on the set $\{x, y\}$, the solution to this problem is far from obvious. (See, e.g., [115], [205], [325], and [327].)

In classical invariant theory, the problem of finding generators for the invariants of the representations $k\tau$, where τ is the defining representation of the general linear group itself, viz., GL(n, $\mathbb{C}$), played a central role. (See, e.g, [451], where the solution to this problem is referred to as a **first fundamental theorem**[19] for GL(n, $\mathbb{C}$).) In [451] one also finds such a theorem for the symmetric group Σ_n starting from its defining representation as a permutation representation on the set $\{x_1, \ldots, x_n\}$. Of course, the ground field throughout [451] is $\mathbb{C}$, the complex numbers. Problem 1 simply poses this type of question for some fixed interesting representation of some interesting group, e.g., a pseudoreflection group in one of its pseudoreflection representations. This type of problem seems to have fallen out of favor, except as a means of providing nasty examples, e.g., to show that $\mathbb{F}[\oplus_k V]^G$ does not have some desirable property if k is large enough, e.g., is not generated by forms of degree less than some fixed bound depending only on $|G|$, or is not Cohen-Macaulay; see Proposition 5.8.1.

1.7 What Makes Rings of Invariants Special?

If $\mathbb{F}$ is a field, then the polynomial ring $\mathbb{F}[z_1, \ldots, z_n]$ over $\mathbb{F}$ in a finite number of variables has a number of pleasant properties. A short list of properties possessed by $\mathbb{F}[z_1, \ldots, z_n]$ would probably include the following:

❶ It is a commutative, graded, connected integral domain.
❷ It is Noetherian.
❸ It is integrally closed.
❹ It has Krull dimension equal to n.
❺ It is factorial.
❻ It is Cohen-Macaulay.
❼ It is finite global dimension, so is a polynomial algebra.
❽ Its field of fractions $\mathbb{F}(z_1, \ldots, z_n) = \mathbb{FF}(\mathbb{F}[V])$ is purely transcendental over the ground field $\mathbb{F}$.

These properties certainly make polynomial algebras special when seen against the background of general commutative graded algebras over a field. It is only natural to ask to what extent these properties are inherited by a ring of invariants $\mathbb{F}[V]^G$. What makes rings of invariants special is that most of these properties are inherited, but not all, and not all the time, i.e., not for all groups, or not in all situations. Obviously, the properties listed in ❶ are inherited by all graded subalgebras of $\mathbb{F}[V]$, hence also by every subring of invariants. Less immediate is that properties ❷, ❸, and ❹ are also inherited by rings of invariants; these, indeed, are classical results due to Emmy Noether; see Theorem 2.1.4 and Corollary 2.1.5.

[19] A **second** fundamental theorem would be an answer to the same question for the relations, i.e., for the first syzygy modules, and so on. One also sometimes uses the expression **first main theorem**, etc.

THEOREM 1.7.1 (Emmy Noether): *Let $\rho : G \hookrightarrow \mathrm{GL}(n, \mathbb{F})$ be a faithful representation of a finite group over the field $\mathbb{F}$. Then the ring of invariants $\mathbb{F}[V]^G$ is an integrally closed Noetherian algebra over $\mathbb{F}$ whose Krull dimension is equal to n.* □

A basic fact used to prove this and other properties of rings of invariants is the following:

PROPOSITION 1.7.2: *Let $\rho : G \hookrightarrow \mathrm{GL}(n, \mathbb{F})$ be a faithful representation of a finite group G over the field $\mathbb{F}$. Then the extension of algebras $\mathbb{F}[V]^G \subseteq \mathbb{F}[V]$ is integral, i.e., every element of $f \in \mathbb{F}[V]$ is the root of a monic polynomial with coefficients in $\mathbb{F}[V]^G$.*

PROOF: If $f \in \mathbb{F}[V]$, then the monic polynomial

$$\Phi_f(X) = \prod_{g \in G}(X - gf)$$

lies in $\mathbb{F}[V]^G[X]$ and has f as a root. □

Note that Proposition 1.7.2 describes a property inherited from $\mathbb{F}[V]$ by every ring of invariants, but not by every subalgebra, namely that $\mathbb{F}[V]^G \subseteq \mathbb{F}[V]$ is an integral extension. This, in turn, implies that the classical lying-over, going-up and going-down theorems, as well as ❹, hold for $\mathbb{F}[V]^G \subseteq \mathbb{F}[V]$; see Section A.2.

Properties ❻ and ❼ belong in the area of Problems 2 of Section 1.3 and Problem 4 of Section 1.4, and are usually not inherited; see Example 3 in Section 1. The investigation of what conditions imply that $\mathbb{F}[V]^G$ is a polynomial algebra has enjoyed an immense popularity; see Section 7.1. These results have found applications to diverse areas of mathematics; see e.g., [73], [1], [78] Chapter 7 Section 3.

Property ❽ amounts to asking for a solution of the classical Noether problem: When is $\mathbb{F}(V)^G = \mathbb{F}\mathbb{F}(\mathbb{F}[V]^G)$ purely transcendental over $\mathbb{F}$? This leads to a completely different area of mathematics, using different tools and methods, and will not be treated in this book. See the survey articles by M. Kervaire and T. Vust, [203], and R. Swan, [419], for more information and references.

So, finally, we come to Property ❺: Not every ring of invariants is factorial, as we see in the following example.

EXAMPLE 1: Let $\mathbb{F}$ be a field of characteristic not equal to 2. The matrix $\mathbf{T}$ may be regarded as implementing a representation of the cyclic group $\mathbb{Z}/2$ of order 2. From this representation we obtain an action of $\mathbb{Z}/2$ on the polynomial algebra

$$\mathbf{T} = \begin{bmatrix} -1 & 0 \\ 0 & -1 \end{bmatrix}$$

$\mathbb{F}[x, y]$ in 2 variables. The action of $\mathbf{T}$ on $\mathbb{F}[x, y]$ sends monomials to monomials, and so a polynomial is T-invariant if and only if it is a sum of T-invariant monomials. A monomial $x^i y^j$ is T-invariant if and only if $i+j$ is even. From this it follows that $\mathbb{F}[x, y]^{\mathbb{Z}/2}$ is generated as an algebra by x^2, y^2, and xy:

$$\mathbb{F}[x, y]^{\mathbb{Z}/2} = \mathbb{F}[x^2, y^2, xy]/\left((xy)^2 - x^2y^2\right).$$

Although this is a nice ring, indeed a hypersurface, it is not factorial, since x^2y^2 has the two distinct factorizations $(x^2) \cdot (y^2) = x^2y^2 = (xy) \cdot (xy)$.

As a positive result we have the following:

PROPOSITION 1.7.3: *Let $\rho : G \hookrightarrow \mathrm{GL}(n, \mathbb{F})$ be a representation of a finite group. If there is no nontrivial homomorphism $G \longrightarrow \mathbb{F}^{\times}$, then $\mathbb{F}[V]^G$ is a unique factorization domain.*

PROOF: Let $f \in \mathbb{F}[V]^G$ and write f as a product of powers of irreducible factors $f = f_1^{k_1} \cdots f_m^{k_m}$ that are pairwise nonassociates in $\mathbb{F}[V]$. The ideals $(f_1), \ldots, (f_m)$ generated by the irreducible factors in $\mathbb{F}[V]$ are permuted by G. Let $B_1, \ldots, B_k$ be the orbits of G acting on the set of ideals $\{(f_1), \ldots, (f_m)\}$ and set

$$F_i = \prod_{f_j \in B_i} f_j^{k_i}, \qquad i = 1, \ldots, k.$$

The ideals $(F_1), \ldots, (F_k)$ are stable under the action of G, i.e., if $h \in (F_i)$ and $g \in G$, then $gh \in (F_i)$. In particular,

$$gF_i = \lambda(g)F_i \qquad \forall\, g \in G, \quad \lambda(g) \in \mathbb{F}^{\times}.$$

The function $\lambda : G \longrightarrow \mathbb{F}^{\times}$ is a representation, and since there are no nontrivial homomorphisms $G \longrightarrow \mathbb{F}^{\times}$, it follows that $\lambda(g) = 1$ for all $g \in G$ and F_i is invariant for $i = 1, \ldots, k$. No factor of F_i can be invariant, since F_i is a product over a G-orbit in $\mathbb{F}[V]$. Therefore, $F_i \in \mathbb{F}[V]^G$ is an irreducible polynomial and

$$f = F_1 \cdots F_k$$

is a representation in $\mathbb{F}[V]^G$ as a product of pairwise nonassociate irreducible factors. The uniqueness of this decomposition (in the usual sense) is readily established, completing the proof. □

It is worthwhile noting that Proposition 1.7.3 is applicable in many cases of interest, such as the following:

(i) G is a simple nonabelian group (e.g., $G = A_n$ for $n \geq 5$),
(ii) G is perfect,
(iii) $\mathbb{F}$ is a field of characteristic p and G is a finite p-group,
(iv) $\mathbb{F}$ is a finite field with q elements and G a finite group whose order is prime to $q - 1$,
(v) $\mathbb{F} = \mathbb{Q}$ and G is a group of odd order,

(vi) $\mathbb{F} = \mathbb{F}_2$ and G is finite, or $\mathbb{F} = \mathbb{F}_3$ and G has odd order.
The conclusion also holds in many other cases, such as pseudoreflection groups (see [94]), but requires a more elaborate proof. One situation of special importance is that of p-groups over a field of characteristic p, which we will need in the sequel, so we record it for future reference.

COROLLARY 1.7.4: *If $\rho : P \hookrightarrow \mathrm{GL}(n, \mathbb{F}_p)$ is a representation of a finite p-group P over a field of characteristic p, then $\mathbb{F}[V]^P$ is a unique factorization domain.* □

This lists the most important special cases of the following characterization of representations with factorial rings of invariants due to H. Nakajima, [275].

THEOREM 1.7.5 (H. Nakajima): *Let $\rho : G \hookrightarrow \mathrm{GL}(n, \mathbb{F})$ be a representation of a finite group G. Denote by $H \subseteq G$ the subgroup generated by all pseudoreflections in G. Then the following are equivalent:*

(1) *$\mathbb{F}[V]^G$ is a unique factorization domain.*
(2) *G/H has no nontrivial one-dimensional representation.*[20]
(3) *$\mathbb{F}[V]^G_\lambda$ is a free $\mathbb{F}[V]^G$-module for every one-dimensional representation $\lambda : G \longrightarrow \mathbb{F}^\times$.* □

[20] Note carefully, that H is always normal in G, see, e.g., [372] Section 7.1.

Chapter 2
Algebraic Finiteness

THIS chapter is devoted to a discussion of the algebraic finiteness property of rings of invariants of finite groups. Recall from Section 1.3 that algebraic finiteness is concerned with the finite generation of rings of invariants. Combinatorial finiteness, on the other hand, is concerned with counting, in an appropriate sense, the number of invariant forms of a given degree. These two problems are interconnected in a number of unexpected ways, which we will explain in Section 3.3, after we have established the basic combinatorial finiteness results.

The solution of the algebraic finiteness problem has led to far-reaching developments in commutative algebra. D. Hilbert's original paper, [163], on the finite generation of rings of invariants of $\mathrm{SL}(n, \mathbb{C})$ contained, as Theorem I, what is usually referred to today as Hilbert's basis theorem: *if* $\mathbb{F}$ *is a field and* $I \subseteq \mathbb{F}[z_1, \ldots, z_n]$ *an ideal, then there exists a finite number of polynomials* $f_1, \ldots, f_m \in I$ *that generate* I, *i.e.,* $I = (f_1, \ldots, f_m)$. Emmy Noether's paper [303] is no less remarkable: It introduced fundamental arguments for finite ring extensions. We will try to arrange these proofs to bring out their general character.

Two important constructions appear in this chapter: the transfer homomorphism, which we introduced in a simplified form in Section 1.4, and the Noether map. These are basic tools for constructing invariant forms. Using these tools we will show in the nonmodular case that a ring of invariants $\mathbb{F}[V]^G$ is always generated by forms of degree at most $|G|$; see Section 2.3. This solves Problem 1 from Section 1.4 in the nonmodular case. We will also make use of the transfer homomorphism to prove Molien's theorem in Chapter 3, which solves the combinatorial finiteness problem in the nonmodular case, and the Noether map in Chapter 4 to construct the fine Chern classes and show that they generate the ring of invariants in many cases.

Since the only tools to construct invariants available to us in this chapter are the transfer homomorphism and the Noether map, there are rather few examples that we can compute. We therefore postpone a serious study of examples until Chapters 3, 5, and 7.

2.1 Emmy Noether's Finiteness Theorem

Let $\rho : G \hookrightarrow \mathrm{GL}(n, \mathbb{F})$ be a representation of a finite group over the field $\mathbb{F}$, $V = \mathbb{F}^n$ the representation space, and $\mathbb{F}[V]^G$ the ring of invariants. Can we find a finite number of invariant forms, say $f_1, \dots, f_m \in \mathbb{F}[V]^G$, such that every element of $\mathbb{F}[V]^G$ can be written as a polynomial in $f_1, \dots, f_m$?

It is probably best to begin with a simple example to show that there really is something to be proved, namely that not every subalgebra of $\mathbb{F}[V]$ need be finitely generated in the above sense. Consider the subalgebra of $\mathbb{F}[x, y]$ generated by the elements

$$1, xy, xy^2, \dots, xy^n, \dots .$$

Clearly, no generator xy^n can be in the subalgebra generated by the remaining generators, since the product of any two other generators, and hence any polynomial in the other generators that is divisible by y^n, must be divisible by x^2. So there are subalgebras of $\mathbb{F}[x, y]$ that are not finitely generated.

In this book, we shall need to make use of a considerable amount of commutative algebra, some of which is special to the graded case, for which we will do our best to provide proofs, or references. As general references for commutative algebra we recommend the books [24] and [25], or one of the standard textbooks [20], [104], [243], and [465]. Some of the most often cited results are collected in Appendix A. We begin by reviewing some terminology and basic facts about Noetherian rings and modules.

A module M over a commutative ring A is said to be **Noetherian** if every ascending chain of submodules $M_0 \subseteq M_1 \subseteq \cdots \subseteq M_k \subseteq \cdots \subseteq M$ is eventually constant. A finite direct sum of Noetherian modules, as well as every submodule and quotient module of a Noetherian module, is again Noetherian. The ring A itself is said to be a **Noetherian ring** if it is so as a module over itself; in other words, if every ascending chain of ideals is eventually constant. The following results can be found in any of the standard references cited above.

PROPOSITION 2.1.1: *A commutative ring A is Noetherian if and only if every ideal is finitely generated. A module M over a Noetherian ring A is Noetherian if and only if it is finitely generated.* □

A key result in the history of commutative algebra is Theorem I from [163].

THEOREM 2.1.2 (Hilbert's Basis Theorem): *If A is a commutative Noetherian ring, then so is the polynomial ring $A[x]$.* □

COROLLARY 2.1.3: *If A is a finitely generated commutative algebra over a field $\mathbb{F}$, then A is Noetherian.*

PROOF: A is a quotient of some polynomial ring over $\mathbb{F}$, which is Noetherian by the Hilbert basis theorem. □

Emmy Noether's proof of the finite generation of rings of invariants of finite groups is based on the concept of integrality. Recall that if $A \supseteq B$ is an extension of rings, we say that an element of A is **integral** over B if it is a zero of a monic polynomial (i.e., a polynomial whose leading coefficient is one) with coefficients in B. This is the same as saying that the element lies in a subring of A that contains B and is finitely generated as a B-module. So, the sum and product of integral elements are again integral. We say that A is an **integral extension** of B if every element of A is integral over B. If A is an integral extension of B and finitely generated over B as an algebra, we say that A is a **finite extension** of B. In this case A is finitely generated as a module over B.

THEOREM 2.1.4 (E. Noether): *Suppose that $\mathbb{F}$ is a field, and G is a finite group acting as automorphisms of a finitely generated commutative $\mathbb{F}$-algebra A. Then A^G is also a finitely generated commutative $\mathbb{F}$-algebra, and A is finitely generated as a module over A^G.*

PROOF: A is an integral extension of A^G by Proposition 1.7.2. Let B be the subalgebra of A^G generated by the coefficients of the monic polynomials satisfied by a finite set of $\mathbb{F}$-algebra generators of A. Then B is a finitely generated $\mathbb{F}$-algebra, and hence Noetherian. A is a finitely generated B-module, and hence so is the B-submodule A^G. Thus A^G is a finitely generated $\mathbb{F}$-algebra. □

REMARK: The proof actually shows much more than is stated, viz.; *If $A \supseteq B$ is a finite extension, and B is a Noetherian ring, then so is A.*

As elegant as this proof is, it still has a flaw: It is nonconstructive.

(i) We need monic polynomials satisfied by generators of A over A^G. *But* we do not know A^G yet.[1]
(ii) A more serious problem is that we need a set of B-module generators for A^G. Since we do not know A^G we have no clue as to how to find the needed generators, not even a bound on their degrees.

COROLLARY 2.1.5: *Let $\rho : G \hookrightarrow \mathrm{GL}(n, \mathbb{F})$ be a representation of a finite group over the field $\mathbb{F}$. Then $\mathbb{F}[V]^G$ is a finitely generated $\mathbb{F}$-algebra, and $\mathbb{F}[V]$ is a finite integral extension of $\mathbb{F}[V]^G$.* □

[1] Actually, this can be remedied to some extent with the aid of an adequate system of parameters for A^G, constructed, e.g., with the aid of a Dade basis (see Section 4.3) or the Dickson polynomials, if $\mathbb{F}$ is finite (see Section 6.1).

As a bonus from the proof we see that for any element $z \in V^*$, there is an integer k, depending on z, such that z^k belongs to the Hilbert ideal $\mathfrak{h}(\rho) = (\mathbb{F}[V]^G)^e$, which is the ideal of $\mathbb{F}[V]$ generated by the invariants of strictly positive degree. In particular, this holds for a basis $z_1, \ldots, z_n$ for V^*. Therefore, choosing k to be maximal among such integers, we see that the coinvariants $\mathbb{F}[V]_G$ are zero in all homogeneous degrees above nk. Hence we have proven the following corollary:

COROLLARY 2.1.6: *If G is a finite group and $G \hookrightarrow \mathrm{GL}(n, \mathbb{F})$ is a representation of G. Then the ring of coinvariants $\mathbb{F}[V]_G$ is a finite-dimensional G-representation.* □

If the characteristic of the ground field $\mathbb{F}$ does not divide the order of G and $\mathbb{F}$ contains enough roots of unity, then every finite-dimensional G-representation $\rho : G \hookrightarrow \mathrm{GL}(n, \mathbb{F})$ decomposes into a direct sum of irreducible G-representations. This is true in particular for the G-representation $\mathrm{Tot}(\mathbb{F}[V]_G)$, obtained from the graded representation $\mathbb{F}[V]_G$ by totalizing; see Section A.1. It therefore follows that $\mathbb{F}[V]_G$ is fixed-point free in the sense that $(\mathbb{F}[V]_G)^G = \mathbb{F}$ is the homogeneous component of degree 0.

On the other hand, if the characteristic of $\mathbb{F}$ divides the order of G, this need no longer be the case; for example, if $\rho : P \hookrightarrow \mathrm{GL}(n, \mathbb{F})$ is a representation of a finite p-group over a field of characteristic p, then the action of P on $\mathbb{F}[V]_P$ must have a nonzero fixed-point set in every homogeneous degree that is nonzero; see Lemma 2.4.2.

Apart from these facts and a few special cases, not a great deal is known about these representations. See, however, Theorem 7.2.1, and Section 7.2 here, as well as the book [69], in particular, Theorem 2.4.6 and Chapter 11.

The iteration of this construction of taking fixed points in the algebra of coinvariants leads to a number of interesting developments; see, e.g., [187] and [285].

In a different vein, coherent rings and modules provide a very satisfactory generalization of Noetherian rings and modules (see, e.g., [358]). It might be expected that the Noetherian finiteness theorems would generalize to coherent finiteness theorems. This is not the case.

EXAMPLE 1 (S. Glaz [131]): Consider the action of $\mathbb{Z}/2$ on the polynomial algebra $\mathbb{F}[x, y, z_1, \ldots, z_n, \ldots]$, where $\mathbb{F}$ is a field of characteristic not equal to 2, given by sending a variable x, y, or z_i to its negative. The polynomial algebra $\mathbb{F}[x, y, z_1, \ldots, z_n, \ldots]$ is the quintessential example of a coherent graded algebra. The fixed subalgebra $\mathbb{F}[x, y, z_1, \ldots, z_n, \ldots]^{\mathbb{Z}/2}$ is, however, not coherent: The ideal

$$(xy{:}x^2) = (y^2, xy, yz_1, yz_2, \ldots)$$

is not finitely generated. One could, of course, take the viewpoint that the $\mathbb{Z}/2$-action lacks a certain coherence property in this example. But what should it be? If $\mathbb{Z}/2$ acts on $\mathbb{F}[x_1, \ldots, y_1, \ldots]$ by interchanging the x-variables and the y-variables, then the ring of invariants is $\mathbb{F}[x_1 + y_1, x_1y_1, x_2 + y_2, x_2y_2, \ldots]$, the prototypical example of a coherent ring that is not Noetherian.

2.2 The Transfer Homomorphism

A basic tool used in the invariant theory of finite groups is the **transfer homomorphism**, which was introduced in Section 1.4. For many applications it is advantageous to have a relative version, which we define next. If H is a subgroup of a finite group G and V is a finite-dimensional G-module, we define the **relative transfer from H to G**

$$\mathrm{Tr}_H^G : \mathbb{F}[V]^H \longrightarrow \mathbb{F}[V]^G$$

by

$$\mathrm{Tr}_H^G(f)(x) = \sum_{gH \in G/H} g(f)(x) = \sum_{gH \in G/H} f(\rho(g)^{-1}(x)).$$

The notation means that the sum runs over a set of left coset representatives of H in G. It is an easy calculation to show that for $f \in \mathbb{F}[V]^H$ and $g'H = g''H \in G/H$ then $g'f = g''f \in \mathbb{F}[V]^G$, so the right-hand side makes sense independent of the choice of elements in G representing the cosets G/H. Since

$$g' \cdot \mathrm{Tr}_H^G(f) = \sum_{g''H \in G/H} g'g''(f) = \sum_{g'g''H \in G/H} g'g''(f) = \mathrm{Tr}_H^G(f)$$

(if $\{g''H\}$ runs through all cosets exactly once, so does $\{g'g''H\}$ for a fixed g'), we see that $\mathrm{Tr}_H^G(f) \in \mathbb{F}[V]^G$. Note that the composite

$$\mathbb{F}[V]^G \hookrightarrow \mathbb{F}[V]^H \xrightarrow{\mathrm{Tr}_H^G} \mathbb{F}[V]^G$$

is equal to multiplication by $|G : H|$, the **index** of H in G. In particular, if $|G : H|$ is invertible in $\mathbb{F}$, then the transfer is surjective, and the map

$$\pi_H^G = \frac{1}{|G : H|} \mathrm{Tr}_H^G : \mathbb{F}[V]^H \longrightarrow \mathbb{F}[V]^G \subseteq \mathbb{F}[V]^H$$

is an idempotent projection whose image is equal to $\mathbb{F}[V]^G$. In this case $\mathbb{F}[V]^H$ is a direct sum of $\mathbb{F}[V]^G$ and the kernel of π_H^G. If $H = 1$, we just write π^G for $\pi_1^G = \frac{1}{|G|} \sum_{g \in G} g$, so that $\mathbb{F}[V] = \mathbb{F}[V]^G \oplus \ker(\pi^G)$ when $|G| \in \mathbb{F}^\times$. So we have shown the following:

PROPOSITION 2.2.1: *Let $\rho : G \hookrightarrow \mathrm{GL}(n, \mathbb{F})$ be a representation of a finite group G over the field $\mathbb{F}$ and $H \leq G$ a subgroup. If $|G : H|$ is invertible in the ground field $\mathbb{F}$, then the transfer homomorphism $\mathrm{Tr}_H^G : \mathbb{F}[V]^H \longrightarrow \mathbb{F}[V]^G$ is an epimorphism.* □

COROLLARY 2.2.2: *Let G be a finite group and $H \leq G$ a subgroup with $|G:H| \in \mathbb{F}^{\times}$. Suppose that V', V'' are representations of G over the field $\mathbb{F}$ and $\varphi : V' \longrightarrow V''$ a G-equivariant map. If the induced map $\varphi^* : \mathbb{F}[V'']^H \longrightarrow \mathbb{F}[V']^H$ is an epimorphism, then so is the induced map $\varphi^* : \mathbb{F}[V'']^G \longrightarrow \mathbb{F}[V']^G$.*

PROOF: This follows from the commutative square

$$\begin{array}{ccc} \mathbb{F}[V'']^H & \xrightarrow{\varphi^*} & \mathbb{F}[V']^H \\ \downarrow {\scriptstyle \mathrm{Tr}^G_H} & & \downarrow {\scriptstyle \mathrm{Tr}^G_H} \\ \mathbb{F}[V'']^G & \xrightarrow[\varphi^*]{} & \mathbb{F}[V']^G \end{array}$$

and Proposition 2.2.1. □

REMARK: A primary application of Corollary 2.2.2 is to the case where H is the trivial subgroup, $|G| \in \mathbb{F}^{\times}$, and the dual map $\varphi^* : V''^* \longrightarrow V'^*$ is an epimorphism. See, e.g., [370], [376], and [379].

In general, the transfer behaves badly with respect to products in $\mathbb{F}[V]^H$. However, note that $\mathbb{F}[V]^G$ being a subalgebra of $\mathbb{F}[V]^H$ means that $\mathbb{F}[V]^H$ is an $\mathbb{F}[V]^G$-module. Moreover,

$$\left.\begin{array}{l} \mathrm{Tr}^G_H(f) = |G:H| \cdot f \\ \mathrm{Tr}^G_H(f \cdot h) = f \cdot \mathrm{Tr}^G_H(h) \end{array}\right\} \quad \forall\, f \in \mathbb{F}[V]^G,\, h \in \mathbb{F}[V]^H,\, \deg(h) > 0.$$

Therefore, Tr^G_H is an $\mathbb{F}[V]^G$-module homomorphism, and if $|G:H| = 0 \in \mathbb{F}$, then the transfer is zero on $\mathbb{F}[V]^G$.

Proposition 2.2.1 assures us that the transfer homomorphism is nonzero in the nonmodular case. To explore what happens in the modular case we require the following version of R. Dedekind's independence lemma. For a proof see [182] Chapter I Section 3 Theorem 3.

LEMMA 2.2.3 (R. Dedekind): *Let A be a commutative integral domain containing a field $\mathbb{F}$ and $\alpha_1, \ldots, \alpha_n \in \mathrm{Aut}(A)$ distinct automorphisms of A. Then $\alpha_1, \ldots, \alpha_n$ are A-linearly independent in $\mathrm{Hom}_{\mathbb{F}}(A, A)$.* □

PROPOSITION 2.2.4: *Let $\rho : G \hookrightarrow \mathrm{GL}(n, \mathbb{F})$ be a representation of a finite group G over the field $\mathbb{F}$. Then $\mathrm{Tr}^G : \mathbb{F}[V] \longrightarrow \mathbb{F}[V]^G$ is not zero.*

PROOF: If $\mathrm{Tr}^G = 0$, then the set $\{\rho(g) \mid g \in G\} \subseteq \mathrm{Aut}(\mathbb{F}[V])$ of distinct automorphisms is linearly dependent in $\mathrm{End}_F(\mathbb{F}[V]) = \mathrm{End}_{\mathbb{F}}(V)$, since

$$0 = \mathrm{Tr}^G = \sum_{g \in G} 1 \cdot g.$$

This contradicts Lemma 2.2.3, the lemma of R. Dedekind. □

So the transfer homomorphism is never zero, not even in the modular case. The image of the transfer $\mathrm{Tr}^G : \mathbb{F}[V] \hookrightarrow \mathbb{F}[V]^G$ is an ideal, and one measure of how large it is is its height. In Theorem 2.4.5 we will show that the height of $\mathrm{Im}(\mathrm{Tr}^G) \subseteq \mathbb{F}[V]^G$ is at most $n-1$ if the characteristic of $\mathbb{F}$ actually divides the order of G.

EXAMPLE 1 (K. Kuhnigk [215]): Consider the action of the symmetric group Σ_n on the polynomial algebra $\mathbb{F}[x_1, \ldots, x_n]$ by permutation of the variables, where $\mathbb{F}$ is a field of characteristic 2. We know that $\mathbb{F}[x_1, \ldots, x_n]^{\Sigma_n} = \mathbb{F}[e_1, \ldots, e_n]$, where $e_1, \ldots, e_n$ are the elementary symmetric polynomials. $\mathbb{F}[x_1, \ldots, x_n]$ is a free finitely generated module over $\mathbb{F}[e_1, \ldots, e_n]$, and is generated by monomials of degree less than or equal to $\binom{n}{2}$. This is rather easy to see using the fact that $\mathbb{F}[z_1, \ldots, z_n]$ is a Cohen-Macaulay algebra and $e_1, \ldots, e_n \in \mathbb{F}[z_1, \ldots, z_n]$ is a regular sequence. See Chapter 5. It therefore suffices to compute the transfer on monomials of degree at most $\binom{n}{2}$, since these will generate $\mathrm{Im}(\mathrm{Tr}^{\Sigma_n})$ as an $\mathbb{F}[e_1, \ldots, e_n]$-module, and hence also as an ideal.

Let $x^K = x_1^{k_1} \cdots x_n^{k_n}$ be a monomial. If two distinct exponents among $k_1, \ldots, k_n$ are equal, then the isotropy group $(\Sigma_n)_{x^K}$ of x^K contains the involution switching those two indices. Let $\mathbb{Z}/2 \leq \Sigma_n$ be the subgroup generated by this involution. Then

$$\mathrm{Tr}^{\Sigma_n}(x^K) = \mathrm{Tr}^{\Sigma_n}_{\mathbb{Z}/2}(\mathrm{Tr}^{\mathbb{Z}/2}(x^K)) = \mathrm{Tr}^{\Sigma_n}_{\mathbb{Z}/2}(x^K + x^K) = 0,$$

since $\mathbb{F}$ has characteristic 2. The only monomials of degree less than or equal to $\binom{n}{2}$ with all indices distinct are the $n!$ monomials in the Σ_n-orbit of the monomial $x_1^1 \cdot x_2^2 \cdots x_{n-1}^{n-1}$. Therefore, the image of the transfer is the principal ideal generated by

$$\mathrm{Tr}^{\Sigma_n}(x_1^1 \cdot x_2^2 \cdots x_{n-1}^{n-1}) = \sum_{\sigma \in \Sigma_n} x_{\sigma(1)}^1 \cdot x_{\sigma(2)}^2 \cdots x_{\sigma(n-1)}^{n-1}.$$

With the aid of the Leibniz rule for expanding determinants applied to the Vandermonde determinant, and the fact that $+1 = -1 \in \mathbb{F}$, one sees that

$$\prod_{1 \leq i < j \leq n} (x_i + x_j) = \det \begin{bmatrix} 1 & x_1^1 & \ldots & x_1^{n-1} \\ \vdots & \vdots & & \vdots \\ 1 & x_n^1 & \ldots & x_n^{n-1} \end{bmatrix} = \sum_{\sigma \in \Sigma_n} x_{\sigma(1)}^1 \cdots x_{\sigma(n-1)}^{n-1}.$$

Thus the image of the transfer in this case is the principal ideal generated by the **discriminant**.

2.3 Emmy Noether's Bound

Recall that for a representation $\rho : G \hookrightarrow \mathrm{GL}(n, \mathbb{F})$ of a finite group over the field $\mathbb{F}$ we write $\beta(\rho)$ for the maximum degree of a generator in a minimal algebra generating set for $\mathbb{F}[V]^G$: $\beta_{\mathbb{F}}(G)$ denotes the maximum of $\beta(\rho)$ for representations ρ of G defined over $\mathbb{F}$. In the remarkable three and one half page paper [301] Emmy Noether gave two complete proofs of the finiteness theorem over $\mathbb{C}$, a correction to an argument of H. Weber's, and showed that $\beta_{\mathbb{C}}(G) \leq |G|$. A careful reading of her proof yields the inequality $\beta_{\mathbb{F}}(G) \leq |G|$ if $|G|!$ is invertible in $\mathbb{F}$. The problem of extending this upper bound, Emmy Noether's bound, to all nonmodular cases was finally achieved, after many partial solutions, e.g., [327], [370], [376], and [379], to name but a few, independently by P. Fleischmann [120] and J. Fogarty [122].

This section is based on an elegant lemma of D. J. Benson [32] in the spirit of one of the arguments due to Emmy Noether, [301]. Both Emmy Noether's proofs of the finiteness theorem have been reworked many times; in addition to the sources already mentioned see, e.g., [26], [387], and [451]. Recall that for a ring extension $B \subseteq A$ and an ideal $I \subset B$ that I^{e} denotes the extended ideal in A. If G acts on A and an ideal $I \subset A$ is stable under the G-action, i.e., mapped into itself by all the elements of G, then I^G denotes the subset of elements fixed by G. This should not be confused with $I^{|G|}$ which denotes the $|G|$th power of the ideal I.

LEMMA 2.3.1 (D. J. Benson): *Let A be a commutative ring with 1 and $\rho : G \hookrightarrow \mathrm{Aut}(A)$ a representation of G by automorphisms of A. If $|G|$ is invertible in A and $I \subset A$ is a G-stable ideal then $I^{|G|} \subset (I^G)^{\mathrm{e}}$.*

PROOF: Choose $|G|$ elements in I and index them by the elements of G, say $\{f_g \mid g \in G\}$. Then

$$\prod_{g \in G} (hg(f_g) - f_g) = 0$$

for any $h \in G$. Multiplying this out and summing over all $h \in G$ gives

$$\sum_{S \subseteq G} (-1)^{|G \setminus S|} \left(\sum_{h \in G} \prod_{g \in S} h(gf_g) \right) \cdot \left(\prod_{g \in G \setminus S} f_g \right) = 0,$$

where S runs through all subsets of G. The term on the left corresponding to the empty set is $\pm|G| \cdot \prod_{g \in G} f_g$ and all the other terms lie in $(I^G)^{\mathrm{e}} = A \cdot I^G$, because $\sum_{h \in G} \prod_{g \in S} h(gf_g)$ is invariant. So the result is proved by rearranging terms and dividing by $|G|$. □

Recall from Section 1.2 that the Hilbert ideal associated with the representation $\rho : G \hookrightarrow \mathrm{GL}(n, \mathbb{F})$ is the ideal $\mathfrak{h}(\rho)$ in $\mathbb{F}[V]$ generated by the G-

invariant forms of positive degree, i.e., $\mathfrak{h}(\rho)$ is the extended ideal $(\overline{\mathbb{F}[V]^G})^e$ of the augmentation ideal $\overline{\mathbb{F}[V]^G}$. From the preceding lemma we therefore obtain the following consequence.

PROPOSITION 2.3.2 (P. Fleischmann, J. Fogarty): *Let $\rho : G \hookrightarrow \mathrm{GL}(n, \mathbb{F})$ be a representation of a finite group over the field $\mathbb{F}$ and suppose that the characteristic of $\mathbb{F}$ does not divide the order of G. Then the Hilbert ideal $\mathfrak{h}(\rho)$ is generated as an ideal by G-invariant forms of degree at most $|G|$.*

PROOF: Let $z_1, \ldots, z_n$ be a basis for the space V^* of linear forms and z^A a monomial in them of degree $|G|$ or larger. Such a monomial is a product of $|G|$ or more linear forms, so by Lemma 2.3.1 applied to the augmentation ideal of $\mathbb{F}[V]$ the monomial z^A lies in $\mathfrak{h}(\rho)$. If $\deg(z^A) > |G|$ write $z^A = z^B \cdot z^C$ where $\deg(z^B) = |G|$. Then $z^B \in \mathfrak{h}(\rho)$, so z^A is not part of a minimal generating set for $\mathfrak{h}(\rho)$. □

From this result, by making use of the original argument of D. Hilbert [163] to prove the finiteness theorem, one obtains Emmy Noether's bound in the nonmodular case.

THEOREM 2.3.3 (P. Fleischmann, J. Fogarty): *Let $\rho : G \hookrightarrow \mathrm{GL}(n, \mathbb{F})$ be a representation of a finite group over the field $\mathbb{F}$ and suppose that the characteristic of $\mathbb{F}$ does not divide the order of G. Then $\mathbb{F}[V]^G$ is generated by G-invariant forms of degree at most $|G|$.*

PROOF: Recall the projection operator $\pi^G = \frac{1}{|G|} \cdot \mathrm{Tr}^G : \mathbb{F}[V] \longrightarrow \mathbb{F}[V]^G$ from Section 2.2. Denote by $\overline{\mathbb{F}[V]^G}$ the augmentation ideal of $\mathbb{F}[V]^G$. Since $\overline{\mathbb{F}[V]^G} \subset \mathfrak{h}(G)$, it follows that $\pi^G : \mathfrak{h}(G) \longrightarrow \overline{\mathbb{F}[V]^G}$ is an epimorphism in positive degrees. Since it is also an $\mathbb{F}[V]^G$-module homomorphism, the ideal generators for $\mathfrak{h}(G)$ are mapped to module generators for $\overline{\mathbb{F}[V]^G}$, but module generators for $\overline{\mathbb{F}[V]^G}$ generate $\mathbb{F}[V]^G$ as an algebra (see Appendix A). □

By a simple linearization argument one obtains a result useful in proving a relative version of Emmy Noether's bound.

COROLLARY 2.3.4: *Let G be a finite group acting by grading preserving automorphisms on the commutative connected graded Noetherian algebra A over the field $\mathbb{F}$. If $|G|$ is invertible in $\mathbb{F}$ then $\beta(A^G) \leq |G| \beta(A)$.*

PROOF: Let W be the graded vector space defined by

$$W = \begin{cases} \mathbb{F}[V]_i & \text{if } 1 \leq i \leq \beta(A) \\ 0 & \text{otherwise} \end{cases}$$

and $\mathbb{F}[W] \longrightarrow A$ the map induced by the inclusion $W \subseteq A$. There is the commutative diagram

$$\begin{array}{ccc} \mathbb{F}[W] & \longrightarrow & A \\ \Big\downarrow {\scriptstyle \mathrm{Tr}^G} & & \Big\downarrow {\scriptstyle \mathrm{Tr}^G} \\ \mathbb{F}[W]^G & \longrightarrow & A^G \end{array}$$

where both transfer maps are epimorphisms. Therefore $\mathbb{F}[W]^G \longrightarrow A^G$ is also an epimorphism and the result follows from Theorem 2.3.3. □

COROLLARY 2.3.5: *Let $\varrho : G \hookrightarrow \mathrm{GL}(n, \mathbb{F})$ be a representation of the finite group G over the field $\mathbb{F}$ and $H \triangleleft G$ a normal subgroup with $|G:H|$ invertible in $\mathbb{F}$. Then $\beta(\varrho) \leq |G:H| \cdot \beta(\varrho\,|_H)$.*

PROOF: This follows from Corollary 2.3.4 and the equality $\mathbb{F}[V]^G = \left(\mathbb{F}[V]^H\right)^{G/H}$. □

Here is a simple example to illustrate the ideas and results developed so far.

EXAMPLE 1: The cyclic group of order 3 acts on the vector space $\mathbb{F}^3$ by cyclic permutation of the standard basis vectors E_1, E_2, E_3. The 2-dimensional subspace V spanned by the vectors with coordinate sum 0 is $\mathbb{Z}/3$-invariant and has as a basis the vectors $E_1 - E_2, E_3 - E_2$. The matrix of the generator of $\mathbb{Z}/3$ with respect to this basis for V is

$$\mathbf{A} = \begin{bmatrix} 0 & -1 \\ 1 & -1 \end{bmatrix} \in \mathrm{GL}(2, \mathbb{F}).$$

If $\{x, y\}$ is the dual basis of V^*, then

$$\mathbf{A}x = -y, \quad \mathbf{A}y = x - y.$$

If the characteristic of $\mathbb{F}$ is different from 3, then by Theorem 2.3.3, $\mathbb{F}[x, y]^{\mathbb{Z}/3}$ is generated as an algebra by polynomials of degree at most 3. The polynomials

$$\begin{aligned} x, y &\in S^1(V^*), \\ x^2, xy, y^2 &\in S^2(V^*), \\ x^3, x^2y, xy^2, y^3 &\in S^3(V^*) \end{aligned}$$

are a basis for the homogeneous components of $\mathbb{F}[x, y]$ of degree 1, 2, and 3 respectively. The proof of Emmy Noether's theorem shows that a set of algebra generators for $\mathbb{F}[x, y]^{\mathbb{Z}/3}$ may be obtained by applying the projection

$$\pi^{\mathbb{Z}/3} : \mathbb{F}[x, y] \longrightarrow \mathbb{F}[x, y]^{\mathbb{Z}/3}$$

to these nine polynomials. Table 2.3.1 summarizes this computation.

φ	$\pi^{\mathbb{Z}/3}(\varphi)$
x	0
y	0

φ	$\pi^{\mathbb{Z}/3}(\varphi)$
x^2	$\frac{1}{3}\left(2x^2 - 2xy + 2y^2\right)$
xy	$\frac{1}{3}\left(x^2 - xy + y^2\right)$
y^2	$\frac{1}{3}\left(2x^2 - 2xy + 2y^2\right)$

φ	$\pi^{\mathbb{Z}/3}(\varphi)$
x^3	$x^2y - xy^2$
x^2y	$\frac{1}{3}\left(-x^3 + 3x^2y - y^3\right)$
xy^2	$\frac{1}{3}\left(-x^3 + 3xy^2 - y^3\right)$
y^3	$-x^2y + xy^2$

TABLE 2.3.1: The values of $\pi^{\mathbb{Z}/3}(\varphi)$ for $\deg(\varphi) \leq 3$

From this we see that the polynomials

$$\begin{aligned} f &= x^2 - xy + y^2, \\ h &= x^2y - xy^2, \\ k &= x^3 - 3xy^2 + y^3 \qquad (\text{or } x^3 - 3x^2y + y^3) \end{aligned}$$

are a set of algebra generators for $\mathbb{F}[x, y]^{\mathbb{Z}/3}$. Since h and k have degree 3, the upper bound on the degrees given by Emmy Noether's theorem is sharp. Note, however, that to obtain these 3 fundamental invariants we had to consider 9 polynomials.

In the modular case Emmy Noether's bound may fail for the ring of invariants. Perhaps the simplest such example, though not the smallest with respect to the dimension,[2] is the representation of $\mathbb{Z}/2$ on the set $\{x_1, x_2, x_3, y_1, y_2, y_3\}$ by simultaneous permutation of x_1, x_2, x_3 with y_1, y_2, y_3.

EXAMPLE 2: Let $\mathbb{F}$ be a field of characteristic 2 and denote by $\sigma_3 : \mathbb{Z}/2 \hookrightarrow \mathrm{GL}(6, \mathbb{F})$ the representation of $\mathbb{Z}/2$ implemented by the permutation matrix

$$\mathbf{S} = \begin{bmatrix} \begin{smallmatrix} 0 & 1 \\ 1 & 0 \end{smallmatrix} & 0 & 0 \\ 0 & \begin{smallmatrix} 0 & 1 \\ 1 & 0 \end{smallmatrix} & 0 \\ 0 & 0 & \begin{smallmatrix} 0 & 1 \\ 1 & 0 \end{smallmatrix} \end{bmatrix} \in \mathrm{GL}(6, \mathbb{F}).$$

The Poincaré series of $\mathbb{F}[x_1, x_2, x_3, y_1, y_2, y_3]^{\mathbb{Z}/2}$ will be computed in Sec-

[2] This is M.-J. Bertin's example of the ring of invariants of $\mathbb{Z}/4$ in characteristic 2 acting by cyclic permutation of x_1, x_2, x_3, x_4, Section 1.1, Example 3. A minimal generating set must contain a form of degree 5.

tion 3.2, Example 1. Here it is:

$$P(\mathbb{F}[x_1, x_2, x_3, y_1, y_2, y_3]^{\mathbb{Z}/2}, t) = \frac{1}{2}\left[\frac{(1+t)^3+(1-t)^3}{(1-t)^3(1-t^2)^3}\right]$$
$$= \frac{1+3t^2}{(1-t)^3(1-t^2)^3} = 1 + 3t + 12t^2 + 28t^3 + \cdots .$$

From this we see that the space of invariant linear forms has dimension 3, the space of invariant quadratic forms dimension 12, the space of invariant cubic forms dimension 28, etc. To compute requisite bases, use the monomial basis in $x_1, x_2, x_3, y_1, y_2, y_3$ and Lemma 3.2.1. This gives

$$\ell_i = x_i + y_i, \qquad i = 1, 2, 3,$$

as a basis for the invariant linear forms, and the 6 products

$$\ell_i \ell_j, \qquad 1 \leq i \leq j \leq 3,$$

together with the 6 quadratic polynomials

$$\begin{aligned} q_i &= x_i y_i, \qquad i = 1, 2, 3, \\ Q_3 &= x_1x_2 + y_1y_2, \\ Q_2 &= x_1x_3 + y_1y_3, \\ Q_1 &= x_2x_3 + y_2y_3 \end{aligned}$$

as a basis for the space of invariant quadratic forms. From these, at most 28 linearly independent invariant cubic polynomials can be generated as products: 18 products of a linear and a quadratic polynomial, and 10 products of three linear polynomials, so the space of cubic forms in the subalgebra generated by these polynomials has at most dimension 28. The crucial observation, [297] or [373] §4 Example 2, is that

$$\ell_1\ell_2\ell_3 = Q_1\ell_1 + Q_2\ell_2 + Q_3\ell_3 + 2(x_1x_2x_3 + y_1y_2y_3),$$

and hence in characteristic 2 this space of cubic forms has dimension at most 27. In characteristic different from 2 it has the required dimension 28. Therefore, the algebra of invariants $\mathbb{F}[x_1, x_2, x_3, y_1, y_2, y_3]^{\mathbb{Z}/2}$ contains an indecomposable[3] cubic form, in particular $x_1x_2x_3 + y_1y_2y_3$ is one such, and Noether's bound, $|\mathbb{Z}/2| = 2$, does not hold in this example.

2.4 Feshbach's Transfer Theorem

The transfer homomorphism played a decisive role in the proof of Emmy Noether's bound in the previous section, and it is an important tool in many other nonmodular invariant-theoretic situations. In the modular case

[3] Indeed, it can be shown, see, e.g., [327], that the 3 linear forms and the 6 quadratic forms together with the cubic form $c = x_1x_2x_3 + y_1y_2y_3$ form a complete set of algebra generators of the ring of invariants in characteristic 2.

the image of the transfer $\mathrm{Tr}^G : \mathbb{F}[V] \longrightarrow \mathbb{F}[V]^G$ is a proper ideal. In [114] M. Feshbach investigated, among other things, the height of this ideal. The height of an ideal provides one measure of how large it is. This section is devoted to a proof of M. Feshbach's transfer theorem taken from the unpublished manuscript [216].

Let $g \in \mathrm{GL}(n, \mathbb{F})$ and define the element $\partial_g = 1 - g \in \mathrm{Mat}_{n,n}(\mathbb{F})$, where $\mathrm{Mat}_{n,n}(\mathbb{F})$ denotes the algebra of $n \times n$ matrices over $\mathbb{F}$. The element ∂_g acts on the algebra $\mathbb{F}[V]$ as a linear **twisted differential**, i.e.,

$$\begin{aligned} \partial_g(f+h) &= \partial_g(f) + \partial_g(h), \\ \partial_g(f \cdot h) &= \partial_g(f) \cdot h + (gf) \cdot \partial(h), \end{aligned}$$

as the following computation shows:

$$\begin{aligned} \partial_g(f+h) &= f + h - gf - gh = f - gf + h - gh = \partial_g(f) + \partial_g(h), \\ \partial_g(f \cdot h) &= f \cdot h - g(f \cdot h) f \cdot h - (gf) \cdot h + (gf) \cdot h - (gf) \cdot (gh) \\ &= (f - gf) \cdot h + (gf)(h - gh) = \partial_g(f) \cdot h + (gf) \cdot \partial_g(h). \end{aligned}$$

Let $I_g \subseteq \mathbb{F}[V]$ denote the ideal generated by $\partial_g(V^*)$, where as usual V^* is the dual space of $V = \mathbb{F}^n$, regarded as the homogeneous component of $\mathbb{F}[V]$ of degree 1, i.e., as the homogeneous linear forms in $\mathbb{F}[V]$.

LEMMA 2.4.1: *Let $g \in \mathrm{GL}(n, \mathbb{F})$ and let $\mathrm{Im}(\partial_g)$ be the image of the twisted differential $\partial_g : \mathbb{F}[V] \longrightarrow \mathbb{F}[V]$. Then $\mathrm{Im}(\partial_g) \subseteq I_g$.*

PROOF: Choose a basis $z_1, \ldots, z_n$ for V^* and identify $\mathbb{F}[V]$ with $\mathbb{F}[z_1, \ldots, z_n]$ as usual. Both the set $\mathrm{Im}(\partial_g)$ and the ideal I_g are graded subsets of $\mathbb{F}[z_1, \ldots, z_n]$, and they agree in grading 1 by definition. Therefore, we may proceed inductively and suppose that $\mathrm{Im}(\partial_g)$ is contained I_g in degree $< d$. If $f \in \mathbb{F}[z_1, \ldots, z_n]$ is a form of degree d, then it is a sum of monomials $z^E = z_1^{e_1} \cdots z_n^{e_n}$, and so, by linearity of ∂_g, it is enough to show that $\partial_g(z^E) \in I_g$. Without loss of generality we may suppose that $e_1 > 0$, so we may write $z^E = z_1 z^F$, where $F = (e_1 - 1, e_2, \ldots, e_n)$. Since the monomial z^F has degree $d - 1$, we know by induction that $\partial_g(z^F) \in I_g$. By the twisted derivation formula,

$$\partial_g(z^E) = \partial_g(z_1) \cdot z^F + (gz_1) \cdot \partial_g(z^F).$$

By definition $\partial_g(z_1) \in I_g$. Since I_g is an ideal, it follows that both terms on the right-hand side of the preceding equation are in I_g, thus completing the induction step, and hence the proof. □

LEMMA 2.4.2: *If $\mathbb{F}$ is a field of characteristic p and $u \in \mathrm{GL}(n, \mathbb{F})$ is an element of order p, then $V^u \neq \{0\} \neq (V^*)^u$.*

PROOF: Let $\mathbb{Z}/p < \mathrm{GL}(n, \mathbb{F})$ be the subgroup generated by u. Since $\mathbb{F}$ is of characteristic p, we have $0 = 1 - u^p = (1-u)^p$. Let k be the largest integer such that $(1-u)^k \neq 0$ and $x \in V$ an element such that $y = (1-u)^k(x) \neq 0$.

Then $(1-u)(y)=0$, so $y \in V^{\mathbb{Z}/p}$ is nonzero. The same argument applies to the dual space V^*. □

If $\mathfrak{p}$ is a prime ideal in the commutative ring A with 1 then the **height of** $\mathfrak{p}$ is defined to be the length of the longest chain of prime ideals

$$\mathfrak{p}_0 \subsetneqq \mathfrak{p}_1 \subsetneqq \cdots \subsetneqq \mathfrak{p}_k = \mathfrak{p} \subset A$$

ending at $\mathfrak{p}$. Prime ideals of height zero are exactly the minimal prime ideals in A. For an arbitrary ideal $I \subset A$ its height is by definition the minimal height of a prime ideal containing it. In a Noetherian ring A ideals always have finite height, and the maximum height of a prime ideal is called the **Krull dimension** of A. For more information about the height of ideals and the Krull dimension see one of the standard text books on commutative algebra, such as, [24], [104], or [465].

PROPOSITION 2.4.3 (M. Feshbach): *Let $\mathbb{F}$ be a field of characteristic p, $u \in \mathrm{GL}(n, \mathbb{F})$ an element of order p, and $P < \mathrm{GL}(n, \mathbb{F})$ the subgroup generated by u. Then $\mathrm{Im}(\mathrm{Tr}^P) \subseteq I_u$ and*

$$\mathit{ht}\left(\sqrt{\mathrm{Im}(\mathrm{Tr}^P)}\right) \leq n - \dim_{\mathbb{F}}((V^*)^u) < n.$$

PROOF: Note that in the group ring $\mathbb{F}(P)$ we have

$$\mathrm{Tr}^P = 1 + u + \cdots + u^{p-1} = (1-u)^{p-1} = \partial_u^{p-1}.$$

Hence $\mathrm{Im}(\mathrm{Tr}^P) = \mathrm{Im}(\partial_u^{p-1}) \subseteq \mathrm{Im}(\partial_u) \subseteq I_u$ by Lemma 2.4.1. The ideal I_u is a prime ideal in $\mathbb{F}[V]$, since it is generated by linear forms. The height of I_u is equal to $\dim_{\mathbb{F}}(\partial_u(V^*)) = n - \dim_{\mathbb{F}}(\ker(\partial_u : V^* \longrightarrow V^*))$ and $\ker(\partial_u : V^* \longrightarrow V^*) = (V^*)^u$. Hence I_u has height $n - \dim_{\mathbb{F}}((V^*)^u)$. Since $I_u \subset \mathbb{F}[V]$ is prime, the contracted ideal $\mathbb{F}[V]^P \cap I_u$ is also prime. By the going-up theorem (see Theorem A.2.3) $\mathit{ht}(\mathbb{F}[V]^P \cap I_u) = \mathit{ht}(I_u)$. Since $\mathrm{Im}(\mathrm{Tr}^P) \subseteq \mathbb{F}[V]^P \cap I_u$ we get the first inequality

$$\mathit{ht}\left(\sqrt{\mathrm{Im}(\mathrm{Tr}^P)}\right) \leq n - \dim_{\mathbb{F}}((V^*)^u).$$

Finally, since u has order p and $\mathbb{F}$ has characteristic p, $(V^*)^u \neq \{0\}$ by Lemma 2.4.2, and the result follows. □

LEMMA 2.4.4: *Let $\rho : G \hookrightarrow \mathrm{GL}(n, \mathbb{F})$ be a representation of a finite group G over the field $\mathbb{F}$, and $H \leq G$ a subgroup. Then $\mathrm{Im}(\mathrm{Tr}^G) \subseteq \mathrm{Im}(\mathrm{Tr}^H)$.*

PROOF: Choose a right transversal $g_1, \ldots, g_t$ for H in G. Then $G = Hg_1 \sqcup \cdots \sqcup Hg_t$, and hence for any $f \in \mathbb{F}[V]$,

$$\mathrm{Tr}^G(f) = \sum_{g \in G} gf = \sum_{h \in H} \sum_{i=1}^{t} hg_i f = \sum_{h \in H} h(g_1 f + \cdots + g_t f) = \mathrm{Tr}^H(g_1 f + \cdots + g_t f),$$

and the result follows. □

THEOREM 2.4.5 (M. Feshbach): *Let $\rho : G \hookrightarrow \mathrm{GL}(n, \mathbb{F})$ be a representation of a finite group over the field $\mathbb{F}$. Then*

(i) $\mathrm{Im}(\mathrm{Tr}^G) \subseteq \bigcap_{g \in G \mid |g| = p} I_g$,

(ii) $\mathit{ht}\left(\sqrt{\mathrm{Im}(\mathrm{Tr}^G)}\right) \leq n - \max_{|g|=p} \{\dim_{\mathbb{F}}(V^g)\}$.

PROOF: For each element $g \in G$ of order p we have by Lemma 2.4.4 that $\mathrm{Im}(\mathrm{Tr}^G) \subseteq \mathrm{Im}(\mathrm{Tr}^{<g>})$, where $<g>$ denotes the subgroup of G generated by g, and the result follows from Proposition 2.4.3. □

This theorem has many consequences, which will appear at various points throughout the remainder of the text. Specifically, we apply it in Section 6.4 to determine the affine variety defined by the ideal $\mathrm{Im}(\mathrm{Tr}^G)^e \subset \mathbb{F}[V]$ in the modular case. In particular, in Theorem 6.4.7 we show that the inequality *(ii)* of M. Feshbach's theorem is always an equality.

We close this chapter by applying our new knowledge about the transfer to two of the examples we have already encountered.

EXAMPLE 1: In Example 3 of Section 1.1 we indicated some properties of the invariants of the regular representation of $\mathbb{Z}/4$ over a field of characteristic 2. Here we show how to calculate the height of the ideal $\mathrm{Im}(\mathrm{Tr}^{\mathbb{Z}/4})$ with the aid of Theorem 2.4.5.

The group $\mathbb{Z}/4 = <g>$ contains precisely one subgroup of order 2, which is generated by g^2. So applying M. Feshbach's transfer theorem yields

$$\mathit{ht}\left(\mathrm{Im}(\mathrm{Tr}^{\mathbb{Z}/4})\right) \leq 4 - \dim_{\mathbb{F}}(V^{g^2}) = 4 - 2 = 2.$$

By R. Dedekind's independence lemma, Proposition 2.2.4, the transfer is not zero, so

$$1 \leq \mathit{ht}\left(\mathrm{Im}(\mathrm{Tr}^{\mathbb{Z}/4})\right) \leq 2.$$

Indeed, $\mathit{ht}\left(\mathrm{Im}(\mathrm{Tr}^{\mathbb{Z}/4})\right) = 2$. One way to see this is as follows: Assume to the contrary that the height of the image of the transfer were 1. Note that by Remark 2.5 in [349] or Theorem 2.4 in [292] the ideal

$$\mathrm{Im}(\mathrm{Tr}^{\mathbb{Z}/4}) \subset \mathbb{F}[V]^{\mathbb{Z}/4}$$

is prime. By Corollary 1.7.4 the ring of invariants $\mathbb{F}[V]^{\mathbb{Z}/4}$ is a unique factorization domain. Therefore, if the height of $\mathrm{Im}(\mathrm{Tr}^{\mathbb{Z}/4})$ were 1, this ideal would have to be principal. This in turn implies that $\mathbb{F}[V]^{\mathbb{Z}/4}$ would be Cohen-Macaulay by Proposition 3.7 in [216]. This is a contradiction, since the ring of invariants has depth 3, but Krull dimension 4.

EXAMPLE 2: Reconsider the 3-fold regular representation of $\mathbb{Z}/2$, Example 2 in Section 2.3. Assume that the ground field has characteristic 2.

Then Theorem 2.4.5 combined with Proposition 2.2.4 gives

$$1 \leq ht\left(\operatorname{Im}(\operatorname{Tr}^{\mathbb{Z}/2})\right) \leq 3.$$

Using the same notation as in Section 2.3 we see that the algebra generators

$$\ell_i, Q_i, \quad i = 1, 2, 3,$$

and the cubic form

$$c = x_1x_2x_3 + y_1y_2y_3$$

are each the transfer of one of their monomials, so

$$I = (\ell_1, \ell_2, \ell_3, Q_1, Q_2, Q_3, c) \subseteq \operatorname{Im}\left(\operatorname{Tr}^{\mathbb{Z}/2}\right).$$

Since

$$\mathbb{F}[V]^{\mathbb{Z}/2}/I = \mathbb{F}[q_1, q_2, q_3]$$

is an integral domain of Krull dimension 3, we have that I is a prime ideal of height 3. Therefore, $I = \operatorname{Im}(\operatorname{Tr}^{\mathbb{Z}/2})$, because if the image of the transfer were strictly larger, it would have height at least 4, a contradiction to Theorem 2.4.5.

Chapter 3
Combinatorial Finiteness

AN important class of problems for rings of invariants involve combinatorial finiteness (see Section 1.3, Problem 4), such as counting, in an appropriate sense, the number of invariant forms of a given degree. The way we have chosen to present this material is with the aid of the Poincaré series of a graded vector space.

DEFINITION: *A graded vector space M over the field $\mathbb{F}$ is said to be of* **finite type** *if* $\dim_{\mathbb{F}}(M_k)$ *is finite for all k. For such a graded vector space M, the* **Poincaré series** *of M is the formal series*

$$P(M, t) = \sum_{-\infty}^{\infty} \dim_{\mathbb{F}}(M_k) t^k.$$

The Poincaré series of a graded vector space of finite type is an adequate extension of the dimension of a finite-dimensional vector space, and enjoys many similar properties, [391]. For example, it is additive: In other words, if M' and M'' are graded vector spaces of finite type, then so is $M' \oplus M''$, and

$$P(M' \oplus M'', t) = P(M' t) + P(M'', t).$$

If M' and M'' are **bounded below**, which means that $M'_{k'} = 0$ and $M''_{k''} = 0$ for k', k'' large and negative, then the tensor product $M' \otimes M''$ is of finite type and

$$P(M' \otimes M'', t) = P(M' t) \cdot P(M'', t).$$

We are primarily concerned with modules that are **positively graded**, i.e., that are zero in negative degrees.[1] The Poincaré series for such modules is really just a Taylor series, and it is advantageous to think of the variable t as

[1] Mathematical terminology should be precise! Here it is not. Positively graded ought to be called **nonnegatively graded**, but it is hard to change a half century of usage.

being a complex variable. This will let us speak of the Laurent expansion, zeros, and poles of $P(M, t)$, and as we shall see, these often have surprising interpretations.

There are a number of hidden connections between algebraic finiteness, combinatorial finiteness, Noetherian finiteness, and homological finiteness. We will examine one of these in this chapter in Section 3.3.

Another combinatorial theme in invariant theory is the study of the invariants of permutation representations. This is probably the oldest part of invariant theory, and one might expect that nothing new remained to be discovered. This is not the case; see, e.g., [129] or [128]. One of the most interesting new result is M. Göbel's theorem [134] that the ring of invariants of a finite group G acting on a finite set X is always generated by forms of degree at most the maximum of 2 and $\binom{|X|}{2}$. We will examine this in Section 3.4.

The Poincaré series of permutation representations are also special, in that they are independent of the ground field. We develop several combinatorial formulae for these Poincaré series in Section 3.2.

3.1 Molien's Theorem on Poincaré Series

In many cases it is possible to compute the Poincaré series of a commutative graded connected algebra directly. For example, for the polynomial algebra $\mathbb{F}[z]$ on a single generator z of degree d, we find that

$$P(\mathbb{F}[z], t) = \sum_{i=0}^{\infty} t^{di} = \frac{1}{1-t^d},$$

and hence from the tensor product rule

$$P(\mathbb{F}[z_1, \dots, z_n], t) = \prod_{i=1}^{n} \frac{1}{(1-t^{d_i})}, \qquad \deg(z_i) = d_i,\ i = 1, \dots, n.$$

The following example, while simple in its formulation, has had wide-ranging consequences in unexpected areas; see, e.g., [359], [368] and the extremely elegant paper [130].

EXAMPLE 1: Consider the representation $\rho : \mathbb{Z}/k \hookrightarrow \mathrm{GL}(2, \mathbb{C})$ given by the matrix

$$\mathbf{T} = \begin{bmatrix} \lambda & 0 \\ 0 & \lambda^{-1} \end{bmatrix},$$

where $\lambda = \exp(2\pi \mathbf{i}/k)$. The action of $\mathbb{Z}/k$ on $\mathbb{C}[x, y]$ sends monomials to monomials, and one easily checks (see Lemma 7.3.5) that $\mathbb{C}[x, y]^{\mathbb{Z}/k}$ has a $\mathbb{C}$-basis consisting of monomials $x^{ak}y^{bk}(xy)^c$, where $a, b, c \in \mathbb{N}_0$, and

$0 \le c \le k-1$. Hence

$$\mathbb{C}[x, y]^{\mathbb{Z}/k} = \bigoplus_{i=0}^{k-1} \mathbb{C}[x^k, y^k] \cdot (xy)^i.$$

The Poincaré series of $\mathbb{C}[x^k, y^k] \cdot (xy)^i$ is

$$\frac{t^{2i}}{(1-t^k)^2},$$

so by the additivity of Poincaré series

$$\begin{aligned} P(\mathbb{C}[x, y]^{\mathbb{Z}/k}, t) &= \frac{1+t^2+\cdots+t^{2k-2}}{(1-t^k)^2} = \frac{(1+t^2+\cdots+t^{2k-2})(1+t^k)}{(1-t^k)^2(1+t^k)} \\ &= \frac{(1+t^2+\cdots+t^{2k-2})(1+t^k)}{(1-t^{2k})(1-t^k)} = \frac{1+t^k}{(1-t^2)(1-t^k)} \\ &= \frac{1-t^{2k}}{(1-t^2)(1-t^k)^2}. \end{aligned}$$

The form of this Poincaré series suggests (see Proposition 3.3.2) that there ought to be three generators, whose degrees are 2, k, and k, and a single relation of degree $2k$. This is indeed the case, suitable generators being xy, x^k, and y^k, and the single relation being $r = x^k y^k - (xy)^k$. In other words, the map

$$\varphi : \mathbb{C}[f, h', h'']/(f^k - h'h'') \longrightarrow \mathbb{C}[x, y]^{\mathbb{Z}/k}$$

defined by

$$\varphi(f) = xy, \quad \varphi(h') = x^k, \quad \varphi(h'') = y^k$$

is an isomorphism of algebras.

Note that this discussion remains unchanged if we replace the complex numbers $\mathbb{C}$ by any other field $\mathbb{F}$ containing a kth root of unity; see, e.g., Section 22.1 in [297].

Recall from Section 2.2 that when $\mathbb{F}$ is a field of characteristic zero and $\rho : G \hookrightarrow \mathrm{GL}(n, \mathbb{F})$ a representation of the finite group G, then the averaging map $\pi^G = \frac{1}{|G|}\sum_{g\in G} g$ is a projection operator on $\mathbb{F}[V]$ with image $\mathbb{F}[V]^G$. So the dimension of $\mathbb{F}[V]_j^G$ is equal to the trace of the matrix representing π^G on $\mathbb{F}[V]_j$, and this gives

$$P(\mathbb{F}[V]^G, t) = \frac{1}{|G|}\sum_{g\in G}\sum_{j=0}^{\infty} \mathrm{tr}(g, \mathbb{F}[V]_j)t^j.$$

If $\mathbb{F}$ has nonzero characteristic, then the trace of a matrix is in $\mathbb{F}$, so cannot be interpreted as a dimension. However, there is a standard way to deal with this if the characteristic of $\mathbb{F}$ does not divide the order of G. Before explaining this we note that the following formula for the trace actually works independently of the characteristic of $\mathbb{F}$.

PROPOSITION 3.1.1 (Trace Formula): *Let $\rho : G \hookrightarrow \mathrm{GL}(n, \mathbb{F})$ be a representation of a finite group G over a field $\mathbb{F}$. If $g \in G$, then the trace of the action of G on $\mathbb{F}[V]$ is given by*

$$\sum_{j=0}^{\infty} \mathrm{tr}(g, \mathbb{F}[V]_j)t^j = \frac{1}{\det(1-g^{-1}t)}.$$

PROOF: Extending the field does not affect either side of this equation, so we may assume that $\mathbb{F}$ is algebraically closed. Then there is a basis in which g is upper triangular on V (if $\mathbb{F}$ has characteristic zero, it is even diagonalizable), say with eigenvalues $\lambda_1, \dots, \lambda_n$. The eigenvalues on V^* are $\lambda_1^{-1}, \dots, \lambda_n^{-1}$, so the eigenvalues on $\mathbb{F}[V]_j$ are the products of j not necessarily distinct λ_i^{-1}'s. Using a basis in which g is upper triangular we have

$$\sum_{j=0}^{\infty} \mathrm{tr}(g, \mathbb{F}[V]_j)t^j = \Big(\sum_{j=0}^{\infty} \lambda_1^{-j}t^j\Big) \cdots \Big(\sum_{j=0}^{\infty} \lambda_n^{-j}t^j\Big) = \prod_{i=1}^{n} \frac{1}{1-\lambda_i^{-1}t} = \frac{1}{\det(1-g^{-1}t)},$$

which proves the proposition. □

REMARK: Apart from a factor of $\det(g^{-1})$ the term on the right-hand side of the formula in Proposition 3.1.1 is nothing but the reciprocal of the characteristic polynomials of the element $\rho(g)$ for $g \in G$. One should think of the right-hand side of this formula as a power series in t with coefficients certain sums of determinants of the action of $g^{-1} \in G$ on V. Note that $\mathrm{Mat}_{n,n}(\mathbb{F})[t] \cong \mathrm{Mat}_{n,n}(\mathbb{F}[t])$, so det(-) makes sense on $\mathrm{Mat}_{n,n}(\mathbb{F})[t]$.

To explain how to proceed in nonzero characteristic we need to recall some facts from the representation theory of finite groups in the nonmodular case; see, e.g., [110], [111], or [345] Part III. The following discussion applies only to finite fields, (see also [354] Section 12 and [162]) but is adequate for our purposes. For the general theory see one of the references just cited.

Let G be a finite group of order d and $\zeta = \exp(2\pi i/d)$ a primitive dth root of unity. Choose a prime ideal $\mathfrak{p}$ containing p in the ring of cyclotomic integers[2] $\mathbb{Z}(\zeta) = \{a_0 + a_1\zeta + \cdots + a_{d-1}\zeta^{d-1} \mid a_0, \dots, a_{d-1} \in \mathbb{Z}\}$. Let $\mathfrak{D}$ be the localization of $\mathbb{Z}(\zeta)$ at $\mathfrak{p}$. A complex representation $\rho : G \longrightarrow \mathrm{GL}(n, \mathbb{C})$ is said to be **defined over** $\mathfrak{D}$ if it is possible to choose a basis $x_1, \dots, x_n$ for $V_{\mathbb{C}} = \mathbb{C}^n$ such that for all $g \in G$ the entries of the matrices $\rho(g)$ with respect to the basis $x_1, \dots, x_n$ lie in $\mathfrak{D}$. In this case we have an action of G on the free $\mathfrak{D}$-module $V_{\mathfrak{D}} = \bigoplus_n \mathfrak{D}$ via the matrices $\{\rho(g) \in \mathrm{GL}(n, \mathfrak{D})\}$ representing $\rho(g)$, $g \in G$, with respect to the basis $x_1, \dots, x_n$.

[2] See, e.g., [226] for an introduction to cyclotomic algebra.

Given a complex representation ρ that is defined over $\mathfrak{D}$ we can construct a representation ρ_p of G in characteristic p as follows: Let $\mathfrak{m} \subset \mathfrak{D}$ be the unique maximal ideal. Then $\mathfrak{D}/\mathfrak{m}$ is the finite field $\mathbb{F}_p(\widehat{\zeta})$ of characteristic p obtained from the prime field $\mathbb{F}_p$ by adjoining a primitive dth root of unity $\widehat{\zeta}$, and $V_p = \mathbb{F}_p(\widehat{\zeta}) \otimes_{\mathfrak{D}} V_{\mathfrak{D}}$ is a representation of G over $\mathbb{F}_p(\widehat{\zeta})$. It is easy to see that $\dim_{\mathbb{C}}(V^G) = \dim_{\mathbb{F}_p(\widehat{\zeta})}(V_p^G)$. Conversely, there is the well-known result from representation theory, [110] Section 4, which provides the requisite construction starting with a representation over a finite field and lifting it to characteristic zero.

PROPOSITION 3.1.2: *Let G be a finite group of order d and p a prime integer not dividing d. Set $\mathbb{F}_q = \mathbb{F}_p(\widehat{\zeta})$, the extension of the prime field $\mathbb{F}_p$ of characteristic p by a primitive dth root of unity $\widehat{\zeta}$. If $\rho : G \longrightarrow \mathrm{GL}(n, \mathbb{F}_q)$ is a representation of G over $\mathbb{F}_q$, then there exists a complex representation $\widetilde{\rho} : G \longrightarrow \mathrm{GL}(n, \mathbb{C})$, defined over $\mathfrak{D} = \mathbb{Z}(\zeta)_{\mathfrak{p}}$, where $\zeta \in \mathbb{C}$ is a primitive dth root of unity, such that $\widetilde{\rho}_p = \rho$.* □

A representation $\widetilde{\rho}$ whose existence is guaranteed under the hypotheses of Proposition 3.1.2 is referred to as a **characteristic 0 lift of** ρ. Let $\mathbb{F}_p(\widehat{\zeta})$ be the Galois field with $q = p^{\nu}$ elements. If we choose, once and for all, an identification of $\mathbb{F}_p(\widehat{\zeta})^{\times}$ with the group of qth roots of unity, and identify $0 \in \mathbb{F}_q$ with $0 \in \mathbb{C}$, then we can interpret the trace of matrices over $\mathbb{F}_q$ as complex numbers. With this interpretation we have the following result:

THEOREM 3.1.3 (T. Molien): *Let $\rho : G \hookrightarrow \mathrm{GL}(n, \mathbb{F})$ be a representation of a finite group G over a field $\mathbb{F}$ of characteristic prime to the order of G. Then the Poincaré series of the ring of invariants is given by*

$$P(\mathbb{F}[V]^G, t) = \frac{1}{|G|} \sum_{g \in G} \frac{1}{\det(1 - g^{-1}t)} = \frac{1}{|G|} \sum_{g \in G} \frac{1}{\det(1 - g\, t)}.$$

PROOF: This follows by using Proposition 3.1.1, and the additivity of the trace to compute the trace of π^G. □

REMARK: Even in noncoprime characteristic, Molien's formula still makes sense, so long as one uses a **Brauer lift**, that is, an identification of $\mathbb{F}^{\times}$ with a group of roots of unity in $\mathbb{C}$, to interpret the trace, determinant, and eigenvalues as complex numbers, and sums only over the p-regular elements of G. However, it no longer calculates the Poincaré series of the invariants, but rather the Poincaré series for the multiplicity of the trivial module as a composition factor; cf. [248], [257].

As is often the case in mathematics, turning things around, by which we mean a reversal of viewpoint, is often quite profitable. Consider again the cyclic group of Example 1, only this time in the following way.

EXAMPLE 2: Let $\lambda \in \mathbb{C}$ be a primitive kth root of unity. We pose the problem of evaluating the sum

$$\sum_{i=0}^{k-1} \frac{1}{(1-\lambda^i t)(1-\lambda^{-i} t)}.$$

The Atiyah–Bott fixed-point theorem yields expressions similar to the one above when studying the defect of a $\mathbb{Z}/k$-action on a closed surface at an isolated fixed point (see [171], [462], [359], and the references there). To evaluate this sum note that it is exactly the sum that arises by applying Molien's theorem to compute the Poincaré series of the ring of invariants of the cyclic group $\mathbb{Z}/k$ studied in Example 1. In that example we computed the Poincaré series by other means, so we obtain

$$\sum_{i=0}^{k-1} \frac{1}{(1-\lambda^i t)(1-\lambda^{-i} t)} = k\left[\frac{1+t^2+\cdots+t^{2k-2}}{(1-t^k)^2}\right],$$

a result that is far from obvious; see, e.g., [368], [130], as well as the references already cited.

EXAMPLE 3: Consider the dihedral group D_{2k} of order $2k$ represented in $\mathrm{GL}(2, \mathbb{R})$ as the group of symmetries of a regular k-gon centered at the origin. This representation is orthogonal, so we may identify it with its own dual. In this representation the group D_{2k} is generated by the matrices

$$\mathbf{D} = \begin{bmatrix} \cos\frac{2\pi}{k} & -\sin\frac{2\pi}{k} \\ \sin\frac{2\pi}{k} & \cos\frac{2\pi}{k} \end{bmatrix} \quad \text{and} \quad \mathbf{S} = \begin{bmatrix} 1 & 0 \\ 0 & -1 \end{bmatrix},$$

where $\mathbf{D}$ is a rotation through $2\pi/k$ radians and $\mathbf{S}$ is a reflection in an axis. Thus the elements of D_{2k} are the identity, the $k-1$ rotations $\mathbf{D}^i$, $i = 1, \ldots, k-1$, and the k reflections $\mathbf{S}\cdot\mathbf{D}^i$, $i = 0, \ldots, k-1$. Over the complex numbers the rotation $\mathbf{D}$ is diagonalizable with complex conjugate eigenvalues $\lambda = \exp(2\pi\mathbf{i}/k)$, $\overline{\lambda} = \lambda^{-1}$. Thus $\mathbf{D}^i$ is diagonalizable with eigenvalues λ^i, λ^{-i}, and hence

$$\det(1-\mathbf{D}^i t) = (1-\lambda^i t)(1-\lambda^{-i} t).$$

The reflections $\mathbf{S}\cdot\mathbf{D}^i$ are all diagonalizable with eigenvalues $1, -1$, so

$$\det(1-\mathbf{S}\cdot\mathbf{D}^i t) = (1-t)(1+t) = 1-t^2.$$

By Molien's theorem (Theorem 3.1.3) we obtain

$$(\star) \qquad P(\mathbb{R}[x,y]^{D_{2k}}, t) = \frac{1}{2k}\left\{\frac{k}{1-t^2} + \sum_{i=0}^{k-1} \frac{1}{(1-\lambda^i t)(1-\lambda^{-i} t)}\right\}.$$

We evaluated the second sum in Example 2 and found

$$\sum_{i=0}^{k-1} \frac{1}{(1-\lambda^i t)(1-\lambda^{-i} t)} = k\left[\frac{1+t^2+\cdots+t^{2k-2}}{(1-t^k)^2}\right].$$

Substituting this into (★) yields

$$\begin{aligned} P(\mathbb{R}[x,y]^{D_{2k}}, t) &= \frac{1}{2k}\left\{k\left[\frac{1+t^2+\cdots+t^{2k-2}}{(1-t^k)^2}\right] + k\frac{1}{1-t^2}\right\} \\ &= \frac{1}{2}\left[\frac{1-t^{2k}+(1-t^k)^2}{(1-t^k)^2(1-t^2)}\right] = \frac{1}{2}\left[\frac{(1-t^k)(1+t^k)+(1-t^k)^2}{(1-t^k)^2(1-t^2)}\right] \\ &= \frac{1}{(1-t^k)(1-t^2)}, \end{aligned}$$

and we see that $\mathbb{R}[x,y]^{D_{2k}}$ has the same Poincaré series as a polynomial algebra on two generators, one of degree 2 and one of degree k.

The action of D_{2k} on $\mathbb{R}^2$ is orthogonal and hence preserves the square of the norm, giving the invariant $q = x^2 + y^2 \in \mathbb{R}[x,y]^{D_{2k}}$. The vertices of the regular k-gon form an orbit and have the coordinates

$$\left(\cos\left(\frac{2\pi i}{k}\right), \sin\left(\frac{2\pi i}{k}\right)\right) \qquad i = 0, \ldots, k-1.$$

The kth elementary symmetric polynomial in these orbit elements is the invariant

$$h = \prod_{i=0}^{k-1}\left[\cos\left(\frac{2\pi i}{k}\right)\cdot x + \sin\left(\frac{2\pi i}{k}\right)\cdot y\right].$$

If $k \not\equiv 0 \bmod 4$, then the coefficient of x^k in this polynomial is nonzero, and one sees that q and h are algebraically independent. If $k \equiv 0 \bmod 4$, then the coefficient of y^2x^{k-2} is nonzero and q and h are algebraically independent in this case also. Therefore, $\mathbb{R}[q,h] \subseteq \mathbb{R}[x,y]^{D_{2k}}$. Since

$$P(\mathbb{R}[x,y]^{D_{2k}}, t) = \frac{1}{(1-t^2)(1-t^k)} = P(\mathbb{R}[q,h], t),$$

it follows that $\mathbb{R}[x,y]^{D_{2k}} = \mathbb{R}[q,h]$.

If we sort the terms in the sum occurring in Molien's theorem according to the multiplicity of the eigenvalue 1, then we see that the Laurent series expansion of $P(\mathbb{F}[V]^G, t)$ begins with $\frac{1}{|G|}/(1-t)^n$, and that the coefficient of $1/(1-t)^{n-1}$ in the Laurent expansion is determined by the elements $g \in G$ such that 1 is an eigenvalue of multiplicity $n-1$. Such an element is called a **pseudoreflection**. (See Sections 6.2, 7.1, and 7.2 for more about pseudoreflections and the groups that they generate.) The set of pseudoreflections in G will be denoted by $s(G)$. It, of course, depends on the representation ϱ, but we do not wish to encumber the notation to make this explicit.

None of the group elements $g \neq 1 \in G$ can have 1 as an eigenvalue of order n, because G has order prime to the characteristic of $\mathbb{F}$. Therefore,

$$\sum_{g \neq 1 \in G} \frac{1}{\det(1 - g^{-1}t)}$$

has a pole of order at most $n-1$ at $t=1$. Thus the value at $t=1$ of the holomorphic function $(1-t)^n P(\mathbb{F}[V]^G, t)$ is equal to $1/|G|$. Elaborating on this argument we obtain the following:

PROPOSITION 3.1.4: *Let $\mathbb{F}$ be a field of characteristic zero and $\varrho : G \hookrightarrow \mathrm{GL}(n, \mathbb{F})$ a finite-dimensional representation of G on $V = \mathbb{F}^n$. Then the Laurent expansion of the Poincaré series $P(\mathbb{F}[V]^G, t)$ begins as follows:*

$$P(\mathbb{F}[V]^G, t) = \frac{\frac{1}{|G|}}{(1-t)^n} + \frac{\frac{|s(G)|}{2 \cdot |G|}}{(1-t)^{n-1}} + \cdots,$$

where $|s(G)|$ is the number of pseudoreflections in $\varrho(G)$.

PROOF: Consider the difference

$$P(\mathbb{F}[V]^G, t) - \frac{1}{|G|}\frac{1}{(1-t)^n} = \frac{1}{|G|} \sum_{g \in G, g \neq 1} \frac{1}{\det(1 - g^{-1}t)}.$$

This is a rational function with a pole of order at most $n-1$ at $t=1$. Multiplying this function by $(1-t)^{n-1}$ yields a holomorphic function with a finite value at $t=1$. This value is the coefficient of $1/(1-t)^{n-1}$ in the Laurent expansion of the Poincaré series of $P(\mathbb{F}[V], t)$ at $t=1$. An element $g \in G$ contributes to this value if and only if $\dim_{\mathbb{F}}(V^g) = n-1$, i.e., the only elements contributing to the value of the coefficient of $1/(1-t)^{n-1}$ are the pseudoreflections. This means that the Poincaré series may be written in the form

$$\frac{1}{|G|}\left\{\frac{1}{(1-t)^n} + \sum_{g \in s(G)} \frac{1}{\det(1 - g^{-1}t)} + \cdots\right\},$$

where $\cdots$ indicates a rational function with a pole of order at most $n-2$ at $t=1$. If g is a pseudoreflection, let $\lambda_g \neq 1$ be the unique eigenvalue of $\varrho(g)$ different from 1. For a pseudoreflection the reciprocal of the characteristic polynomial is

$$\frac{1}{\det(1 - g^{-1}t)} = \frac{1}{(1-t)^{n-1}(1 - \lambda_g^{-1}t)},$$

and after multiplying by $(1-t)^{n-1}$, the value at $t=1$ is $1/(1-\lambda_g^{-1})$. Observe that if g is a pseudoreflection, so is g^{-1}, and $g = g^{-1}$ precisely when g has order 2. For the pseudoreflections of order different from 2, $\lambda_g^{-1} = \lambda_{g^{-1}}$, and the identity

$$\frac{1}{1-\lambda^{-1}}+\frac{1}{1-\lambda}=\frac{(1-\lambda)+(1-\lambda^{-1})}{1-\lambda^{-1}-\lambda+1}=1$$

shows that if we pair together each of these pseudoreflections with its inverse, then their total contribution to the coefficient of $1/(1-t)^{n-1}$ in $P(\mathbb{F}[V]^G, t)$ is equal to one-half of their number. For the reflections of order 2 we have $\lambda=-1$, so that their contribution is also one-half their number, and the result follows. □

COROLLARY 3.1.5: *Let $G \hookrightarrow \mathrm{GL}(n, \mathbb{F})$ be a faithful representation of a finite group G. If the characteristic of $\mathbb{F}$ does not divide the order of G and*

$$\mathbb{F}[V]^G \cong \mathbb{F}[f_1, \dots, f_n], \quad \deg(f_i)=d_i,\ i=1,\dots,n,$$

then

$$|G| = d_1\cdots d_n \text{ and } |s(G)| = \sum_{i=1}^{n}(d_i-1).$$

PROOF: This follows by equating the formula for the Poincaré series of a polynomial algebra with that obtained from Proposition 3.1.4. □

REMARK: These results can be extended to the modular context: The leading coefficient of the Poincaré series of a ring of invariants is always $\frac{1}{|G|}$ (see Theorem 4.5.3). If $\mathbb{F}[V]^G \cong \mathbb{F}[f_1, \dots, f_n]$ and $\deg(f_i)=d_i$, $i=1,\dots,n$, then $d_1\cdots d_n=|G|$ (Corollary 4.5.4). A formula for the second coefficient, when the ground field is the prime field $\mathbb{F}_p$, has been obtained by D. J. Benson and W. W. Crawley-Boevey [31] (see also [283]), and the ramification formula, and attendant inequality of [31] has been extended to arbitrary fields by J. Hartmann [158].

If $\rho : G \hookrightarrow \mathrm{SL}(n, \mathbb{F})$ is a representation into the special linear group, then the Poincaré series of $\mathbb{F}[V]^G$ satisfies a very useful functional equation used by R. P. Stanley in [400].

COROLLARY 3.1.6 (R. P. Stanley): *Let $\rho : G \hookrightarrow \mathrm{SL}(n, \mathbb{F})$ be a representation of a finite group over the field $\mathbb{F}$ whose characteristic is prime to the order of G. Then*

$$P\left(\mathbb{F}[V]^G, \frac{1}{t}\right) = (-1)^n t^n P(\mathbb{F}[V]^G, t).$$

PROOF: By Molien's theorem we have

$$\begin{aligned} P\left(\mathbb{F}[V]^G, \frac{1}{t}\right) &= \frac{1}{|G|}\sum_{g\in G}\frac{1}{\det(1-g^{-1}t^{-1})} \\ &= \frac{t^n}{|G|}\sum_{g\in G}\frac{1}{\det(t-g^{-1})} = \frac{t^n}{|G|}\sum_{g\in G}\frac{1}{\det(gt-1)\det(g^{-1})} \end{aligned}$$

$$= \frac{(-1)^n t^n}{|G|} \sum_{g \in G} \frac{1}{\det(1 - gt)} = (-1)^n t^n P(\mathbb{F}[V]^G, t),$$

since $\det(g) = 1$ for all $g \in G$. □

REMARK: This implies by another theorem of Stanley, [400] Theorem 4.4, that rings of invariants of nonmodular representations into the special linear group are Gorenstein. A result previously obtained by K. Watanabe [443]; see Theorem 5.7.6.

EXAMPLE 4 (T. Molien): Consider the group $\mathcal{I}$ of rotations of a regular icosahedron embedded in $\mathbb{R}^3$ centered at the origin; see, for example, Chapter 2 in [153]. The icosahedron has 20 faces, each one an equilateral triangle. It has 12 vertices and 30 edges. The group $\mathcal{I} \subset \mathbb{SO}(3) \subset \mathrm{GL}(3, \mathbb{R})$ has order 60 and is isomorphic to the alternating group A_5. Every element T of $\mathcal{I}$ is a rotation, so has a polar axis L_{T} by Euler's theorem (see, e.g., Theorem 2.3.1 in [153]) and an angle of rotation ϑ_{T} in the plane orthogonal to the polar axis. The points of intersection of L_{T} with the unit sphere of $\mathbb{R}^3$ are called the **poles** of T. The polar axes of the icosahedron consist of the 6 lines joining opposite vertices, 15 lines joining midpoints of opposite edges, and 10 lines joining the centers of opposite faces. Thus there are $6 + 15 + 10 = 31$ polar axes and 62 poles. The rotations about the lines joining opposite vertices are through an angle $2\pi k/5$, where $k = 1, 2, 3, 4$, those about polar axes joining midpoints of opposite edges through an angle π, and those about a polar axis through the centers of opposite sides are rotations through an angle $2\pi k/3$, where $k = 1, 2$. Since

$$1 + 6 \cdot 4 + 15 \cdot 1 + 10 \cdot 2 = 60,$$

this accounts for all the elements of $\mathcal{I}$.

As an aid to computing the Poincaré series $P(\mathbb{R}[x, y, z]^{\mathcal{I}}, t)$ Table 3.1.1 displays the diagonal forms of the elements in $\mathcal{I}$ and their characteristic polynomials. In the table $\zeta = \exp(2\pi \mathbf{i}/3)$ and $\omega = \exp(2\pi \mathbf{i}/5)$.

Substituting from this table into the formula of Molien gives for the Poincaré series

$$\frac{1}{60}\left[\frac{1}{(1-t)^3} + \frac{15}{(1-t)(1-t^2)} + \frac{20}{1-t^3} + 6\sum_{i=1}^{4} \frac{1}{(1-\omega^i t)(1-\omega^{-i} t)(1-t)}\right].$$

The sum[3]

$$\sum_{i=1}^{4} \frac{1}{(1-\omega^i t)(1-\omega^{-i} t)(1-t)} = \frac{1}{1-t}\sum_{i=1}^{4} \frac{1}{(1-\omega^i t)(1-\omega^{-i} t)}$$

[3] T. Molien, by the way, gives no clue how he made this calculation.

ORDER OF ROTATION	DIAGONAL FORM OF THE MATRIX	$\det(1-gt)$	NUMBER OF TERMS
1	$\begin{bmatrix}1&0&0\\0&1&0\\0&0&1\end{bmatrix}$	$(1-t)^3$	1
2	$\begin{bmatrix}0&1&0\\1&0&0\\0&0&1\end{bmatrix}$	$(1-t)(1-t^2)$	15
3	$\begin{bmatrix}\zeta^{\pm1}&0&0\\0&\zeta^{\mp1}&0\\0&0&1\end{bmatrix}$	$1-t^3$	20
5	$\begin{bmatrix}\omega^i&0&0\\0&\omega^{-i}&0\\0&0&1\end{bmatrix}$ $i=1,2,3,4$	$(1-(\omega^i+\omega^{-i})t+t^2)(1-t)$ $i=1,2,3,4$	24

TABLE 3.1.1: The icosahedral group $\mathcal{I}$.

may be evaluated with the aid of Example 2. We obtain

$$\begin{aligned}\sum_{i=1}^{4}\frac{1}{(1-\omega^i t)(1-\omega^{-i}t)} &= \sum_{i=0}^{4}\frac{1}{(1-\omega^i t)(1-\omega^{-i}t)}-\frac{1}{(1-t)^2}\\ &= 5\frac{1+t^2+t^4+t^6+t^8}{(1-t^5)^2}-\frac{1}{(1-t)^2}.\end{aligned}$$

Combining these results and simplifying gives

$$\begin{aligned}&P(\mathbb{R}[x,y,z]^{\mathcal{I}},t)\\ &\quad=\frac{1}{60}\left[\frac{15}{(1-t)(1-t^2)}+\frac{20}{1-t^3}+30\frac{1+t^2+t^4+t^6}{(1-t^4+t^8)^2(1-t)}-\frac{6}{(1-t)^3}\right]\\ &\quad=\frac{1+t^{15}}{(1-t^2)(1-t^6)(1-t^{10})}.\end{aligned}$$

This formula suggests that $\mathbb{R}[x,y,z]^{\mathcal{I}}$ contains a polynomial subalgebra generated by three polynomials f, h, k of degrees 2, 6, 10, and a polynomial r of degree 15, such that

$$\mathbb{R}[x,y,z]^{\mathcal{I}}=\mathbb{R}[f,h,k]\oplus\mathbb{R}[f,h,k]\cdot r.$$

This is indeed the case, and suitable polynomials f, h, k, and r may be found as follows.

The group $\mathcal{G}$ acts on the icosahedron by rotations of $\mathbb{R}^3$ so preserves the inner product, and hence for suitably chosen x, y, z the quadratic polynomial $f = x^2 + y^2 + z^2$ is invariant. The set of polar axes $\mathcal{L}$ of $\mathcal{G}$ is permuted by the action of $\mathcal{G}$ on $\mathbb{R}^3$. Each polar axis is represented by a linear functional $\ell : \mathbb{R}^3 \longrightarrow \mathbb{R}$ whose kernel is the plane through the origin orthogonal to the polar axis. The action of $\mathcal{G}$ on $\mathcal{L}$ separates $\mathcal{L}$ into orbits. Two polar axes ℓ' and ℓ'' belong to the same orbit if and only if the corresponding angles of rotation $\vartheta_{\ell'}$ and $\vartheta_{\ell''}$ are multiples of the same fraction of 2π, e.g., $2\pi/5$, $6\pi/5$, etc. Thus there are three orbits $\mathcal{A}, \mathcal{B}, \mathcal{C} \subset \mathcal{L}$ of the $\mathcal{G}$-action on $\mathcal{L}$ consisting of 6, 10, and 15 axes corresponding to the rotations of order 5, 3, and 2, respectively. If we take the product of all the elements in each of these orbits, we obtain polynomials h, k, and r suitable to the above description of $\mathbb{R}[x, y, z]^{\mathcal{G}}$.

Since $\mathcal{G}$ acts by rotations, the representation has image in $SL(3, \mathbb{R})$, so Corollary 3.1.6 applies. Indeed, a little computation shows that

$$P\left(\mathbb{R}[x, y, z]^{\mathcal{G}}, \frac{1}{t}\right) = \frac{1 + \frac{1}{t^{15}}}{(1 - \frac{1}{t^2})(1 - \frac{1}{t^6})(1 - \frac{1}{t^{10}})} = \frac{t^{15} + 1}{(t^2 - 1)(t^6 - 1)(t^{10} - 1)} \frac{t^{18}}{t^{15}}$$
$$= -t^3 P(\mathbb{R}[x, y, z]^{\mathcal{G}}, t).$$

EXAMPLE 5 (R. P. Stanley): Consider[4] the two matrices

$$\begin{bmatrix} -1 & 0 & 0 \\ 0 & -1 & 0 \\ 0 & 0 & 1 \end{bmatrix}, \quad \begin{bmatrix} 1 & 0 & 0 \\ 0 & 1 & 0 \\ 0 & 0 & \boldsymbol{i} \end{bmatrix} \in GL(3, \mathbb{C}).$$

Together they generate a subgroup isomorphic to $\mathbb{Z}/2 \times \mathbb{Z}/4$. Since the group contains $-\mathbf{I}$, the invariants are all of even degree. The induced action on $\mathbb{C}[x, y, z]$ preserves monomials, and it may be easily verified that

$$x^2, \ xy, \ y^2, \ z^4$$

are a minimal set of algebra generators for the invariants $\mathbb{C}[x, y, z]^{\mathbb{Z}/2\times\mathbb{Z}/4}$. The invariants contain $\mathbb{C}[x^2, y^2, z^4]$ as a subalgebra and $\mathbb{C}[x, y, z]^{\mathbb{Z}/2\times\mathbb{Z}/4}$ is a free $\mathbb{C}[x^2, y^2, z^4]$-module on the two generators[5] $1, xy$. In other words,

$$\mathbb{C}[x, y, z]^{\mathbb{Z}/2\times\mathbb{Z}/4} = \mathbb{C}[x^2, y^2, z^4] \oplus \mathbb{C}[x^2, y^2, z^4] \cdot (xy)$$

as a $\mathbb{C}[x^2, y^2, z^4]$-module. Thus, we obtain for the Poincaré series

$$P(\mathbb{C}[x, y, z]^{\mathbb{Z}/2\times\mathbb{Z}/4}, t) = \frac{1}{(1 - t^2)^2(1 - t^4)} + \frac{t^2}{(1 - t^2)^2(1 - t^4)} = \frac{1}{(1 - t^2)^3}.$$

---

[4] For a complete discussion of this example see Example 23.2.3 in [297].

[5] This is a special case of a theorem of J. A. Eagon and M. Hochster, [172], which we will discuss in Chapter 5.

Therefore, the invariants $\mathbb{C}[x, y, z]^{\mathbb{Z}/2\times\mathbb{Z}/4}$ have the same Poincaré series as a polynomial algebra $\mathbb{C}[f_1, f_2, f_3]$ on three generators f_1, f_2, f_3, all of degree 2. However, $\mathbb{C}[x, y, z]^{\mathbb{Z}/2\times\mathbb{Z}/4}$ is not isomorphic to such a polynomial algebra. Even worse, the three matrices

$$\begin{bmatrix} -1 & 0 & 0 \\ 0 & 1 & 0 \\ 0 & 0 & 1 \end{bmatrix}, \quad \begin{bmatrix} 1 & 0 & 0 \\ 0 & -1 & 0 \\ 0 & 0 & 1 \end{bmatrix}, \quad \begin{bmatrix} 1 & 0 & 0 \\ 0 & 1 & 0 \\ 0 & 0 & -1 \end{bmatrix} \in \mathrm{GL}(3, \mathbb{C})$$

define an action of $(\mathbb{Z}/2)^3$ on $\mathbb{C}^3$, and the ring of invariants is $\mathbb{C}[x^2, y^2, z^2]$, which is a polynomial algebra, and also has Poincaré series $1/(1-t^2)^3$. Thus, the form of the Poincaré series $P(\mathbb{F}[V]^G, t)$ can be an aid in studying the algebra structure of $\mathbb{F}[V]^G$, but certainly does not determine it.

3.2 Poincaré Series of Permutation Representations

Suppose that a finite group G acts as permutations on a finite set X. We then refer to X together with the G-action as a **finite G-set**. A subset $B \subset X$ is called an **orbit** if G permutes the elements of B among themselves and the induced permutation action of G on B is transitive. If $B \subseteq X$ is an orbit and $b \in B$ then $B = \{gb \mid g \in G\}$ so as a G-set $B = G/G_b$, where $G_b \leq G$ is the subgroup fixing b, which is called the **isotropy group** of b. If b', b'' belong to the same orbit B of G on X then $G_{b'}, G_{b''} \leq G$ are conjugate subgroups.

If X is a finite G-set, then X decomposes into a disjoint union

$$X = X^G \sqcup X_1 \sqcup \cdots \sqcup X_s,$$

where $X_1, \ldots, X_s$ are the orbits of G on X with more than one element, and the orbits with a single element have been collected together in the fixed point set X^G. For eaxh orbit X_i choose a point $x_i \in X_i$ and denote by G_i the isotropy group of x_i. Then

$$|X| = |X^G| + \sum_{i=1}^{s} |G : G_i|$$

is the **class equation**, which counts the elements of X in a way appropriate to the G-action.

NOTATION: *If X is a G-set and R a commutative ring, denote by $\mathscr{F}_R(X)$ the free R-module on the set X, and regard it as an $R(G)$-module via the action of G by permutation of the basis X. If $R = \mathbb{F}$ is a field, we also write V_X for $\mathscr{F}_{\mathbb{F}}(X)$. It is the underlying vector space of the G-representation over $\mathbb{F}$ corresponding to the permutation representation of G on X.*

DEFINITION: *If G is a finite group acting on a module M over the ring R and $B \subseteq M$ is a G-orbit, then $\sum_{b\in B} b$ is called the* **orbit sum of** B, *and denoted by* $\mathfrak{S}_B$. *If* X *is a* G*-set and* $x \in X$ *with* G*-orbit* $[x]$, *then we also write* $\mathfrak{S}_x$ *instead of* $\mathfrak{S}_{[x]}$.

LEMMA 3.2.1: *Let* X *be a finite* G*-set and* R *a commutative ring. Then the orbit sums*

$$\mathfrak{S}_B = \sum_{b \in B} b, \qquad B \subset X \text{ an orbit,}$$

are a basis for the fixed-point set $\mathscr{F}_R(X)^G$.

PROOF: Since $\mathfrak{S}_B$ is a sum over an entire orbit, it belongs to the fixed-point set $\mathscr{F}_R(X)^G$. The orbits of G on X are disjoint. Let

$$X = X_1 \sqcup \cdots \sqcup X_s$$

be the decomposition of X into orbits and write $\mathfrak{S}_i$ for $\mathfrak{S}_{X_i}$. Denote the elements of X_i by $y_{i,1}, \ldots, y_{i,m_i}$ for $i = 1, \ldots, s$. In this notation $\{y_{i,j} \mid 1 \le j \le m_i, i = 1, \ldots, s\}$ is a basis for $\mathscr{F}_R(X)$. The action of G on X, and hence on $\mathscr{F}_R(X)$, maps each X_i bijectively to itself for $i = 1, \ldots, s$. Suppose that

$$y = \sum a_{i,j} y_{i,j} \in \mathscr{F}_R(X)^G.$$

Rewrite this in the form

$$y = \sum_{j=1}^{m_1} a_{1,j} y_{1,j} + \cdots + \sum_{j=1}^{m_s} a_{s,j} y_{s,j}.$$

If $y_{k,j'}, y_{k,j''} \in X_k$, then there exists an element $g \in G$ such that $g y_{k,j'} = y_{k,j''}$, so applying g to y and equating coefficients, we see that $a_{k,j'} = a_{k,j''}$. Since this holds for all $1 \le j', j'' \le m_k$, $k = 1, \ldots, s$, we may write

$$y = a_1 \sum_{j=1}^{m_1} y_{1,j} + \cdots + a_s \sum_{j=1}^{m_s} y_{s,j} = a_1 \mathfrak{S}_1 + \cdots + a_s \mathfrak{S}_s,$$

so the orbit sums $\mathfrak{S}_1, \ldots, \mathfrak{S}_s$ generate $\mathscr{F}_R(X)^G$.

To see that the orbit sums are linearly independent, suppose

$$a_1 \mathfrak{S}_1 + \cdots + a_s \mathfrak{S}_s = 0.$$

Rewrite this as

$$\begin{aligned} 0 &= a_1 \sum_{j=1}^{m_1} y_{1,j} + \cdots + a_s \sum_{j=1}^{m_s} y_{s,j} \\ &= a_1 y_{1,1} + \cdots + a_1 y_{1,m_1} + \cdots + a_s y_{s,1} + \cdots + a_s y_{s,m_s}. \end{aligned}$$

Since the set $\{y_{i,j} \mid 1 \le j \le m_i, i = 1, \ldots, s\}$ is a basis for $\mathscr{F}_R(X)$, it follows that $a_1 = \cdots = a_s = 0$ as required. □

If X is a finite G-set, R is a commutative ring, and $R[X]$ the polynomial ring over R with generators X, then the action of G on $R[X]$ sends monomials to monomials. The monomials of degree k are a basis for $R[X]_k$, and hence $R[X]_k$ is the linear representation associated to the permutation representation of G on the monomial basis.

PROPOSITION 3.2.2: *Let G be a finite group acting on the finite set X. If R is a commutative ring, then $R[X]_k^G = R \otimes_{\mathbb{Z}} \mathbb{Z}[X]_k^G$ is a free R-module, whose rank is independent of R. In particular, the Poincaré series of $\mathbb{F}[V]^G$ is independent of the ground field $\mathbb{F}$.*

PROOF: By Lemma 3.2.1, $R[X]_k^G$ has an R-basis consisting of the orbit sums. The number of such orbit sums does not depend on whether we count them in $R[X]_k$ or $\mathbb{Z}[X]_k$. □

This proposition can be combined with Molien's theorem to provide a formula for the Poincaré series of rings of invariants of permutation representations in the modular case. Here is an example, promised in Section 2.3, Example 2, to illustrate this.

EXAMPLE 1: Let $\mathbb{F}$ be a field and denote by $\sigma_k : \mathbb{Z}/2 \hookrightarrow \mathrm{GL}(2k, \mathbb{F})$ the k-fold sum of the regular representation of $\mathbb{Z}/2$. This is a permutation representation so Poincaré series of $\mathbb{F}[x_1, \dots, x_k, y_1, \dots, y_k]^{\mathbb{Z}/2}$ can be computed, no matter what $\mathbb{F}$ is, by using Molien's theorem to compute the Poincaré series over $\mathbb{C}$. This is easy, since the representation is implemented by the permutation matrix

$$\mathbf{S} = \begin{bmatrix} \begin{smallmatrix} 0 & 1 \\ 1 & 0 \end{smallmatrix} & 0 & 0 \\ 0 & \ddots & 0 \\ 0 & 0 & \begin{smallmatrix} 0 & 1 \\ 1 & 0 \end{smallmatrix} \end{bmatrix} \in \mathrm{GL}(2k, \mathbb{F}).$$

So the characteristic polynomial of the generator of $\mathbb{Z}/2$ is $(1-t^2)^k$. We obtain

$$\begin{aligned} P(\mathbb{F}[x_1, \dots, x_k, y_1, \dots, y_k]^{\mathbb{Z}/2}, t) &= \frac{1}{2}\left[\frac{1}{(1-t)^{2k}} + \frac{1}{(1-t^2)^k}\right] \\ &= \frac{1}{2}\left[\frac{(1+t)^k + (1-t)^k}{(1-t)^k(1-t^2)^k}\right] = \frac{1}{2}\left[\frac{\sum_{i=0}^{k}\binom{k}{i}t^i + \sum_{i=0}^{k}(-1)^i\binom{k}{i}t^i}{(1-t)^k(1-t^2)^k}\right] \\ &= \frac{\sum_{i=0}^{\left[\frac{k}{2}\right]}\binom{k}{2i}t^{2i}}{(1-t)^k(1-t^2)^k}. \end{aligned}$$

For example, when $k = 3$ this formula gives

$$P(\mathbb{F}[x_1, x_2, x_3, y_1, y_2, y_3]^{\mathbb{Z}/2}, t) = \frac{1}{2}\left[\frac{(1+t)^3 + (1-t)^3}{(1-t)^3(1-t^2)^3}\right]$$
$$= \frac{1+3t^2}{(1-t)^3(1-t^2)^3} = 1 + 3t + 12t^2 + 28t^3 + \cdots,$$

as was used in Example 2 of Section 2.3. Proposition 3.2.2 suggests that there must be a combinatorial formula for the Poincaré series $P(\mathbb{F}[X]^G, t)$ of a permutation representation. To find the dimension of $\mathbb{F}[X]_k^G$ over $\mathbb{F}$ we need to find the number of G-orbits on the monomial basis of $\mathbb{F}[X]_k$. The answer to this problem is provided by a result of A. Cauchy and G. Frobenius, which is often called Burnside's lemma. (For a discussion of the history of this result and its (mis)attribution see [402], Volume II, page 404.)

PROPOSITION 3.2.3 (A. Cauchy, G. Frobenius): *Let G be a finite group and X a finite G set. Then*

$$|X/G| = \frac{1}{|G|}\sum_{g\in G}|X^g|.$$

PROOF: Consider the set

$$S = \{(g, x) \mid g \in G, x \in X \text{ and } gx = x\}.$$

We count the elements in S in two different ways. First, denoting by G_x the isotropy group of x and $[x]$ the orbit of $x \in X$, we find

$$\sum_{g\in G}|X^g| = |S| = \sum_{x\in X}|G_x| = \sum_{x\in X}\frac{|G|}{|[x]|} = |G|\sum_{x\in X}\frac{1}{|[x]|}.$$

If we note that $|[x']| = |[x'']|$ whenever x', x'' lie in the same orbit, then we can convert the last sum into a sum over the orbits $B \subseteq X$, which are the points of X/G, and doing so, we obtain that

$$\sum_{x\in X}\frac{1}{|[x]|} = \sum_{B\in X/G}\sum_{x\in B}\frac{1}{|[x]|} = \sum_{B\in X/G}\Big(\underbrace{\frac{1}{|B|} + \cdots + \frac{1}{|B|}}_{|B|}\Big) = \sum_{B\in X/G} 1 = |X/G|,$$

so altogether, we have shown that

$$\sum_{g\in G}|X^g| = |G|\cdot|X/G|$$

as was claimed. □

To obtain more information about the Poincaré series of $\mathbb{F}[X]^G$ we choose to study the action of G on the monomial basis using[6] some modern com-

[6] The combinatorics could be replaced by a modest dose of character theory. However, since

binatorial tools (see, e.g., [201], [403], and [402]). To this end we introduce some new concepts.

For a set X denote by $\underset{n}{\times}X$ the n-fold Cartesian product $X\times\cdots\times X$ of X with itself. The symmetric group Σ_n acts on $\underset{n}{\times}X$ by permutation of the factors. If X is a G-set, then $\underset{n}{\times}X$ is made into a G-set by setting

$$g\cdot(x_1,\ldots,x_n)=(gx_1,\ldots,gx_n).$$

The Σ_n and G-actions commute with each other, i.e.,

$$\sigma g(x_1,\ldots,x_n)=g\sigma(x_1,\ldots,x_n),\ \ \forall\, g\in G,\ \sigma\in\Sigma_n,\ (x_1,\ldots,x_n)\in\underset{n}{\times}X.$$

The n-**fold symmetric product of** X, $\mathbb{SP}^n(X)$, is defined to be the orbit space of $\underset{n}{\times}X$ by the Σ_n-action, i.e.,

$$\mathbb{SP}^n(X)=\underset{n}{\times}X/\Sigma_n.$$

The elements of $\mathbb{SP}^n(X)$ may be thought of as unordered n-tuples of elements $[x_1,\ldots,x_n]$ of X.

EXAMPLE 2: Let $X=\{x,y,z\}$ be a set with 3 elements. Then the first three symmetric powers of X are

$$\mathbb{SP}^1(X)=\{x,y,z\},$$
$$\mathbb{SP}^2(X)=\left\{\begin{matrix}[x,x],[y,y],[z,z]\\ [x,y],[y,z],[z,x]\end{matrix}\right\},$$
$$\mathbb{SP}^3(X)=\left\{\begin{matrix}[x,x,x],[y,y,y],[z,z,z]\\ [x,x,y],[y,y,x],[z,z,x]\\ [x,x,z],[y,y,z],[z,z,y]\\ [x,y,z]\end{matrix}\right\}.$$

Since the G and Σ_n-actions on $\underset{n}{\times}X$ commute, the action of G on $\underset{n}{\times}X$ passes down to $\mathbb{SP}^n(X)$, i.e.,

$$\alpha:G\times\mathbb{SP}^n(X)\longrightarrow\mathbb{SP}^n(X),$$
$$g\cdot[x_1,\ldots,x_n]\longmapsto[gx_1,\ldots,gx_n],$$

for $g\in G$ and $[x_1,\ldots,x_n]\in\mathbb{SP}^n(X)$, defines an action of G on $\mathbb{SP}^n(X)$, and the quotient map

$$\underset{n}{\times}X\xrightarrow{\ q\ }\mathbb{SP}^n(X)$$

is a G-map. The following lemma explains our interest in the functor $\mathbb{SP}^n(-)$ and why it is relevant to the discussion of invariants of permutation representations.

---

this chapter is about combinatorics we prefer to use combinatorial methods.

LEMMA 3.2.4: *Let X be a finite set. Then the map*

$$M : \mathbb{SP}^n(X) \longrightarrow S^n(V_X^*)$$

defined by

$$M([x_1, \ldots, x_n]) = x_1 \cdots x_n$$

is a bijection between the elements of $\mathbb{SP}^n(X)$ and the monomial basis of $S^n(V_X)$. If G is a finite group and X is a G-set, then M is a G-map.

PROOF: This is clear, since to a monomial $x_1^{e_1} \cdots x_m^{e_m}$ of degree n we may associate the unordered n-tuple $[x_1, \ldots, x_1, \ldots, x_m, \ldots, x_m]$, where each x_i is repeated e_i times. This correspondence is an inverse to M. □

Combining Lemma 3.2.4 with Proposition 3.2.3, allows us to write a combinatorial formula for the Poincaré series of the ring of invariants of a permutation group:

PROPOSITION 3.2.5: *Let G be a finite group and X a finite G-set. If $\mathbb{F}$ is any field, then*

$$P(\mathbb{F}[X]^G, t) = \frac{1}{|G|} \sum_{g \in G} \sum_{k=0}^{\infty} |\mathbb{SP}^k(X)^g| t^k. \quad \square$$

This formula does not depend on $\mathbb{F}$ as predicted by Proposition 3.2.2. To make some actual computations, note that Lemma 3.2.4 allows us to count the number of elements of $\mathbb{SP}^n(X)$, as the following lemma shows:

LEMMA 3.2.6: *Let X be a finite set. Then*

$$|\mathbb{SP}^n(X)| = \binom{|X| + n - 1}{n}. \quad \square$$

DEFINITION: *Let X be a finite G-set, G a finite group. Define*

$$A(n, X, G) = |\mathbb{SP}^n(X)/G|,$$

i.e., $A(n, X, G)$ is the number of G-orbits on $\mathbb{SP}^n(X)$.

From Lemmas 3.2.1 and 3.2.4 we obtain the following:

PROPOSITION 3.2.7: *Let G be a finite group and X be a finite G-set. Then the Poincaré series of the ring of invariants $\mathbb{F}[X]^G$ is independent of $\mathbb{F}$ and is given by*

$$P(\mathbb{F}[X]^G, t) = \sum_{n=0}^{\infty} A(n, X, G) t^n. \quad \square$$

To study the numbers $A(n, X, G)$ it is convenient to introduce another way to visualize the elements of $\mathbb{SP}^n(X)$, which is suggested by Lemma 3.2.4.

DEFINITION: *A* **multiset of size** n *consists of a pair* (Y, μ), *where* Y *is a set and* $\mu : Y \longrightarrow \mathbb{N}$ *is a function, such that* $\sum_{y \in Y} \mu(y) = n$. *The set* Y *is called the* **underlying set** *of the multiset,* $\mu(y)$ *the* **multiplicity** *of the element* $y \in Y$, *and* μ *the* **multiplicity function** *of the multiset.*

Since multiplicities are strictly positive, the underlying set of a multiset of size n has cardinality at most n.

To a point $[x] \in \mathbb{SP}^n(X)$ we associate a multiset of size n as follows. The point $[x]$ is an unordered n-tuple $[x_1, \ldots, x_n]$ of elements of X. For the underlying set of the multiset $E_{[x]}$ we take $\{x_1, \ldots, x_n\}$. N.b., in dealing with sets repetitions may be struck from the list of elements, i.e., $\{x, x\} = \{x\}$. In other words, $E_{[x]}$ is the set $\{z_1, \ldots, z_m\}$, where $z_1, \ldots, z_m$ are the *distinct* elements of the n-tuple $[x] = [x_1, \ldots, x_n]$. The multiplicity function $\mu_{[x]}$ is defined by setting $\mu_{[x]}(z)$ equal to the number of occurrences of z in the unordered n-tuple $[x_1, \ldots, x_n]$. This procedure associates a multiset $(E_{[x]}, \mu_{[x]})$, with underlying set a subset of X, to every point $[x]$ of $\mathbb{SP}^n(X)$. Conversely, if (Y, μ) is a multiset of size n whose underlying set is a subset of X, we may associate to (Y, μ) a point $[x]$ in $\mathbb{SP}^n(X)$ by taking the unordered n-tuple of points consisting of the elements of Y, each repeated as often as its multiplicity. Thus points of $\mathbb{SP}^n(X)$ may be viewed as the multisets of size n whose underlying set is a subset of X. In the language of multisets the symmetric square, i.e., $\mathbb{SP}^2(X)$, of $X = \{x, y, z\}$ may be described by Table 3.2.1. Each row of the table gives the multiplicity of the elements of a point of $\mathbb{SP}^2(\{x, y, z\})$ viewed as a multiset of size 2.

UNDERLYING SET	$\mu(x)$	$\mu(y)$	$\mu(z)$
$\{x\}$	2		
$\{y\}$		2	
$\{z\}$			2
$\{x, y\}$	1	1	
$\{x, z\}$	1		1
$\{y, z\}$		1	1

TABLE 3.2.1: $\mathbb{SP}^2(\{x, y, z\})$

DEFINITION: *If* (Y, μ) *is a multiset of size* n, *then the* **type** *of* (Y, μ) *is the* $|Y|$*-tuple of numbers* $\{\mu(y) \mid y \in Y\}$ *arranged in decreasing order.*

The type of (Y, μ) is a partition of n with parts the numbers $\{\mu(y) \mid y \in Y\}$.

LEMMA 3.2.8: *Let* G *be a finite group and* X *a finite* G*-set. Then for every* $[x] \in \mathbb{SP}^n(X)$ *and* $g \in G$ *the multisets* $E_{[x]}$ *and* $E_{g[x]}$ *have the same type.*

PROOF: Suppose $[x] = [x_1, \ldots, x_n]$. Since the order does not matter, we may write $[x]$ in the form

$$[\underbrace{y_1, \dots, y_1}_{\leftarrow \mu(y_1) \rightarrow}, \underbrace{y_2, \dots, y_2}_{\leftarrow \mu(y_2) \rightarrow}, \dots, \underbrace{y_s, \dots, y_s}_{\leftarrow \mu(y_s) \rightarrow}],$$

where $\mu(y_1) \geq \mu(y_2) \geq \cdots \geq \mu(y_s)$ is the type of $[x]$. By definition of the G-action

$$g \cdot [x] = [\underbrace{gy_1, \dots, gy_1}_{\longleftarrow \mu(y_1) \longrightarrow}, \underbrace{gy_2, \dots, gy_2}_{\longleftarrow \mu(y_2) \longrightarrow}, \dots, \underbrace{gy_s, \dots, gy_s}_{\longleftarrow \mu(y_s) \longrightarrow}],$$

and the type has not changed. □

The proof of the preceding lemma shows for finite G-set X that the action of G on $\mathbb{SP}^n(X)$ sends a point y in the multiset $E_{[x]} \in \mathbb{SP}^n(X)$ of multiplicity k to a point of the same multiplicity. Classifying multisets by their type can be of use in determining the orbit structure of $\mathbb{SP}^n(X)$. Since the type of a multiset of size n is just a partition of n, we can use standard partition theory notation to indicate the type. (See, for example, [329] Chapter 6.) One such standard notation indicates the type by listing the number of occurances of each integer k in the partitions an exponent: for example, for partitions of 2 the type 2^1 is the partition 2 of 2, and 1^2 the partition 1, 1 of 2. This is the notation we will employ in the sequel.

Here is a simple example to illustrate these ideas. Apart from the actual numbers that appear it is a special case of a general phenomena for permutation actions of cyclic groups of prime order (see Corollary 3.2.10).

EXAMPLE 3: Let $G = \mathbb{Z}/3$ acting on $\{x, y, z\}$ by cyclic permutation. Let us compute the orbits of the action on the the first three symmetric products $\mathbb{SP}^1(X)$, $\mathbb{SP}^2(X)$, and $\mathbb{SP}^3(X)$. $X = \mathbb{SP}^1(X)$ consists of a single $\mathbb{Z}/3$ orbit because the action is transitive. $\mathbb{SP}^2(X)$ consists of two $\mathbb{Z}/3$ orbits corresponding to the two types of multisets of size 2, viz., type 1^2 and 2^1. The orbits are

$$\begin{array}{ll} \{[x,x],[y,y],[z,z]\} & \text{of type } 2^1, \\ \{[x,y],[y,z],[z,x]\} & \text{of type } 1^2, \end{array}$$

The three types of multisets of size 3 are 3^1, $2^1 1^1$, and 1^3, and to each of these there correspond $\mathbb{Z}/3$ orbits in $\mathbb{SP}^3(X)$. These are

$$\begin{array}{ll} \{[x,y,z]\} & \text{of type } 1^3, \\ \{[x,x,y],[y,y,z],[z,z,x]\} & \text{of type } 2^1 1^1, \\ \{[x,x,z],[y,y,x],[z,z,y]\} & \text{of type } 2^1 1^1, \\ \{[x,x,x],[y,y,y],[z,z,z]\} & \text{of type } 3^1, \end{array}$$

so that there are four orbits of the $\mathbb{Z}/3$ action on $\mathbb{SP}^3(X)$. The orbit sums are listed in Table 3.2.2.

deg	ORBIT SUM	NOTATION
1	$x + y + z$	e_1
2	$x^2 + y^2 + z^2$	s_2
2	$xy + yz + zx$	e_2
3	xyz	e_3
3	$x^2y + y^2z + z^2x$	f
3	$x^2z + y^2x + z^2y$	h
3	$x^3 + y^3 + z^3$	s_3

TABLE 3.2.2: $\mathbb{Z}/3$-orbit sum polynomials

By Theorem 2.3.3, $\mathbb{F}[x, y, z]^{\mathbb{Z}/3}$ is generated by elements of degree at most 3, provided that 3 is invertible in $\mathbb{F}$. From this table we see that the polynomials e_1, e_2, e_3, and f (or h) may be chosen as generators for $\mathbb{F}[x, y, z]^{\mathbb{Z}/3}$. This should be no surprise, since $\mathbb{Z}/3$ acting cyclically on $\{x, y, z\}$ is just the alternating group A_3 in its tautological representation. So the invariants[7] are generated by e_1, e_2, e_3, and $\nabla = h$.

PROPOSITION 3.2.9: *Let the finite group G act transitively on the set $\{x_1, \ldots, x_m\} = X$. Then*

$$\mathbb{SP}^n(X)^G = \begin{cases} \varnothing, & n \not\equiv 0 \bmod m, \\ [\underbrace{x_1, \ldots, x_m, \ldots, x_1, \ldots, x_m}_{k}], & n = km. \end{cases}$$

PROOF: The indicated point is indeed a fixed point. If $[x] \in \mathbb{SP}^n(X)^G$, then the underlying set of $[x]$ must contain all the elements $x_1, \ldots, x_m$ because G acts transitively on X. Moreover, invariance and transitivity imply that each pair of elements $x_{i'}, x_{i''}$ must have the same multiplicity. This is possible if and only if n is a multiple of m and $[x]$ is the indicated element. □

If G is cyclic of prime order, then the only subgroups of G are G itself and the trivial subgroup. This greatly simplifies the orbit structure of $\mathbb{SP}^n(X)$, since the only possible orbits are fixed points and orbits with p elements.

COROLLARY 3.2.10: *Let the cyclic group $\mathbb{Z}/p$ of prime order act by cyclic permutation on the set $\{x_1, \ldots, x_p\} = X$. Then*

$$|\mathbb{SP}^n(X)/\mathbb{Z}/p| = \begin{cases} \frac{1}{p}\binom{p+n-1}{n}, & n \not\equiv 0 \bmod p, \\ 1 + \frac{1}{p}\left[\binom{p+n-1}{n} - 1\right], & n \equiv 0 \bmod p. \end{cases}$$

PROOF: The result follows from Lemma 3.2.6, Proposition 3.2.9, and Proposition 3.2.3. □

We are now in a position to compute the Poincaré series of the ring of invariants of the regular representation of a cyclic group of prime order.

[7] Once we have proved Theorem 3.4.2 we will see that this holds also in characteristic 3.

PROPOSITION 3.2.11: *Let $\rho : \mathbb{Z}/p \hookrightarrow \mathrm{GL}(p, \mathbb{F})$ denote the regular representation of the cyclic group of prime order p. Then the Poincaré series of the ring of invariants $\mathbb{F}[x_1, \ldots, x_p]^{\mathbb{Z}/p}$ is given by*

$$P(\mathbb{F}[x_1, \ldots, x_p]^{\mathbb{Z}/p}, t) = \frac{1}{p}\left[\frac{1}{(1-t)^p} + \frac{p-1}{(1-t^p)}\right]$$
$$= \frac{1}{p}\left[\frac{(1+t+\cdots+t^{p-1})^p + (p-1)(1-t^p)^{p-1}}{(1-t^p)^p}\right].$$

PROOF: The regular representation is the permutation representation associated to the action of $\mathbb{Z}/p$ on itself given by left translation. In other words, ρ is the linear representation associated to the action of $\mathbb{Z}/p$ on the set $X = \{x_1, \ldots, x_p\}$ by cyclic permutation. By Corollary 3.2.10, and the definitions introduced in this section, we have

$$A(n, X, \mathbb{Z}/p) = \begin{cases} \frac{1}{p}\binom{p+n-1}{n}, & n \not\equiv 0 \bmod p, \\ 1 + \frac{1}{p}\left[\binom{p+n-1}{n} - 1\right], & n \equiv 0 \bmod p. \end{cases}$$

By Proposition 3.2.7 we obtain

$$P\left(\mathbb{F}[x_1, \ldots, x_p]^{\mathbb{Z}/p}, t\right) = \sum_{n=0}^{\infty} A(n, X, \mathbb{Z}/p)\, t^n$$
$$= \frac{1}{p}\left[\sum_{n=0}^{\infty}\binom{p+n-1}{n} t^n + (p-1)\sum_{k=0}^{\infty} t^{kp}\right] = \frac{1}{p}\left[\frac{1}{(1-t)^p} + \frac{p-1}{1-t^p}\right],$$

and the result follows by combining the fractions over their common denominator. □

3.3 The Hilbert–Serre Theorem on Poincaré Series

A rather surprising result concerning the Poincaré series of the polynomial invariants of a finite group is that it is a rational function. This is immediate from Molien's theorem, Theorem 3.1.3, in the nonmodular case, but far from apparent otherwise. In point of fact, this is a special case of a more general theorem for finitely generated graded modules over graded Noetherian rings.

Recall that for a polynomial algebra A over $\mathbb{F}$ generated by elements $x_1, \ldots, x_n$ of degrees $d_1, \ldots, d_n$ that the Poincaré series is given by

$$P(A, t) = \prod_{i=1}^{n} \frac{1}{1 - t^{d_i}}.$$

This is a rational function with a pole at $t = 1$ of order n. The theorem of

D. Hilbert and J.-P. Serre generalizes this.[8]

THEOREM 3.3.1 (D. Hilbert, J.-P. Serre): *Suppose that A is a graded connected commutative $\mathbb{F}$-algebra finitely generated over $\mathbb{F}$ by homogeneous elements $x_1, \ldots, x_s$ in positive degrees. Suppose that M is a finitely generated graded A-module. Then the Poincaré series $P(M, t)$ is of the form*

$$\frac{f(t)}{\prod_{j=1}^{s}(1 - t^{\deg(x_j)})},$$

where $f(t)$ is a polynomial in t with integer coefficients.

PROOF: The proof is by induction on s. If $s = 0$, then M is just a finite-dimensional vector space over $\mathbb{F}$, so $P(M, t)$ is a polynomial. So suppose $s > 0$. Let $\lambda : M \longrightarrow M$ denote left multiplication by the element $x_s \in A$. We have exact sequences

$$0 \longrightarrow \ker(\lambda) \longrightarrow M \longrightarrow M/\ker(\lambda) \longrightarrow 0,$$

$$0 \longrightarrow \operatorname{Im}(\lambda) \longrightarrow M \longrightarrow \operatorname{coker}(\lambda) \longrightarrow 0,$$

yielding

$$(\text{✣}) \qquad P(M, t) = P(\ker(\lambda), t) + P(M/\ker(\lambda), t),$$
$$(\text{✠}) \qquad P(M, t) = P(\operatorname{Im}(\lambda), t) + P(\operatorname{coker}(\lambda), t).$$

The map λ is a morphism of degree $\deg(x_s)$, so

$$(M/\ker(\lambda))_j \cong (\operatorname{Im}(\lambda))_{j+\deg(x_s)},$$

and hence

$$P(M/\ker(\lambda), t) = t^{-\deg(x_s)} P(\operatorname{Im}(\lambda), t).$$

Substituting in the equations (✣), (✠), we obtain

$$P(M, t) = P(\ker(\lambda), t) + t^{-\deg(x_s)} P(\operatorname{Im}(\lambda), t),$$
$$P(M, t) = P(\operatorname{Im}(\lambda), t) + P(M/\operatorname{Im}(\lambda), t).$$

Multiplying the first of these equations by $t^{\deg(x_s)}$ and subtracting from the second equation yields

$$P(M, t) = \frac{-t^{\deg(x_s)}}{1 - t^{\deg(x_s)}} P(\ker(\lambda), t) + \frac{1}{1 - t^{\deg(x_s)}} P(M/\operatorname{Im}(\lambda), t).$$

The modules $\ker(\lambda)$ and $M/\operatorname{Im}(\lambda)$ both have trivial x_s-action, and are thus finitely generated over the subalgebra of A generated by $x_1, \ldots, x_{s-1}$. By the inductive hypothesis the right-hand side of the preceding equation is a rational function of the stated form. □

---

[8] We will see a different proof of this result at the end of Section 5.2.

The Poincaré series of a graded module is of course unique, but the number of ways to write it as a rational function as in Theorem 3.3.1 is alarmingly large. Still, once written in this form one could wonder what the significance of the integer s, the exponents $\deg(x_1), \ldots, \deg(x_s)$, and the coefficients of the numerator $f(t)$ are; see Theorem 5.5.1. The following result fits in this context. Recall that a commutative graded algebra H over the field $\mathbb{F}$ is called a **graded complete intersection** if $H \cong \mathbb{F}[f_1, \ldots, f_m]/(h_1, \ldots, h_k)$, where $h_1, \ldots, h_k \in \mathbb{F}[f_1, \ldots, f_m]$ are algebraically independent and $\mathbb{F}[f_1, \ldots, f_m]$ is a free $\mathbb{F}[h_1, \ldots, h_k]$-module.

PROPOSITION 3.3.2: *Suppose that H is a complete intersection over the field $\mathbb{F}$, so $H \cong \mathbb{F}[f_1, \ldots, f_m]/(h_1, \ldots, h_k)$, where the forms $h_1, \ldots, h_k$ in $\mathbb{F}[f_1, \ldots, f_m]$ are algebraically independent and $\mathbb{F}[f_1, \ldots, f_m]$ is a free $\mathbb{F}[h_1, \ldots, h_k]$-module. Then*

$$P(H, t) = \frac{\prod_{i=1}^{k}(1 - t^{\deg(h_i)})}{\prod_{j=1}^{m}(1 - t^{\deg(f_j)})}.$$

PROOF: Since $\mathbb{F}[f_1, \ldots, f_m]$ is a free $\mathbb{F}[h_1, \ldots, h_k]$-module, it follows that

$$\mathbb{F}[f_1, \ldots, f_m] \cong \mathbb{F}[h_1, \ldots, h_k] \otimes_{\mathbb{F}} H$$

as $\mathbb{F}[h_1, \ldots, h_k]$-modules. The Poincaré series is multiplicative over tensor products, and solving for $P(H, t)$ yields the desired conclusion. □

If we think of $P(M, t)$ as a function of a complex variable, then the following proposition shows that we may interpret the order of the pole at $t = 1$ as telling us about the rate of growth of the dimension of the graded pieces.

PROPOSITION 3.3.3: *Suppose that*

$$P(M, t) = \frac{f(t)}{\prod_{j=1}^{s}(1 - t^{k_j})} = \sum_{r \geq 0} a_r t^r,$$

where $f(t)$ is a polynomial with integer coefficients and the a_r are nonnegative integers. Let $d(M)$ be the order of the pole of $P(M, t)$ at $t = 1$. Then

(i) *there exists a constant $\kappa > 0$ such that $a_n \leq \kappa \cdot n^{d(M)-1}$ for $n > 0$, but*

(ii) *there does not exist a constant $\kappa > 0$ such that $a_n \leq \kappa \cdot n^{d(M)-2}$ for $n > 0$.*

PROOF: The hypothesis and conclusion remain unaltered if we replace $P(M, t)$ by $P(M, t) \cdot (1 + t + \cdots + t^{k_j - 1})$, and so without loss of generality we may assume each $k_j = 1$. Hence, we may suppose that the Poincaré series

$P(M, t)$ is equal to $f(t)/(1-t)^{d(M)}$ with $f(1) \neq 0$. Let $f(t) = \alpha_m t^m + \cdots + \alpha_0$. We have

$$a_n = \alpha_0 \binom{n+d(M)-1}{d(M)-1} + \alpha_1 \binom{n+d(M)-2}{d(M)-1} + \cdots + \alpha_m \binom{n+d(M)-m-1}{d(M)-1}$$

and the condition $f(1) \neq 0$ implies that $\alpha_0 + \cdots + \alpha_m \neq 0$, so this expression is a polynomial of degree exactly $d(M) - 1$ in n. □

NOTATION: *We write $d(M)$ for the order of the pole at $t = 1$ of $P(M, t)$.*

LEMMA 3.3.4: *Suppose that $A \supseteq B$ is a finite extension of graded algebras, each of which is finitely generated as an algebra over $\mathbb{F}$ by homogeneous elements of positive degrees. Then $d(A) = d(B)$.*

PROOF: As a B-module, A is finitely generated, so it is a quotient of a finitely generated free graded B-module M. Then $P(M, t) = f(t)P(B, t)$ for some polynomial $f(t)$ with $f(1) > 0$. Thus as power series[9] we have

$$P(B, t) \leq P(A, t) \leq f(t)P(B, t),$$

which proves the lemma. □

We apply this lemma to overrings of invariants and get the following:

PROPOSITION 3.3.5: $d(\mathbb{F}[V]^G) = \dim_{\mathbb{F}}(V)$.

PROOF: By Theorem 2.1.4, the ring $\mathbb{F}[V]$ is a finite extension of $\mathbb{F}[V]^G$, so by the lemma, it suffices to look at the pole at $t = 1$ of

$$P(\mathbb{F}[V], t) = 1/(1-t)^n,$$

where $n = \dim_{\mathbb{F}}(V)$. □

3.4 Göbel's Theorem on Permutation Invariants

The study of permutation-invariant polynomials is probably one of the oldest subjects in invariant theory. Surprisingly, there are still many things we do not know, and lots of new theorems remain to be discovered. One such theorem was discovered by M. Göbel; it appeared in his Diplomarbeit in 1992 and forms the subject of this section. The presentation of the material is taken from [205]; see also [322]. To formulate M. Göbel's theorem we need some definitions.

DEFINITION: *If $K = (k_1, \ldots, k_n)$ is an n-tuple of nonnegative integers, then K is called an* **exponent sequence**. *The* **associated partition of** *K is the ordered set consisting of the n numbers $k_1, \ldots, k_n$ rearranged in*

[9] The stated inequalities hold coefficient for coefficient.

weakly decreasing order. We denote by $\lambda(K)$ *the partition associated to* K*, so*

$$\lambda(K) = (\lambda_1(K) \geq \lambda_2(K) \geq \cdots \geq \lambda_n(K))$$

and the n*-tuple* $(\lambda_1(K), \ldots, \lambda_n(K))$ *is a permutation of* $k_1, \ldots, k_n$*. The monomial* x^K *is called* **special** *if the associated partition* $\lambda(K)$ *of the exponent sequence* K *satisfies*

(i) $\lambda_i(K) - \lambda_{i+1}(K) \leq 1$ *for all* $i = 1, \ldots, n-1$*, and*
(ii) $\lambda_n(K) = 0$.

Notice that if two exponent sequences A and B are permutations of each other, then $\lambda(A) = \lambda(B)$.

M. Göbel's theorem says that the rings of invariants of permutation representations are generated as an algebra by the orbit sums of the special monomials and the elementary symmetric polynomial $e_n = x_1 \cdots x_n$. From Lemma 3.2.1 we know that the orbit sums of monomials are a basis for the invariants of a permutation representation. Thus, what we need to show is that such orbit sums belong to the subalgebra generated by the orbit sums of special monomials and e_n. The idea for the proof of this is similar to the one used in the proof of the fundamental theorem of symmetric polynomials, [372] Theorem 1.1.1. We start with an orbit sum and by a succession of reductions rewrite it in the desired form. To be sure that this procedure comes to an end we need a suitable ordering on the monomials in addition to the ordering by the degree of the monomial. Here is the definition of this ordering.

DEFINITION: *If* $A = (a_1, \ldots, a_n)$ *and* $B = (b_1, \ldots, b_n)$ *are sequences of nonnegative integers, then we say that the monomial* x^A *is smaller in the* **dominance order** *than* x^B*, and denote this by* $x^A \leq_{\text{dom}} x^B$*, if the associated partitions satisfy*

$$\begin{aligned} \lambda_1(A) &\leq \lambda_1(B), \\ \lambda_1(A) + \lambda_2(A) &\leq \lambda_1(B) + \lambda_2(B), \\ &\vdots \\ \lambda_1(A) + \cdots + \lambda_n(A) &\leq \lambda_1(B) + \cdots + \lambda_n(B). \end{aligned}$$

If both $x^A \leq_{\text{dom}} x^B$ *and* $x^B \leq_{\text{dom}} x^A$ *then we write* $x^A =_{\text{dom}} x^B$*.*

Note that $x^A =_{\text{dom}} x^B$ means that A and B are permutations of each other.

To reduce the monomials we need a rewriting construction. Here it is:

DEFINITION: *Let* x^K *be a monomial and* $\lambda(K)$ *the partition associated to the exponent sequence* K*. Set* $t_K = \min\{i \mid \lambda_i(K) - \lambda_{i+1}(K) > 1\}$*, and define the* **reduced monomial of** x^K *to be* $x^{\widetilde{K}}$*, where the exponent sequence* $\widetilde{K}$ *is obtained from* K *by lowering the* t_K *largest exponents by* 1*.*

Note that in the notation of this definition t_K is the first index where the first condition to be special fails in the associated partition: Passing to the reduced monomial lowers each of the t_K largest exponents by 1, so $x^{\widetilde{K}}$ is closer to being special than x^K was.

Let us have a look at an example. The monomial

$$x^K = x_1^3 x_2 x_3 \in \mathbb{F}[x_1, x_2, x_3]$$

is not special, because the associated partition is $(3 \geq 1 \geq 1)$, which does not satisfy either one of the two conditions for a special monomial. The associated reduced monomial is

$$x^{\widetilde{K}} = x_1^2 x_2 x_3 \in \mathbb{F}[x_1, x_2, x_3]$$

(with $t_K = 1$). It is closer to being special than x^K was, because by the reduction we were able to assure that the first condition to be special holds. Note also that $x^{\widetilde{K}} = x_1 \cdot x_1 x_2 x_3$ is the product of a special monomial and $e_3 = x_1 x_2 x_3$. This is, in fact, the key to the proof of Göbel's theorem, and the content of the following:

LEMMA 3.4.1: *Let $A = (a_1, \ldots, a_n)$ be an n-tuple of nonnegative integers. Then every monomial x^B occurring in $\mathfrak{S}_{[x^{\tilde{A}}]} \cdot e_{t_A}$, where e_{t_A} is the t_Ath elementary symmetric polynomial, is lower in the dominance order than x^A, with equality if and only if x^B is a term in $\mathfrak{S}_{[x^A]}$.*

PROOF: Set $t = t_A$. Let x^B be a monomial occurring in $\mathfrak{S}_{[x^{\tilde{A}}]} \cdot e_t$. We may write

$$x^B = x^C x_{i_1} \cdots x_{i_t},$$

where C is a permutation of $\widetilde{A}$. Note that $\lambda(C) = \lambda(\widetilde{A})$ and

$$(\because) \qquad b_i = \begin{cases} c_i & \text{if } i \notin \{i_1, \ldots, i_t\}, \\ c_i + 1 & \text{if } i \in \{i_1, \ldots, i_t\}. \end{cases}$$

To show that $B \leq_{\text{dom}} A$ we need to express $\lambda(B) = (\lambda_1(B) \geq \cdots \geq \lambda_n(B))$ in terms of $\lambda(A) = (\lambda_1(A) \geq \cdots \geq \lambda_n(A))$. To this end let χ_I be the characteristic function of the subset $I = \{i_1, \ldots, i_t\} \subseteq \{1, \ldots, n\}$. In other words, $\chi_I(j) = 1$ if $j \in I$ and $\chi_I(j) = 0$ if $j \in \{1, \ldots, n\} \setminus I$. Since C is a permutation of $\widetilde{A}$, we may write

$$c_i = \widetilde{a}_{\tau(i)}, \; i = 1, \ldots, n,$$

for some permutation $\tau \in \Sigma_n$. Note that τ permutes $\{1, \ldots, n\}$, preserving the partition consisting of I and $\{1, \ldots, n\} \setminus I$. Then $(\because)$ says that

$$(\maltese) \qquad b_i = \begin{cases} \widetilde{a}_{\tau(i)} & \text{if } i \notin \{i_1, \ldots, i_t\}, \\ \widetilde{a}_{\tau(i)} + 1 & \text{if } i \in \{i_1, \ldots, i_t\}. \end{cases}$$

Since

$$\tilde{a}_i = \begin{cases} a_i - 1 & \text{if } 1 \le i \le t, \\ a_i & \text{if } t+1 \le i \le n, \end{cases}$$

we may rewrite (⊕) to get

$$(\bigstar) \qquad b_i = \begin{cases} \tilde{a}_{\tau(i)} - 1 + \chi_I(\tau(i)) & \text{if } 1 \le i \le t, \\ \tilde{a}_{\tau(i)} + \chi_I(\tau(i)) & \text{if } t+1 \le i \le n. \end{cases}$$

The partition $\lambda(B)$ is the n-tuple B arranged in weakly decreasing order, so $(\bigstar)$ says that

$$\lambda_i(B) = \begin{cases} \lambda_{\tau(i)}(A) - 1 + \chi_I(\tau(i)) & \text{if } 1 \le i \le t, \\ \lambda_{\tau(i)}(A) + \chi_I(\tau(i)) & \text{if } t+1 \le i \le n. \end{cases}$$

This has prepared the way for us to show that $B \le_{\text{dom}} A$, for which we need to compare the sums

$$\sum_{i=1}^{s} \lambda_i(B) \quad \text{and} \quad \sum_{i=1}^{s} \lambda_i(A) \quad \forall\, s = 1, \dots, n.$$

First consider the case $s \le t$. Then

$$\begin{aligned} \sum_{i=1}^{s} \lambda_i(B) &= \sum_{i=1}^{s} \Big(\lambda_{\tau(i)}(A) - 1 + \chi_I(\tau(i)) \Big) \\ &= \Big(\sum_{i=1}^{s} \lambda_{\tau(i)}(A) \Big) - s + \sum_{i=1}^{s} \chi_I(\tau(i)) \le \sum_{i=1}^{s} \lambda_i(A), \end{aligned}$$

because

$$\sum_{i=1}^{s} \lambda_{\tau(i)}(A) \le \sum_{i=1}^{s} \lambda_i(A) \text{ and } \sum_{i=1}^{s} \chi_I(\tau(i)) \le s,$$

since any reordering of $\lambda(A) = (\lambda_1(A) \ge \cdots \ge \lambda_n(A))$ can only lead to a smaller sum for the first s terms.

If $s > t$, then

$$\begin{aligned} \sum_{i=1}^{s} \lambda_i(B) &= \sum_{i=1}^{t} \lambda_i(B) + \sum_{i=t+1}^{s} \lambda_i(B) \\ &= \sum_{i=1}^{t} \Big(\lambda_{\tau(i)}(A) - 1 + \chi_I(\tau(i)) \Big) + \sum_{i=t+1}^{s} \Big(\lambda_{\tau(i)}(A) + \chi_I(\tau(i)) \Big) \\ &= \Big(\sum_{i=1}^{t} \lambda_{\tau(i)}(A) \Big) - t + \sum_{i=1}^{s} \chi_I(\tau(i)) + \Big(\sum_{i=t+1}^{s} \lambda_{\tau(i)}(A) \Big) \\ &= \sum_{i=1}^{t} \lambda_{\tau(i)}(A) + \sum_{i=t+1}^{s} \lambda_{\tau(i)}(A) - t + \sum_{i=1}^{s} \chi_I(\tau(i)). \end{aligned}$$

Again, since τ preserves the partition of $\{1, \ldots, n\}$ consisting of I and its complement, we have

$$\sum_{i=1}^{t} \lambda_{\tau(i)}(A) = \sum_{i=1}^{t} \lambda_i(A).$$

Reordering the weakly decreasing sequence $\lambda(A)$ can only lead to a smaller sum for the first s terms. It follows that

$$\sum_{i=t+1}^{s} \lambda_{\tau(i)}(A) \leq \sum_{i=t+1}^{s} \lambda_i(A).$$

Finally, $\sum_{i=1}^{s} \chi_I(\tau(i)) \leq t$, so

$$\sum_{i=1}^{s} \lambda_i(B) = \sum_{i=1}^{t} \lambda_{\tau(i)}(A) + \sum_{i=t+1}^{s} \lambda_{\tau(i)}(A) - t + \sum_{i=1}^{s} \chi_I(\tau(i))$$
$$\leq \sum_{i=1}^{t} \lambda_i(A) + \sum_{i=t+1}^{s} \lambda_i(A) = \sum_{i=1}^{s} \lambda_i(A)$$

as was required to be shown.

If x^B is a term in $\mathfrak{S}_{x^A}$, then A and B are permutations of each other, so $\lambda(A) = \lambda(B)$, which implies $x^A =_{\text{dom}} x^B$.

Conversely, if $x^A =_{\text{dom}} x^B$, then $\lambda(A) = \lambda(B)$. The monomial x^B occurs in the orbit sum $\mathfrak{S}_{[x^{\tilde{A}}]} \cdot e_{t_A}$. Hence, we get x^B by

first subtracting 1 from t_A of the largest exponents $a_1, \ldots, a_n$ in the exponent sequence A,

second permuting the result by an element g of G to get a monomial in the orbit of $x^{\tilde{A}}$,

third adding 1 to the exponents in the result that correspond to the exponents of some monomial in $\mathfrak{S}_{[x^{\tilde{A}}]} \cdot e_{t_A}$.

Since the resulting partition is B, and $\lambda(B) = \lambda(A)$, exactly the same gap occurs in B as in A, so the exponents to which we added 1 in the third step must be exactly the ones we subtracted 1 from in the first step. This says that g permutes A to B, so x^B is a term of $\mathfrak{S}_{[x^A]}$. □

THEOREM 3.4.2 (M. Göbel): *Let G be a finite group, X a finite G-set, and R a commutative ring. Then the ring of invariants $R[X]^G$ is generated as an algebra by $e_{|X|} = \prod_{x \in X} x$, the top degree elementary symmetric polynomial in the elements of X, and the orbit sums of special monomials.*

PROOF: Set $n = |X|$. By Lemma 3.2.1 we need to show that the orbit sum of any monomial belongs to the subalgebra S generated by e_n and the orbit sums of the special monomials. So, let $x^K \in R[X]$ be a nonspecial monomial. We use induction on the degree and the dominance order of the monomial.

If $\lambda_n(K) \neq 0$, then x^K is divisible by e_n, so $x^K = x^{K'} \cdot e_n$, and by repeated division by e_n we may suppose that $\lambda_n(K) = 0$. This being the case there is a gap in the associated partition $\lambda(K)$, and we write

$$\mathfrak{S}_{[x^K]} = \mathfrak{S}_{[x^{\tilde{K}}]} \cdot e_{t_K} - \Delta(x^K),$$

where t_K is the first index for which the condition to be a special monomial fails in $\lambda(K)$. The t_Kth elementary symmetric polynomial is the sum of the elements in the Σ_n orbit of $x_1 \cdots x_{t_K}$. This monomial is a special monomial, since $t_K < n$. Since the Σ_n-orbit of $x_1 \cdots x_{t_K}$ is a disjoint union of G-orbits, it follows that $e_{t_K} \in S$. The orbit sum $\mathfrak{S}_{[x^{\tilde{K}}]}$ belongs to S by induction on the degree of the monomial x^K. Therefore, $\mathfrak{S}_{[x^{\tilde{K}}]} \cdot e_{t_K}$ belongs to S. Finally, write

$$\Delta(x^K) = \mathfrak{S}_{[x^{\tilde{K}}]} \cdot e_{t_K} - \mathfrak{S}_{[x^K]},$$

and note by Lemma 3.4.1 that all the monomials that occur in $\Delta(x^K)$ are lower in the dominance order than x^K. Hence $\Delta(x^K)$ is in S by induction, and therefore so is $\mathfrak{S}_{[x^K]}$. □

COROLLARY 3.4.3 (M. Göbel): *Let G be a finite group, X a finite G-set, and R a commutative ring. Then the ring of invariants $R[X]^G$ is generated as an algebra by forms of degree at most* $\max\left\{|X|, \binom{|X|}{2}\right\}$.

PROOF: The maximal degree of a special monomial is $\binom{|X|}{2}$, and $e_{|X|}$ of course has degree $|X|$. □

The inequality $\beta_{\mathbb{F}}(\pi) \le \max\left\{n, \binom{n}{2}\right\}$ for permutation representations π on a set X of cardinality n, independent of the group G or the ground field $\mathbb{F}$, is known as **Göbel's bound**. These results together with a *pullback technique*, which will also be used in Sections 4.1, 4.2, and 7.3, [376], and [379], as well as other contexts, can be used to advantage here to yield a bound for $\beta(\rho)$ when ρ is an irreducible representation in the nonmodular case. The idea is quite simple.

Suppose that $\rho : G \hookrightarrow \mathrm{GL}(n, \mathbb{F})$ is an irreducible representation of a finite group G over the field $\mathbb{F}$. If $B \subset V^*$ is the G-orbit of a nonzero vector, then $\mathrm{Span}_{\mathbb{F}}\{B\} = V^*$, for otherwise, V^* would contain the nontrivial proper G-invariant subspace $\mathrm{Span}_{\mathbb{F}}\{B\}$, contrary to the assumption that ρ is irreducible. The group G acts on B by permutations and hence also on the polynomial algebra $\mathbb{F}[B]$, where the elements of B are regarded as formal variables (indeterminates over $\mathbb{F}$). The inclusion $B \subset V^*$ induces an epimorphism of algebras

$$\lambda : \mathbb{F}[B] \longrightarrow S(V^*) = \mathbb{F}[V]$$

that is compatible with the G-actions. If in addition we assume that $|G| \in \mathbb{F}^\times$, then λ induces an epimorphism upon taking invariants. Hence we derive a corollary to Göbel's theorem:

COROLLARY 3.4.4: *Let $\varrho : G \hookrightarrow \mathrm{GL}(n, \mathbb{F})$ be an irreducible representation of a finite group G over the field $\mathbb{F}$ and $B \subset V^* \setminus \{0\}$ a G-orbit. If $|G| \in \mathbb{F}^\times$, then $\mathbb{F}[V]^G$ is generated as an algebra by orbit sums of special monomials in the elements of the orbit B. In particular, $\beta(\varrho) \leq \max\left\{|B|, \binom{|B|}{2}\right\}$.* □

The relation between the orbit sum $\mathfrak{S}_{[x^K]}$ of a monomial and the transfer $\mathrm{Tr}^G(x^K)$ is quite simple: $\mathrm{Tr}^G(x^K) = |G : G_{x^K}| \cdot \mathfrak{S}_{[x^K]}$, where G_{x^K} is the isotropy subgroup of x^K. If X is a finite G-set and the ground field has characteristic p dividing the order $|G|$ of G, then the image of the transfer $\mathrm{Im}(\mathrm{Tr}^G) \subset \mathbb{F}[X]^G$ is generated as an ideal by the transfers of the special monomials x^K with $|G : G_{x^K}|$ relatively prime to p; see [295] Theorem 1.1. This is proved with suitably modified versions of Lemma 3.4.1 and Theorem 3.4.2, in which the transfer is replaced by the orbit sum.

The example of the defining representation of the symmetric group Σ_n over a field of characteristic 2 (see Section 2.2, Example 1) is of this type. Here is another such example taken from [295].

EXAMPLE 1: Consider the alternating group A_n, $n \geq 4$, in its defining representation on $X = \{x_1, \dots, x_n\}$, and let the ground field $\mathbb{F}$ have characteristic 2. As in Section 2.2, Example 1, $e_1, \dots, e_n \in \mathbb{F}[x_1, \dots, x_n]^{A_n}$, where $e_1, \dots, e_n$ are the elementary symmetric polynomials, and, as was noted there, it follows from Corollary 5.5.1 that likewise $\mathrm{Im}(\mathrm{Tr}^{A_n})$ is generated as an ideal by the transfers of monomials of degree at most $\binom{n}{2}$.

If $x^K = x_1^{k_1} \cdots x_n^{k_n} \in \mathbb{F}[x_1, \dots, x_n]$ is a monomial and there exist four pairwise distinct exponents i, j, r, s, and $k_i = k_j$, $k_r = k_s$, then the product of the two transpositions $i \leftrightarrow j$ and $r \leftrightarrow s$ lies in A_n, has order two, and fixes x^K. Hence the isotropy group of x^K has even order. As $\mathbb{F}$ has characteristic 2

$$\mathrm{Tr}^{A_n}(x^K) = |(A_n)_{x^K}| \cdot \mathfrak{S}_{x^K} = 0 \in \mathbb{F}[x_1, \dots, x_n]^{A_n}.$$

The only special monomials in $\mathbb{F}[x_1, \dots, x_n]$ of degree at most $\binom{n}{2}$ where no such index pairs occur all lie in the Σ_n-orbit of one of the following:

$$x_1^1 x_2^2 \cdots x_{n-1}^{n-1},$$
$$x_1^a x_2^1 x_3^2 \cdots x_{n-1}^{n-2} \quad \text{with } a \leq n-2,$$
$$x_1^b x_2^b x_3^1 \cdots x_{n-1}^{n-3} \quad \text{with } b \leq n-3.$$

The special monomials x^K that appear in this list are those where the associated partition $\lambda(K)$ has at most 3 of the terms of $\lambda_i(K)$ equal, so that

$\mathrm{Im}(\mathrm{Tr}^{A_n})$ is generated by the orbit sums of special monomials $\mathfrak{S}_{[x^K]}$ where at most three equalities occur in the associated partition $\lambda(K) = (\lambda_1(K) \geq \cdots \geq \lambda_{n-1}(K) \geq \lambda_n(K) = 0)$.

Finally, we calculate the height of $\mathrm{Im}(\mathrm{Tr}^{A_n})$, for $n \geq 4$, in the modular situation. The product of the transpositions (12)(34) belongs to A_n, so by Feshbach's transfer theorem (Theorem 2.4.5, or Theorem 2.4 in [292], or Theorem 1 of [378])

$$\mathit{ht}\left(\mathrm{Im}(\mathrm{Tr}^{\mathbb{Z}/2})\right) \leq 2.$$

We claim the ideal $\mathrm{Im}(\mathrm{Tr}^{\mathbb{Z}/2})$ has height precisely 2. Since the image of the transfer is nontrivial, its height is positive, and we assume to the contrary that

$$\mathit{ht}\left(\mathrm{Im}(\mathrm{Tr}^{A_n})\right) = 1.$$

Then the ideals

$$\mathrm{Im}(\mathrm{Tr}^{A_n}) \cap \mathbb{F}[V]^{\Sigma_n} \supseteq \mathrm{Im}(\mathrm{Tr}^{\Sigma_n})$$

have height 1. Recall that the image of the transfer of the symmetric group Σ_n is the principal prime ideal generated by the discriminant, Δ_n; see Section 2.2, Example 1. This implies

$$\mathrm{Im}(\mathrm{Tr}^{A_n}) \cap \mathbb{F}[V]^{\Sigma_n} = \mathrm{Im}(\mathrm{Tr}^{\Sigma_n}).$$

However, let

$$\nabla'_n,\ \nabla''_n \in \mathrm{Im}(\mathrm{Tr}^{A_n})$$

be the orbit sums of $x_1^1 x_2^2 \cdots x_{n-1}^{n-1}$ and $x_1^2 x_2^1 x_3^3 \cdots x_{n-1}^{n-1}$, respectively. The Σ_n-orbit of $x_1^1 x_2^2 \cdots x_{n-1}^{n-1}$ splits into the disjoint union of the A_n-orbits of $x_1^1 x_2^2 \cdots x_{n-1}^{n-1}$ and $x_1^2 x_2^1 x_3^3 \cdots x_{n-1}^{n-1}$, so

$$\Delta_n = \nabla'_n + \nabla''_n$$

and

$$\nabla'_n \cdot \nabla''_n \in \mathbb{F}[V]^{\Sigma_n}.$$

This latter polynomial is not in the image of the transfer of Σ_n because it is irreducible in $\mathbb{F}[V]^{\Sigma_n}$, so cannot be divisible by the discriminant Δ_n in $\mathbb{F}[V]^{\Sigma_n}$. Since

$$\nabla'_n \cdot \nabla''_n \in \mathrm{Im}(\mathrm{Tr}^{A_n}) \cap \mathbb{F}[V]^{\Sigma_n} = \mathrm{Im}(\mathrm{Tr}^{\Sigma_n}),$$

this is a contradiction, so $\mathit{ht}\left(\mathrm{Im}(\mathrm{Tr}^{A_n})\right) = 2$, for $n \geq 4$, when the ground field $\mathbb{F}$ has characteristic 2.

Chapter 4
Noetherian Finiteness

LOOKING at the general form of the Poincaré series of a commutative graded connected finitely generated algebra over a field given by Theorem 3.3.1 (the theorem of D. Hilbert and J.-P. Serre) suggests that such an algebra might contain a polynomial subalgebra over which it is finitely generated as a module. This is the content of the Noether normalization theorem[1] (see, e.g., [372] Theorem 5.3.3). In the case of a ring of invariants $\mathbb{F}[V]^G$ of a finite group G one might hope to be able to find such a subalgebra by using properties of the representation $\varrho : G \hookrightarrow \mathrm{GL}(n, \mathbb{F})$. This is indeed the case. A construction of E. Dade, see Section 4.3, provides a means of doing this.

The central purpose of this chapter is to create tools for constructing invariant forms in a systematic way. We start by discussing the orbit Chern classes, which were briefly introduced in Section 1.4. These, and their refinements, have proven to be invaluable when it comes to performing computations with rings of invariants of finite groups. In particular, they provide us with a large family of natural invariants associated with the orbit structure of the G-action on V^*. The regular representation of G plays an important role here. More generally, the action of G on the cosets of a subgroup H of G lead to a refinement of the orbit Chern classes. These invariants, constructed from the combinatorics of the regular representation, also play a role in the theory. For example, we show that it is always possible to construct a system of parameters for a ring of invariants consisting of Chern class (Proposition 4.1.1), and that under appropriate conditions one can always find algebra generators which are refinements of Chern classes (Theorem 4.1.2). This leads naturally to the conclusion that the ring of invariants of the regular representation of a group G of order d over a field $\mathbb{F}$ of characteristic p,

[1] A statement of the version that we use appears in Appendix A as Theorem A.3.1.

where $p > d$ or $p = 0$ realizes $\beta_{\mathbb{F}}(G)$: This is Theorem 4.1.4.

As we develop these tools we illustrate their use with a number of elementary examples. We wish to emphasize that the examples have not been chosen for their complexity, but rather for their transparency.

4.1 Orbit Chern Classes

A basic tool for constructing invariants are the orbit Chern classes. Their first formal appearance would seem to be in [387], where they were motivated by the splitting principal for Chern classes of vector bundles, and introduced for the purposes of [366]. The idea, however, is very simple.

Let V be a finite-dimensional representation of a finite group G, and $B \subset V^*$ be a G-invariant subset. Set

$$(*) \qquad \varphi_B(X) = \prod_{b \in B} (X + b),$$

which we regard as an element of the ring $\mathbb{F}[V][X]$, where X is an additional indeterminate. If $B = B' \sqcup B''$ decomposes into the disjoint union of two G-invariant subsets, then $\varphi_B(X) = \varphi_{B'}(X) \cdot \varphi_{B''}(X)$. Since any G-invariant subset is a disjoint union of orbits, we may, for the most part, restrict attention to subsets B that are orbits. If B is a G-orbit, the polynomial $\varphi_B(X)$ is called the **orbit polynomial** of B.

We let G act on the ring $\mathbb{F}[V][X]$ via its action on V and by fixing X. Since the product $(*)$ is taken over an invariant subset of V^*, it is clear that $\varphi_B(X) \in \mathbb{F}[V]^G[X] = (\mathbb{F}[V][X])^G$. Expand $\varphi_B(X)$ in powers of X to obtain a polynomial of degree $|B|$ in X, viz.,

$$\varphi_B(X) = \sum_{i+j=|B|} c_i(B) \cdot X^j.$$

The elements $c_i(B) \in \mathbb{F}[V]^G$, $i = 0, \ldots, |B|$, are called the **orbit Chern classes** of the orbit B. Note that $\mathbb{F}[V]$ is a finite integral extension of $\mathbb{F}[V]^G$, and for $v \in V^*$, the orbit polynomial $\varphi_{G\cdot v}(X)$ is nothing but the minimal polynomial of the element $-v$ over $\mathbb{F}[V]^G$.

The following generalizes a result of H. Weber, [372] Corollary 3.1.9. It provides us with a means of finding a system of parameters for a ring of invariants by working with the structure of V^* as a G-representation.

PROPOSITION 4.1.1 (L. Smith–R. E. Stong): *Let $\rho : G \hookrightarrow \mathrm{GL}(n, \mathbb{F})$ be a representation of a finite group over the field $\mathbb{F}$. Let $S \subseteq \mathbb{F}[V]^G$ be the subalgebra generated by all orbit Chern classes. Then S is Noetherian and $\mathbb{F}[V]^G \supseteq S$ is a finite ring extension. Hence S contains a system of parameters for $\mathbb{F}[V]^G$.*

PROOF: Let $K \subseteq \mathbb{F}[V]$ be the subalgebra generated by the coefficients of the orbit polynomials of a basis $z_1, \ldots, z_n \in V^*$. K is finitely generated, hence Noetherian. Since $z_1, \ldots, z_n$ generate $\mathbb{F}[V]$ as an algebra, $\mathbb{F}[V]$ is a finite K-module. Therefore, we have $K \subseteq S \subseteq \mathbb{F}[V]$, with $\mathbb{F}[V]$ a finite K-module. Hence S is also a finite K-module. Since $S \supseteq K$, $\mathbb{F}[V]$ is a finite S-module. □

The first, or **bottom, orbit Chern class**, $c_1(B)$, is the sum of the orbit elements, and hence $c_1(B) = \mathrm{Tr}^{G/G_b}(b)$, where $b \in B$ is arbitrary and $G_b \leq G$ is the isotropy group of b. The last orbit Chern class $c_{|B|}(B)$ is the product of all the elements in the orbit B and referred to as the **top Chern class** of the orbit, and is often denoted by $c_{\mathrm{top}}(B)$. If $b \in B$, then the **norm** of b is the product $\prod_{g \in G} gb$ and is equal to $c_{|B|}(B)^{|G:G_b|}$. Applied to arbitrary finite G-invariant sets the first Chern class is additive, the top Chern class is multiplicative, and all the Chern classes together satisfy

$$c_k(B' \sqcup B'') = \sum_{i+j=k} c_i(B') \cdot c_j(B''),$$

which is referred to as the **Whitney sum formula**. This terminology comes from algebraic topology.

The Chern classes of the orbit are nothing but the elementary symmetric functions in the elements of the orbit. Denote by Fun(B, $\mathbb{F}$) the $\mathbb{F}$-vector space of functions from B to $\mathbb{F}$. If B is a G-orbit the group G acts on B, since B is a G-orbit, and hence also on Fun(B, $\mathbb{F}$) by $(g \cdot f)(b) = f(g^{-1} \cdot b)$, where $f \in$ Fun(B, $\mathbb{F}$), $b \in B$, and $g \in G$ are arbitrary. There is a natural linear map

$$\eta_B : \mathrm{Fun}(B, \mathbb{F}) \longrightarrow V^*, \qquad \eta_B(f) = \sum_{b \in B} f(b) \cdot b, \qquad \forall\, f \in \mathrm{Fun}(B, \mathbb{F}),$$

which is G-equivariant, since

$$\eta_B(gf) = \sum_{b \in B} f(g^{-1} \cdot b)b = \sum_{b \in B} f(b)gb = g\left(\sum_{b \in B} f(b)b\right) = g \cdot \eta_B(f).$$

The vector space $V_B^* = \mathrm{Fun}(B, \mathbb{F})$ has a basis consisting of the **delta functions**

$$\delta_b : B \longrightarrow \mathbb{F}, \qquad \delta_b(x) = \begin{cases} 1 & \text{for } x = b, \\ 0 & \text{otherwise,} \end{cases}$$

which allows us to identify B with a basis for V_B^*. The map η_B induces a map $\mathbb{F}[V_B]^{\Sigma_{|B|}} \longrightarrow \mathbb{F}[V]^G$ that sends the kth elementary symmetric polynomial in the delta function basis, $e_k \in \mathbb{F}[V_B]^{\Sigma_{|B|}}$, into the kth orbit Chern class of B, $c_k(B) \in \mathbb{F}[V]^G$. By studying $\mathbb{F}[V_B]^{\Sigma_{|B|}}$ we can therefore deduce information about $\mathbb{F}[V]^G$. This is connected with Problem 1 in Section 1.6, since as a $\Sigma_{|B|}$-representation, V_B is the $|B|$-fold direct sum of the defining permutation representation of $\Sigma_{|B|}$. This will be examined in more detail in Section 4.2.

EXAMPLE 1 (L. E. Dickson): The group $\mathrm{GL}(2, \mathbb{F}_p)$ on $V = \mathbb{F}_p^2$ acts on $\mathbb{F}_p^2$ in its defining representation. The only orbits of $\mathrm{GL}(2, \mathbb{F}_p)$ on V^* are $\{0\}$ and $\widetilde{V} = V^* \setminus \{0\}$. To compute the Chern classes of the orbit $\widetilde{V}^*$ it is convenient to consider instead

$$\varphi_{V^*}(t) = \prod_{v \in V^*} (t + v) = t\varphi_{\widetilde{V}^*}(t).$$

Let $\{x, y\}$ be a basis for the dual vector space V^*, and set $L = \mathrm{Span}_{\mathbb{F}_p}\{x\} \subset V^*$. Note that

$$\varphi_L(s) = \prod_{w \in L} (s + w) = s^p - x^{p-1}s$$

is $\mathbb{F}_p$-linear in s. Thus

$$\begin{aligned}
\varphi_{V^*}(t) &= \prod_{v \in V^*} (t + v) = \prod_{\lambda, \mu \in \mathbb{F}_p} (t + \lambda x + \mu y) = \prod_{\mu \in \mathbb{F}_p} \prod_{\lambda \in \mathbb{F}_p} (t + \lambda x + \mu y) \\
&= \prod_{\mu \in \mathbb{F}_p} \varphi_L(t + \mu y) = \prod_{\mu \in \mathbb{F}_p} (\varphi_L(t) + \mu \varphi_L(y)) = \varphi_L(t)^p - \varphi_L(y)^{p-1}\varphi_L(t) \\
&= (t^p - x^{p-1}t)^p - (y^p - x^{p-1}y)^{p-1}(t^p - x^{p-1}t) \\
&= t^{p^2} - (x^{p(p-1)} + y^{p-1}(y^{p-1} - x^{p-1})^{p-1})t^p + (xy^p - x^p y)^{p-1}t,
\end{aligned}$$

and hence

$$\begin{aligned}
c_{p^2-1}(\widetilde{V}^*) &= (xy^p - x^p y)^{p-1}, \\
c_{p^2-p}(\widetilde{V}^*) &= (x^{p(p-1)} + y^{p-1}(y^{p-1} - x^{p-1})^{p-1}) = \frac{xy^{p^2} - x^{p^2}y}{xy^p - x^p y} \\
&= (x^{p-1})^p + (x^{p-1})^{p-1}(y^{p-1}) + \cdots + (x^{p-1})(y^{p-1})^{p-1} + (y^{p-1})^p.
\end{aligned}$$

We will see when we examine the Dickson algebra in Chapter 6 that in fact these two polynomials are algebraically independent and generate $\mathbb{F}_p[x, y]^{\mathrm{GL}(2, \mathbb{F}_p)}$ as an algebra.

The following result from [387] shows that the orbit Chern classes are often sufficient to generate the ring of invariants as an algebra.

THEOREM 4.1.2: *Let $\rho : G \hookrightarrow \mathrm{GL}(n, \mathbb{F})$ be a representation of a finite group G over a field $\mathbb{F}$. Suppose either the field $\mathbb{F}$ is of characteristic zero or that the order of G is less than the characteristic of $\mathbb{F}$. Then $\mathbb{F}[V]^G$ is generated by orbit Chern classes. If b is the size of the largest orbit of G acting on V^*, then $\mathbb{F}[V]^G$ is generated by forms of degree at most b.*

PROOF: By Theorem 2.3.3 $\mathbb{F}[V]^G$ is generated by forms of degree at most $|G|$. If $f \in \mathbb{F}[V]^G$ is a polynomial of degree at most $|G|$, then the formula

$$(-1)^j j!\, u_1 \cdots u_j = \sum_{I \subseteq \{1,\ldots,n\}} (-1)^{|I|} \left(\sum_{i \in I} u_i \right)^j$$

(see Lemma 7.3.1) shows that we may write f in the form

$$f = \sum_{\ell \in V^*} \ell^d,$$

where $d = \deg(f)$, and the sum extends over a finite set of homogeneous linear polynomials $\ell \in V^*$.

Since the characteristic of $\mathbb{F}$ is prime to the order of $|G|$, or $\mathbb{F}$ has characteristic zero, the transfer $\mathrm{Tr}^G : \mathbb{F}[V] \longrightarrow \mathbb{F}[V]^G$ is surjective. Therefore, we may suppose that $\mathbb{F}[V]^G$ is generated by elements $\mathrm{Tr}^G(f)$ of degree at most $|G|$. If $f \in \mathbb{F}[V]$ has degree $d \le |G|$, then f may be expressed as a sum of dth powers, so we obtain

$$\mathrm{Tr}^G(f) = \sum_{\ell \in V^*} \sum_{g \in G} (g\,\ell)^d.$$

Therefore, it will suffice to show that

$$\sum_{g \in G} (g\,\ell)^d$$

may be written as a polynomial in orbit Chern classes for any linear polynomial $\ell \in V^*$.

To do this, let $\ell \in V^*$ with isotropy group G_ℓ. Denote by $[\ell]$ the orbit of ℓ. The map

$$G \longrightarrow [\ell], \qquad g \longmapsto g\,\ell$$

induces an isomorphism of G-sets $G/G_\ell \xrightarrow{\cong} [\ell]$. Therefore,

$$\sum_{g \in G} (g\,\ell)^d = |G : G_\ell| \sum_{h \in [\ell]} (h)^d.$$

Notice that $\sum_{h \in [\ell]} (h)^d$ is the dth power sum polynomial, which is a symmetric polynomial in the elements of the orbit $[\ell]$. Hence it may be rewritten as a polynomial in the elementary symmetric polynomials[2] $e_1, \ldots, e_{|G:G_\ell|}$ of the elements of the orbit, which by definition are nothing but the orbit Chern classes of the orbit $[\ell]$. □

---

[2] The actual formula being used is

$$k e_k = \sum_{i=1}^{k} (-1)^{i-1} p_i e_{k-i} \quad k \in \mathbb{N}_0,$$

with the convention that $e_0 = 1$, and is usually called **Newton's formula**.

EXAMPLE 2: Consider the dihedral group D_8 of order 8. This group has a real representation of dimension 2 as the group of isometries of a square, as pictured in Figure 4.1.1. The action on the dual vector space with dual standard basis $\{x, y\}$ is generated by the matrices

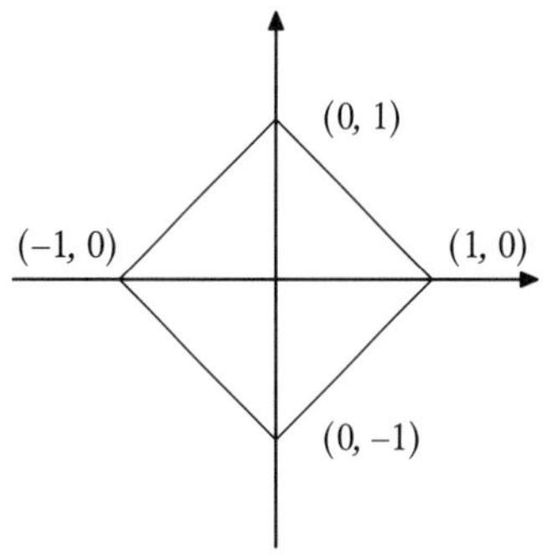

FIGURE 4.1.1

$$\mathbf{D} = \begin{bmatrix} 0 & -1 \\ 1 & 0 \end{bmatrix}, \qquad \mathbf{S} = \begin{bmatrix} -1 & 0 \\ 0 & 1 \end{bmatrix},$$

where $\mathbf{D}$ is a rotation through 90° and $\mathbf{S}$ a reflection in the y-axis. By Theorem 4.1.2, $\mathbb{R}[x, y]^{D_8}$ is generated by orbit Chern classes. Each orbit is invariant under the involution $v \longmapsto -v$, so the Chern classes, being symmetric functions of the elements of the orbit, are even functions, and hence there are no odd-degree invariants. The action being orthogonal the norm must be preserved, so $x^2 + y^2$ is an invariant. The orbit of x is $\{\pm x, \pm y\}$ and has as top Chern class x^2y^2. A short computation with the transfer shows that these forms span the invariants of degree 2 and 4 respectively, and hence

$$\mathbb{R}[x, y]^{D_8} = \mathbb{R}[x^2 + y^2, x^2y^2].$$

In fact, as long as the characteristic of $\mathbb{F}$ is not 2, the ring of invariants $\mathbb{F}[x, y]^{D_8}$ is a polynomial algebra on $x^2 + y^2$ and x^2y^2. This can be seen from Theorem 4.5.5. In characteristic 2 the representation is no longer faithful.

The following example is the other side of the coin: The orbit Chern classes will often fail miserably to generate the ring of invariants as an algebra.

EXAMPLE 3 (L. E. Dickson [86]): Consider the subgroup of $GL(2, \mathbb{F}_3)$ generated by the matrices

$$\mathbf{A} = \begin{bmatrix} 0 & 1 \\ -1 & 0 \end{bmatrix}, \quad \mathbf{B} = \begin{bmatrix} -1 & 1 \\ 1 & 1 \end{bmatrix} \in GL(2, \mathbb{F}_3).$$

It is not hard to see that these matrices generate a quaternion group of order 8. Just set

$$\mathbf{C} = \mathbf{AB} = \begin{bmatrix} 1 & 1 \\ 1 & -1 \end{bmatrix},$$

and check that

$$\mathbf{A}^2 = \mathbf{B}^2 = \mathbf{C}^2 = -\mathbf{I},$$

where $\mathbf{I}$ is the identity matrix. Hence the subgroup of $GL(2, \mathbb{F}_3)$ generated by $\mathbf{A}$ and $\mathbf{B}$ is isomorphic to the quaternion group Q_8 of order 8.

The elements of Q_8 are $\pm\mathbf{I}, \pm\mathbf{A}, \pm\mathbf{B}, \pm\mathbf{C}$. Inspection of these matrices shows that every nonzero vector in $\mathbb{F}_3^2$ occurs exactly once as a first column, so Q_8 acts transitively on $\mathbb{F}_3^2$. Thus the orbits of Q_8 acting on V^* are $\{0\}$ and $V^* \setminus \{0\}$, and the only Chern classes are therefore (see Example 1)

$$\frac{xy^9 - x^9y}{xy^3 - x^3y}, \quad (xy^3 - x^3y)^2,$$

where $\{x, y\}$ is the dual of the canonical basis of $\mathbb{F}_3^2$. These polynomials are of degree 6 and 8. However, $x^4 + y^4$ is invariant, and therefore $\mathbb{F}_3[x, y]^{Q_8}$ cannot be generated by orbit Chern classes alone.

This example is by no means an isolated instance: It recurs whenever G acts transitively on the lines in V; see, e.g., the doubly infinite family of such subgroups of $GL(2, \mathbb{F}_q)$ in Section 7.4, Example 10. In Section 4.2 we will introduce a refinement of orbit Chern classes that will let us deal with this example, and indeed with a whole class of similar examples (see also [379]).

It follows from Noether's bound (Theorem 2.3.3) that in the nonmodular case there is a representation ϱ for which $\beta(\varrho) = \beta_{\mathbb{F}}(G) \leq |G|$. If the field $\mathbb{F}$ has characteristic zero or larger than $|G|$, then this is the regular representation of G over $\mathbb{F}$, [381]. This was originally proved by B. Schmid in [333] for $\mathbb{F} = \mathbb{C}$ by completely different methods. We require a preliminary result for our proof that explains why the regular representation has this property.

PROPOSITION 4.1.3: *Let $\mathbb{F}$ be a field, G a finite group, and Ω a finite transitive G-set. Then Ω occurs as an orbit in the regular representation of G defined over $\mathbb{F}$.*

PROOF: Identify the regular representation of G over $\mathbb{F}$ with the action of G from the left on the group ring $\mathbb{F}(G)$ of G over $\mathbb{F}$. If $H \leq G$ is the isotropy group of a point of Ω, then we may identify Ω as a G-set with the left action of G on the left cosets of H in G, so $\Omega \cong G/H$ as G-sets. Let

$$\sigma_H = \sum_{h \in H} h \in \mathbb{F}(G).$$

Then the isotropy group of σ_H in $\mathbb{F}(G)$ is H, so the orbit of σ_H is also isomorphic to G/H as a left G-set. □

THEOREM 4.1.4: *Let G be a finite group of order d and $\mathbb{F}$ a field of characteristic p. If $p = 0$ or $p > d$, then $\beta(\varrho) \leq \beta(\mathrm{reg}_{\mathbb{F}}(G))$ for any representation $\varrho : G \hookrightarrow GL(n, \mathbb{F})$, where $\mathrm{reg}_{\mathbb{F}}(G) : G \hookrightarrow GL(d, \mathbb{F})$ is the regular representation of G over the field $\mathbb{F}$.*

PROOF: Since $p = 0$ or $d = |G| < p$, it follows from Theorem 4.1.2 that $\mathbb{F}[V]^G$ is generated by orbit Chern classes. If $B \subset V^*$ is an orbit of G in the space of linear forms on V, then by Proposition 4.1.3, we may find a G-equivariant embedding $B \hookrightarrow \mathbb{F}(G)$. This in turn induces an epimorphism

$$\alpha_B : \mathbb{F}[\mathbb{F}(G)] \longrightarrow \mathbb{F}[B],$$

where $\mathbb{F}[B]$ is the polynomial algebra in the formal variables $\{b \in B\}$. Since $p \nmid d$, passing to fixed subalgebras yields an epimorphism by Corollary 2.2.2,

$$\alpha_B^G : \mathbb{F}[\mathbb{F}(G)]^G \longrightarrow \mathbb{F}[B]^G.$$

If k denotes the number of elements in B, then the action of G on B, being by permutations, leaves the elementary symmetric polynomials $e_1, \ldots, e_k$ in the elements $b \in B$ invariant, i.e.,

$$\mathbb{F}[e_1(B), \ldots, e_k(B)] \subseteq \mathbb{F}[B]^G.$$

The inclusion $B \subset V^*$ induces a G-equivariant map

$$\beta_B : \mathbb{F}[B] \longrightarrow \mathbb{F}[V],$$

and by definition of the orbit Chern classes

$$\beta_B(e_i(B)) = c_i(B), \; i = 1, \ldots, k,$$

where $c_1(B), \ldots, c_k(B) \in \mathbb{F}[V]^G$ are the Chern classes of the orbit B. Therefore, the composite

$$\gamma_B : \mathbb{F}[\mathbb{F}(G)]^G \xrightarrow{\alpha_B^G} \mathbb{F}[B]^G \xrightarrow{\beta_B^G} \mathbb{F}[V]^G$$

contains the subalgebra of $\mathbb{F}[V]^G$ generated by the orbit Chern classes $c_1(B), \ldots, c_k(B) \in \mathbb{F}[V]^G$ of the orbit B. Since $\mathbb{F}[V]^G$ is generated by orbit Chern classes, we can find a finite number $B_1, \ldots, B_m$ of orbits of G in V^* such that the map

$$\bigotimes_{i=1}^{m} : \mathbb{F}[\mathbb{F}(G)]^G \xrightarrow{\bigotimes_{i=1}^{m} \gamma_{B_i}} \bigotimes_{i=1}^{m} \mathbb{F}[V]^G \xrightarrow{\mu} \mathbb{F}[V]^G$$

is an epimorphism, where μ is the multiplication map. Since $\bigotimes_{i=1}^{m} \mathbb{F}[\mathbb{F}(G)]^G$ is generated by forms of degree at most $\beta_{\mathbb{F}}(\mathrm{reg}_{\mathbb{F}}(G))$, the result follows. □

In the next section we will see how to extend Theorem 4.1.4 to cover the case where G contains a chain of subgroups

$$\{1\} = G_k < G_{k-1} < \cdots < G_1 < G_0 = G,$$

and $\mathbb{F}$ is a field of characteristic p, which satisfies

$$p > \max\left\{ |G_{k-1} : G_k|, \ldots, |G_0 : G_1| \right\}.$$

4.2 A Refinement of Orbit Chern Classes

Although Theorem 4.1.1 shows that we can use the orbit Chern classes to find a system of parameters for a ring of invariants, Example 3 in the previous section shows that the orbit Chern classes may fail to provide enough invariant forms to generate the ring of invariants as an algebra. In this section we will describe a refinement of the orbit Chern classes based on [379]. For an alternative approach see [211].

Let $\rho : G \hookrightarrow \mathrm{GL}(n, \mathbb{F})$ be a representation of a finite group G over the field $\mathbb{F}$, $V = \mathbb{F}^n$ the corresponding G-module. Denote by $\mathbb{F}(G)$ the group algebra of G over $\mathbb{F}$, and consider the map

$$\alpha : \mathbb{F}(G) \otimes V^* \longrightarrow V^*$$

defined by

$$\alpha(g \otimes z) = gz, \qquad g \in G, z \in V^*.$$

If we let G act on $\mathbb{F}(G) \otimes V^*$ by

$$h(g \otimes z) = hg \otimes z, \qquad h, g \in G, z \in V^*,$$

then the map α is G-equivariant. Denote by W the vector space dual of $\mathbb{F}(G) \otimes V^*$. The chain of isomorphisms

$$\begin{aligned} W = (\mathbb{F}(G) \otimes_{\mathbb{F}} V^*)^* &\cong (\mathbb{F}(G))^* \otimes_{\mathbb{F}} V^{**} \\ &\cong (\mathbb{F}(G))^* \otimes_{\mathbb{F}} V = \mathrm{Hom}_{\mathbb{F}}(\mathbb{F}(G), V) = \mathrm{map}(G, V) \end{aligned}$$

shows that we can identify W with $\mathrm{map}(G, V)$, the space of all maps from the set G to the vector space V with the vector space operations given by pointwise addition of functions and multiplication of a function by a scalar. Under this identification the action of G on W corresponds to the G-action on $\mathrm{map}(G, V)$ given by

$$(g \cdot f)(h) = f(gh), \qquad f \in \mathrm{map}(G, V), g, h \in G.$$

The dual of the map α, namely

$$\eta : V \longrightarrow \mathrm{map}(G, V) = W,$$

is called the **Noether map**. It is given explicitly by the formula

$$\eta(v)(g) = gv, \qquad g \in G, v \in V.$$

From η (or equivalently α) we obtain a G-equivariant map

$$\eta : \mathbb{F}[W] \longrightarrow \mathbb{F}[V],$$

which we also call the Noether map.[3] Note that since the action of G on $W = \mathrm{map}(G, V)$ is via left translation of G, it extends to an action of the full symmetric group Σ_d, $d = |G|$.

---

[3] This multiple use of η and the term Noether map should cause no problems.

Choose a basis $z_1, \ldots, z_n$ for V^*. Then $\{g \otimes z_i \mid i = 1, \ldots, n, g \in G\}$ may be identified with a basis for W^*. Define the operator

$$\mathsf{E} : \mathbb{F}[V] \longrightarrow \mathbb{F}[W]$$

to be the $\mathbb{F}$-linear extension of the map on the monomial basis given by

$$\mathsf{E}(z^A) = \sum_{g \in G} (g \otimes z_1)^{a_1} \cdots (g \otimes z_n)^{a_n},$$

where $A = (a_1, \ldots, a_n) \in \mathbb{N}_0 \times \cdots \times \mathbb{N}_0$ is a multi-index and $z^A \in \mathbb{F}[V]$ the corresponding monomial. The following lemma is a direct consequence of the definition of E and Tr^G.

LEMMA 4.2.1: *With the preceding notation, the operator* E *satisfies*
(i) $\mathsf{E}(f) \in \mathbb{F}[W]^{\Sigma_d}$ *for all* $f \in \mathbb{F}[V]$, *and*
(ii) $\eta(\mathsf{E}(f)) = \mathrm{Tr}^G(f)$ *for all* $f \in \mathbb{F}[V]$. □

As a consequence we obtain the following result:

PROPOSITION 4.2.2: *Let* $\mathbb{F}$ *be a field of characteristic* p *and* G *a group of order* $d = |G|$. *Suppose that* p *does not divide* d. *Regard* G *as a subgroup of* Σ_d *via its regular representation. Then, the composition*

$$\varphi : \mathbb{F}[W]^S \hookrightarrow \mathbb{F}[W]^G \xrightarrow{\ \eta^G\ } \mathbb{F}[V]^G$$

is surjective for any subgroup S *of* Σ_d *containing* G.

PROOF: It is enough to show that φ is onto for $S = \Sigma_d$, which follows from the commutative triangle

$$\begin{array}{ccc} \mathbb{F}[V] & \xrightarrow{\ \mathsf{E}\ } & \mathbb{F}[W]^{\Sigma_d} \\ & {\scriptstyle \mathrm{Tr}^G} \searrow & \downarrow {\scriptstyle \eta} \\ & & \mathbb{F}[V]^G \end{array}$$

and the fact that the transfer is onto in the nonmodular case. □

Already at this point we can exploit the first main theorem of invariant theory for the symmetric group ([372] Section 3.4, or [451] Chapter 2) to refine the orbit Chern classes. Here is how this works.

Denote by Ω the underlying set of G regarded as a G-set via the left translation action of G on itself. In this notation

$$W = \mathrm{map}(\Omega, V) = \bigoplus_n \mathrm{map}(\Omega, \mathbb{F}).$$

The group Σ_d, $d = |G|$, acts on Ω, extending the action of G, and $X = \mathrm{map}(\Omega, \mathbb{F})$ is the linearization of the defining permutation representation of Σ_d over $\mathbb{F}$. The elements of Ω may be identified with linear forms on $\mathrm{map}(\Omega, \mathbb{F})$ by setting $\omega(f) = f(\omega)$, $\quad \omega \in \Omega$, and $f \in \mathrm{map}(\Omega, \mathbb{F})$.

Let $X(1), \ldots, X(n)$ be copies of X and $\Omega(1), \ldots, \Omega(n)$ copies of Ω. Elements of $X(i)$ will be written in the form $x(i)$ and similarly $\omega(j)$ for elements of $\Omega(j)$. Then

$$W = X(1) \oplus \cdots \oplus X(n) = \bigoplus_n X,$$

so, as a Σ_d-module, W is the n-fold direct sum of the tautological representation X of Σ_d. This is a much studied example, see, e.g., the discussion in [451] of the first fundamental theorem of invariant theory for the symmetric group. Its computation goes back to the roots of invariant theory in the 19th century see e.g. [284] Band II sechsunddreissigte Vorlesung, and it is at the center of Emmy Noether's first proof of the basic finiteness theorem for finite groups [301].

From the preceding discussion it follows that if $p > d$, the ring of invariants $\mathbb{F}[W]^{\Sigma_d}$ is generated by the **polarized elementary symmetric polynomials** in d vector variables of dimension n. Explicitly these are defined as follows: Let $I = (i_1, \ldots, i_n) \in \mathbb{N}_0 \times \cdots \times \mathbb{N}_0$ be a multi-index satisfying $i_1 + \cdots + i_n \leq d$. For $r = 1, \ldots, n$ choose i_r distinct elements of $\Omega(r)$, say $\omega(r)_1, \ldots, \omega(r)_{i_r}$, and let $\omega(I) \in \mathbb{F}[W]_{|I|}$, where $|I| = i_1 + \cdots + i_n$, be their product, so

$$\omega(I) = \omega(1)_1 \cdots \omega(1)_{i_1} \cdot \omega(2)_1 \cdots \omega(2)_{i_2} \cdots \omega(n)_1 \cdots \omega(n)_{i_n} \in \mathbb{F}[W].$$

The orbit sum of the Σ_d-orbit of $\omega(I)$ is the I**th polarized elementary symmetric polynomial** in d vector variables of dimension n, and is denoted by

$$e(I) = \mathfrak{S}_{\Sigma_d \cdot \omega(I)} \in \mathbb{F}[W]^{\Sigma_d}.$$

Table 4.2.1 shows the polarized elementary symmetric polynomials in 2 vector variables $z_{i,j}$, $i = 1, 2$ and $j = 1, \ldots, 4$, of dimension 4.

In fact, in the definition of $e(I)$ it is sufficient to consider only those choices where the same element is never chosen from different blocks. If we restrict ourselves to only such choices, then, introducing indeterminates $t_1, \ldots, t_n$, we will have the identity

$$\prod_{I=(i_1,\ldots,i_n),\, |I| \leq d} (1 + \omega(1)_{i_1} t_1 + \cdots + \omega(n)_{i_n} t_n) = \sum_{I,\, |I| \leq d} e(I) t^I$$

in $\mathbb{F}[W][t_1, \ldots, t_n]$, which could be used to define $e(I)$. The image $\eta(e(I)) \in \mathbb{F}[V]^G$ is the I**th polarized Chern class** and is denoted by $c(I)$.

Before turning to an example to illustrate these ideas, we emphasize that this is not the only way to generalize the orbit Chern classes. Here we have adopted the approach of [379]. In [211], the idea is to start with a G-invariant subset B of V^*, and use multipartitions of B associated to the cosets of subgroups of G to introduce polarized generalizations of the ordinary orbit Chern classes.

$$
\begin{aligned}
\sigma_{(0,0)} &= 1\\
\sigma_{(1,0)} &= z_{11}+z_{12}+z_{13}+z_{14}\\
\sigma_{(0,1)} &= z_{21}+z_{22}+z_{23}+z_{24}\\
\sigma_{(2,0)} &= z_{11}z_{12}+z_{11}z_{13}+z_{11}z_{14}+z_{12}z_{13}+z_{12}z_{14}+z_{13}z_{14}\\
\sigma_{(0,2)} &= z_{21}z_{22}+z_{21}z_{23}+z_{21}z_{24}+z_{22}z_{23}+z_{22}z_{24}+z_{23}z_{24}\\
\sigma_{(1,1)} &= z_{11}z_{22}+z_{11}z_{23}+z_{11}z_{24}+z_{12}z_{21}+z_{12}z_{23}+z_{12}z_{24}\\
&\quad +z_{13}z_{21}+z_{13}z_{22}+z_{13}z_{24}+z_{14}z_{21}+z_{14}z_{22}+z_{14}z_{23}\\
\sigma_{(3,0)} &= z_{11}z_{12}z_{13}+z_{11}z_{12}z_{14}+z_{11}z_{13}z_{14}+z_{12}z_{13}z_{14}\\
\sigma_{(0,3)} &= z_{21}z_{22}z_{23}+z_{21}z_{22}z_{24}+z_{21}z_{23}z_{24}+z_{22}z_{23}z_{24}\\
\sigma_{(2,1)} &= z_{11}z_{12}z_{23}+z_{11}z_{12}z_{24}+z_{11}z_{13}z_{22}+z_{11}z_{13}z_{24}\\
&\quad +z_{11}z_{14}z_{22}+z_{11}z_{14}z_{23}+z_{12}z_{13}z_{21}+z_{12}z_{13}z_{24}\\
&\quad +z_{12}z_{14}z_{21}+z_{12}z_{14}z_{23}+z_{13}z_{14}z_{21}+z_{13}z_{14}z_{22}\\
\sigma_{(1,2)} &= z_{11}z_{22}z_{23}+z_{11}z_{22}z_{24}+z_{11}z_{23}z_{24}+z_{12}z_{21}z_{23}\\
&\quad +z_{12}z_{21}z_{24}+z_{12}z_{23}z_{24}+z_{13}z_{21}z_{22}+z_{13}z_{21}z_{24}\\
&\quad +z_{13}z_{22}z_{24}+z_{14}z_{21}z_{22}+z_{14}z_{21}z_{23}+z_{14}z_{22}z_{23}\\
\sigma_{(4,0)} &= z_{11}z_{12}z_{13}z_{14}\\
\sigma_{(0,4)} &= z_{21}z_{22}z_{23}z_{24}\\
\sigma_{(3,1)} &= z_{11}z_{12}z_{13}z_{24}+z_{11}z_{12}z_{14}z_{23}+z_{11}z_{13}z_{14}z_{22}+z_{12}z_{13}z_{14}z_{21}\\
\sigma_{(1,3)} &= z_{11}z_{22}z_{23}z_{24}+z_{12}z_{21}z_{23}z_{24}+z_{13}z_{21}z_{22}z_{24}+z_{14}z_{21}z_{22}z_{23}\\
\sigma_{(2,2)} &= z_{11}z_{12}z_{23}z_{24}+z_{11}z_{13}z_{22}z_{24}+z_{11}z_{14}z_{22}z_{23}\\
&\quad +z_{12}z_{13}z_{21}z_{24}+z_{12}z_{14}z_{21}z_{23}+z_{13}z_{14}z_{21}z_{22}
\end{aligned}
$$

TABLE 4.2.1: The polarized elementary symmetric functions $\dim_{\mathbb{F}}(V) = n = 2,\ d = 4$

EXAMPLE 1: Consider the representation of $\mathbb{Z}/4$ implemented by the matrix

$$\mathrm{T} = \begin{bmatrix} 0 & -1 \\ 1 & 0 \end{bmatrix} \in GL(2, \mathbb{F}),$$

where $\mathbb{F}$ is a field of characteristic different from 2. Then $n = 2$ and $d = 4$.

$$\begin{aligned}
\eta(\sigma_{(0,0)}) &= 1\\
\eta(\sigma_{(1,0)}) &= 0\\
\eta(\sigma_{(0,1)}) &= 0\\
\eta(\sigma_{(2,0)}) &= -(x^2+y^2)\\
\eta(\sigma_{(0,2)}) &= -(x^2+y^2)\\
\eta(\sigma_{(1,1)}) &= 0\\
\eta(\sigma_{(3,0)}) &= 0\\
\eta(\sigma_{(0,3)}) &= 0\\
\eta(\sigma_{(2,1)}) &= 0\\
\eta(\sigma_{(1,2)}) &= 0\\
\eta(\sigma_{(4,0)}) &= x^2y^2\\
\eta(\sigma_{(0,4)}) &= x^2y^2\\
\eta(\sigma_{(3,1)}) &= 2(xy^3-x^3y)\\
\eta(\sigma_{(1,3)}) &= 2(x^3y-xy^3)\\
\eta(\sigma_{(2,2)}) &= x^4+y^4-4x^2y^2
\end{aligned}$$

TABLE 4.2.2: Polarized Chern Classes

From these we obtain by setting $z_{1,j} = \mathbf{T}^j(x)$ and $z_{2,j} = \mathbf{T}^j(y)$ the corresponding polarized Chern classes, which are displayed in Table 4.2.2. Apparently, there is a great deal of redundancy after evaluation. In any case, we obtain the three polynomials

$$\begin{aligned}
f_1 &= x^2+y^2,\\
f_2 &= x^2y^2,\\
f_3 &= x^3y-xy^3
\end{aligned}$$

that are generators for $\mathbb{F}[x, y]^{\mathbb{Z}/4}$, and

$$\mathbb{F}[x, y]^{\mathbb{Z}/4} \cong \mathbb{F}[f_1, f_2, f_3]/(f_3^2 - f_2f_1^2 - 4f_2^2).$$

The first two of these invariants are ordinary orbit Chern classes, to wit, $f_1 = c_2([x])$ and $f_2 = c_4([x])$. However, the algebraic relation

$$12f_3 = c_4([x+2y]) - 4c_2([x])^2 + 7c_4([x])$$

shows that the subalgebra generated by the ordinary orbit Chern classes is the full ring of invariants $\mathbb{F}[x, y]^{\mathbb{Z}/4}$ if and only if the characteristic of $\mathbb{F}$ is not 3, i.e., greater than 4, the order of the group. However, in characteristic 3 one needs in addition the polarized orbit Chern class f_3.

The polarized Chern classes are the first step toward obtaining more invariants out of the action of G on V^*. To describe these we need to review some facts about permutation representations, cf. [452].

DEFINITION: *Let G be a finite group and Ω a transitive G-set. A* **system of imprimitivity** *for Ω is a decomposition of Ω*

$$\Omega = \Omega_1 \sqcup \cdots \sqcup \Omega_e$$

into disjoint subsets, called the **blocks** *of the system of imprimitivity, such that for every $g \in G$ and every $i \in \{1, \dots, e\}$ the image $g(\Omega_i)$ is one of the, perhaps other, blocks, Ω_j.*

So a system of imprimitivity is a partition of Ω into blocks that are permuted by the elements of G. Notice that we assumed that G acts transitively, so the blocks must all have the same size, say c, and G is a subgroup of $\Sigma_c \wr \Sigma_e \leq \Sigma_d$, where $|\Omega| = d = c \cdot e$. The embedding of the wreath product corresponds to the the system of imprimitivity. This means that each $g \in G$ acts by an e-tuple $(\pi_1, \dots, \pi_e)$, with $\pi_i \in \Sigma_c$ permuting the elements of the ith block Ω_i, and an element $\tilde{g} \in \Sigma_e$ that permutes the blocks. The following lemma shows how to find systems of imprimitivity. The proof is elementary and left to the reader (see [452] for a discussion of systems of imprimitivity).

LEMMA 4.2.3: *Let G be a finite group and Ω a transitive G-set. Then $\Omega \cong G/G_\omega$ as G-sets, where $G_\omega \leq G$ is the isotropy subgroup of some point in $\omega \in \Omega$. If H is a subgroup of G such that $G_\omega \leq H \leq G$, then the decomposition*

$$G/H = H \cdot \omega \sqcup g_2 \cdot H \cdot \omega \sqcup \cdots \sqcup g_e \cdot H \cdot \omega,$$

where $1 = g_1, g_2, \ldots, g_e \in G$ is a transversal for H in G, is a system of imprimitivity for Ω. The number e of blocks is $|G : H|$, and their size is $c = |H|$. Every system of imprimitivity for Ω with e blocks of size c arises in this way. □

In the situation of Lemma 4.2.3 we say that

$$\Omega = H \cdot \omega \sqcup g_2 \cdot H \cdot \omega \sqcup \cdots \sqcup g_e \cdot H \cdot \omega$$

is the **system of imprimitivity corresponding to the subgroup** H. Note that there is a nontrivial system of imprimitivity, i.e., one with more than a single block, precisely when the isotropy group G_ω of a point $\omega \in \Omega$ is not a maximal subgroup of G.

Let us reconsider a linear representation $\rho : G \hookrightarrow \mathrm{GL}(n, \mathbb{F})$ of a finite group over the field $\mathbb{F}$, and as before, let Ω denote the underlying set of G regarded as a G-set via left translation, and $W = \mathrm{map}(\Omega, V)$, with

$$\eta : \mathbb{F}[W] \longrightarrow \mathbb{F}[V]$$

the Noether map. We again need n copies $\Omega(1), \ldots, \Omega(n)$ of Ω. If $H < G$ is a subgroup, $|H| = c$, the corresponding system of imprimitivity for $\Omega(i)$, $1 \leq i \leq n$, is denoted by

$$\Omega(i) = \Omega(i)_1 \sqcup \cdots \sqcup \Omega(i)_e,$$

where $e = |G : H|$. If $X(1), \ldots, X(n)$ are copies of $X = \mathrm{map}(\Omega, \mathbb{F})$, there is a corresponding decomposition

$$X(i) = X(i)_1 \oplus \cdots \oplus X(i)_e, \qquad 1 \leq i \leq n,$$

where $X(i)_j = \mathrm{map}(\Omega(i)_j, \mathbb{F})$, and $j = 1, \ldots, e$. Therefore, we may write

$$\mathbb{F}[W] = \mathbb{F}[X(1) \oplus \cdots \oplus X(n)] = \mathbb{F}[\bigoplus_{\substack{1 \leq i \leq n \\ 1 \leq j \leq e}} X(i)_j].$$

This finer decomposition of W is preserved by the action of the wreath product $\Sigma_c \wr \Sigma_e < \Sigma_d$ acting on $W = \mathrm{map}(\Omega, \mathbb{F})$ via its action on Ω. This is because $\Sigma_c \wr \Sigma_e$ is precisely the subgroup of Σ_d that preserves the system of imprimitivity $\Omega = \Omega_1 \sqcup \cdots \sqcup \Omega_e$ corresponding to $H < G$. We therefore obtain a map

$$\varphi : \mathbb{F}[W]^{\Sigma_c \wr \Sigma_e} \longrightarrow \mathbb{F}[V]^G,$$

which, by Proposition 4.2.2, is onto if $d = c \cdot e$ is invertible in $\mathbb{F}$. If the characteristic of $\mathbb{F}$ is actually strictly larger than either of c or e, then the ring of invariants $\mathbb{F}[W]^{\Sigma_c \wr \Sigma_e}$ is generated by certain doubly polarized elementary symmetric polynomials, which we next describe; see [379].

We arrange the blocks $\Omega(i)_j$, $1 \le i \le n$, $1 \le j \le e$, in a rectangular array

$$\begin{matrix} \Omega(1)_1 & \Omega(1)_2 & \cdots & \Omega(1)_e \\ \vdots & \vdots & \vdots & \vdots \\ \Omega(n)_1 & \Omega(n)_2 & \cdots & \Omega(n)_e. \end{matrix}$$

For each $n \times e$-matrix $\mathbf{K} = [k_{i,j}]$ with entries in $\mathbb{N}_0$ that satisfies

(1) row sums are at most c, and
(2) the sum of all the entries is at most d,

select $k_{i,j}$ distinct elements from $\Omega(i)_j$, say $\omega(i)_{j,1}, \ldots, \omega(i)_{j,k_{i,j}}$; set $k_i = k_{i,1} + \cdots + k_{i,e}$ and take the product

$$\omega(\mathbf{K}) = \omega(1)_1 \cdots \omega(1)_{k_1} \cdot \omega(2)_1 \cdots \omega(2)_{k_2} \cdots \omega(n)_1 \cdots \omega(n)_{k_n} \in \mathbb{F}[W].$$

Let $e(\mathbf{K})$ denote the orbit sum $\mathfrak{S}_{\Sigma_c \wr \Sigma_e \cdot \omega(\mathbf{K})}$ of the $\Sigma_c \wr \Sigma_e$-orbit of $\omega(\mathbf{K}) \in \mathbb{F}[W]^{\Sigma_c \wr \Sigma_e}$. Again, it is possible to avoid some duplications by restricting the choices so that the same element is not chosen from $\Omega(i')_j$ and $\Omega(i'')_j$ for different values of i' and i''. We call the forms $\omega(\mathbf{K})$ **doubly polarized elementary symmetric polynomials of $\Sigma_c \wr \Sigma_e$**. The image, $\eta(e(\mathbf{K})) = c(\mathbf{K}) \in \mathbb{F}[V]^G$, is called the **Kth doubly polarized Chern class** of G.

This whole procedure can be iterated for a chain of subgroups in G,

$$(\because) \qquad \{1\} = G_m < G_{m-1} < \cdots < G_1 < G_0 = G,$$

to provide **multipolarized Chern classes** in $\mathbb{F}[V]^G$ indexed by multidimensional arrays. Collectively we refer to these as **fine Chern classes**: Here is how this works.

The chain of subgroups $(\because)$ leads to a decomposition of Ω into fine blocks:

$$\Omega = \Omega_{j_{m-2}, \ldots, j_0}, \qquad 1 \le j_k \le e_k,\ k = 0, \ldots, m-2.$$

Each of these fine blocks contains e_{m-1} elements. The basic fine block is G_{m-1} itself. The first step to obtain the remaining fine blocks is to consider the e_{m-2} cosets of G_{m-1} in G_{m-2}, which partition G_{m-2} into e_{m-2} copies of G_{m-1}, and thereby into $e_{m-1} \cdot e_{m-2}$ fine blocks. The process continues with the cosets of G_{m-2} in G_{m-3}, and so on. The result is the multidimensional array

$$\left\{ \Omega_{j_{m-2}, \ldots, j_0} \mid 1 \le j_{m-k} \le e_{m-k},\ k = 2, \ldots, m \right\}.$$

For example, if $G = Q_8 = \{\pm 1, \pm \boldsymbol{i}, \pm \boldsymbol{j}, \pm \boldsymbol{k}\}$ is the quaternion group of order 8, then there is the chain of subgroups $\{1\} < \mathbb{Z}/2 < \mathbb{Z}/4 < Q_8$, where $\mathbb{Z}/4$ is generated by $\boldsymbol{i}$, and $\mathbb{Z}/2$ by -1. This means that there are 4 fine blocks of 2 elements each as shown in the following table:

	$\mathbb{Z}/4$	$\boldsymbol{j}\cdot\mathbb{Z}/4$
$\mathbb{Z}/2$	$\Omega_{1,1}=\{\pm 1\}$	$\Omega_{1,2}=\{\pm \boldsymbol{j}\}$
$\boldsymbol{i}\mathbb{Z}/2$	$\Omega_{2,1}=\{\pm \boldsymbol{i}\}$	$\Omega_{2,2}=\{\pm \boldsymbol{k}\}$.

The first column is the partition of $\mathbb{Z}/4$ into the two cosets of $\mathbb{Z}/2$, and the second column arises from the first by multiplying the first column by $\boldsymbol{j}$. The columns correspond to the cosets of $\mathbb{Z}/4$ in Q_8, and the rows inside columns to the partition of the cosets of $\mathbb{Z}/4$ into cosets of $\mathbb{Z}/2$.

We denote by

$$\Omega(i)_{j_{m-2},\dots,j_0}, \qquad i=1,\dots,n,\ 1\le j_k\le e_k,\ k=0,\dots,m-2,$$

the corresponding decomposition of the copies of Ω. Then $\sqcup\Omega(i)_{j_{m-2},\dots,j_0}$ is a basis for W^*. The corresponding linear decomposition of W^* into a direct sum is preserved by the action of the iterated wreath product $(\Sigma_{e_{m-1}} \wr (\Sigma_{e_{m-2}} \cdots \wr \Sigma_{e_1}) \wr \Sigma_{e_0}$ on Ω, and hence on $W=\mathrm{map}(\Omega, G)$. Set $d_i = |G_i|$ and $e_i=|G_i:G_{i+1}|$ for $i=0,\dots,m$. We consider m-dimensional arrays $\mathsf{A}=[a(i;j_{m-2},\dots,j_0)]$ of nonnegative integers with $1\le i\le n$ satisfying

$$a(i;j_{m-2},\dots,j_0)\le e_{m-1},$$

$$\sum_{j_{m-2}=1}^{e_{m-2}} a(i;j_{m-2},\dots,j_0)\le e_{m-2},$$

$$\ddots$$

$$\sum_{j_0=1}^{e_0} a(i;j_{m-2},\dots,j_0)\le e_0,$$

$$a(i';j_{m-2},\dots,j_0)\cdot a(i'';j_{m-2},\dots,j_0)>0 \Rightarrow i'=i''.$$

So the sum of all the elements in the array is at most $d=e_{m-1}\cdot e_{m-2}\cdots e_0$, and for fixed $(j_{m-2},\dots,j_0)$ there is at most one i with $a(i;j_{m-2},\dots,j_0)\neq 0$. We choose $a(i;j_{m-2},\dots,j_0)$ distinct elements of $\Omega(i)_{j_{m-2},\dots,j_0}$, form the product of all of these, and let $e(\mathsf{A})\in\mathbb{F}[W]^{(\Sigma_{e_{m-1}}\wr(\Sigma_{e_{m-2}}\cdots\wr\Sigma_{e_1})\wr\Sigma_{e_0}}$ be the orbit sum of the $(\Sigma_{e_{m-1}} \wr (\Sigma_{e_{m-2}} \cdots \wr \Sigma_{e_1}) \wr \Sigma_{e_0}$-orbit of this product. We call $e(\mathsf{A})\in\mathbb{F}[W]^{(\Sigma_{e_{m-1}}\wr(\Sigma_{e_{m-2}}\cdots\wr\Sigma_{e_1})\wr\Sigma_{e_0}}$ the **multipolarized elementary symmetric polynomials** and the image $\eta(e(\mathsf{A}))\in\mathbb{F}[V]^G$ the **multipolarized**, or **fine, Chern classes**. Note that $e(\mathsf{A})$ depends on the choice of the chain (⁘) of subgroups used to define it. This is not indicated in the notation, which is already complicated enough. As we will see, this dependence on the chain can be of importance.

The justification for all this is to be found in [379], whose main result may be stated in the following form:

THEOREM 4.2.4: *Let $\mathbb{F}$ be a field of characteristic p, G a finite group, and*

$$\{1\} = G_m < G_{m-1} < \cdots < G_1 < G_0 = G$$

a chain of subgroups in G. Set $e_i = |G_i : G_{i+1}|$ for $i = 0, \ldots, m-1$. If

$$p > \max\{e_{m-1}, \ldots, e_0\},$$

then $\mathbb{F}[V]^G$ is generated as an algebra by the fine Chern classes associated to this chain and all its subchains.

PROOF: Let Ω be the underlying set of G regarded as a G-set and

$$\Omega = \Omega_1 \sqcup \cdots \sqcup \Omega_{e_0}$$

the decomposition of Ω into the left cosets of G_1 in $G_0 = G$. Let

$$\Omega_i = \Omega_{i,1} \sqcup \cdots \sqcup \Omega_{i,e_1} \quad i = 1, \ldots, e_0$$

be the decomposition of Ω_i into blocks consisting of the left cosets of G_2 in G_1. Continuing in this way we obtain partitions

$$\Omega = \bigsqcup_{i_{k-1}=1}^{e_{k-1}} \cdots \bigsqcup_{i_1=1}^{e_0} \Omega_{i_1, \ldots, i_k}$$

of Ω into blocks using the successive left cosets of G_{k+1} in G_k for $k = 1, \ldots, m-1$. The action of G on Ω must preserve the partitions

$$\Omega = \bigsqcup \Omega_i,\ \Omega = \bigsqcup \Omega_{i_1, i_2}, \ldots,\ \Omega = \bigsqcup \Omega_{i_1, \ldots, i_m}.$$

The largest subgroup of Σ_d preserving these partitions is the iterated wreath product

$$(\Sigma_{e_{m-1}} \wr (\Sigma_{e_{m-2}} \cdots \wr \Sigma_{e_1}) \wr \Sigma_{e_0},$$

so $G \leq (\Sigma_{e_{m-1}} \wr (\Sigma_{e_{m-2}} \cdots \wr \Sigma_{e_1}) \wr \Sigma_{e_0}$ (see e.g., the discussion of primitive permutation groups in [452]).

By hypothesis $p \nmid d = e_0 \cdots e_{m-1}$ so $p \nmid |(\Sigma_{e_{m-1}} \wr (\Sigma_{e_{m-2}} \cdots \wr \Sigma_{e_1}) \wr \Sigma_{e_0}|$ and hence the restriction of the Noether map

$$\eta : \mathbb{F}[\Omega]^{(\Sigma_{e_{m-1}} \wr (\Sigma_{e_{m-2}} \cdots \wr \Sigma_{e_1}) \wr \Sigma_{e_0}} \longrightarrow \mathbb{F}[V]^G$$

is an epimorphism. By Corollary 2.3.4 the desired conclusion will follow if we can show

$$(*) \qquad \beta\left(\mathbb{F}[\Omega]^{(\Sigma_{e_{m-1}} \wr (\Sigma_{e_{m-2}} \cdots \wr \Sigma_{e_1}) \wr \Sigma_{e_0}}\right) \leq e_0 \cdots e_{m-1}.$$

This is done with an iterated application of the first fundamental theorem of invariant theory and induction on m.

For $m = 1$ $(*)$ is a consequence of the first fundamental theorem for the symmetric group Σ_{e_0}. So suppose $m > 1$ and we have shown that for the defining representation Ω of $(\Sigma_{e_{m-2}} \wr (\Sigma_{e_{m-3}} \cdots \wr \Sigma_{e_1}) \wr \Sigma_{e_0}$ and any $k \in \mathbb{N}$

that $\beta\left(\mathbb{F}[\bigsqcup_k \Omega]^{(\Sigma_{e_{m-2}} \wr (\Sigma_{e_{m-3}} \cdots \wr \Sigma_{e_1}) \wr \Sigma_{e_0}}\right) \leq e_0 \cdots e_{m-2}$.. Note that

$$\mathbb{F}[\Omega]^{(\Sigma_{e_{m-1}} \wr (\Sigma_{e_{m-2}} \cdots \wr \Sigma_{e_1}) \wr \Sigma_{e_0}} = \left(\mathbb{F}[\Omega]^{\left(\times_{e_0}(\Sigma_{e_{m-1}} \wr (\Sigma_{e_{m-2}} \cdots \wr \Sigma_{e_2}) \wr \Sigma_{e_1}))\right)}\right)^{\Sigma_{e_0}}.$$

Let $c = d/e_0$. If Γ is the natural permutation representation for Σ_c, then

$$\Omega = \underbrace{\Gamma \sqcup \cdots \sqcup \Gamma}_{e_0}$$

as Σ_c-set. Therefore

$$\mathbb{F}[\Omega]^{\left(\times_{e_0}(\Sigma_{e_{m-1}} \wr (\Sigma_{e_{m-2}} \cdots \wr \Sigma_{e_2}) \wr \Sigma_{e_1}))\right)} = \bigotimes_{e_0} \mathbb{F}[\bigsqcup_k \Gamma]^{(\Sigma_{e_{m-1}} \wr (\Sigma_{e_{m-2}} \cdots \wr \Sigma_{e_2}) \wr \Sigma_{e_1}}.$$

Γ regarded as a $(\Sigma_{e_{m-1}} \wr (\Sigma_{e_{m-2}} \cdots \wr \Sigma_{e_2}) \wr \Sigma_{e_1}$-set is the defining representation so $\bigsqcup_k \Gamma$ is a $(\Sigma_{e_{m-1}} \wr (\Sigma_{e_{m-2}} \cdots \wr \Sigma_{e_2}) \wr \Sigma_{e_1}$-set to which our induction hypothesis can be applied. We therefore conclude that

$$\beta\left(\mathbb{F}[\bigsqcup_k \Gamma]^{(\Sigma_{e_{m-1}} \wr (\Sigma_{e_{m-2}} \cdots \wr \Sigma_{e_2}) \wr \Sigma_{e_1}}\right) \leq e_1 \cdots e_{k-1}.$$

Set

$$A = \mathbb{F}[\bigsqcup_k \Gamma]^{(\Sigma_{e_{m-1}} \wr (\Sigma_{e_{m-2}} \cdots \wr \Sigma_{e_2}) \wr \Sigma_{e_1}}.$$

Then

$$\mathbb{F}[\Omega]^{\left(\times_{e_0}(\Sigma_{e_{m-1}} \wr (\Sigma_{e_{m-2}} \cdots \wr \Sigma_{e_2}) \wr \Sigma_{e_1}\right)} = \bigotimes_{e_0} \mathbb{F}[\bigsqcup_k \Gamma]^{(\Sigma_{e_{m-1}} \wr (\Sigma_{e_{m-2}} \cdots \wr \Sigma_{e_2}) \wr \Sigma_{e_1}} = \bigotimes_{e_0} A$$

and the action of Σ_{e_0} on $\otimes_{e_0} A$ is by permutation of the factors. Therefore by Corollary 2.3.4 and the inductive hypothesis we obtain

$$\begin{aligned}\beta\left(\mathbb{F}[\Omega]^{(\Sigma_{e_{m-1}} \wr (\Sigma_{e_{m-2}} \cdots \wr \Sigma_{e_1}) \wr \Sigma_{e_0}}\right) &= \beta\left(\left(\mathbb{F}[\Omega]^{\left(\times_{e_0}(\Sigma_{e_{m-1}} \wr (\Sigma_{e_{m-2}} \cdots \wr \Sigma_{e_2}) \wr \Sigma_{e_1}))\right)}\right)^{\Sigma_{e_0}}\right) \\ &= \beta\left(\left(\otimes_{e_0} A\right)^{\Sigma_{e_0}}\right) \leq e_0 \beta(A) = e_0 \cdot e_1 \cdots e_{k-1}\end{aligned}$$

as was to be shown. □

Here is an example to illustrate these ideas.

EXAMPLE 2 (D. Krause [211]): The quaternion group Q_8 of order 8 has a representation over $\mathbb{F} = \mathbb{F}_3$ given by the matrices

$$\boldsymbol{i} = \mathbf{A} = \begin{bmatrix} 0 & 1 \\ -1 & 0 \end{bmatrix}, \quad \boldsymbol{j} = \mathbf{B} = \begin{bmatrix} -1 & 1 \\ 1 & 1 \end{bmatrix} \in \mathrm{GL}(2, \mathbb{F}_3).$$

We saw in Example 3 of the previous section that the invariants are not generated by ordinary orbit Chern classes, of which there are only the two

$$\frac{xy^9 - x^9y}{xy^3 - x^3y}, \quad (xy^3 - x^3y)^2.$$

Consider in Q_8 the chain of subgroups

$$(\maltese) \qquad \{1\} < \mathbb{Z}/2 < \mathbb{Z}/4 < Q_8,$$

where $\mathbb{Z}/4$ is generated by $\mathbf{A}$ and $\mathbb{Z}/2$ by its square $-\mathbf{I}$. The corresponding decomposition of the underlying set Ω of Q_8 consists of the four blocks as shown in the accompanying table. The second column of the table results from the first by multiplying the first column by $\boldsymbol{j}$, and the second row from the first by multiplying by $\boldsymbol{i}$.

	$\mathbb{Z}/4$	$\boldsymbol{j}\cdot\mathbb{Z}/4$
$\mathbb{Z}/2$	$\{\pm 1\}$	$\{\pm\boldsymbol{j}\}$
$\boldsymbol{i}\mathbb{Z}/2$	$\{\pm\boldsymbol{i}\}$	$\{\pm\boldsymbol{k}\}$.

The representation of Q_8 is 2-dimensional, so we have $n = 2$. For the chain ($\maltese$) we obtain $d = 8$, $e_2 = e_1 = e_0 = 2$. To compute the fine Chern classes associated to this chain of subgroups we need two copies of the multidimensional array given in the preceding table, say,

(☯)

	$\mathbb{Z}/4$	$\boldsymbol{j}\cdot\mathbb{Z}/4$
$\mathbb{Z}/2$	$\Omega(1)_{1,1} = \{\pm 1\}$	$\Omega(1)_{1,2} = \{\pm\boldsymbol{j}\}$
$\boldsymbol{i}\mathbb{Z}/2$	$\Omega(1)_{2,1} = \{\pm\boldsymbol{i}\}$	$\Omega(1)_{2,2} = \{\pm\boldsymbol{k}\}$

	$\mathbb{Z}/4$	$\boldsymbol{j}\cdot\mathbb{Z}/4$
$\mathbb{Z}/2$	$\Omega(2)_{1,1} = \{\pm 1\}$	$\Omega(2)_{1,2} = \{\pm\boldsymbol{j}\}$
$\boldsymbol{i}\mathbb{Z}/2$	$\Omega(2)_{2,1} = \{\pm\boldsymbol{i}\}$	$\Omega(2)_{2,2} = \{\pm\boldsymbol{k}\}$.

To compute the threefold polarized elementary symmetric polynomials we need to work in the polynomial algebra in $16 = 2 \cdot 8 = n \cdot d$ variables, which we choose to call $z_{i;j,k,\pm}$, where $1 \le i, j, k \le 2$ and both values of the sign $\pm$ are used. The threefold polarized elementary symmetric polynomials are indexed by 3-dimensional arrays of nonnegative integers $\mathbb{A} = [a(i;j,k)]$ satisfying

$$\begin{aligned} &1 \le i, j, k \le 2, \\ &a(i;j,k) \le e_2 = 2, \\ &a(i;1,k) + a(i;2,k) \le e_1 = 2,\ k = 1, 2, \\ &a(i;j,1) + a(i;j,2) \le e_0 = 2,\ j = 1, 2, \\ &a(1;j,k) \cdot a(2;j,k) = 0. \end{aligned}$$

For each such array we obtain a threefold polarized elementary symmetric polynomial $e(\mathbb{A}) \in \mathbb{F}_3[z_{i;j,k,\pm}]^{(\Sigma_2 \wr \Sigma_2)\wr \Sigma_2}$. To do so we choose $a(i;j,k)$ distinct elements from $\Omega(i;j,k)$, form their product in $\mathbb{F}_3[z_{i;j,k,\pm}]$, and take the orbit sum of the $\big((\Sigma_2 \wr \Sigma_2) \wr \Sigma_2\big)$-orbit of this product. This leads to a long list. To obtain the corresponding threefold polarized Chern class $c(\mathbb{A}) \in \mathbb{F}_3[x, y]^{Q_8}$ we need to substitute according to the scheme depicted in Table 4.2.3. This scheme arises from the tables (🅢) by applying the Noether map η.

$$z_{i;j,k,\pm} \longrightarrow \begin{cases} \pm x & \text{for } i=1 \text{ and } (j,k)=(1,1) \\ \pm y & \text{for } i=1 \text{ and } (j,k)=(2,1) \\ \pm(-x+y) & \text{for } i=1 \text{ and } (j,k)=(1,2) \\ \pm(x+y) & \text{for } i=1 \text{ and } (j,k)=(2,2) \\ \pm y & \text{for } i=2 \text{ and } (j,k)=(1,1) \\ \pm x & \text{for } i=2 \text{ and } (j,k)=(2,1) \\ \pm(x+y) & \text{for } i=2 \text{ and } (j,k)=(1,2) \\ \pm(x-y) & \text{for } i=2 \text{ and } (j,k)=(2,2) \end{cases}$$

TABLE 4.2.3

As we have already seen in the preceding example, there is a great deal of redundancy after the substitution, so rather than present a full table (n.b., the group $(\Sigma_2 \wr \Sigma_2) \wr \Sigma_2$ has order 128, so orbit sums are possibly quite large) we present the result of judicious choices for the arrays.

If $\mathbb{A}$ is the array $[a(i;j,k)]$ and $\mathbb{B}$ the array $[b(i;j,k)]$ whose only nonzero entries are

$$a(1;1,1) = 2 = a(1;2,1),$$
$$b(1;i,j) = 1, \qquad 1 \leq i,j \leq 2,$$

respectively, then

$$f_1 = c(\mathbb{A}) = 2(x^4 - x^2y^2 + y^4),$$
$$f_2 = c(\mathbb{B}) = 2(xy^3 - x^3y).$$

To these we can add the ordinary Chern class of the orbit $V^* \setminus \{0\}$, namely

$$f_3 = c_6(V^* \setminus \{0\}) = \frac{xy^9 - x^9y}{xy^3 - x^3y} = x^4y^2 - x^2y^4.$$

We next compute the Poincaré series of $\mathbb{F}_3[x,y]^{Q_8}$. To do so we avail ourselves of a characteristic zero lift of the representation $Q_8 \hookrightarrow \mathrm{GL}(2,\mathbb{F}_3)$ (see [372] Chapter 7, Section 3, Example 1). The matrices

$$\widetilde{\mathbf{A}} = \begin{bmatrix} 0 & 1 \\ -1 & 0 \end{bmatrix}, \qquad \widetilde{\mathbf{B}} = \frac{1}{\sqrt{-2}} \begin{bmatrix} -1 & 1 \\ 1 & 1 \end{bmatrix} \in \mathrm{GL}(2, \mathbb{Q}(\sqrt{-2}))$$

have product

$$\widetilde{\mathbf{C}} = \widetilde{\mathbf{A}} \cdot \widetilde{\mathbf{B}} = \frac{1}{\sqrt{-2}} \begin{bmatrix} 1 & 1 \\ 1 & -1 \end{bmatrix}.$$

Direct computation shows that

$$\widetilde{\mathbf{A}}^2 = \widetilde{\mathbf{B}}^2 = \widetilde{\mathbf{C}}^2 = -\mathbf{I},$$

so that the subgroup of $GL(2, \mathbb{Q}(\sqrt{-2}))$ generated by $\widetilde{\mathbf{A}}$ and $\widetilde{\mathbf{B}}$ is isomorphic to Q_8. Denote by $\mathbb{Z}_{(3)}$ the integers localized at the prime ideal (3) and by $\mathbb{Z}_{(3)}(\sqrt{-2})$ the subring of $\mathbb{Q}(\sqrt{-2})$ consisting of elements of the form $a + b\sqrt{-2}$, where $a, b \in \mathbb{Z}_{(3)}$. The principal ideal $(-1 + \sqrt{-2}) \subset \mathbb{Z}_{(3)}(\sqrt{-2})$ is prime, contains $3 = -(1 + \sqrt{-2})(-1 + \sqrt{-2})$, and the natural map

$$\mathbb{Z}_{(3)}(\sqrt{-2}) \xrightarrow{q} \mathbb{F}_3 \qquad a + b\sqrt{-2} \longmapsto a + b$$

identifies the quotient $\mathbb{Z}_{(3)}(\sqrt{-2})/(-1 + \sqrt{-2})$ with $\mathbb{F}_3$. Under the map q the matrices $\widetilde{\mathbf{A}}$ and $\widetilde{\mathbf{B}}$ map to the matrices $\mathbf{A}$ and $\mathbf{B}$, respectively, which generate the copy of Q_8 in $GL(2, \mathbb{F}_3)$ whose invariants we are studying. With the aid of this lift we can compute the Poincaré series with the help of Table 4.2.4 and Molien's theorem.

g	$\rho(g)$	$\det(1 - g^{-1}t)$	g	$\rho(g)$	$\det(1 - g^{-1}t)$
$\mathbf{I}$	$\begin{bmatrix} 1 & 0 \\ 0 & 1 \end{bmatrix}$	$(1 - t)^2$	$-\mathbf{I}$	$\begin{bmatrix} -1 & 0 \\ 0 & -1 \end{bmatrix}$	$(1 + t)^2$
$\widetilde{\mathbf{A}}$	$\begin{bmatrix} 0 & 1 \\ -1 & 0 \end{bmatrix}$	$1 + t^2$	$-\widetilde{\mathbf{A}}$	$\begin{bmatrix} 0 & -1 \\ 1 & 0 \end{bmatrix}$	$1 + t^2$
$\widetilde{\mathbf{B}}$	$\frac{1}{\sqrt{-2}}\begin{bmatrix} -1 & 1 \\ 1 & 1 \end{bmatrix}$	$1 + t^2$	$-\widetilde{\mathbf{B}}$	$-\frac{1}{\sqrt{-2}}\begin{bmatrix} -1 & 1 \\ 1 & 1 \end{bmatrix}$	$1 + t^2$
$\widetilde{\mathbf{C}}$	$\frac{1}{\sqrt{-2}}\begin{bmatrix} 1 & 1 \\ 1 & -1 \end{bmatrix}$	$1 + t^2$	$-\widetilde{\mathbf{C}}$	$-\frac{1}{\sqrt{-2}}\begin{bmatrix} 1 & 1 \\ 1 & -1 \end{bmatrix}$	$1 + t^2$

TABLE 4.2.4: The quaternion subgroup Q_8 in $GL(2, \mathbb{Q}(\sqrt{-2})$

The result of this computation is

$$P(\mathbb{F}_3[V]^{Q_8}, t) = \frac{1}{8}\left(\frac{1}{(1-t)^2} + \frac{1}{(1+t)^2} + \frac{6}{1+t^2}\right) = \frac{1+t^6}{(1-t^4)^2}$$
$$= 1 + 2t^4 + t^6 + 3t^8 + \cdots.$$

We can convert this into a table of dimensions for the spaces of invariant forms of low degree. We obtain

$$\begin{array}{rccccccccccl} \deg: & 0 & 1 & 2 & 3 & 4 & 5 & 6 & 7 & 8 & \cdots, \\ \dim: & 1 & 0 & 0 & 0 & 2 & 0 & 1 & 0 & 3 & \cdots. \end{array}$$

The two fine Chern classes $f_1 = x^4 - x^2y^2 + y^4$ and $f_2 = x^3y - xy^3$ are linearly independent, so span $\mathbb{F}_3[x, y]_4^{Q_8}$. They are algebraically independent, so their squares and product span $\mathbb{F}_3[x, y]_8^{Q_8}$. The sixth ordinary Chern class f_3 is equal to $c_6(V^* \setminus \{0\}) = (xy^9 - x^9y)/(xy^3 - x^3y) = x^4y^2 - x^2y^4$, so cannot be in the subalgebra generated by f_1 and f_2. Since $\mathbb{F}_3[x, y]^{Q_8}$ is generated by forms of degree at most 8 by Theorem 2.3.3, it follows that we have found algebra generators for $\mathbb{F}_3[x, y]^{Q_8}$, and

$$\mathbb{F}_3[x, y]^{Q_8} \cong \mathbb{F}_3[f_1, f_2, f_3]/\mathfrak{p},$$

where $\mathfrak{p}$ is a prime ideal of height 1. Since $\mathbb{F}_3[f_1, f_2, f_3]$ is a unique factorization domain, $\mathfrak{p}$ must be a principal ideal. In fact, it is generated by a polynomial of degree 12 expressing f_3^2 as a polynomial in f_1 and f_2 (see Section 7.4, Example 10).

We already commented that the fine Chern classes depend on the choice of the chain of subgroups (⁘) used in their construction. Here is an example to show how this can be exploited: It is the by now familiar example of $Q_8 < \mathrm{GL}(2, \mathbb{F}_3)$, but worked out in a different manner.

EXAMPLE 3: Consider the quaternion subgroup Q_8 of $\mathrm{GL}(2, \mathbb{F}_3)$ from the previous example. Denote by Ω the orbit $V^* \setminus \{0\}$ of G. This orbit is imprimitive, and the system of imprimitivity corresponding to the subgroup $\mathbb{Z}/4$ in Q_8 generated by $\mathbf{A}$ consists of the two blocks

$$\Omega_1 = \{\pm x, \pm y\},$$
$$\Omega_2 = \{\pm(x+y), \pm(x-y)\}.$$

The elementary symmetric polynomials in the block entries are called **block Chern classes**. In this example each block has two nonzero block Chern classes, viz.,

$$c_2(\Omega_1) = -(x^2+y^2),$$
$$c_4(\Omega_1) = x^2y^2,$$
$$c_2(\Omega_2) = x^2+y^2,$$
$$c_4(\Omega_2) = (x^2-y^2)^2.$$

Taking symmetric functions of these will give us invariants of Q_8. Such invariants we call **fine block Chern classes**. These are

$$\Phi = c_4(\Omega_1) + c_4(\Omega_2) = (x^2+y^2)^2,$$
$$\Theta = c_4(\Omega_1)c_2(\Omega_2) + c_2(\Omega_1)c_4(\Omega_2) = -(x^2+y^2)(x^4+y^4) = \mathbf{d}_{2,1}.$$

There is another system of imprimitivity for Ω corresponding to the subgroup $\mathbb{Z}/4$ generated by $\mathbf{B}$, whose blocks are

$$\Lambda_1 = \{\pm x, \pm(x-y)\},$$
$$\Lambda_2 = \{\pm y, \pm(x+y)\}.$$

From the block Chern classes of this system of imprimitivity we obtain the fine block Chern class

$$\Psi = c_4(\Lambda_1) + c_4(\Lambda_2) = x^2(x-y)^2 + y^2(x+y)^2.$$

The forms $\Phi, \Psi \in \mathbb{F}_3[x, y]^{Q_8}$ form a system of parameters, and (see the discussion in [372] of Example 1 in Section 3.1)

$$\mathbb{F}_3[x, y]^{Q_8} = \mathbb{F}_3[\Phi, \Psi] \oplus \mathbb{F}_3[\Phi, \Psi] \cdot \Theta.$$

Therefore, the fine block Chern classes Φ, Ψ, and Θ generate $\mathbb{F}_3[x, y]^{Q_8}$ as an algebra.

4.3 Dade Bases and Systems of Parameters

The subject of this section is the construction of systems of parameters for rings of invariants of finite groups. Proposition 4.1.1 tells us that we can find polynomials in the orbit Chern classes that form a system of parameters. The following shows that in fact, we can find top orbit Chern classes that do the trick. The basic idea is attributed by R. Stanley, [401], to E. Dade (private correspondence) in the case where the ground field is of characteristic zero. The following account is taken from the unpublished manuscript [323].

PROPOSITION 4.3.1: *Let $\rho : G \hookrightarrow \mathrm{GL}(n, \mathbb{F})$ be a representation of a finite group G over the field $\mathbb{F}$, and set $V = \mathbb{F}^n$. Suppose that there is a basis $z_1, \ldots, z_n$ for the dual representation V^* that satisfies the condition*

$$(*) \qquad z_i \notin \bigcup_{g_1, \ldots, g_{i-1} \in G} \mathrm{Span}_{\mathbb{F}}\{g_1 \cdot z_1, \ldots, g_{i-1} \cdot z_{i-1}\}, \qquad i = 2, \ldots, n.$$

Then the top Chern classes

$$c_{\mathrm{top}}([z_1]), \ldots, c_{\mathrm{top}}([z_n]) \in \mathbb{F}[V]^G$$

of the orbits $[z_1], \ldots, [z_n]$ of the basis elements are a system of parameters.

PROOF: Let t_i denote the cardinality $|[z_i]|$ of the G-orbit $[z_i]$ of z_i for $i = 1, \ldots, n$. Since the extension of rings $\mathbb{F}[V]^G \subseteq \mathbb{F}[V]$ is finite and integral, it is equivalent to show that $c_{t_1}([z_1]), \ldots, c_{t_n}([z_n]) \in \mathbb{F}[V]$ are a system of parameters. Passing to an algebraic closure $\overline{\mathbb{F}}$ of $\mathbb{F}$ we need to show by Theorem A.3.6 that the variety defined by the ideal $(c_{t_1}([z_1]), \ldots, c_{t_n}([z_n])) \subset \mathbb{F}[V]$ consists of the origin $0 \in \overline{V} = \overline{\mathbb{F}}^n$ alone. Note that the variety $\mathfrak{V}(c_{t_i}([z_i]))$ consists of the union of the hyperplanes $\ker\{g \cdot z_i\}$, where g ranges over all the elements of G. Since $z_1, \ldots, z_n$ satisfy condition $(*)$, the linear forms $g_1 z_1, \ldots, g_n z_n$ are linearly independent for any $g_1, \ldots, g_n \in G$, and therefore the intersection of their kernels is zero. Hence for the variety $\mathfrak{V}(c_{t_1}([z_1]), \ldots, c_{t_n}([z_n]))$ defined by the ideal $(c_{t_1}([z_1]), \ldots, c_{t_n}([z_n])) \subset \mathbb{F}[V]$ we have

$$\begin{aligned} \mathfrak{V}(c_{t_1}([z_1]), \ldots, c_{t_n}([z_n])) &= \bigcap_{j=1}^{n} \mathfrak{V}(c_{t_j}([z_j])) = \bigcap_{j=1}^{n} \bigcup_{g_j \in G} \ker\{g_j \cdot z_j\} \\ &= \bigcup_{g_1, \ldots, g_n \in G} \ker\{g_1 z_1\} \cap \cdots \cap \ker\{g_n z_n\} = \{0\} \end{aligned}$$

as required. □

Given a representation $\rho : G \hookrightarrow \mathrm{GL}(n, \mathbb{F})$, a basis $z_1, \ldots, z_n \in V^*$ that satisfies the condition $(*)$ of Proposition 4.3.1 will be called a **Dade basis** for V^*, and we refer to $(*)$ as **Dade's condition.**

PROPOSITION 4.3.2: *Let G be a finite group, $n \in \mathbb{N}$, and $\mathbb{F}$ a field. If $|G|^{n-1} < |\mathbb{F}|$, then for any representation of G, $\rho : G \hookrightarrow \mathrm{GL}(n, \mathbb{F})$, the dual representation V^* of $V = \mathbb{F}^n$ admits a Dade basis.*

PROOF: Choose $z_1 \neq 0 \in V^*$ and proceed inductively, assuming that we have already chosen linearly independent elements $z_1, \ldots, z_i \in V^*$ satisfying Dade's condition. If $i = n$, we are done. Otherwise, to choose z_{i+1} note that the subspace

$$\mathrm{Span}_{\mathbb{F}}\{g_1 \cdot z_1, \ldots, g_i \cdot z_i\},$$

with $g_1, \ldots, g_i \in G$ contains $|\mathbb{F}|^i$ elements. If $\mathbb{F}$ is not finite, then V cannot be the union of a finite number of proper subspaces, so we can choose z_{i+1} not in any of these subspaces. If $\mathbb{F}$ is finite, the vector space V^* contains $|\mathbb{F}|^n$ elements. Since $|G|^i \leq |G|^{n-1} < |\mathbb{F}|$, we have

$$\left| \bigcup_{g_1, \ldots, g_i \in G} \mathrm{Span}_{\mathbb{F}}\{g_1 \cdot z_1, \ldots, g_i \cdot z_i\} \right| \leq |G|^{n-1} \cdot |\mathbb{F}|^i < |\mathbb{F}| \cdot |\mathbb{F}|^{n-1} = |\mathbb{F}|^n = |V^*|.$$

Hence there is an element $z_{i+1} \in V^*$ for us to choose such that

$$z_{i+1} \notin \bigcup_{g_1, \ldots, g_i \in G} \mathrm{Span}_{\mathbb{F}}\{g_1 \cdot z_1, \ldots, g_i \cdot z_i\}.$$

In particular, $z_{i+1} \notin \mathrm{Span}\{z_1, \ldots, z_i\}$, so $z_1, \ldots, z_i, z_{i+1} \in V^*$ are linearly independent and satisfy Dade's condition. This constructs inductively a Dade basis for V^*. □

These results have the following neat consequence.

COROLLARY 4.3.3: *Let $\rho : G \hookrightarrow \mathrm{GL}(n, \mathbb{F})$ be a representation of a finite group G over the field $\mathbb{F}$. Suppose the dual representation V^* has a Dade basis. Then $\mathbb{F}[V]^G$ contains a system of parameters $h_1, \ldots, h_n \in \mathbb{F}[V]^G$ such that $\deg(h_1) = \cdots = \deg(h_n) = |G|$. If moreover $\mathbb{F}[V]^G$ is Cohen-Macaulay,[4] then the Poincaré series of $\mathbb{F}[V]^G$ has the form*

$$P(\mathbb{F}[V]^G, t) = (1 - t^{|G|})^{-n} \sum_{j=0}^{s} t^{m_j},$$

for some $s \in \mathbb{N}_0$, $m_0, \ldots, m_s \in \mathbb{N}$.

[4] This is the case if, for example, $|G| \in \mathbb{F}^\times$: see Theorem 5.5.2.

PROOF: By Proposition 4.3.1 the top orbit Chern classes of a Dade basis $z_1, \ldots, z_n$ form a system of parameters. The cardinality t_i of the G-orbit $[z_i]$ of z_i divides $|G|$ for $i = 1, \ldots, n$, and therefore, setting

$$h_i = c_{t_i}([z_i])^{\frac{|G|}{t_i}}, \qquad i = 1, \ldots, n,$$

yields a system of parameters for $\mathbb{F}[V]^G$ with $\deg(h_1) = \cdots = \deg(h_n) = |G|$.

If $\mathbb{F}[V]^G$ is Cohen–Macaulay, then it is a finitely generated free module over the subalgebra $\mathbb{F}[h_1, \ldots, h_n]$ generated by the algebraically independent elements $h_1, \ldots, h_n \in \mathbb{F}[V]$. If $1 = F_0, \ldots, F_s \in \mathbb{F}[V]^G$ is a basis for $\mathbb{F}[V]^G$ over $\mathbb{F}[h_1, \ldots, h_n]$, then

$$P(\mathbb{F}[V]^G, t) = \sum_{j=0}^{s} P(\mathbb{F}[h_1, \ldots, h_n], t)t^{\deg(F_j)} = (1 - t^{|G|})^{-n} \sum_{j=0}^{s} t^{\deg(F_j)}$$

as claimed. □

For representations of finite p-groups in characteristic p there is always a Dade basis. In fact, we give a specific construction of such a basis, using iterated fixed-point sets. We begin by defining the ideas involved.

Let M be a finite-dimensional $\mathbb{F}(G)$-module. The fixed-point set of G acting on M is denoted by M^G as usual. Define the **ith iterated fixed-point set**, denoted by $M^{\overset{G\cdots G}{\leftarrow i \rightarrow}}$ or M^{G^i}, inductively by

$$M^{G^i} = \begin{cases} M^G & \text{if } i = 1, \\ \{x \in M \mid x - gx \in M^{G^{i-1}} \ \forall\, g \in G\} & \text{for } i > 1. \end{cases}$$

It is easy to see that these form a chain of G-submodules, and

$$\{0\} \subseteq M^G \subseteq M^{G \cdot G} \subseteq \cdots \subseteq M^{\overset{G\cdots G}{\leftarrow i \rightarrow}} \subseteq \cdots \subseteq M$$

is called the **iterated fixed-point filtration** of M. Since M is finite-dimensional this filtration stabilizes after at most $\dim_{\mathbb{F}}(M)$ steps, and the integer $j(m)$ where it stabilizes is called the **iterated fixed-point length**. If $|G| \in \mathbb{F}^{\times}$, then $M^G \subseteq M$ is a direct summand, so M/M^G is fixed-point free and $M^G = M^{G \cdot G} = \cdots = M^{G^i}$ for all $i \in \mathbb{N}$. If the characteristic of $\mathbb{F}$ divides the order of G, then this need not be the case. (Recall the discussion of the stable invariants from Chapter 1, Section 1.2.) It is helpful to set $M^{G^0} = \{0\}$ and to define

$$\operatorname{cofix}_G(M) = M/M^G.$$

Note that G acts naturally on $\operatorname{cofix}_G(M)$ and that the quotient map

$$\varphi : M \longrightarrow \operatorname{cofix}_G(M)$$

is an $\mathbb{F}(G)$-module homomorphism. Hence we can iterate the functor

cofix_G also and obtain a chain of epimorphisms

$$M \longrightarrow \mathrm{cofix}_G(M) \longrightarrow \cdots \longrightarrow \mathrm{cofix}^i_G(M) \longrightarrow \cdots.$$

Since

$$\dim_{\mathbb{F}}(M) \geq \dim_{\mathbb{F}}(\mathrm{cofix}_G(M)) \geq \cdots \geq \dim_{\mathbb{F}}(\mathrm{cofix}^i_G(M)) \geq \cdots,$$

this sequence stabilizes also after at most $\dim_{\mathbb{F}}(M)$ steps. The iterated fixed-point set M^{G^i} is the kernel of the epimorphism $M \longrightarrow \mathrm{cofix}^i_G(M)$.

If P is a finite p-group and $\mathbb{F}$ a field of characteristic p, then Lemma 2.4.2 shows for any nonzero $\mathbb{F}(P)$-module N that $N^P \neq \{0\}$. The augmentation ideal of $\mathbb{F}(P)$ is nilpotent and coincides with the nilradical of $\mathbb{F}(P)$. In this case the iterated fixed-point filtration is called the **socle filtration** (see, e.g., [16]), and the iterated fixed-point length is called the **socle length.** The sequence $N \longrightarrow \mathrm{cofix}_P(N) \longrightarrow \cdots \longrightarrow \mathrm{cofix}^i_P(N) \longrightarrow \cdots$ terminates with the zero module.

Conversely, if the augmentation ideal of $\mathbb{F}(G)$ acts nilpotently, then the map $1 - g : N \longrightarrow N$ is nilpotent for all $g \in G$, and therefore 1 is the only eigenvalue of g. Looking at the Jordan form of g, we see that g has p-power order. Hence we have shown the following:

LEMMA 4.3.4: *Let P be a finite p-group and M be a finite-dimensional $\mathbb{F}(P)$-module. Then the iterated fixed-point filtration*

$$\{0\} \subset M^P \subset M^{P \cdot P} \subset \cdots \subset M^{P^j} = M$$

terminates at M, and the iterated fixed-point length is $j = j(M) \leq \dim_{\mathbb{F}}(M)$. □

In this case the integers

$$k_i = \dim_{\mathbb{F}}(M^{P^i}), \qquad i = 1, \ldots, j(m),$$

are called the **ifp-dimensions,**[5] and we may choose a basis $z_1, \ldots, z_n$ for M such that

$$z_{k_i+1}, \ldots, z_{k_{i+1}} \in M^{G^{i+1}} \qquad i = 0, \ldots, j-1,$$

and the residue classes in $M^{G^{i+1}}/M^{G^i}$ are a basis over $\mathbb{F}$. We call such a basis for M an **ifp-basis.**

THEOREM 4.3.5: *Let P be a finite p-group and M an $\mathbb{F}(P)$-module. Then the augmentation ideal of $\mathbb{F}(P)$ acts nilpotently on M and an ifp-basis for M satisfies Dade's condition.*

PROOF: Since the augmentation ideal of $\mathbb{F}(G)$ acts nilpotently on M, the iterated fixed-point filtration of M terminates at M after at most $\dim_{\mathbb{F}}(M)$ terms. Let j be the iterated fixed-point length of M, $(k_1, \ldots, k_j)$

[5] Of course, *ifp* stands for *iterated fixed point.*

the ifp-dimensions, and $z_1, \ldots, z_n \in M$ an ifp-basis. The vectors $z_1, \ldots, z_{k_1}$ are fixed points of P, and hence they satisfy Dade's condition trivially, since they are linearly independent. Suppose we have shown inductively that $z_1, \ldots, z_{k_i}$ satisfy Dade's condition. Consider the elements $z_{k_i+1}, \ldots, z_{k_{i+1}}$. Since

$$M^{G^i} = \mathrm{Span}_{\mathbb{F}}\{z_1, \ldots, z_{k_i}\} = \mathrm{Span}_{\mathbb{F}}\{g_1 \cdot z_1, \ldots, g_{k_i} \cdot z_{k_i} \mid g_1, \ldots, g_{k_i} \in G\}$$

and $z_{k_i+1}, \ldots, z_{k_{i+1}} \in (M/M^{G^i})^G$ are a basis, the elements $z_1, \ldots, z_{k_{i+1}}$ satisfy Dade's condition. □

If M is an $\mathbb{F}(G)$-module, then the augmentation ideal of $\mathbb{F}(G)$ acts nilpotently on M if and only if each element $g - 1 \in \mathbb{F}(G)$ acts nilpotently. This is the case if, for example, $G = P$ is a finite p-group and $\mathbb{F}$ has characteristic p, or if G is generated by transvections. (For a discussion of transvections see Section 6.2)

COROLLARY 4.3.6: *Let $\rho : P \hookrightarrow \mathrm{GL}(n, \mathbb{F})$ be a representation of a finite p-group over the field $\mathbb{F}$ of characteristic p. Let $z_1, \ldots, z_n$ be an ifp-basis for the dual representation V^* of $V = \mathbb{F}^n$ and let t_i be the cardinality of the P-orbit $[z_i]$ of z_i for $i = 1, \ldots, n$. Then the top Chern classes*

$$c_{t_1}([z_1]), \ldots, c_{t_n}([z_n]) \in \mathbb{F}[V]^P$$

are a system of parameters.

PROOF: Apply Proposition 4.3.1 and Theorem 4.3.5. □

4.4 Euler Classes and Related Constructions

In addition to the orbit Chern classes and their refinements, another construction adapted from characteristic class theory is that of the Euler class; see [387]. Again, the idea is quite simple, but it has powerful consequences.

Suppose that $\rho : G \hookrightarrow \mathrm{GL}(n, \mathbb{F})$ is a representation of a finite group over the field $\mathbb{F}$. A **configuration of hyperplanes** in V is a set $\mathcal{C}$ of codimension-1 linear subspaces of V. The configuration $\mathcal{C}$ is said to be G-invariant if for each hyperplane $W \in \mathcal{C}$ and each $g \in G$ the hyperplane $g(W)$ is also in $\mathcal{C}$. Dual to the configuration $\mathcal{C}$ of hyperplanes in V is a **configuration of lines** $\mathcal{L}_{\mathcal{C}}$ in the dual space V^* defined by $\ell \in \mathcal{L}_{\mathcal{C}}$ if and only if for every nonzero $z \in \ell$ one has $\ker(z) \in \mathcal{C}$. If we choose one nonzero linear form $z_W \in \ell_W = W^{\perp} \subset V^*$ for each $W \in \mathcal{C}$, then the polynomial $\mathbf{e}_{\mathcal{C}} = \prod_{W \in \mathcal{C}} z_W \in \mathbb{F}[V]$ defines the affine variety associated to $\mathcal{C}$, i.e.,

$$\mathfrak{V}(\mathcal{C}) = \bigcup_{W \in \mathcal{C}} W = \{x \in V \mid \mathbf{e}_{\mathcal{C}}(x) = 0\};$$

see, e.g., [313] Chapter 1. If $\mathcal{C}$ is G-invariant, then for any g in G, $g\mathbf{e}_{\mathcal{C}}$ also defines this variety. Therefore, for $g \in G$, one has $g\mathbf{e}_{\mathcal{C}} = \lambda(g)\mathbf{e}_{\mathcal{C}}$ for some nonzero $\lambda(g) \in \mathbb{F}$. The function $\lambda : G \longrightarrow \mathbb{F}^\times$ is a 1-dimensional representation of G, and the form $\mathbf{e}_C$ is a G λ-relative invariant; see Section 1.1. We call $\mathbf{e}_C$ the **pre-Euler class** of the G-invariant configuration C. Since G is finite, some power of λ is the trivial representation, and hence some power of $\mathbf{e}_C$ is a G-invariant form. The smallest such power is called the **Euler class** of the configuration and is denoted by E_C. Note that when there are no 1-dimensional representations of G over $\mathbb{F}$ apart from the trivial one, then the pre-Euler class is already a G-invariant form, and so equal to the Euler class. This will happen for any field $\mathbb{F}$ when G is a nonabelian simple group or perfect; see Section 1.7.

For example, if $\mathbb{F}$ is a finite field with q elements, then the configuration of all hyperplanes in V is invariant under any group. The pre-Euler class of this configuration is a form L_n of degree $(q^n - 1)/(q - 1)$ and figures in the computation of the ring of invariants of the special linear group $\mathrm{SL}(n, \mathbb{F})$. Since $\mathrm{SL}(n, \mathbb{F})$ has no nontrivial 1-dimensional representations, this pre-Euler class is already an $\mathrm{SL}(n, \mathbb{F})$-invariant. We denote it simply by L_n, or if n is clear from context, simply by L. Since it is a universal Euler class for representations $\rho : G \hookrightarrow \mathrm{SL}(n, \mathbb{F})$ the notations E_n and E are also sometimes employed.

Here is an example to illustrate this.

EXAMPLE 1 (L. E. Dickson): We consider the invariants of the special linear group $\mathrm{SL}(2, \mathbb{F}_q)$ of the finite field $\mathbb{F}_q$ in its defining two-dimensional representation. Let $\{x, y\}$ be a basis for the dual vector space V^*. The only orbits of $\mathrm{SL}(2, \mathbb{F}_q)$ on V^* are $\{0\}$ and $\widetilde{V} = V^* \setminus \{0\}$. The computation of the Chern classes is therefore the same as for $\mathrm{GL}(2, \mathbb{F}_p)$, apart from the substitution of q for p, that we made in Section 4.1, Example 1. The result in this instance is

$$\begin{aligned} c_{q^2-1}(\widetilde{V}^*) &= (xy^q - x^q y)^{q-1}, \\ c_{q^2-q}(\widetilde{V}^*) &= (x^{q(q-1)} + y^{q-1}(y^{q-1} - x^{q-1})^{q-1}) \\ &= \frac{xy^{q^2} - x^{q^2} y}{xy^q - x^q y}. \end{aligned}$$

The pre-Euler class associated to the configuration of all hyperplanes, i.e., lines, in $V = \mathbb{F}_q^2$ is $\mathrm{SL}(2, \mathbb{F}_q)$ invariant, and denoted by L. If $\ell \subset V^*$ is a line and $z \in \ell$ is a nonzero form, then the other nonzero forms in ℓ are the multiples αz, where $\alpha \in \mathbb{F}_q^\times$. Since $\alpha^{q-1} = 1$ for any nonzero α in $\mathbb{F}_q$ it follows that $\mathsf{L}^{q-1} = c_{q^2-1}(\widetilde{V}^*)$. It is not hard to see that the two forms $c_{q^2-q}(\widetilde{V}^*)$ and L are a system of parameters in $\mathbb{F}_q[x, y]$, see, e.g., [372] Section 5.6, Example 4. From Theorem 4.5.5, which is proved in the next section, it therefore follows that

$$\mathbb{F}_q[x, y]^{\mathrm{SL}(2, \mathbb{F}_q)} = \mathbb{F}_q[c_{q^2-q}(\widetilde{V}^*), \mathsf{L}].$$

The Chern classes of the orbit $\widetilde{V}^*$ are called the Dickson polynomials, and are usually denoted by $\mathbf{d}_{2,0} = c_{q^2-1}(\widetilde{V}^*)$ and $\mathbf{d}_{2,1} = c_{q^2-q}(\widetilde{V}^*)$. The choice of the double indexing will become clear in Section 6.1 when we study the invariants of $\mathrm{GL}(n, \mathbb{F}_q)$ in general.

4.5 The Degree Theorem

According to the Noether normalization theorem (Theorem A.3.1) the Krull dimension of a commutative graded connected algebra A over a field $\mathbb{F}$ is equal to the order of the pole of the rational function $P(A, t)$ at $t = 1$. If the Krull dimension is n, and $a_1, \ldots, a_n \in A$ is a homogeneous system of parameters, then

$$P(A, t) = \frac{f(t)}{\prod_{i=1}^{n}(1 - t^{|a_i|})}$$

where $f(t)$ is a polynomial with integral coefficients. So the product $(1-t)^n P(A, t)$ is a rational function, which when evaluated at $t = 1$ gives a nonzero rational number, called the **degree** of A and written $\deg(A)$, [391]. If $A = \mathbb{F}[V]^G$, and $\mathbb{F}$ has characteristic prime to $|G|$, then, as a consequence of Molien's theorem, (Theorem 3.1.3) $\deg(\mathbb{F}[V]^G) = \frac{1}{|G|}$. If $B \supseteq A$ is a finite extension of finitely generated graded integral domains over $\mathbb{F}$, then we shall show that the ratio of their degrees is equal to the degree of the corresponding extension of their graded fields of fractions. The degree of this extension of graded fields coincides with the degree of the extension of the usual fields of fractions, which, being Galois, lets us conclude that $\deg(\mathbb{F}[V]^G) = \frac{1}{|G|}$ regardless of the characteristic of $\mathbb{F}$. We call this the degree theorem and it is the subject of this section.

Let A be a graded integral domain over the field $\mathbb{F}$. We define the **graded field of fractions** of A to be the $\mathbb{Z}$-graded object $\mathit{FF}(A)$, where for $j \in \mathbb{Z}$ the component $\mathit{FF}(A)_j$ consists of the fractions x/y, $y \neq 0$, with $x \in A_{m+j}$, $y \in A_m$, and $m \geq \max\{0, -j\}$. These fractions are added and multiplied in the usual way. $\mathit{FF}(A)$ is a **graded field** in the sense that every nonzero homogeneous element has a homogeneous inverse. It is not a field in the usual sense, nor is its totalization, but the degree zero component $\mathit{FF}(A)_0$ is a field. If A has nonzero elements of nonzero degree, it is easy to see that $\mathit{FF}(A)$ may be identified with the Laurent polynomials $\mathit{FF}(A)_0[X, X^{-1}]$, with X some nonzero element of minimal strictly positive degree in A. The *full* field of fractions of A, i.e., the field of fractions of its totalization $\mathrm{Tot}(A)$ is, of course, $\mathit{FF}(A)_0(X)$.

If $\mathbb{K} \supset \mathbb{L}$ are graded fields, with $\mathbb{K} = \mathbb{K}_0[X, X^{-1}]$ finitely generated as an $\mathbb{L}$-module, where $\mathbb{L} = \mathbb{L}_0[Y, Y^{-1}]$, then $\mathbb{K}$ is clearly free as an $\mathbb{L}$-module, and we write $|\mathbb{K} : \mathbb{L}|$ for the number of generators. This is easily seen to be equal to the degree of the ungraded field extension $\mathbb{F}\mathbb{F}(\mathrm{Tot}(\mathbb{K})) \supset \mathbb{F}\mathbb{F}(\mathrm{Tot}(\mathbb{L}))$. If moreover $\deg(X) = \deg(Y)$, this number also conincides with the degrees $|\mathbb{K}_0 : \mathbb{L}_0|$ and $|\mathbb{K}_0(X) : \mathbb{L}_0(Y)|$ regarded as ungraded field extensions.

LEMMA 4.5.1: *Suppose that $B \supseteq A$ are graded integral domains over a field with B integral over A. Let their graded fields of fractions be $\mathbb{K} = \mathbb{F}\mathbb{F}(B)$ and $\mathbb{L} = \mathbb{F}\mathbb{F}(A)$, and assume that $\mathbb{K} \supset \mathbb{L}$ is a finite extension. Then there is a basis of $\mathbb{K}$ over $\mathbb{L}$ consisting of elements of B. The A-submodule of B generated by this basis is free.*

PROOF: Since B is integral over A, every $b \in B$ has a minimal polynomial, so satisfies some equation of the form

$$b^r + a_{r-1}b^{r-1} + \cdots + a_0 = 0$$

with $a_i \in A$. Since B is an integral domain, we may suppose that $a_0 \neq 0$. Thus there is an element of B, namely $\widetilde{b} = -b^{r-1} - a_{r-1}b^{r-2} - \cdots - a_1$, with the property that $\widetilde{b}b = a_0$ is a nonzero element of A.

Let $|\mathbb{K} : \mathbb{L}| = m$, and choose elements $y_1, \ldots, y_m$ forming a basis of $\mathbb{K}$ as a vector space over $\mathbb{L}$. Let $y_i = b_i'/b_i''$ with $b_i', b_i'' \in B$. Choose $\widetilde{b}_i''$ as above for $i = 1, \ldots, m$. Then the elements $\widetilde{y}_i = \widetilde{b}_i''b_i' \in B$ for $i = 1, \ldots, m$ also form a basis of $\mathbb{K}$ over $\mathbb{L}$. These elements are linearly independent in B regarded as an A-module (since they are linearly independent over the field of fractions), so the A-submodule of B generated by them is free. □

PROPOSITION 4.5.2: *Suppose that $B \supseteq A$ is a finite extension of finitely generated connected graded integral domains over $\mathbb{F}$. Then*

$$\deg(B) = |\mathbb{F}\mathbb{F}(B) : \mathbb{F}\mathbb{F}(A)| \deg(A).$$

PROOF: Without loss of generality we may suppose that A contains a nonzero element of positive degree, for otherwise, $A = \mathbb{F} = B$ and there is nothing to prove. Then $\dim(A) = n = \dim(B)$ is strictly positive since A and B are domains. According to Lemma 4.5.1 we may choose a basis $x_1, \ldots, x_m$ of $\mathbb{F}\mathbb{F}(B)$ over $\mathbb{F}\mathbb{F}(A)$ consisting of homogeneous elements of B such that the A-submodule of B generated by $x_1, \ldots, x_m$ is a free A-module, of rank m, say, $Ax_1 \oplus \cdots \oplus Ax_m \subseteq B$. Hence[6]

$$P(A, t)\Big(\sum_{i=1}^{m} t^{\deg(x_i)}\Big) \leq P(B, t).$$

Since B is a finitely generated A-module, we may choose a nonzero element $a \in A$, e.g., a common multiple of the denominators of $x_1, \ldots, x_m$, with

[6] If $\sum_{i=0}^{\infty} a_i$ and $\sum_{i=0}^{\infty} b_i$ are power series with real coefficients, we write $\sum_{i=0}^{\infty} a_i \leq \sum_{i=0}^{\infty} b_i$ to mean $a_i \leq b_i$ for $i = 0, \ldots$.

inverse a^{-1} in $I\!F\!I\!F(A)$, such that $B \subseteq A[a^{-1}]x_1 \oplus \cdots \oplus A[a^{-1}]x_m$. This implies

$$P(B, t) \leq P(A, t)t^{-\deg(a)}\Big(\sum_{i=1}^{m} t^{\deg(x_i)}\Big).$$

Therefore, we have

$$P(A, t)\Big(\sum_{i=1}^{m} t^{\deg(x_i)}\Big) \leq P(B, t) \leq P(A, t)t^{-\deg(a)}\Big(\sum_{i=1}^{m} t^{\deg(x_i)}\Big).$$

We multiply this inequality by $(1-t)^n$, put $t = 1$, and get

$$\deg(A) \cdot m \leq \deg(B) \leq \deg(A) \cdot m$$

as desired. □

THEOREM 4.5.3 (Degree Theorem): *Let $\rho : G \hookrightarrow \mathrm{GL}(n, \mathbb{F})$ be a representation of a finite group G over the field $\mathbb{F}$. Then $\deg(\mathbb{F}[V]^G) = \frac{1}{|G|}$. Hence the Laurent expansion of the Poincaré series of $\mathbb{F}[V]^G$ about $t = 1$ begins*

$$P(\mathbb{F}[V]^G, t) = \frac{\frac{1}{|G|}}{(1-t)^n} + \cdots.$$

PROOF: $\mathbb{F}[V]$ is a finite extension of $\mathbb{F}[V]^G$, so we may apply Proposition 4.5.2. At the level of (ungraded) fields of fractions, $\mathbb{F}(V)$ is a Galois extension of $\mathbb{F}(V)^G$ with Galois group G. So we conclude from Proposition 4.5.2 that

$$\deg(\mathbb{F}[V]) = |G| \cdot \deg(\mathbb{F}[V]^G).$$

Since $P(\mathbb{F}[V], t) = 1/(1-t)^n$, we have $\deg(\mathbb{F}[V]) = 1$, and solving for $\deg(\mathbb{F}[V]^G)$ yields the result. □

The degree theorem has a number of important consequences for the polynomial algebra problem (see Section 1.4, Problem 5). We first note the following corollary:

COROLLARY 4.5.4: *Suppose that $\rho : G \hookrightarrow \mathrm{GL}(n, \mathbb{F})$ is a representation of a finite group G over the field $\mathbb{F}$. If $\mathbb{F}[V]^G = \mathbb{F}[f_1, \ldots, f_n]$, where $\deg(f_i) = d_i$ for $i = 1, \ldots, n$, then $d_1 d_2 \cdots d_n = |G|$.*

PROOF: Direct computation gives

$$\begin{aligned} \deg(\mathbb{F}[f_1, \ldots, f_n]) &= (1-t)^n \prod_{i=1}^{n} \frac{1}{1-t^{d_i}}\Bigg|_{t=1} \\ &= \prod_{i=1}^{n} \frac{1}{1+t+t^2+\cdots+t^{d_i-1}}\Bigg|_{t=1} = \frac{1}{d_1 d_2 \cdots d_n}, \end{aligned}$$

and by Theorem 4.5.3, $\deg(\mathbb{F}[V]^G) = \frac{1}{|G|}$, so the result follows. □

From the degree theorem (Theorem 4.5.3) we deduce furthermore that a sort of converse to Corollary 4.5.4 holds:

PROPOSITION 4.5.5: *Let $G \hookrightarrow \mathrm{GL}(n, \mathbb{F})$ be a representation of a finite group G over the field $\mathbb{F}$. Suppose $\mathbb{F}[V]^G$ contains a system of parameters $f_1, \ldots, f_n$, such that $\deg(f_1) \cdots \deg(f_n) = |G|$. Then $\mathbb{F}[V]^G \cong \mathbb{F}[f_1, \ldots, f_n]$.*

PROOF: Note that $f_1, \ldots, f_n$ is also a system of parameters for $\mathbb{F}[V]$ so they are algebraically independent by Theorem A.3.1. If A denotes the subalgebra of $\mathbb{F}[V]^G$ generated by $f_1, \ldots, f_n$, then $A \cong \mathbb{F}[f_1, \ldots, f_n]$, and hence

$$P(A, t) = \prod_{i=1}^{n} \left(\frac{1}{1 - t^{d_i}} \right) = \frac{\frac{1}{d_1 \cdots d_n}}{(1-t)^n} + \cdots ,$$

where $d_i = \deg(f_i)$ $i = 1, \ldots, n$. Therefore, $\deg(A) = d_1 \cdots d_n$. Also by Theorem A.3.1 the Poincaré series $P(\mathbb{F}[V]^G, t)$ has a pole of order n at $t = 1$. By Theorem 4.5.3 $\deg(\mathbb{F}[V]^G) - \frac{1}{|G|}$. Hence A and $\mathbb{F}[V]^G$ have the same field of fractions $\mathbb{K}$. Since $\mathbb{F}[V]^G$ is finite over A, the extension $A \subseteq \mathbb{F}[V]^G$ is integral. However, $A \cong \mathbb{F}[f_1, \ldots, f_n]$ is integrally closed in its field of fractions. Therefore, we have $\mathbb{F}[V]^G \subseteq A \cong \mathbb{F}[f_1, \ldots, f_n]$, and the result follows. □

In the nonmodular case we were able to compute not only the leading coefficient of the Laurent expansion of the Poincaré series of a ring of invariants, but the next coefficient as well; see Proposition 3.1.4. The corresponding computation for modular representations is considerably more difficult; see, e.g., [30] Theorem 3.13.2, [62], [283] and [158].

For further applications of the degree theorem see, e.g., [377], [217], [385], and the correction [2] to the penultimate section of [1]. The following example is a typical illustration of how to use this theorem.

EXAMPLE 1 (L. Flatto [117]): We reconsider the example of the dihedral group D_{2k} of order $2k$ represented in $\mathrm{GL}(2, \mathbb{R})$ as the group of symmetries of a regular k-gon centered at the origin, Section 3.1, Example 3. The group $D_{2k} < \mathrm{GL}(2, \mathbb{R})$ is generated by the matrices

$$\mathbf{D} = \begin{bmatrix} \cos\frac{2\pi}{k} & -\sin\frac{2\pi}{k} \\ \sin\frac{2\pi}{k} & \cos\frac{2\pi}{k} \end{bmatrix}, \qquad \mathbf{S} = \begin{bmatrix} 1 & 0 \\ 0 & -1 \end{bmatrix}$$

where $\mathbf{D}$ is a rotation through $2\pi/k$ radians and $\mathbf{S}$ a reflection in an axis. These matrices are orthogonal, so they preserve the norm, and hence $f_1 = x^2 + y^2$ is an invariant quadratic polynomial. If we introduce the complex variable $z = x + \boldsymbol{i}y$, then with respect to z the action of D_{2k} is given by

$$\mathbf{S}(z) = \overline{z}, \qquad \mathbf{D}(z) = \exp\frac{2\pi\boldsymbol{i}}{k} z.$$

Thus $f_2 = \mathfrak{Re}(z^k)$, where $\mathfrak{Re}(-)$ denotes the real part of $-$, is a polynomial in

x and y that is invariant under the action of D_{2k}. The system of equations

$$x^2 + y^2 = 0,$$
$$\mathfrak{Re}(z^k) = 0$$

has only the trivial solution $(0, 0)$ in $\mathbb{C}$, and hence by Proposition A.3.6 f_1, f_2 is a system of parameters. Moreover, $\deg(f_1)\deg(f_2) = 2k = |D_{2k}|$, so by Proposition 4.5.5

$$\mathbb{R}[x, y]^{D_{2k}} = \mathbb{R}[f_1, f_2].$$

We close this section with an important application of Proposition 4.5.5 which is particularly useful in studying the invariants of finite p-groups in characteristic p (see Section 6.2 and the references there).

PROPOSITION 4.5.6 (H. Nakajima, R. E. Stong): *Let G be a finite group and $\varrho : G \hookrightarrow GL(n, \mathbb{F})$ a representation. Suppose there is a chain of G-invariant subspaces*

$$V^* = V_n^* \supset V_{n-1}^* \supset \cdots \supset V_0^* = \{0\}$$

with $\dim_{\mathbb{F}}(V_i^*) = i$ and vectors $z_i \in V_i^* \setminus V_{i-1}^*$ with G-orbits $[z_i]$ for $i = 1, \ldots, n$ satisfying the following condition:

$$\prod_{i=1}^{n} |[z_i]| = |G|.$$

Then $\mathbb{F}[V]^G = \mathbb{F}[f_1, \ldots, f_n]$, where

$$f_i = c_{d_i}([z_i]), \quad d_i = |[z_i]| \quad \text{for } i = 1, \ldots, n$$

are the top Chern classes of the orbits $[z_1], \ldots, [z_n]$.

PROOF: The polynomial f_i has degree d_i, so we may apply Proposition 4.5.5 to obtain the conclusion, provided that $f_1, \ldots, f_n$ is a system of parameters. Therefore, it suffices to show that $\mathbb{F}[V]$ is integral over the subalgebra $A_0 \subseteq \mathbb{F}[V]^G$ generated by $f_1, \ldots, f_n$. Let $A_i \subseteq \mathbb{F}[V]$ be the subalgebra generated by $f_1, \ldots, f_n$ together with $z_1, \ldots, z_i$. Thus we have

$$A_0 \subseteq A_1 \subseteq \cdots \subseteq A_n = \mathbb{F}[V].$$

Every element of $[z_i]$ is of the form $\lambda z_i + u$, where $\lambda \in \mathbb{F}$ is nonzero and $u \in V_{i-1}^*$. Therefore,

$$f_i = \prod_{z \in [v_i]} z = \sum_{j=1}^{d_i} \varphi_{i,j}(z_1, \ldots, z_{i-1}) z_i^j,$$

where the leading coefficient $\varphi_{i,d_i}(z_1, \ldots, z_{i-1})$ is a nonzero field element.

Hence z_i is a root of the monic polynomial

$$\frac{1}{\varphi_{i,d_i}(z_1, \ldots, z_{i-1})}\left[f_i - \sum_{j=1}^{d_i} \varphi_{i,j}(z_1, \ldots, z_{i-1})X^j\right] \in A_{i-1}[X].$$

Therefore, A_i is integral over A_{i-1} for $i = 1, \ldots, n$, and hence by transitivity of integrality $\mathbb{F}[V] = A_n$ is integral over A_0. □

As an example of the applicability of this result we consider the unipotent subgroup of $\mathrm{GL}(n, \mathbb{F}_q)$.

EXAMPLE 2 ($\mathrm{Uni}(n, \mathbb{F}_q)$): Consider $\mathrm{Uni}(n, \mathbb{F}_q) \subset \mathrm{GL}(n, \mathbb{F}_q)$, the unipotent subgroup, which consists of the upper triangular matrices with 1's on the main diagonal, viz.,

$$\mathrm{Uni}(n, \mathbb{F}_q) = \left\{ \begin{bmatrix} 1 & & * \\ & \ddots & \\ 0 & & 1 \end{bmatrix} \in \mathrm{GL}(n, \mathbb{F}_q) \mid * \in \mathbb{F}_q \right\}.$$

It is just a p-Sylow subgroup of $\mathrm{GL}(n, \mathbb{F}_q)$, and as such is a finite p-group of order $q^{n-1} \cdot q^{n-2} \cdots q = q^{\binom{n}{2}}$. The elements of $\mathrm{Uni}(n, \mathbb{F}_q)$ all have determinant 1, so $\mathrm{Uni}(n, \mathbb{F}_q)$ is also a p-Sylow subgroup of $\mathrm{SL}(n, \mathbb{F}_q)$ and any subgroup between $\mathrm{SL}(n, \mathbb{F}_q)$ and $\mathrm{GL}(n, \mathbb{F}_q)$.

The group $\mathrm{Uni}(n, \mathbb{F}_q)$ preserves the complete flag of subspaces

$$\{0\} = V_0 \subsetneq V_1 \subsetneq \cdots \subsetneq V_n = V = \mathbb{F}_q^n,$$

where V_i is the subspace of V consisting of vectors whose last $n - i$ coordinates are zero. This means that the standard basis for V is a Dade basis, and hence by Proposition 4.3.1 the top Chern classes of the standard basis vectors are a system of parameters. To verify that the condition of the theorem of Nakajima and Stong (Theorem 4.5.6) holds for this chain of subspaces we need to compute the size of the orbits of the standard dual basis vectors.

To this end, let $\mathbf{E}_{i,j}$ denote the elementary $n \times n$ matrix with a 1 in the ith row and the jth column, and otherwise 0. The matrices $\mathbf{u}_{i,j} = \mathbf{I} + \mathbf{E}_{i,j}$ belong to $\mathrm{Uni}(n, \mathbb{F}_q)$ for $i < j$. If $\{x_1, \ldots, x_n\}$ is the standard basis for $V = \mathbb{F}_q^n$, then

$$\mathbf{E}_{i,j}(x_k) = \begin{cases} 0, & j \neq k, \\ x_i, & j = k, \end{cases}$$

$$\mathbf{u}_{i,j}(x_k) = \begin{cases} x_k, & j \neq k, \\ x_k + x_i, & j = k. \end{cases}$$

Passing to the dual vector space V^* and the dual basis $\{z_1, \ldots, z_n\}$ we obtain

$$\mathbf{E}_{i,j}(z_k) = \begin{cases} 0, & i \neq k, \\ z_j, & i = k, \end{cases}$$

$$\mathbf{u}_{i,j}(z_k) = \begin{cases} z_k, & i \neq k, \\ z_k + z_j, & i = k. \end{cases}$$

The elements of $\mathrm{Uni}(n, \mathbb{F}_q)$ can be written uniquely in the form

$$\mathbf{u} = \mathbf{I} + \sum_{i<j} a_{i,j}\mathbf{E}_{i,j}, \qquad a_{i,j} \in \mathbb{F}_q,$$

so we have

$$\mathbf{u}(z_k) = \Big(\mathbf{I} + \sum_{i<j} a_{i,j}\mathbf{E}_{i,j}\Big)(z_k) = \begin{cases} z_k + \sum_{k<j} a_{k,j}z_j, & k < n, \\ z_n, & k = n. \end{cases}$$

Therefore, the orbit of z_k under the action of the group $\mathrm{Uni}(n, \mathbb{F}_q)$ is

$$[z_k] = \begin{cases} \{z_k + b_{k+1}z_{k+1} + \cdots + b_n z_n \mid b_{k+1}, \ldots, b_n \in \mathbb{F}_q\}, & k < n, \\ \{z_n\}, & k = n, \end{cases}$$

which contains q^{n-k} elements. If we set $\mathbf{h}_{n,n-k} = c_{q^{n-k}}([z_k])$, then we obtain

$$\mathbb{F}_q[z_1, \ldots, z_n]^{\mathrm{Uni}(n,\mathbb{F}_q)} = \mathbb{F}_q[\mathbf{h}_{n,0}, \ldots, \mathbf{h}_{n,n-1}]$$

by Theorem 4.5.6.

Chapter 5
Homological Finiteness

NOETHERIAN commutative graded connected algebras over a field are finitely generated. This means that for such an algebra A we can choose elements $a_1, \ldots, a_k \in A$ such that the map of algebras $\mathbb{F}[u_1, \ldots, u_k] \longrightarrow A$, where $u_i \longmapsto a_i$, for $i = 1, \ldots, k$, is an epimorphism. If this map is called ε, then the kernel of ε, say I_0, is an ideal in $\mathbb{F}[u_1, \ldots, u_k]$ and as such, it is finitely generated. To be finitely generated as an ideal means that I_0 is finitely generated as an $H = \mathbb{F}[u_1, \ldots, u_k]$-module. If we choose a finite set $h_1, \ldots, h_l \in I_0$ of elements that generate I_0 as an H-module, then we can find an epimorphism of the free $\mathbb{F}[u_1, \ldots, u_k]$-module generated by say $w_1, \ldots, w_l$ onto I_0 by sending $w_i \longmapsto h_i$ for $i = 1, \ldots, l$. The kernel of this map in turn is again finitely generated, and so on.

We think it is clear that any mathematician has to ask: Does this process come to an end in a finite number of steps? This is the question from which homological algebra arose, and also the starting point for this chapter. It leads directly to Hilbert's syzygy theorem, and its converse, which solves Problem 2 from Section 1.3. A natural byproduct are the notions of homological dimension and codimension which can be used to classify commutative graded connected algebras into types of varying homological structure. In this chapter we introduce tools, such as the Koszul complex, to study the interplay of homological algebra with with invariant theory. We examine many basic homological notions and their ring theoretic interpretations with a view towards deciding under what conditions a ring of invariants has the indicated property.

5.1 The Koszul Complex

Although Hilbert's syzygy theorem was conceived for rings of invariants, it is a central theorem of homological algebra, and best placed in this more general context. For a polynomial algebra $\mathbb{F}[z_1, \ldots, z_n]$ (n.b. we do not require that $\deg(z_i) = 1$, only that $\deg(z_i) > 0$) we aim to describe for each module M over $\mathbb{F}[z_1, \ldots, z_n]$ a functorial, minimal resolution of M as an $\mathbb{F}[z_1, \ldots, z_n]$-module. As a reference for the missing proofs the reader may consult [372] Chapter 6.

DEFINITION: *Let A be a commutative graded connected algebra over a field $\mathbb{F}$. A sequence $a_1, \ldots, a_n \in \overline{A}$ is called a* **regular sequence** *if a_1 is not a zero divisor in A and a_i is not a zero divisor in $A/(a_1, \ldots, a_{i-1})$ for $i = 2, \ldots, n$, where as usual, $\overline{A}$ is the augmentation ideal of A.*

For example, the elementary symmetric polynomials $e_1, \ldots, e_n$ regarded as elements of $\mathbb{F}[z_1, \ldots, z_n]$ are a regular sequence. A regular sequence is always algebraically independent, [372] Theorem 6.2.1. The converse is false: $x, xy \in \mathbb{F}[x, y]$ are algebraically independent but not a regular sequence in $\mathbb{F}[x, y]$. However, because they are algebraically independent, they are a regular sequence in the subalgebra $\mathbb{F}[x, xy] \subset \mathbb{F}[x, y]$. Therefore, regularity does not behave well under extension. Nor does it behave well under restriction. Consider the subalgebra $A \subset \mathbb{F}[x, y]$ generated by

$$x^2, x^2y, y^2, y^3 \in \mathbb{F}[x, y].$$

The elements $x^2, y^2 \in A$ regarded as elements of $\mathbb{F}[x, y]$ are a regular sequence. However,

$$y^2(x^2y) = x^2y^3 = 0 \in A/(x^2),$$

and since x^2y does not belong to the ideal generated by x^2 in A, it follows that y^2 is a zero divisor in $A/(x^2)$. Hence x^2, y^2 is not a regular sequence in A.

For pairs of elements in rings of invariants we do however have a positive result.

PROPOSITION 5.1.1: *Let $\rho : G \hookrightarrow \mathrm{GL}(n, \mathbb{F})$ be a representation of a finite group over the field $\mathbb{F}$ and $f_1, f_2 \in \mathbb{F}[V]^G$. Then f_1, f_2 are a regular sequence in $\mathbb{F}[V]^G$ if and only if they are also a regular sequence in $\mathbb{F}[V]$.*

PROOF: Suppose that $f_1, f_2 \in \mathbb{F}[V]^G$ are a regular sequence in $\mathbb{F}[V]^G$. If h_1, h_2 belong to $\mathbb{F}[V]$ and satisfy

$$h_1f_1 + h_2f_2 = 0.$$

It is enough to show that $h_1 = Fh_2$ for some $F \in \mathbb{F}[V]$, since then $0 = h_1f_1 + h_2f_2 = Fh_2f_1 + h_2f_2$ implies $Ff_1 + f_2 = 0$ and hence $f_2 \in (f_1)$. If $h_2 = 0$, then since $\mathbb{F}[V]$ is an integral domain $h_1 = 0$ also, so $F = 1$ will work, and we may suppose that $h_2 \neq 0$. Then, in the field of fractions $\mathbb{F}(V)$ of $\mathbb{F}[V]$

we find

$$\frac{h_1}{h_2} = -\frac{f_2}{f_1} \in \mathbb{F}(V)^G.$$

If we multiply h_1/h_2 by 1 in the form

$$1 = \frac{\prod_{1 \neq g \in G} gh_2}{\prod_{1 \neq g \in G} gh_2}$$

we obtain

$$\frac{H_1}{H_2} = -\frac{f_2}{f_1} \in \mathbb{F}(V)^G,$$

where H_2 is G-invariant. Cross-multiplying shows that H_1 is also G-invariant. Therefore

$$H_1 f_1 + H_2 f_2 = 0 \in \mathbb{F}[V]^G$$

with both H_1 and H_2 in $\mathbb{F}[V]^G$. Since f_1, f_2 are a regular sequence in $\mathbb{F}[V]^G$ it follows that $H_1 = FH_2$ for some $F \in \mathbb{F}[V]^G$. Both sides of this equality are divisible by $\prod_{1 \neq g \in G} gh_2$ and performing the division then yields $h_1 = Fh_2$ as required.

Conversely, let $f_1, f_2 \in \mathbb{F}[V]^G$ be a regular sequence in $\mathbb{F}[V]$. Assume to the contrary that they are not a regular sequence in $\mathbb{F}[V]^G$. Then f_1 is a zero divisor in $\mathbb{F}[V]^G/(f_2)$. This means that we can choose forms $h_1, h_2 \in \mathbb{F}[V]^G$, such that $h_1 \notin (f_2) \subset \mathbb{F}[V]^G$, and $h_1 f_1 = h_2 f_2$. Since f_1, f_2 are a regular sequence in $\mathbb{F}[V]$ the form h_1 must belong to the extended ideal $(f_2)^e \subset \mathbb{F}[V]$. This means we can find a form $h \in \mathbb{F}[V]$ such that $h_1 = hf_2$. We then have $h_2 f_2 = h_1 f_1 = hf_2 f_1$. Since $\mathbb{F}[V]$ is an integral domain and $f_2 \neq 0$ this implies that $h_2 = hf_1$. In the field of fractions $\mathbb{F}(V)$ this says that $h = \frac{h_2}{f_1}$, so $h \in \mathbb{F}(V)^G$, as both h_2 and f_1 are. Since h is also in $\mathbb{F}[V]$ it must be a G-invariant polynomial, i.e., $h \in \mathbb{F}[V]^G$. The equation $h_1 = hf_2$ implies that h_1 belongs to the ideal $(f_2) \subset \mathbb{F}[V]^G$, which is a contradiction to how we chose h_1. □

A fact of basic importance is the inequality expressed in the next proposition; see also [372] Theorem 6.2.2.

PROPOSITION 5.1.2: *Let A be a commutative graded connected algebra over a field $\mathbb{F}$ and $a_1, \ldots, a_k \in A$ a regular sequence. Then $k \leq \dim(A)$.* □

So, in a Noetherian algebra the length of the longest regular sequence is bounded above by the Krull dimension. Algebras for which the length of the longest regular sequence they contain is equal to their Krull dimension are called **Cohen–Macaulay algebras**. If A is Cohen-Macaulay and $a_1, \ldots, a_n \in A$ a regular sequence of length $n = \dim(A)$ then $A/(a_1, \ldots, a_n)$

has Krull dimension zero, hence is totally finite, so $a_1, \ldots, a_n \in A$ are a system of parameters.

The role of Cohen-Macaulay algebras in invariant theory will be examined in more detail in Section 5.5. At this point note however that Proposition 5.1.1 says that rings of invariants in two variables are always Cohen-Macaulay.

NOTATION: *If V is a graded vector space over $\mathbb{F}$, we denote by $E[V]$ the exterior algebra on the dual vector space V^*. If $u_1, \ldots, u_n$ is a basis for V^*, then $E[V] = E[u_1, \ldots, u_n]$ has as a basis the elements*

$$u_{i_1} \cdots u_{i_m}, \qquad i_1 < i_2 < \cdots < i_m,$$

of degree $\deg(u_{i_1}) + \cdots + \deg(u_{i_m})$. Multiplication is by juxtaposition and subject to the rules

$$u_i u_j = \begin{cases} -u_j u_i & \text{for } i \neq j, \\ 0 & \text{for } i = j. \end{cases}$$

The elements of $E[V]$ may be thought of as alternating homogeneous, multilinear functions from V to $\overline{\mathbb{F}}$, where $\overline{\mathbb{F}}$ is the algebraic closure of $\mathbb{F}$. The homogeneous component of $E[V]$ of degree k is denoted by $E[V]_k$, $\Lambda^k(V^*)$, or $\Lambda^k(u_1, \ldots, u_n)$, and is isomorphic to the kth exterior power of V^*.

DEFINITION: *Let A be a commutative graded connected algebra over a field $\mathbb{F}$ and $a_1, \ldots, a_n \in A$. The* **Koszul complex** *of A with respect to $a_1, \ldots, a_n \in A$ is the differential graded commutative algebra*

$$\mathcal{K} = \mathcal{K}(a_1, \ldots, a_n) = A \otimes E[sa_1, \ldots, sa_n],$$

where[1] $\deg(sa_i) = 1 + \deg(a_i)$, *and the differential ∂ is defined by requiring*

$$\begin{aligned} \partial\,|_A &= 0, \\ \partial(sa_i) &= a_i \quad \text{for } i = 1, \ldots, n, \\ \partial(x \cdot y) &= \partial(x)y + (-1)^{\deg(x)} x \partial(y), \qquad \forall\, x, y \in \mathcal{K}. \end{aligned}$$

The syzygy problem is concerned with finding finite free resolutions of modules over a polynomial algebra $\mathbb{F}[z_1, \ldots, z_n]$. The Koszul complex provides a solution to this problem by **bigrading** it in the way we describe next. In $\mathcal{K}_A(a_1, \ldots, a_n) = A \otimes E[sa_1, \ldots, sa_n]$ we introduce a bigrading by demanding

$$\operatorname{bideg}(a \otimes sa_{i_1} sa_{i_2} \cdots sa_{i_k}) = (k, \deg(a) + \deg(a_{i_1}) + \cdots + \deg(a_{i_k})),$$

where $a \in A$. The first component of the bidegree is called the **homological degree**, or **resolution degree**, and the second component the **internal**

[1] The s is meant to suggest **suspension** or **shift**, since the grading of sa is shifted by 1 from the corresponding element of A. This kind of shifting of degrees arose originally in algebraic topology to relate the cohomology of a space to that of its suspension.

degree. The homogeneous component of homological degree k is isomorphic to the tensor product $A \otimes \Lambda^k(sa_1, \ldots, sa_n)$. The differential in the Koszul complex lowers the homological degree by 1, and so we may arrange the Koszul complex in the form of a sequence of graded A-modules

$$0 \longleftarrow \mathcal{K}_{0,*} \xleftarrow{\partial_1} \mathcal{K}_{1,*} \xleftarrow{\partial_2} \cdots \xleftarrow{\partial_n} \mathcal{K}_{n,*} \longleftarrow 0,$$

where $\mathcal{K}_{i,*} \cong A \otimes \Lambda^i(V^*)$. The homology of the Koszul complex is the homology of this chain complex of A-modules. The fundamental theorems about Koszul complexes are Theorems 6.2.3 and 6.2.4 in [372]. Taken together they give us the following result.

THEOREM 5.1.3 (J.-L. Koszul): *Let A be a commutative graded connected algebra over a field $\mathbb{F}$ and $a_1, \ldots, a_n \in A$. The Koszul complex $\{\mathcal{K}_A(a_1, \ldots, a_n), \partial\}$ is acyclic if and only if $a_1, \ldots, a_n$ is a regular sequence. In this case the Koszul complex $\mathcal{K}_A(a_1, \ldots, a_n)$ provides a free acyclic resolution of $A/(a_1, \ldots, a_n)$ as an A-module.* □

For example, the Koszul complex

$$\mathcal{K} = \mathbb{F}[z_1, \ldots, z_n] \otimes_{\mathbb{F}} E[sz_1, \ldots, sz_n]$$

with differential

$$\partial(z_i) = 0, \qquad \partial(sz_i) = z_i, \qquad i = 1, \ldots, n,$$

is acyclic and provides a free acyclic resolution of $\mathbb{F}$ as an $\mathbb{F}[z_1, \ldots, z_n]$-module. Likewise,

$$\mathcal{L} = \mathbb{F}[z_1, \ldots, z_n] \otimes_{\mathbb{F}} E[sz_1, \ldots, sz_n] \otimes_{\mathbb{F}} \mathbb{F}[z_1, \ldots, z_n]$$

with differential

$$\partial(1 \otimes z_i) = 0 = \partial(z_i \otimes 1), \qquad \partial(sz_i) = z_i \otimes 1 - 1 \otimes z_i$$

is acyclic and provides a free resolution of $\mathbb{F}[z_1, \ldots, z_n]$ as a bimodule over $\mathbb{F}[z_1, \ldots, z_n]$, i.e., as an $\mathbb{F}[z_1, \ldots, z_n] \otimes \mathbb{F}[z_1, \ldots, z_n]$-module. This can be put to use to solve the syzygy problem for modules over a polynomial algebra.

PROPOSITION 5.1.4: *Let M be a module over $\mathbb{F}[z_1, \ldots, z_n]$. Then the complex*

$$\mathcal{L}(M) = \mathcal{L} \otimes_{\mathbb{F}[z_1, \ldots, z_n]} M$$

is a free resolution of M as an $\mathbb{F}[z_1, \ldots, z_n]$-module. □

The complex $\mathcal{K}$ is often referred to simply as the Koszul complex for $\mathbb{F}[z_1, \ldots, z_n]$, and $\mathcal{L}$ as the **two-sided Koszul complex** for $\mathbb{F}[z_1, \ldots, z_n]$ (see [28], where the two-sided Koszul complex was introduced). These complexes are useful even when they are not acyclic. See e.g., [344], [104], [47], and the many references in these sources. We make just such a use of them in Section 6.5 to study the depth of rings of invariants.

5.2 Hilbert's Syzygy Theorem

With the aid of the Koszul complex it is a simple matter to solve the problem of syzygies (Section 1.3, Problem 2). Here is one version of the result:

THEOREM 5.2.1 (Hilbert's Syzygy Theorem): *A graded module M over the graded polynomial algebra $\mathbb{F}[z_1, \ldots, z_n]$ has a finite free resolution of length at most n, i.e., there is a finite free resolution*

$$(\mathcal{F}) \qquad 0 \longrightarrow F_m \longrightarrow \cdots \longrightarrow F_1 \longrightarrow F_0 \longrightarrow M \longrightarrow 0$$

of M as an $\mathbb{F}[z_1, \ldots, z_n]$-module with $m \leq n$. If M is finitely generated, then $F_0, \ldots, F_m$ may be chosen finitely generated as well.

PROOF: Choose $\mathcal{F} = \mathcal{L}(M)$. □

COROLLARY 5.2.2: *Let $G \hookrightarrow \mathrm{GL}(V)$ be a finite-dimensional representation of a finite group G and $f_1, \ldots, f_s \in \mathbb{F}[V]^G$ a system of algebra generators. Then $\mathbb{F}[V]^G$ has at most s syzygies.*

PROOF: The homomorphism of $\mathbb{F}[F_1, \ldots, F_s]$-modules

$$\mathbb{F}[F_1, \ldots, F_s] \longrightarrow \mathbb{F}[V]^G, \qquad F_i \longmapsto f_i,$$

is surjective so $\mathbb{F}[V]^G$ admits a finite free resolution over $\mathbb{F}[F_1, \ldots, F_s]$ of length at most s. □

DEFINITION: *Let A be a commutative graded connected algebra over a field $\mathbb{F}$ and M an A-module. Let*

$$(\mathcal{F}) \qquad \cdots \longrightarrow F_s \longrightarrow \cdots \longrightarrow F_1 \longrightarrow F_0 \longrightarrow M \longrightarrow 0$$

be a free resolution of M as an A-module. We say that $\mathcal{F}$ has **finite length** *n if $F_n \neq 0$ and $F_s = 0$ for $s > n$. If M admits a free resolution of finite length, then the minimal such length is called the* **projective dimension** *or* **homological dimension** *of M, and is denoted by $\text{proj-dim}_A(M)$ or $\text{hom-dim}_A(M)$. The algebra A is said to have* **finite global (projective) dimension** *d if $\sup_{M \in \mathrm{MOD}/A}\{\text{proj-dim}_A(M)\}$ is finite and equal to d, where MOD/A denotes the category of graded modules over A.*

The global dimension of A is denoted by $\text{gl-dim}(A)$ or $\text{proj-dim}(A)$. As usual, when the suprema are not finite, we set $\text{hom-dim}_A(M) = \infty$, respectively $\text{gl-dim}(A) = \infty$.

We assume familiarity with the basic facts about resolutions and derived functors. There is no lack of references for the results we need, and we intend to use them freely. Among the numerous books on homological algebra are [68], [232], [306], and [448], and there are many others containing substantial amounts of homological algebra, such as [104] and [372], as well as the two foundational articles [126] and [151].

One of the less well known and interesting connections between homological algebra and commutative algebra (see, e.g., [163], [391]) is via the Euler characteristic, which links up the Poincaré series of a finitely generated module over a commutative graded connected Noetherian algebra with a resolution of that module over a Noether normalization of the module.

To properly discuss this we need to recall an elementary fact about the Euler characteristic, but in a slightly more general setting than is usual.

Here are the facts we need: The additivity of Poincaré series can be extended from direct sums to a long exact sequence of vector spaces of finite type, viz., if the sequence of graded vector spaces of finite type

$$0 \longrightarrow M_{k'} \longrightarrow M_{k'-1} \longrightarrow \cdots \longrightarrow M_{k''+1} \longrightarrow M_{k''} \longrightarrow 0$$

is exact, then Euler's formula says that

$$(\mathcal{E}) \qquad \sum_{i=k''}^{k'} (-1)^i P(M_i, t) = 0.$$

This leads to an alternative proof of the theorem of D. Hilbert and J.-P. Serre on the rationality of the Poincaré series of finitely generated modules over a commutative graded Noetherian algebra over a field.

Proof of Theorem 3.3.1: Let M be a finitely generated module over the commutative graded connected Noetherian algebra A. Let $\mathbf{N} = \mathbb{F}[z_1, \ldots, z_n] \subseteq A$ be a Noether normalization of A. Then M is also a finitely generated $\mathbb{F}[z_1, \ldots, z_n]$-module. Let

$$(\cdot\!\!\!\!\!+\!\cdot) \qquad 0 \longrightarrow F_m \longrightarrow \cdots \longrightarrow F_1 \longrightarrow F_0 \longrightarrow M \longrightarrow 0$$

be the resolution of M as an $\mathbb{F}[z_1, \ldots, z_n]$-module constructed in Theorem 5.2.1. Then F_i is a free finitely generated $\mathbb{F}[z_1, \ldots, z_n]$-module for $i = 0, \ldots, m$, so

$$P(F_i, t) = \frac{p_i(t)}{(1 - t^{d_1}) \cdots (1 - t^{d_n})}$$

for some polynomial $p_i(t) \in \mathbb{N}[t]$, where $d_j = \deg(z_j) = \deg(a_j)$ for $j = 1, \ldots, n$. If we apply the Euler formula $(\mathcal{E})$ to the resolution $(\cdot\!\!\!\!\!+\!\cdot)$ and solve for $P(M, t)$, we obtain

$$P(M, t) = \sum_{i=0}^{m} (-1)^i P(F_i, t) = \sum_{i=0}^{m} (-1)^i \frac{p_i(t)}{(1 - t^{d_1}) \cdots (1 - t^{d_n})},$$

which is a rational function with integral coefficients. □

5.3 The Converse of Hilbert's Syzygy Theorem

Hilbert's syzygy theorem (Theorem 5.2.1) says that polynomial algebras $\mathbb{F}[V]$ have finite global dimension. The converse of this result is due to J.-P. Serre. It gives us an example of a ring theoretic property (polynomial algebra over a field) with a homological characterization (finite global dimension). We intend to prove it by induction over the Krull dimension, and we start with a lemma that will enable us to prove the theorem in the case of Krull dimension zero.

Let A be a graded A-algebra over the field $\mathbb{F}$ with augmentation ideal $\overline{A}$, and M a graded A-module. Then $Q(M) = M/\overline{A} \cdot M$ is called the **module of indecomposable elements** of M. (See Section A.1 in Appendix A for a discussion of Nakayama's lemma and the module of indecomposables).

LEMMA 5.3.1: *Let A be a commutative graded connected algebra over a field $\mathbb{F}$. Suppose that every element in $\overline{A}$ of positive degree is a zero divisor. If F', F'' are free A-modules, $F' \neq 0$, and $\varphi : F' \longrightarrow F''$ an A-module homomorphism such that $Q(\varphi) : QF' \longrightarrow QF''$ is trivial, then* $\ker(\varphi) \neq 0$.

PROOF: Clearly, it suffices to consider the case where QF' has rank 1. Let $x' \in F'$ be a generator, and $x_1'', \ldots, x_n'' \in F''$ such that

$$\varphi(x') = \sum_{i=1}^{n} a_i x_i''.$$

For degree reasons $\operatorname{Im}(\varphi)$ lies in the free submodule spanned by those x_i'' with $\deg(x_i'') \leq \deg(x')$, and, since $Q(\varphi) = 0$, in fact in the submodule spanned by those x_i'' with $\deg(x_i'') < \deg(x')$. So without loss of generality we may suppose $\deg(x_i'') < \deg(x')$ for $i = 1, \ldots, n$. Since every element of $\overline{A}$ is a zero divisor, $\overline{A}$ is an associated prime ideal of $(0) \subset A$. To see this recall that the set of zero divisors in A is the union of the associated prime ideals of (0) in A. If

$$\overline{A} \subseteq \bigcup_{\mathfrak{p} \in \operatorname{Ass}(A)} \mathfrak{p},$$

then by the prime avoidance lemma (Lemma A.2.1), $\overline{A} \subseteq \mathfrak{p}$ for some $\mathfrak{p} \in \operatorname{Ass}(A)$. The augmentation ideal $\overline{A}$ is maximal, so this entails that $\overline{A} = \mathfrak{p}$, i.e., $\overline{A}$ is an associated prime of (0) in A. A prime ideal is associated to (0) in A if and only if it is the radical of the annihilator ideal of some nonzero element in A, so there is an element $0 \neq b \in A$ such that $\overline{A}$ annihilates b. We then have

$$\varphi(bx') = b\varphi(x') = b\sum_{i=1}^{n} a_i x_i'' = \sum_{i=1}^{n} ba_i x_i'' = \sum_{i=1}^{n} 0 \cdot x_i'' = 0,$$

so $0 \neq bx' \in \ker(\varphi)$ as required. □

The following theorem not only settles the global dimension of graded connected algebras of zero divisors, it also introduces the powerful idea of a **minimal resolution**. This is a resolution

$$\cdots \xrightarrow{d_{s+1}} F_s \xrightarrow{d_s} F_{s-1} \xrightarrow{d_{s-1}} \cdots \xrightarrow{d_1} F_0 \xrightarrow{\varepsilon} M \longrightarrow 0$$

of an A-module M where the induced maps between indecomposable modules $Q(d_s) : Q(F_s) \longrightarrow Q(F_{s-1})$ are all zero.

THEOREM 5.3.2 (S. Eilenberg): *Let A be a commutative graded connected algebra in which every element of positive degree is a zero divisor. If $A \neq \mathbb{F}$ then A has infinite global dimension.*

PROOF: We regard $\mathbb{F}$ as an A-module via the augmentation homomorphism $\varepsilon : A \longrightarrow \mathbb{F}$. We proceed to construct a minimal resolution for $\mathbb{F}$ as an A-module. To this end let $x_1, \ldots, x_n, \ldots$ be a minimal ideal basis for the augmentation ideal $\overline{A} = \ker(\varepsilon)$, i.e., the images of $x_1, \ldots, x_n, \ldots$ in $\mathbb{F} \otimes_A \overline{A}$ form an $\mathbb{F}$-basis, and set

$$\bigoplus_{i \geq 1} A \cdot \overline{x}_i = F_1 \xrightarrow{d_1} F_0 \xrightarrow{\varepsilon} \mathbb{F} \longrightarrow 0,$$

where $F_0 = A$, ε is the augmentation, $A \cdot \overline{x}_i$ is a free A-module on the single generator $\overline{x}_i$ of degree $\deg(\overline{x}_i) = \deg(x_i)$, and d_1 is defined by $d_1(\overline{x}_i) = x_i$ for $i \geq 1$. Note that $Q(d_1) = 0$ and $\ker(d_1) \subset \overline{A} \cdot F_1$. Suppose inductively that we have constructed a partial free resolution

$$F_s \xrightarrow{d_s} F_{s-1} \xrightarrow{d_{s-1}} \cdots \xrightarrow{d_1} F_0 \xrightarrow{\varepsilon} \mathbb{F} \longrightarrow 0$$

with $Q(d_i) = 0$ and $\ker(d_i) \subset \overline{A} \cdot F_i$, $i = 1, \ldots, s$. By the preceding lemma $\ker(d_s) \neq (0)$. Choose generators $x_{s,1}, \ldots, x_{s,r}, \ldots$ projecting in $\mathbb{F} \otimes_A \ker(d_s)$ to an $\mathbb{F}$-basis. Since $\ker(d_s) \subset \overline{A} \cdot F_s$, it follows that the $x_{s,i}$ project to zero in $Q(F_s) = \mathbb{F} \otimes_A F_s$ for $i \geq 1$. Thus we may define

$$\bigoplus_{j \geq 1} A \cdot \overline{x}_{s,j} = F_{s+1} \xrightarrow{d_{s+1}} F_s$$

by $d_{s+1}(\overline{x}_{s,j}) = x_{s,j}$. Then $Q(d_{s+1}) = 0$. Since the elements $\overline{x}_{s,j}$ project to an $\mathbb{F}$-basis of $\mathbb{F} \otimes_A \ker(d_s)$, we have $\ker(d_{s+1}) \subset \overline{A} \cdot F_{s+1}$. Inductively, we obtain a free resolution

$$(\mathcal{F}) \qquad \cdots \xrightarrow{d_{s+1}} F_s \xrightarrow{d_s} \cdots \xrightarrow{d_1} F_0 \xrightarrow{\varepsilon} \mathbb{F} \longrightarrow 0,$$

where

$$Q(F_s) \neq 0 \text{ and } 0 = Q(d_s) : Q(F_s) \longrightarrow Q(F_{s-1}) \quad \forall\, s.$$

We may compute $\operatorname{Tor}_A^*(\mathbb{F}, \mathbb{F})$ from $\mathbb{F} \otimes_A \mathcal{F}$ by taking homology. This is the complex

$$\cdots \xrightarrow{d_{s+1}=0} Q(F_s) \xrightarrow{Q(d_s)=0} \cdots \xrightarrow{Q(d_1)=0} Q(F_0) \longrightarrow 0,$$

and hence

$$\mathrm{Tor}_A^s(\mathbb{F}, \mathbb{F}) = Q(F_s) \neq 0, \quad \forall\, s \geq 0.$$

So A has infinite global dimension. □

Note that S. Eilenberg's theorem is the converse of Hilbert's syzygy theorem for rings of depth zero; see Section 5.5 for the definition of depth. It also says that in the classical context of finite-dimensional algebras over a field, the homological dimension can assume only two values, 0 or ∞, and therefore is not a very sensitive invariant in this context at all. Eilenberg's theorem does provide however the initial step for the proof of the converse of the syzygy theorem.

THEOREM 5.3.3 (J.-P. Serre): *Let A be a finitely generated commutative graded connected algebra over a field $\mathbb{F}$. If A has finite global dimension d, then $A \cong \mathbb{F}[z_1, \ldots, z_d]$.*

PROOF: We proceed by induction on the Krull dimension. In an algebra of Krull dimension zero every element of positive degree is nilpotent, so the case of dimension zero is a special case of S. Eilenberg's Theorem 5.3.2, and we may suppose that A has positive Krull dimension. Again, appealing to Theorem 5.3.2 we may suppose in addition that A contains a nonzero divisor. Choose a set of elements $z_1, \ldots, z_n$ that project to a basis for QA. By Theorem A.3.1 we may suppose that at least one of these elements, say z_1, is a nonzero divisor in A. Then the Koszul complex $\mathcal{K}_A(z_1)$ is acyclic, so

$$\mathrm{Tor}^i_{\mathbb{F}[z_1]}(A, \mathbb{F}) = 0, \quad i > 0,$$

and A is a free $\mathbb{F}[z_1]$-module. Let $B = \mathbb{F} \otimes_{\mathbb{F}[z_1]} A \cong A/(z_1)$. Then B is generated by $z_2, \ldots, z_n$. Therefore, $\dim_{\mathbb{F}}(QB) < n$, and by Corollary A.3.4, $\dim(B) \leq n - 1$.

Assume that A has finite global dimension. We claim that B does also. Let M be a B-module. The quotient map $q : A \longrightarrow B$ allows us to view M as an A-module. As an A-module M has a finite free resolution, say,

$$(\mathcal{F}) \qquad 0 \longrightarrow F_d \longrightarrow \cdots \longrightarrow F_1 \longrightarrow F_0 \longrightarrow M \longrightarrow 0$$

of length at most $d = \text{gl-dim}(A)$. Each F_i is free over A, so by transitivity also over $\mathbb{F}[z_1]$. Hence the sequence

$$0 \longrightarrow \mathbb{F} \otimes_{\mathbb{F}[z_1]} F_d \longrightarrow \cdots \longrightarrow \mathbb{F} \otimes_{\mathbb{F}[z_1]} F_0 \longrightarrow \mathbb{F} \otimes_{\mathbb{F}[z_1]} M \longrightarrow 0$$

is exact. Since z_1 annihilates M, $\mathbb{F} \otimes_{\mathbb{F}[z_1]} M \cong M/(z_1 \cdot M) \cong M$. The modules $\mathbb{F} \otimes_{\mathbb{F}[z_1]} F_i$, $i = 0, \ldots, d$, are free over $\mathbb{F} \otimes_{\mathbb{F}[z_1]} A = B$. Thus we have constructed a free resolution of M as a B-module of length at most $d = \text{gl-dim}(A)$, and hence B has finite global dimension as claimed.

By the inductive hypothesis we therefore have $B \cong \mathbb{F}[\bar{z}_2, \ldots, \bar{z}_n]$, where

$\overline{z}_i = 1 \otimes_{\mathbb{F}[z_1]} z_i \in B$, $i = 2, \ldots, n$. Define

$$\varphi : \mathbb{F}[z_1, \overline{z}_2, \ldots, \overline{z}_n] \longrightarrow A \text{ by } \begin{cases} \varphi(z_1) = z_1, \\ \varphi(\overline{z}_i) = z_i, & \text{for } i = 2, \ldots, n. \end{cases}$$

We claim that φ is an isomorphism. It is surjective by Corollary A.1.2, because $Q(\varphi)$ is an isomorphism. To see that it is also a monomorphism, suppose that $f \neq 0 \in \mathbb{F}[z_1, \overline{z}_2, \ldots, \overline{z}_n]$ is of minimal degree in $\ker(\varphi)$. We may write $f = z_1 q + r$ with $q \in \mathbb{F}[z_1, \overline{z}_2, \ldots, \overline{z}_n]$ and $r \in \mathbb{F}[\overline{z}_2, \ldots, \overline{z}_n]$. Since the composite

$$\mathbb{F}[\overline{z}_2, \ldots, \overline{z}_n] \hookrightarrow \mathbb{F}[z_1, \overline{z}_2, \ldots, \overline{z}_n] \longrightarrow A \longrightarrow B$$

is the identity, we see that $r = 0$. But $z_1 \in A$ was chosen not to be a zero divisor, so $q = 0$, and hence $f = 0$, a contradiction. Thus φ is an isomorphism. □

At this point we can derive a result that comes in handy in the study of the polynomial algebra problem, see, e.g., Section 1.4, Problem 5, and Theorem 7.1.

COROLLARY 5.3.4: *Suppose A is a graded subalgebra of the graded polynomial algebra*[2] *$\mathbb{F}[h_1, \ldots, h_n]$. If the algebra $\mathbb{F}[h_1, \ldots, h_n]$ is free as an A-module, then $A = \mathbb{F}[f_1, \ldots, f_k]$, with $f_1, \ldots, f_k$ a regular sequence in $\mathbb{F}[h_1, \ldots, h_n]$, $k \leq n$.*

PROOF: First we show that A has finite global dimension. Set $S = \mathbb{F}[h_1, \ldots, h_n]$. Suppose that M is a graded A-module. Let

$$0 \longrightarrow K_n \longrightarrow P_{n-1} \longrightarrow \cdots \longrightarrow P_0 \longrightarrow M \longrightarrow 0$$

be a partial projective resolution of M, i.e., all but K_n are projective and the sequence is exact. Since S is free as an A-module, the functor $- \otimes_A S$ is exact and sends projective A-modules to projective S-modules. Thus

$$0 \longrightarrow K_n \otimes_A S \longrightarrow P_{n-1} \otimes_A S \longrightarrow \cdots \longrightarrow M \otimes_A S \longrightarrow 0$$

is an exact sequence of S-modules, where all but perhaps the last term is S-projective. It follows from [372] Theorem 6.3.3 that $K_n \otimes_A S$ is a projective S-module, because the global dimension of S is n. Since S is free as an A-module, $K_n \otimes_A S$ is also projective over A. Moreover, S being a free A module with $1 \in S$ as one of the free generators, there is an A-module splitting $\sigma : S \longrightarrow A$ to the inclusion $i : A \hookrightarrow S$. Thus

$$K_n = K_n \otimes_A A \hookrightarrow K_n \otimes_A S \underset{1\otimes i}{\overset{1\otimes\sigma}{\rightleftarrows}} K_n \otimes_A A$$

represents K_n as an A-module direct summand in the projective A-module $K_n \otimes_A S$, and hence K_n is projective as an A-module, showing that the

[2] We do not require that $h_1, \ldots, h_n$ have degree 1, only positive degree.

global dimension of A is at most n. By Theorem 5.3.3 it follows that $A = \mathbb{F}[f_1, \ldots, f_k]$ is a polynomial algebra with $k \leq n$.

By assumption S is free as a module over A. Therefore, the acyclic Koszul complex

$$\mathcal{K} = \mathcal{K}(f_1, \ldots, f_k) = A \otimes E[sf_1, \ldots, sf_k]$$

of free A-modules remains acyclic when we apply $S \otimes_A -$

$$S \otimes_A \mathcal{K} = S \otimes E[sf_1, \ldots, sa_k].$$

Hence by Theorem 5.1.3 we have that $f_1, \ldots, f_k$ form a regular sequence in $S = \mathbb{F}[h_1, \ldots, h_n]$. □

5.4 Poincaré Duality Algebras

The Koszul complex can be used to derive information about the ring of coinvariants of finite pseudoreflection groups in the nonmodular case, or more generally, of representations whose invariants are a polynomial algebra. As is often the case, it is best to place this result in a setting more general than just that of invariant theory.

DEFINITION: *Let A be a commutative graded connected algebra over a field $\mathbb{F}$. We say that A is a* **Poincaré duality algebra** *of dimension n if*

(i) $A_i = 0$ *for* $i > n$,

(ii) $\dim_{\mathbb{F}}(A_n) = 1$,

(iii) the pairing $A_i \otimes_{\mathbb{F}} A_{n-i} \longrightarrow A_n$ given by multiplication is nonsingular, i.e., an element $a \in A_i$ is zero if and only if $a \cdot b = 0$ for all $b \in A_{n-i}$.

If A is a Poincaré duality algebra of dimension n, a nonzero element $[A]$ of A_n is referred to as a **fundamental class** *for A.*

The notion of Poincaré duality comes from the study of closed manifolds in algebraic topology, and goes back at least to H. Poincaré; see, e.g., [340] §69.

If A is a commutative graded connected algebra over a field $\mathbb{F}$ and $[A] \in A$ is an element of degree n, then it is easy to see that A is a Poincaré duality algebra of dimension n with fundamental class $[A]$ if and only if $A^i = 0$ for $i > n$ and

$$\mathrm{Ann}_A([A]) = \overline{A},$$

$$\mathrm{Ann}_A(\overline{A}) = \mathbb{F} \cdot [A],$$

where $\overline{A}$ is the augmentation ideal of A. A fundamental class is well-defined up to a nonzero multiple by an element of $\mathbb{F}$.

THEOREM 5.4.1: *Let $\mathbb{F}$ be a field, $f_1, \ldots, f_n \in \mathbb{F}[z_1, \ldots, z_n]$ a regular sequence of maximal length, and $A = \mathbb{F}[z_1, \ldots, z_n]/(f_1, \ldots, f_n)$. Write*

$f_i = \sum a_{i,j}z_j$, where $a_{i,j} \in \mathbb{F}[z_1, \ldots, z_n]$, and set $[A] = \det(a_{i,j})$. Then $[A] \in A$ is well-defined, and A is a Poincaré duality algebra with fundamental class $[A]$.

PROOF: A simple calculation with determinants shows that the element $[A] \in A$ is well-defined. Introduce the Koszul complexes

$$\mathcal{E} = \mathbb{F}[z_1, \ldots, z_n] \otimes E[sf_1, \ldots, sf_n] \quad \text{with } \partial(sf_i) = f_i,\ 1 \le i \le n,$$
$$\mathcal{K} = \mathbb{F}[z_1, \ldots, z_n] \otimes E[sz_1, \ldots, sz_n] \quad \text{with } \partial(sz_i) = z_i,\ 1 \le i \le n.$$

By Theorem 5.1.3 $\mathcal{E}$ is a free acyclic resolution of A as an $\mathbb{F}[z_1, \ldots, z_n]$-module, and $\mathcal{K}$ is a free acyclic resolution of $\mathbb{F}$ as an $\mathbb{F}[z_1, \ldots, z_n]$-module. Therefore, $\operatorname{Tor}^*_{\mathbb{F}[z_1,\ldots,z_n]}(A, \mathbb{F})$ is the homology of any of the three complexes

$$A \otimes_{\mathbb{F}[z_1,\ldots,z_n]} \mathcal{K}, \quad \mathcal{E} \otimes_{\mathbb{F}[z_1,\ldots,z_n]} \mathcal{K}, \quad \mathcal{E} \otimes_{\mathbb{F}[z_1,\ldots,z_n]} \mathbb{F},$$

and the augmentation maps provide chain homotopy equivalences between them:

$$A \otimes_{\mathbb{F}[z_1,\ldots,z_n]} \mathcal{K} \xleftarrow{\eta} \mathcal{E} \otimes_{\mathbb{F}[z_1,\ldots,z_n]} \mathcal{K} \xrightarrow{\varepsilon} \mathcal{E} \otimes_{\mathbb{F}[z_1,\ldots,z_n]} \mathbb{F}.$$

From the complex $\mathcal{E} \otimes_{\mathbb{F}[z_1,\ldots,z_n]} \mathbb{F}$ one readily sees that

❶ $$\operatorname{Tor}^*_{\mathbb{F}[z_1,\ldots,z_n]}(A, \mathbb{F}) \cong E[sf_1, \ldots, sf_n].$$

We next redo this computation using the complex $A \otimes_{\mathbb{F}[z_1,\ldots,z_n]} \mathcal{K}$. Let

$$\alpha_i = \sum_{j=1}^{n} a_{i,j} sz_j \in A \otimes_{\mathbb{F}[z_1,\ldots,z_n]} \mathcal{K},$$

and note that

$$\partial(\alpha_i) = f_i = \sum_{j=1}^{n} a_{i,j} z_j = 0 \in A,$$

so α_i is a cycle. Examining the chain homotopy equivalences η and ε we see that no linear combination of $\alpha_1, \ldots, \alpha_n$ is a boundary, and therefore $\alpha_1, \ldots, \alpha_n$ is a basis for $\operatorname{Tor}^1_{\mathbb{F}[z_1,\ldots,z_n]}(A, \mathbb{F})$. From ❶ it then follows that

❷ $$\operatorname{Tor}^*_{\mathbb{F}[z_1,\ldots,z_n]}(A, \mathbb{F}) \cong E[\alpha_1, \ldots, \alpha_n].$$

Next we examine $\operatorname{Tor}^n_{\mathbb{F}[z_1,\ldots,z_n]}(A, \mathbb{F})$ using this complex. A typical chain of homological degree n is $\zeta = a \otimes sz_1 \cdots sz_n$. The boundary of such a chain is

$$\partial(\zeta) = \sum (-1)^i a\bar{z}_i \otimes sz_1 \cdots \widehat{sz_i} \cdots sz_n,$$

where $\bar{z}_i$ denotes the residue class of $z_i \in A$ and $\widehat{\ }$ means that the term under the $\widehat{\ }$ is omitted. Since the elements $sz_1 \cdots \widehat{sz_i} \cdots sz_n$ are linearly independent, ζ is a cycle if and only if $a\bar{z}_i = 0$ for $i = 1, \ldots, n$, i.e., if and only if $a \in \operatorname{Ann}_A(\overline{A})$, so

❸ $$\operatorname{Tor}^n_{\mathbb{F}[z_1,\ldots,z_n]}(A, \mathbb{F}) \cong \operatorname{Ann}_A(\overline{A}).$$

But ❷ provides an alternative computation of $\mathrm{Tor}^n_{\mathbb{F}[z_1,\ldots,z_n]}(A, \mathbb{F})$, namely

❹ $$\mathrm{Tor}^n_{\mathbb{F}[z_1,\ldots,z_n]}(A, \mathbb{F}) \cong \mathrm{Span}_{\mathbb{F}}\{\det(a_{i,j})\},$$

since

$$\left(\sum a_{1,j}sz_j\right)\cdots\left(\sum a_{n,j}sz_j\right) = \det(a_{i,j}) \otimes sz_1\cdots sz_n$$

by the definition of the determinant. By combining ❸ and ❹ we get

❺ $$0 \neq \mathrm{Ann}_A(\overline{A}) = \mathrm{Span}_{\mathbb{F}}\{\det(a_{i,j})\}.$$

Next[3] apply Cramer's rule (see [159] for a short history of this rule) to the linear system in A,

$$\sum_{j=1}^{n} a_{i,j}x_j = 0, \qquad i = 1, \ldots, n,$$

to obtain

$$\mathrm{Ann}_A(\det(a_{i,j})) \supseteq (\overline{z}_1, \ldots, \overline{z}_n),$$

whence $\mathrm{Ann}_A(\det(a_{i,j})) \supseteq \overline{A}$, and since by ❺ $\det(a_{i,j}) \neq 0$, the annihilator ideal cannot be any larger.

To complete the proof note that any element $a \in A$ of maximal degree annihilates $\overline{A}$ for degree reasons. By ❺ it follows that $A^i = 0$ for $i > d = \deg([A])$ and that A^d is a one-dimensional vector space spanned by $[A]$. Moreover, if $a \in A_i$ for $0 < i < d$, then there must exist an element $b \in A$ with $a \cdot b \neq 0$. If b has maximal degree, then it follows that $\deg(b) = d - i$. □

COROLLARY 5.4.2: *Let $\mathbb{F}$ be a field, $f_1, \ldots, f_n \in \mathbb{F}[z_1, \ldots, z_n]$ a regular sequence of maximal length, and $A = \mathbb{F}[z_1, \ldots, z_n]/(f_1, \ldots, f_n)$. If $\deg(f_1)\cdots\deg(f_n)$ is relatively prime to the characteristic of $\mathbb{F}$, then A is a Poincaré duality algebra with fundamental class given by the Jacobian determinant $\det[\partial f_i/\partial z_j]$.*

---

[3] One way to think about this is as follows. If

$$A\begin{bmatrix} y_1 \\ \vdots \\ y_n \end{bmatrix} = \begin{bmatrix} 0 \\ \vdots \\ 0 \end{bmatrix},$$

then multiplying both sides by the cofactor matrix (see e.g., [374] page 242) of A we obtain

$$\begin{bmatrix} 0 \\ \vdots \\ 0 \end{bmatrix} = A^{\mathrm{cof}}A\begin{bmatrix} y_1 \\ \vdots \\ y_n \end{bmatrix} = \det(A)\begin{bmatrix} y_1 \\ \vdots \\ y_n \end{bmatrix}.$$

Therefore, $\det(A) \in \mathrm{Ann}(y_1, \ldots, y_n)$.

PROOF: By Euler's formula

$$\deg(f_i)f_i = \sum \frac{\partial f_i}{\partial z_j} z_j,$$

so the Jacobian determinant is a nonzero multiple of the fundamental class given by Theorem 5.4.1. □

These results yield our first structural result for $\mathbb{F}[V]_G$, namely if $\mathbb{F}[V]^G$ is a polynomial algebra then the algebra of coinvariants is a Poincaré duality algebra. The Jacobian determinant is a $\det^{-1}$-relative invariant, and hence so is the fundamental class of $\mathbb{F}[V]_G$ in the nonmodular case. This remains true in the modular case but requires a completely different proof, see Theorem 7.2.2.

Here is a small example to illustrate this circle of ideas.

EXAMPLE 1: Consider the representation $\rho : D_8 \hookrightarrow \mathrm{GL}(2, \mathbb{R})$ of the dihedral group D_8 of order 8 as the group of isometries of a square. The ring of invariants $\mathbb{R}[x, y]^{D_8}$ is $\mathbb{R}[x^2 + y^2, x^2y^2]$ (see Section 4.1, Example 2). Therefore, the ring of covariants is a Poincaré duality algebra having as fundamental class the form of degree 4

$$[\mathbb{R}[x, y]_{D_8}] = \det \begin{bmatrix} \frac{\partial(x^2+y^2)}{\partial x} & \frac{\partial(x^2+y^2)}{\partial y} \\ \frac{\partial(x^2y^2)}{\partial x} & \frac{\partial(x^2y^2)}{\partial y} \end{bmatrix} = 4(x^3y - xy^3).$$

The elements of $\mathbb{R}[x, y]_{D_8}$ satisfy numerous relations, e.g., $x^2 = -y^2$ and $x^4 = x^2y^2 = -y^4$. The ring $\mathbb{R}[x, y]$ is a free module over $\mathbb{R}[x^2 + y^2, x^2y^2] = \mathbb{R}[x, y]^{D_8}$, and therefore the Poincaré series of $\mathbb{R}[x, y]_{D_8}$ is given by

$$P(\mathbb{R}[x, y]_{D_8}, t) = \frac{(1 - t^2)(1 - t^4)}{(1 - t)^2} = 1 + 2t + 2t^2 + 2t^3 + t^4.$$

So the homogeneous components of $\mathbb{R}[x, y]_{D_8}$ are of dimensions 1, 2, 2, 2, and 1, and $\mathrm{Tot}(\mathbb{R}[x, y]_{D_8})$ has dimension 8. Diagram 5.4.1 is an aid to visualizing this ring of coinvariants. As in Section 1.2, Example 1, the nodes on a horizontal level indicate basis vectors for the elements of degree equal to the height of the node above the node labeled 1, which has degree 0. The dimension of the totalization $\mathrm{Tot}(\mathbb{R}[x, y]_{D_8})$ is 8, which is also the order of D_8. The representation of D_8 on $(\mathbb{R}[x, y]_{D_8})_1$ is of course ρ, whereas the representation on the fundamental class is isomorphic to the determinant representation. If we denote by ψ_i the representation of D_8 on the homoge-

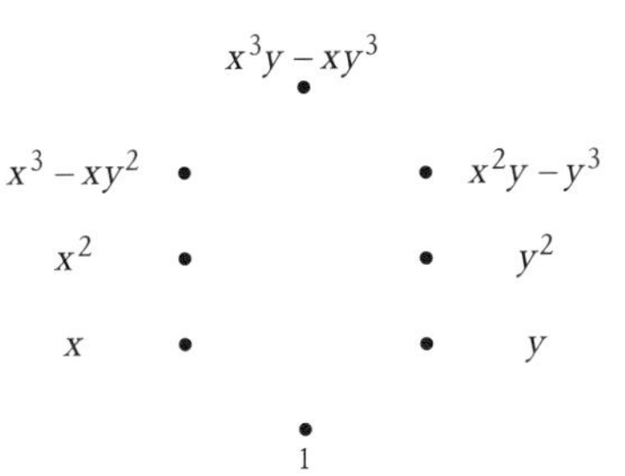

DIAGRAM 5.4.1: $\mathbb{F}[x, y]_{D_8}$

neous component of degree i in this algebra, then Poincaré duality implies that $\psi_1 \otimes \det \cong \psi_3$ and that $\psi_2 \otimes \det \cong \psi_2$. The representation of D_8 on $(\mathbb{R}[x, y]_{D_8})_2$ is the direct sum of two irreducible 1-dimensional representations: On the first node, labeled x, it is the lifted representation $\mathrm{rot}_{\mathbb{Z}/4}^{D_8}$ of $\mathbb{Z}/4$ by rotations on the square over the quotient map $D_8 \longrightarrow \mathbb{Z}/4$ obtained by dividing the normal subgroup $\mathbb{Z}/2$ generated by reflection in the y-axis. On the second node, labeled y^2, it acts via $\mathrm{rot}_{\mathbb{Z}/4}^{D_8} \otimes \det$. Therefore, we have the following table for the representations of D_8 on the homogeneous components of $\mathbb{R}[x, y]_{D_8}$:

i	0	1	2	3	4
ψ_i	1	ρ	$\mathrm{rot}_{\mathbb{Z}/4}^{D_8} \oplus \left(\mathrm{rot}_{\mathbb{Z}/4}^{D_8} \otimes \det\right)$	$\rho \otimes \det$	$\det$

Note that $\rho \otimes \det = \rho$, so that this is a complete list of the irreducible representations of D_8, each of which occurs as often as its dimension; see, e.g., [183] Section 18.3 or [345] Section 5.3. Hence $\mathrm{Tot}\left(\mathbb{R}[x, y]_{D_8}\right)$ is isomorphic to the regular representation of D_8.

Since something similar happened in Example 1 of Section 1.2, it is not surprising that this is no accident, rather an example of a general phenomenon, which will first become apparent when we discuss the invariants of finite pseudoreflection groups in Chapter 7.

If the ring of invariants of $G \hookrightarrow \mathrm{GL}(n, \mathbb{F})$ is a polynomial algebra, say $\mathbb{F}[f_1, \ldots, f_n] = \mathbb{F}[V]^G \subseteq \mathbb{F}[V]$, but the order of the group G is divisible by the characteristic of the ground field $\mathbb{F}$, then the Jacobian determinant of $f_1, \ldots, f_n$ may fail to be a fundamental class in $\mathbb{F}[V]_G$, as the following example illustrates. (In fact, it must fail in this case, as we will prove in Theorem 7.2.8.)

EXAMPLE 2: Consider the representation of $\mathbb{Z}/p$, $p \in \mathbb{N}_0$ a prime, implemented by the matrix

$$\mathbf{T} = \begin{bmatrix} 1 & 1 \\ 0 & 1 \end{bmatrix} \in \mathrm{GL}(n, \mathbb{F}),$$

where $\mathbb{F}$ is a field of characteristic p. The action of $\mathbf{T}$ on the canonical basis for the dual vector space is given by

$$\mathbf{T}(x) = x + y, \qquad \mathbf{T}(y) = y,$$

and the ring of invariants, $\mathbb{F}[x, y]^{\mathbb{Z}/p}$, is a polynomial algebra with generators $x(x^{p-1} - y^{p-1}) = c_{\mathrm{top}}(x)$ and y, as may be verified by Theorem 4.5.5. The Jacobian matrix of the generators is

$$\begin{bmatrix} -y^{p-1} & xy^{p-2} \\ 0 & 1 \end{bmatrix},$$

so the Jacobian determinant is $-y^{p-1}$, and is zero in $\mathbb{F}[x, y]_{\mathbb{Z}/p} = \mathbb{F}[x]/(x^p)$.

We will return to rings of coinvariants of this sort in Section 7.2.

5.5 The Cohen–Macaulay Property

For a representation $\rho : G \hookrightarrow \mathrm{GL}(n, \mathbb{F})$ with ring of invariants $\mathbb{F}[V]^G$, the Noether normalization theorem tells us that $\mathbb{F}[V]^G$ is a finite extension of a polynomial algebra $\mathsf{N} = \mathbb{F}[h_1, \ldots, h_n] \subset \mathbb{F}[V]^G$. Problem 4 in Section 1.4 asks, what kind of a module is $\mathbb{F}[V]^G$ as N-module? For example, is it free? By Theorem 5.1.3 this is the case if and only if $h_1, \ldots, h_n \in \mathbb{F}[V]^G$ is a regular sequence. This problem can be put in a more general context.

The least upper bound of the lengths of regular sequences in a graded connected algebra A over a field $\mathbb{F}$ is a measure of finiteness called the **homological codimension** of A and denoted by hom-codim(A). By Proposition 5.1.2 codim$(A) \leq$ dim(A). (The codimension is sometimes referred to in the literature as the **depth** of the algebra.) If A is a commutative graded connected algebra over a field $\mathbb{F}$, we say that a regular sequence $a_1, \ldots, a_r \in A$ is maximal if it cannot be extended to a regular sequence $a_1, \ldots, a_r, a_{r+1}$ in A with one more element. Although it is not clear that two maximal regular sequences contain the same number of elements, i.e., have the same length, this is the case for graded connected algebras, and a proof may be found for example in [372] Chapter 6, Section 2.

DEFINITION: *Let A be a commutative graded connected $\mathbb{F}$-algebra. Then A is called* **Cohen–Macaulay** *if* hom-codim(A) = dim(A).

THEOREM 5.5.1: *Let A be a commutative graded connected $\mathbb{F}$-algebra and $\mathsf{N} = \mathbb{F}[a_1, \ldots, a_n] \subseteq A$ a Noether normalization of A. Then the following conditions are equivalent:*

(1) *A is Cohen–Macaulay.*
(2) *$a_1, \ldots, a_n \in A$ is a regular sequence.*
(3) *A is a free finitely generated N-module.*
(4) *The Poincaré series of A has the form*

$$P(A, t) = P(\mathsf{N}, t) \cdot P(A//\mathsf{N}, t),$$

where $A//\mathsf{N} = \mathbb{F} \otimes_{\mathsf{N}} A \cong A/(a_1, \ldots, a_n)$.

PROOF: Since we are in a graded connected situation, the properties of A-modules free, projective, and flat coincide (see Proposition A.1.5 in Appendix A.1 for a proper statement, and for example [372] Theorem 6.1.1 for a proof). This yields the equivalence of (2) and (3), whereas the equivalence of (1) with (2), and hence also (3) follows from F. S. Macaulay's theorem (see e.g., [25] Theorem 3.3.5 or [372] Corollary 6.7.7).

Finally we consider condition (4). Since $A//\mathsf{N}$ contains a set of module generators for A over N, there is certainly an epimorphism

$$\varphi : \mathsf{N} \otimes A//\mathsf{N} \longrightarrow A$$

of $\mathbf{N}$-modules. Since the source and the target of φ are of finite type, the map φ is an isomorphism if and only if

$$P(A, t) = P(\mathbf{N}, t) \cdot P(A//\mathbf{N}, t).$$

If this is the case, then A is free as an $\mathbf{N}$-module with basis a vector space basis for $A//\mathbf{N}$. □

Notice that condition (4) of Theorem 5.5.1 says that the Poincaré series of a Cohen-Macaulay algebra has a special form, namely

$$P(A, t) = \frac{p(t)}{\prod_{i=1}^{n} 1 - t^{d_i}},$$

where $n = \dim(A)$, $d_1, \ldots, d_n$ are the degrees of a system of parameters for A, and $p(t)$ is a polynomial with *nonnegative* integral coefficients.

THEOREM 5.5.2 (J. A. Eagon-M. Hochster): *Let $\rho : G \hookrightarrow \mathrm{GL}(n, \mathbb{F})$ be a representation of a finite group G over a field $\mathbb{F}$. If $|G|$ is prime to the characteristic of $\mathbb{F}$, then $\mathbb{F}[V]^G$ is a Cohen-Macaulay algebra.*

PROOF: Let $f_1, \ldots, f_n \in \mathbb{F}[V]^G$ be a system of parameters. Then

$$\mathbb{F}[f_1, \ldots, f_n] \subseteq \mathbb{F}[V]^G \subseteq \mathbb{F}[V]$$

are finite extensions, so $f_1, \ldots, f_n \in \mathbb{F}[V]$ is a system of parameters. $\mathbb{F}[V]$ is Cohen-Macaulay so by Macaulay's theorem, Theorem A.3.5, $f_1, \ldots, f_n \in \mathbb{F}[V]$ is a regular sequence. Hence by Theorem 5.1.3 $\mathbb{F}[V]$ is a free $\mathbb{F}[f_1, \ldots, f_n]$-module. The projection

$$\pi^G : \mathbb{F}[V] \longrightarrow \mathbb{F}[V]^G$$

derived from the transfer (see Section 2.2) represents $\mathbb{F}[V]^G$ as an $\mathbb{F}[V]^G$-module, and a fortiori as an $\mathbb{F}[f_1, \ldots, f_n]$-module direct summand in $\mathbb{F}[V]$. Hence $\mathbb{F}[V]^G$ is a free $\mathbb{F}[f_1, \ldots, f_n]$-module, so by Theorem 5.1.3 $f_1, \ldots, f_n \in \mathbb{F}[V]^G$ is a regular sequence. □

Rings of invariants in two variables are Cohen-Macaulay by Proposition 5.1.1. Here is a slight generalization of this.

PROPOSITION 5.5.3: *Let $\rho : G \hookrightarrow \mathrm{GL}(n, \mathbb{F})$ be a representation of a finite group over the field $\mathbb{F}$. If $n \geq 2$ then* $\text{hom-codim}(\mathbb{F}[V]^G) \geq 2$ *also.*

PROOF: Without loss of generality we may suppose that $\mathbb{F}$ is algebraically closed so there is a Dade basis for V^*, say $z_1, \ldots, z_n$. The top Chern classes $c_{\text{top}}(z_1), \ldots, c_{\text{top}}(z_n)$ are a system of parameters and hence also a regular sequence. By Proposition 5.1.1 any two of them are also a regular sequence in $\mathbb{F}[V]^G$. □

EXAMPLE 1 (I. Böcker [38]): The **generalized quaternion group** Q_{4k} of order $4k$ (See page 253 in [68]) also called the **di-cyclic group** in [80] and much of the classical literature) is defined by two generators $\mathbf{S}, \mathbf{T} \in Q_{4k}$ and two relations

$$\mathbf{S}^k = \mathbf{T}^2, \qquad \mathbf{STS} = \mathbf{T}.$$

By iterating the second relation we obtain $\mathbf{S}^k\mathbf{T}\mathbf{S}^k = \mathbf{T}$, from which it follows that $\mathbf{T}^4 = \mathbf{I} = \mathbf{S}^{2k}$, so a complete list, without repetition, of the elements of Q_{4k} consists of

$$\left\{\mathbf{S}^{\sigma}\mathbf{T}^{\delta} \mid \sigma = 0, \ldots, 2k-1, \delta = 0, 1\right\}.$$

Note for $k = 1$ we have $Q_4 \cong \mathbb{Z}/4$ and for $k = 2$ the group Q_8 is the usual quaternion group.

There is a faithful two-dimensional representation $\rho : Q_{4k} \hookrightarrow \mathrm{GL}(2, \mathbb{C})$ given by

$$\rho(\mathbf{S}) = \begin{bmatrix} \lambda & 0 \\ 0 & \lambda^{-1} \end{bmatrix} \quad \text{and} \quad \rho(\mathbf{T}) = \begin{bmatrix} 0 & -1 \\ 1 & 0 \end{bmatrix},$$

where $\lambda = \exp(\pi \boldsymbol{i}/k)$. Note that $\lambda^k = -1$ and $\lambda^{2k} = 1$. In fact, if we identify $\mathbb{C}^2$ with $\mathbb{H}$, the quaternions, then Q_{4k} may be identified with the subgroup of the unit quaternions generated by

$$\exp\left(\frac{\pi \boldsymbol{i}}{k}\right), \quad \boldsymbol{j},$$

and ρ with the representation induced by left multiplication of Q_{4k} on $\mathbb{C}^2$ regarded as the underlying complex vector space of $\mathbb{H}$.

We begin our study of $\mathbb{C}[x, y]^{Q_{4k}}$ by writing down some invariants. Since $-\mathbf{I} \in Q_{4k}$, the invariants have even degrees, because every polynomial $f(x, y)$ of odd degree is mapped by $-\mathbf{I}$ to $-f(x, y)$. Using the formulae

$$\begin{aligned} \mathbf{S}(x) &= \lambda x, & \mathbf{T}(x) &= -y, \\ \mathbf{S}(y) &= \lambda^{-1} y, & \mathbf{T}(y) &= x, \end{aligned}$$

one readily verifies that

$$\mathbf{S}(x^2y^2) = \lambda^2 x^2 \lambda^{-2} y^2 = x^2y^2,$$
$$\mathbf{T}(x^2y^2) = (-y)^2(x)^2 = x^2y^2,$$

so $f_1 = x^2y^2$ is an invariant of degree 4. Experimenting with small values of k leads to the discovery of further invariants:

$$\mathbf{S}(x^{2k} + y^{2k}) = \lambda^{2k}x^{2k} + \lambda^{-2k}y^{2k} = x^{2k} + y^{2k},$$
$$\mathbf{T}(x^{2k} + y^{2k}) = (-y)^{2k} + x^{2k} = x^{2k} + y^{2k},$$

and

$$\mathbf{S}(x^{2k+1}y - xy^{2k+1}) = \lambda^{2k+1}x^{2k+1}\lambda^{-1}y - \lambda x\lambda^{-(2k+1)}y^{2k+1} = x^{2k+1}y - xy^{2k+1},$$
$$\mathbf{T}(x^{2k+1}y - xy^{2k+1}) = (-y)^{2k+1}x + yx^{2k+1} = x^{2k+1}y - xy^{2k+1}.$$

Set $f_2 = x^{2k} + y^{2k}$ and $h = x^{2k+1}y - xy^{2k+1}$. Then $h^2 = f_1 f_2^2 - 4f_1^{k+1}$.

We next apply T. Molien's theorem (Theorem 3.1.3) to compute the Poincaré series. To this end note that

$$\det(1 - \mathbf{S}^m t) = \det\begin{bmatrix} 1-\lambda^m t & 0 \\ 0 & 1-\lambda^{-m}t \end{bmatrix} = \frac{1}{(1-\lambda^m t)(1-\lambda^{-m}t)}$$

and

$$\det(1 - \mathbf{S}^m\mathbf{T}t) = \det\begin{bmatrix} 1 & \lambda^m t \\ -\lambda^{-m}t & 1 \end{bmatrix} = 1 + t^2.$$

Molien's Theorem 3.1.3 then gives

$$P(\mathbb{C}[x,y]^{Q_{4k}}, t) = \frac{1}{4k}\left[\frac{2k}{1+t^2} + \sum_{m=0}^{2k-1} \frac{1}{(1-\lambda^m t)(1-\lambda^{-m}t)}\right].$$

The second term in the formula is one we evaluated in Section 3.1, Example 2, and substituting from the formula there and simplifying further yields

$$\begin{aligned}
P(\mathbb{C}[x,y]^{Q_{4k}}, t) &= \frac{1}{4k}\left[\frac{2k}{1+t^2} + 2k\frac{1+t^2+\cdots+t^{4k-2}}{(1-t^{2k})^2}\right] \\
&= \frac{1}{2}\left[\frac{1}{1+t^2} + \frac{1+t^2+\cdots+t^{4k-2}}{(1-t^{2k})^2}\right] \\
&= \frac{1}{2}\left[\frac{(1-t^{2k})^2 + (1+t^2)(1+t^2+\cdots+t^{4k-2})}{(1+t^2)(1-t^{2k})^2}\right] \\
&= \frac{1}{2}\left[\frac{1-2t^{2k}+t^{4k}+1+2t^2+\cdots+2t^{4k-2}+t^{4k}}{(1+t^2)(1-t^{2k})^2}\right] \\
&= \frac{1}{2}\left[\frac{2+2t^2+\cdots+\widehat{2t^{2k}}+\cdots+2t^{4k}}{(1+t^2)(1-t^{2k})^2}\right] \\
&= \frac{1+t^2+\cdots+\widehat{t^{2k}}+\cdots+t^{4k}}{(1+t^2)(1-t^{2k})^2}.
\end{aligned}$$

Since $\deg(f_1)$ and $\deg(f_2)$ are even, while $\deg(h)$ is odd, a short computation shows

$$\mathbb{C}[f_1, f_2] \oplus \mathbb{C}[f_1, f_2]\cdot h \subseteq \mathbb{C}[x,y]^{Q_{4k}},$$

and suggest further simplifications of the Poincaré series

$$\begin{aligned}
P(\mathbb{C}[x,y]^{Q_{4k}},t) &= \frac{(1+t^2+\cdots+t^{2k-2})+(t^{2k+2}+\cdots+t^{4k})}{(1+t^2)(1-t^{2k})^2} \\
&= \frac{(1-t^2)(1+\cdots+t^{2k-2})+(1-t^2)(t^{2k+2}+\cdots+t^{4k})}{(1-t^2)(1+t^2)(1-t^{2k})^2} \\
&= \frac{(1-t^{2k})+t^{2k+2}(1-t^{2k})}{(1-t^4)(1-t^{2k})^2} \\
&= \frac{1+t^{2k+2}}{(1-t^4)(1-t^{2k})}.
\end{aligned}$$

Since

$$P(\mathbb{C}[f_1,f_2]\oplus\mathbb{C}[f_1,f_2]\cdot h,t) = \frac{1+t^{2k+2}}{(1-t^4)(1-t^{2k})}$$

as well, we conclude that

$$\mathbb{C}[x,y]^{Q_{4k}} = \mathbb{C}[f_1,f_2]\oplus\mathbb{C}[f_1,f_2]\cdot h.$$

The elements f_1, f_2 are a system of parameters, and $\{1, h\}$ is a basis for $\mathbb{C}[x,y]^{Q_{4k}}$ over $\mathbb{C}[f_1,f_2]$. There is one syzygy provided by the relation $h^2 = f_1(f_2{}^2 - 4f_1{}^k)$.

In the modular case the ring of invariants $\mathbb{F}[V]^G$ may fail to be Cohen-Macaulay. Since rings of invariants in three or fewer variables are Cohen-Macaulay [369] (see Proposition 5.6.10), the lowest-dimensional example is M.-J. Bertin's [97]: The regular representation of $\mathbb{Z}/4$ over a field of characteristic 2. We postpone a discussion of this example to Section 5.8. (See, however, Section 1.1, Example 3, and Section 2.4, Example 1.)

The following example appeared in [54]: The treatment here comes from [297] (see [373] Section 4 Example 2), as the computations there is easier for us to explain.

EXAMPLE 2: Consider the representation $\sigma_3 : \mathbb{Z}/2 \hookrightarrow \mathrm{GL}(6,\mathbb{F})$ given by simultaneous permutation of x_1, x_2, x_3 with y_1, y_2, y_3, where $\mathbb{F}$ is a field of characteristic 2. We have looked at this example a number of times already, to wit in Section 1.6, Example 1, Section 2.3, Example 2, Section 2.4, Example 2, and Section 3.2, Example 1. The polynomials

$$\left.\begin{array}{ll} \ell_i = x_i + y_i, & i = 1,2,3 \\ q_j = x_j y_j, & j = 1,2,3 \end{array}\right\} \in \mathbb{F}[x_1,x_2,x_3,y_1,y_2,y_3]^{\mathbb{Z}/2}$$

form a system of parameters: ℓ_i, q_i are the elementary symmetric functions in x_i, y_i for $i = 1, 2, 3$, and so together are a system of parameters for $\mathbb{F}[x_1,x_2,x_3,y_1,y_2,y_3]$. If the ring of invariants $\mathbb{F}[x_1,x_2,x_3,y_1,y_2,y_3]^{\mathbb{Z}/2}$ were Cohen-Macaulay, then these polynomials would have to be a regular sequence by F. S. Macaulay's theorem; see Theorem A.3.5. But the relation

(compare Section 2.3, Example 2)

$$\ell_1 Q_1 = \ell_2 Q_2 + \ell_3 Q_3 + \ell_1 \ell_2 \ell_3 \in \mathbb{F}[x_1, x_2, x_3, y_1, y_2, y_3]^{\mathbb{Z}/2}$$

shows that $\ell_1 Q_1 \in (\ell_2, \ell_3) \subset \mathbb{F}[x_1, x_2, x_3, y_1, y_2, y_3]^{\mathbb{Z}/2}$. Since the quadratic form Q_1 does not belong to the ideal in $\mathbb{F}[x_1, x_2, x_3, y_1, y_2, y_3]^{\mathbb{Z}/2}$ generated by ℓ_2 and ℓ_3 we conclude that ℓ_1 is a zero divisor modulo (ℓ_2, ℓ_3). Therefore the linear forms ℓ_1, ℓ_2, ℓ_3 do not form a regular sequence in the ring of invariants $\mathbb{F}[x_1, x_2, x_3, y_1, y_2, y_3]^{\mathbb{Z}/2}$, which means that $\mathbb{F}[x_1, x_2, x_3, y_1, y_2, y_3]^{\mathbb{Z}/2}$ cannot be Cohen-Macaulay.

So in the modular case rings of invariants may, or may not, be Cohen-Macaulay. If they are, then A. Broer [43] found a bound for $\beta(\mathbb{F}[V]^G)$ which we explain next. For our proof we require an elementary lemma about integral extensions. For the sake of completness we include the short proof.

LEMMA 5.5.4: *Let $A \subseteq B$ be a integral extension of commutative integral domains. If $I \subseteq B$ is a nonzero ideal then so is $I \cap A \subseteq A$.*

PROOF: Let $0 \neq b \in I$. Since B is integral over A there exists an integer k and elements $a_1, \ldots, a_k \in A$ such that

$$b^k + a_1 b^{k-1} + \cdots + a_{k-1} b + a_k = 0.$$

If in addition we choose k minimal with this property then $a_k \neq 0$ since B is a domain. Solve such an equation for a_k, viz.,

$$a_k = -(b^k + a_1 b^{k-1} + \cdots + a_{k-1} b).$$

The right hand side of this equation is in I and the left hand side is in A. Therefore $0 \neq a_k \in I \cap A$. □

If A is a graded connected algebra over the field $\mathbb{F}$ and M is finitely generated as an A-module then M cannot be generated by fewer than $\dim_{\mathbb{F}}(\mathrm{Tot}(Q_A(M)))$ elements, and the integers (see Section A.1)

$$\sigma_A^+(M) = \max\{k \mid (Q_A(M))_k \neq 0\}$$
$$\sigma_A^-(M) = \min\{k \mid (Q_A(M))_k \neq 0\}$$

are the maximum and minimum degrees of generators in a minimal generating set.

THEOREM 5.5.5 (A. Broer): *Let $\rho: G \hookrightarrow GL(n, \mathbb{F})$ be a representation of a finite group G over the field $\mathbb{F}$ and $\mathbf{N} = \mathbb{F}[h_1, \ldots, h_n] \subseteq \mathbb{F}[V]^G$ a Noether normalization. If $\mathbb{F}[V]^G$ is Cohen-Macaulay then $\mathbb{F}[V]^G$ is generated as an $\mathbf{N}$-module by elements of degree at most*

$$\deg(h_1) + \cdots + \deg(h_n) - n.$$

PROOF: Let $d = \sigma^+_{\mathsf{N}}(\mathbb{F}[V]^G)$. Since $\mathbb{F}[V]^G$ is Cohen-Macaulay it is a free N-module, so it follows from Lemma A.1.4 that there is a nonzero homomorphism of N-modules $\varphi : \mathbb{F}[V]^G \longrightarrow \mathsf{N}$ of degree $-d$. Consider the composition

$$\varphi \circ \mathrm{Tr}^G : \mathbb{F}[V] \xrightarrow{\mathrm{Tr}^G} \mathbb{F}[V]^G \xrightarrow{\varphi} \mathsf{N}.$$

The ideal $\mathrm{Im}(\mathrm{Tr}^G) \subset \mathbb{F}[V]^G$ is nonzero, so by Lemma 5.5.4 there is a nonzero element $h \in \mathsf{N} \cap \mathrm{Im}(\mathrm{Tr}^G)$. Let $h = \mathrm{Tr}^G(f)$. Since φ is nonzero there is also an element $F \in \mathbb{F}[V]^G$ such that $\varphi(F) \neq 0$. Then using that φ is an N-module homomorphism and Tr^G an $\mathbb{F}[V]^G$-module homomorphism we obtain

$$(\varphi \circ \mathrm{Tr}^G)(f \cdot F) = \varphi(\mathrm{Tr}^G(f \cdot F)) = \varphi(\mathrm{Tr}^G(f) \cdot F) = \varphi(h \cdot F) = h \cdot \varphi(F) \neq 0$$

since there are no zero divisors in N. Therefore $\varphi \circ \mathrm{Tr}^G : \mathbb{F}[V] \longrightarrow \mathsf{N}$ is a nonzero N-module homomorphism. By Theorem 5.5.1 the Poincaré series of $Q_{\mathsf{N}}(\mathbb{F}[V])$ is

$$P(Q_{\mathsf{N}}(\mathbb{F}[V]), t) = \frac{(1 - t^{\deg(h_1)}) \cdots (1 - t^{\deg(h_n)})}{(1-t)^n} = \prod_{i=1}^{n} (1 + t + \cdots + t^{\deg(h_i)-1})$$

and has degree

$$\deg(h_1) + \cdots + \deg(h_n) - n.$$

So by Lemma A.1.4

$$-d \geq -(\deg(h_1) + \cdots + \deg(h_n) - n)$$

and multiplying by -1 gives the desired conclusion. □

If $\mathsf{N} \subseteq \mathbb{F}[V]^G$ is a Noether normalization, then module generators for $\mathbb{F}[V]^G$ over N together with algebra generators for N will generate $\mathbb{F}[V]^G$ as an algebra. Therefore we obtain the following bounds for $\beta(\mathbb{F}[V]^G)$ in the Cohen-Macaulay case.

COROLLARY 5.5.6 (A. Broer): *Let $\rho : G \hookrightarrow \mathrm{GL}(n, \mathbb{F})$ be a representation of a finite group G over the field $\mathbb{F}$ and $\mathsf{N} = \mathbb{F}[h_1, \ldots, h_n] \subseteq \mathbb{F}[V]^G$ a Noether normalization. If $\mathbb{F}[V]^G$ is Cohen-Macaulay, then $\mathbb{F}[V]^G$ is generated as an algebra by forms of degree at most*

$$\max\{\deg(h_1), \ldots, \deg(h_n), \deg(h_1) + \cdots + \deg(h_n) - n\}.$$

PROOF: This is immediate from Theorem 5.5.5. □

COROLLARY 5.5.7 (A. Broer): *Let $\rho : G \hookrightarrow \mathrm{GL}(n, \mathbb{F})$ be a representation of a finite group G over the field $\mathbb{F}$ and $\mathsf{N} = \mathbb{F}[h_1, \ldots, h_n] \subseteq \mathbb{F}[V]^G$ a Noether normalization. Let $d = \dim_{\mathbb{F}}((V^*)^G)$ and $c = n - d$. If $\mathbb{F}[V]^G$ is Cohen-Macaulay then $\mathbb{F}[V]^G$ is generated as an algebra by forms of degree* *at most*

$$\max\{|G|, c(|G|-1)\}.$$

PROOF: There is no loss in generality in assuming that $\mathbb{F}$ is algebraically closed. Then a basis $u_1, \ldots, u_d$ for $(V^*)^G$ extends to a Dade basis $u_1, \ldots, u_d, w_1, \ldots, w_c$ for V^* (see Section 4.3). The top Chern classes $c_{\mathrm{top}}(u_1), \ldots, c_{\mathrm{top}}(u_d), c_{\mathrm{top}}(w_1), \ldots, c_{\mathrm{top}}(w_c)$ form a system of parameters for $\mathbb{F}[V]^G$. Let G_w denote the isotropy group of $w \in V^*$. Then $c_{\mathrm{top}}(u_1), \ldots, c_{\mathrm{top}}(u_d), c_{\mathrm{top}}(w_1)^{|G:G_{w_1}|}, \ldots, c_{\mathrm{top}}(w_c)^{|G:G_{w_c}|}$ is also a system of parameters for $\mathbb{F}[V]^G$ and therefore we have the Noether normalization

$$\mathsf{N} = \mathbb{F}[c_{\mathrm{top}}(u_1), \ldots, c_{\mathrm{top}}(u_d), c_{\mathrm{top}}(w_1)^{|G:G_{w_1}|}, \ldots, c_{\mathrm{top}}(w_c)^{|G:G_{w_c}|}] \subseteq \mathbb{F}[V]^G.$$

Since $\deg(c_{\mathrm{top}}(u_i)) = 1$, for $i = 1, \ldots, d$, and $\deg(c_{\mathrm{top}}(w_j))^{|G:G_{w_j}|} = |G|$ for $j = 1, \ldots, c$ the result follows from Theorem 5.5.5. □

The following proposition will be of use when we study the relation between the invariants of the orthogonal group of a finite field and its rotation subgroup. (See Section 7.1, Example 1, and Section 8.1, Example 2.)

PROPOSITION 5.5.8: *Let $\rho : G \hookrightarrow \mathrm{GL}(n, \mathbb{F})$ be a representation of a finite group G over the field $\mathbb{F}$ and $h_1, \ldots, h_n \in \mathbb{F}[V]^G$ a system of parameters. If $\mathbb{F}[V]^G$ is Cohen-Macaulay and*

$$\prod_{i=1}^{n} \deg(h_i) = 2 \cdot |G|,$$

then

$$\mathbb{F}[V]^G = \mathbb{F}[h_1, \ldots, h_n] \oplus \mathbb{F}[h_1, \ldots, h_n] \cdot h,$$

where h is an element of minimal degree in $\mathbb{F}[V]^G$ not in $\mathbb{F}[h_1, \ldots, h_n]$. In other words, $\mathbb{F}[V]^G$ is a hypersurface.

PROOF: Since $\mathbb{F}[V]^G$ is Cohen-Macaulay, its Poincaré series is of the form

$$P(\mathbb{F}[V]^G, t) = \frac{1 + a_1 t + \cdots + a_k t^k}{(1 - t^{\deg(h_1)}) \cdots (1 - t^{\deg(h_n)})},$$

where $a_1, \ldots, a_k \in \mathbb{N}_0$ and $a_k \neq 0$. Multiply this expression by $(1-t)^n$, evaluate the result at $t = 1$, and use the degree theorem (Theorem 4.5.3): The result is

$$\begin{aligned}
\frac{1}{|G|} &= (1-t)^n P(\mathbb{F}[V]^G, t)\Big|_{t=1} \\
&= \frac{1 + a_1 t + \cdots + a_k t^k}{(1 + t + \cdots + t^{\deg(h_1)-1}) \cdots (1 + t + \cdots + t^{\deg(h_n)-1})}\Big|_{t=1} \\
&= \frac{1 + a_1 + \cdots + a_k}{2|G|}.
\end{aligned}$$

Therefore, $a_1 = \cdots = a_{k-1} = 0$ and $a_k = 1$, and the result follows. □

For further information about the Cohen–Macaulay property of modular invariant rings consult the detailed study of G. Kemper [196] and the references there.

5.6 Homological and Cohomological Dimensions

We begin with a short review of some basic facts concerning zero divisors and regular elements on a module in graded form as a preliminary to introducing the homological invariants we need in this section. We adopt[4] the definition that a prime ideal $\mathfrak{p} \subset A$ is an **associated prime of an** A**-module** M if there exists a nonzero element $x_{\mathfrak{p}} \in M$ whose annihilator ideal is exactly $\mathfrak{p}$. Then the set $\mathrm{Ass}_A(M)$ of associated prime ideals of M is nonempty (see Lemma A.4.2).

DEFINITION: *Let A be a commutative graded connected algebra over the field $\mathbb{F}$ and M an A-module. An element $a \in A$ is called a* **zero divisor on** *M if there is a nonzero element $x \in M$ such that $a \in \mathrm{Ann}_A(x)$. Otherwise, a is called* **regular on** *M. The set of regular elements of A on M is denoted by $\mathcal{Reg}_A(M)$, and $\mathcal{Zero}_A(M)$ denotes the set of zero divisors in A on M.*

LEMMA 5.6.1: *Let A be a commutative graded connected Noetherian algebra over the field $\mathbb{F}$ and M an A-module. Then*

$$\mathcal{Zero}_A(M) = \bigcup_{\mathfrak{p} \in \mathrm{Ass}_A(M)} \mathfrak{p}.$$

PROOF: Clearly the union of the associated primes of M is a subset of $\mathcal{Zero}_A(M)$. On the other hand, if $a \in \mathcal{Zero}_A(M)$, choose $0 \neq x \in M$ with $a \in \mathrm{Ann}_A(x)$. Hence a belongs to a maximal element of $\mathrm{Ann}_A(M)$, but as was shown in the proof of Lemma A.4.2 such maximal elements are prime, so $a \in \mathfrak{p} \in \mathrm{Ass}_A(M)$. □

LEMMA 5.6.2: *Let A and B be Noetherian commutative graded connected algebras over the field $\mathbb{F}$ and $\varphi : A \longrightarrow B$ a homomorphism of graded algebras. Let M be a B-module that is finitely generated as an A-module. Then $\mathcal{Zero}_A(M) = \overline{A}$ if and only if $\mathcal{Zero}_B(M) = \overline{B}$, where $\overline{A}, \overline{B}$ denote the augmentation ideals of A and B respectively.*

PROOF: If every element of $\overline{B}$ is a zero divisor on M, then clearly so is every element of $\overline{A}$. Conversely, if $\overline{A} = \mathcal{Zero}_A(M)$, then by Lemma 5.6.1

$$\overline{A} = \mathcal{Zero}_A(M) = \bigcup_{\mathfrak{p} \in \mathrm{Ass}_A(M)} \mathfrak{p}.$$

[4] This is not the only definition possible or in use; see, e.g., [465].

So by prime avoidance (see Lemma A.2.1) we have $\overline{A} \subseteq \mathfrak{p}$ for some $\mathfrak{p} \in \mathrm{Ass}_A(M)$. Since $\overline{A}$ is maximal, we have equality, i.e., $\overline{A} \in \mathrm{Ass}_A(M)$. Choose $x \in M$ with $\overline{A} = \mathrm{Ann}_A(x)$ and let N be the B-submodule of M generated by x. Since A is Noetherian and M is a finitely generated A-module, so is N. By Lemma A.4.2 $\mathrm{Ass}_B(N) \neq \emptyset$. Suppose $\mathfrak{q} \in \mathrm{Ass}_B(N)$. Then $\mathfrak{q} = \mathrm{Ann}_B(bx)$ for some $b \in B$, and the B-submodule of N generated by bx is isomorphic with a degree shift to $B/\mathfrak{q}$. It, too, is a finitely generated A-module. However, note that

$$a(bx) = b(ax) = 0, \qquad \forall\, a \in \overline{A},$$

so $B/\mathfrak{q}$ is a finitely generated $A/\overline{A} = \mathbb{F}$-module. In other words, $B/\mathfrak{q}$ is a finite-dimensional graded vector space over $\mathbb{F}$. Since $\mathfrak{q} \subset B$ is a prime ideal, the algebra $B/\mathfrak{q}$ contains no zero divisors. The only finite-dimensional graded commutative algebra over $\mathbb{F}$ with no zero divisors is $\mathbb{F}$ itself. Hence $\mathfrak{q} = \overline{B}$, and therefore $\overline{B} = \mathcal{Z}ero_B(M)$, as was to be shown. □

The preceding lemma can be reformulated for regular elements instead of zero divisors as follows.

LEMMA 5.6.3: *Let A and B be Noetherian commutative graded connected algebras over the field $\mathbb{F}$ and $\varphi : A \longrightarrow B$ a homomorphism of graded algebras. Let M be a B-module that is finitely generated as an A-module. Then B contains a regular element on M if and only if A does.* □

Lemma 5.6.1 also has the following consequence for regular elements.

LEMMA 5.6.4: *Let A be a commutative graded connected Noetherian algebra over the field $\mathbb{F}$ and M an A-module. Then $\mathcal{R}eg_A(M)$ is nonempty if and only if* $\mathrm{Hom}_A^*(\mathbb{F}, M) = 0$, *where* $\mathrm{Hom}_A^*(-, -)$ *denotes the graded* Hom-*functor (see Section* A.1*).*

PROOF: Suppose that $\varphi : \mathbb{F} \longrightarrow M$ is a nonzero homomorphism of A-modules of degree d. If $x = \varphi(1) \in M_d$ then $\mathrm{Ann}_A(x) = \overline{A}$. Since $\overline{A}$ is a maximal ideal it is prime, so $\overline{A}$ is an associated prime of M and hence by Lemma 5.6.1 every element of positive degree in A is a zero divisor on M, so $\mathcal{R}eg_A(M) = \emptyset$.

Conversely, suppose that $\mathcal{R}eg_A(M) = \emptyset$. Then by Lemma 5.6.1

$$\overline{A} \subseteq \bigcup_{\mathfrak{p} \in \mathrm{Ass}_A(M)} \mathfrak{p},$$

so by the prime avoidance lemma (Lemma A.2.1) $\overline{A}$ must be an associated prime ideal of M. If $0 \neq x \in M$ has $\overline{A}$ as annihilator ideal, then the map $\varphi : \mathbb{F} \longrightarrow M$ defined by requiring $\varphi(1) = x$ is a nonzero homomorphism of A-modules of degree $|x|$, so $\mathrm{Hom}_A^*(\mathbb{F}, M) \neq 0$. □

DEFINITION: *Let A be a commutative graded connected Noetherian algebra over the field $\mathbb{F}$ and M a nontrivial graded A-module. A sequence of elements $a_1, \ldots, a_c \in \overline{A}$ is called a* **regular sequence on** M *if (with the convention that $a_0 = 0$) $a_i \in \mathcal{Reg}_A(M/(a_0 \cdot M + \cdots + a_{i-1} \cdot M)$ for $i = 1, \ldots, c$. The length of the longest regular sequence on M is called the* **codimension of M as an A-module** *and denoted by* $\operatorname{codim}_A(M)$.

The integer $\operatorname{codim}_A(\overline{A})$ is the **codimension of the algebra** A as already defined, and denoted by $\operatorname{codim}(A)$.

If $A = \mathbb{F}[b_1, \ldots, b_n] \subseteq B$ is a Noether normalization of the finitely generated graded algebra B over $\mathbb{F}$, then the homological codimension of B can be computed by restricting our attention to the regular sequences that lie in A. This surprising fact is a consequence of the following change of rings result for codimension (see, e.g., [344] IV Proposition 1.2, or [47] Exercise 1.2.26), which follows from the preceding lemmas.

PROPOSITION 5.6.5: *Let A and B be commutative graded connected Noetherian algebras over the field $\mathbb{F}$ and $\varphi : A \longrightarrow B$ a homomorphism of graded algebras. If M is a B-module that is finitely generated as an A-module, then* $\operatorname{codim}_A(M) = \operatorname{codim}_B(M)$.

PROOF: Let $a_1, \ldots, a_d \in \overline{A}$ be a regular sequence of maximal length. We must show that there is no element $b \in B$ such that $a_1, \ldots, a_d, b \in B$ is a regular sequence on M. But this is equivalent to showing that no element of B can be regular on $M/(a_1 \cdot M + \cdots + a_d \cdot M)$, which follows from Lemma 5.6.3. □

These results have a number of consequences for rings of invariants. In Chapter 4 we presented several methods for constructing systems of parameters for rings of invariants. Using one or the other of these, for example Proposition 5.6.5, allows us to check the codimension of $\mathbb{F}[V]^G$ by examining only regular sequences contained in a Noether normalization of $\mathbb{F}[V]^G$. In the case where $\mathbb{F}$ is a Galois field Proposition 5.6.5 points to a universal source for a regular sequence of maximal length.

NOTATION: *If $\mathbb{F}_q$ is the Galois field with q elements and $n \in \mathbb{N}_0$, then $\mathrm{GL}(n, \mathbb{F}_q)$ is a finite group, and the algebra of invariants $\mathbf{D}(n) = \mathbb{F}_q[z_1, \ldots, z_n]^{\mathrm{GL}(n, \mathbb{F}_q)}$ is called the* **Dickson algebra**.

If $\rho : G \hookrightarrow \mathrm{GL}(n, \mathbb{F}_q)$ is a representation of a finite group G over $\mathbb{F}_q$, then $\mathbf{D}(n) \subseteq \mathbb{F}_q[V]^G$ is a finite extension. In Chapter 6 we will see that $\mathbf{D}(n)$ has a particularly simple structure: It is, in fact, a polynomial algebra,[5] viz., $\mathbf{D}(n) = \mathbb{F}_q[\mathbf{d}_{n,0}, \ldots, \mathbf{d}_{n,n-1}]$, and the generators, which are unique up to nonzero scalars, are a universal system of parameters for $\mathbb{F}_q[V]^G$.

[5] For $n = 2$ this was shown in Section 4.1, Example 1.

THEOREM 5.6.6: *Let $\rho : G \hookrightarrow \mathrm{GL}(n, \mathbb{F})$ be a representation of a finite group G over the field $\mathbb{F}$. Then for any subgroup $H \leq G$*

$$\operatorname{codim}\left(\mathbb{F}[V]^H\right) = \operatorname{codim}_{\mathbb{F}[V]^G}\left(\mathbb{F}[V]^H\right).$$

In particular, if $\mathbb{F} = \mathbb{F}_q$ is a Galois field, then

$$\operatorname{codim}\left(\mathbb{F}_q[V]^G\right) = \operatorname{codim}_{\mathbf{D}(n)}\left(\mathbb{F}_q[V]^G\right).$$

PROOF: The inclusion $\mathbb{F}[V]^G \subseteq \mathbb{F}[V]^H$ turns $\mathbb{F}[V]^H$ into a finitely generated $\mathbb{F}[V]^G$-module. Hence we may apply Proposition 5.6.5 to the extension $A = \mathbb{F}[V]^G \subseteq \mathbb{F}[V]^H = B$ with $M = \mathbb{F}[V]^H$ to get

$$\operatorname{codim}_{\mathbb{F}[V]^G}\left(\mathbb{F}[V]^H\right) = \operatorname{codim}_{\mathbb{F}[V]^H}\left(\mathbb{F}[V]^H\right) = \operatorname{codim}\left(\mathbb{F}[V]^H\right).$$

If we apply this to the pair of groups $\rho(G) \leq \mathrm{GL}(n, \mathbb{F}_q)$ and the tautological representation $\mathrm{id} : \mathrm{GL}(n, \mathbb{F}_q) \hookrightarrow \mathrm{GL}(n, \mathbb{F}_q)$, then we obtain the second assertion. □

THEOREM 5.6.7: *Let $\rho : G \hookrightarrow \mathrm{GL}(n, \mathbb{F})$ be a representation of a finite group over a field of characteristic p. Then*

$$\operatorname{codim}(\mathbb{F}[V]^G) \geq \operatorname{codim}(\mathbb{F}[V]^{\mathrm{Syl}_p(G)}).$$

PROOF: Let $\mathbf{N} = \mathbb{F}[h_1, \ldots, h_n] \subseteq \mathbb{F}[V]^G$ be a Noether normalization. Since $\mathbb{F}[V]^G \subseteq \mathbb{F}[V]^{\mathrm{Syl}_p(G)}$ is a finite extension, $\mathbf{N} \subseteq \mathbb{F}[V]^{\mathrm{Syl}_p(G)}$ is likewise a Noether normalization. By Proposition 5.6.6 it is equivalent to show

$$\operatorname{codim}_{\mathbf{N}}(\mathbb{F}[V]^G) \geq \operatorname{codim}_{\mathbf{N}}(\mathbb{F}[V]^{\mathrm{Syl}_p(G)}).$$

Let

$$\mathrm{Tr}^G_{\mathrm{Syl}_p(G)} : \mathbb{F}[V]^{\mathrm{Syl}_p(G)} \longrightarrow \mathbb{F}[V]^G$$

be the relative transfer (see Section 2.2). Since $|G : \mathrm{Syl}_p(G)|$ is relatively prime to p, the map

$$\pi^G_{\mathrm{Syl}_p(G)} = \frac{1}{|G : \mathrm{Syl}_p(G)|}\mathrm{Tr}^G_{\mathrm{Syl}_p(G)} : \mathbb{F}[V]^{\mathrm{Syl}_p(G)} \longrightarrow \mathbb{F}[V]^G$$

is an $\mathbb{F}[V]^G$-module, and hence a fortiori an $\mathbf{N}$-module homomorphism, splitting the inclusion $\mathbb{F}[V]^G \subseteq \mathbb{F}[V]^{\mathrm{Syl}_p(G)}$.

Suppose that $h'_1, \ldots, h'_c \in \overline{\mathbf{N}}$ is a regular sequence on $\mathbb{F}[V]^{\mathrm{Syl}_p(G)}$. Then $\mathbb{F}[V]^{\mathrm{Syl}_p(G)}$ is a free $\mathbb{F}[\overline{h}'_1, \ldots, \overline{h}'_c]$-module, where multiplication by h'_i defines the action of $\overline{h}'_i$ on $\mathbb{F}[V]^{\mathrm{Syl}_p(G)}$ for $i = 1, \ldots, c$. The map $\pi^G_{\mathrm{Syl}_p(G)}$ is then an $\mathbb{F}[\overline{h}'_1, \ldots, \overline{h}'_c]$-module splitting for the inclusion $\mathbb{F}[V]^G \subseteq \mathbb{F}[V]^{\mathrm{Syl}_p(G)}$, and hence $\mathbb{F}[V]^G$ is a projective $\mathbb{F}[\overline{h}'_1, \ldots, \overline{h}'_c]$-module. By Koszul's theorem (Theorem 5.1.3) it is then free as an $\mathbb{F}[\overline{h}'_1, \ldots, \overline{h}'_c]$-module, so $h'_1, \ldots, h'_c$ is also a regular sequence on $\mathbb{F}[V]^G$. □

The extreme case $(\mathrm{codim}(\mathbb{F}[V]^{\mathrm{Syl}_p(G)}) = \dim(\mathbb{F}[V]) = n)$ leads to the following corollary (see [58], or [372] Proposition 8.3.1).

COROLLARY 5.6.8: *Let $\mathbb{F}$ be a field of characteristic $p \neq 0$ and $\rho : G \hookrightarrow \mathrm{GL}(n, \mathbb{F})$ a representation of a finite group G. If $\mathbb{F}[V]^{\mathrm{Syl}_p(G)}$ is Cohen-Macaulay, then so is $\mathbb{F}[V]^G$.* □

By making use of this we can give another short proof that rings of invariants in two variables are Cohen-Macaulay.

PROPOSITION 5.6.9: *Let $\rho : G \hookrightarrow \mathrm{GL}(2, \mathbb{F})$ be a representation of a finite group over the field $\mathbb{F}$ of characteristic p. Then $\mathbb{F}[V]^G$ is Cohen-Macaulay.*

PROOF: By Corollary 5.6.8 we may suppose that G is a finite p-group. Up to conjugation it is then a subgroup of $\mathrm{Uni}(2, \mathbb{F})$; in other words, the elements of $\rho(G)$ after suitable change of coordinates are of the form

$$\begin{bmatrix} 1 & 0 \\ * & 1 \end{bmatrix}.$$

Hence the standard basis z_1, z_2 for V^* is a Dade basis for V^*, and moreover, the number of elements in the orbit of z_2 is $|G|$, whereas z_1 is a fixed point. By the theorem of H. Nakajima and R. E. Stong (Theorem 4.5.6) it follows that $\mathbb{F}[z_1, z_2]^G = \mathbb{F}[z_1, c_{|G|}(z_2)]$. Polynomial algebras are, of course, Cohen-Macaulay. □

Rings of invariants in at most three variables are known to be Cohen-Macaulay [369]. The following elementary argument proves much more. It is taken from [55] Proposition 17, which will be used again in the proof of Lemma 6.5.5.

PROPOSITION 5.6.10: *Let $\rho : G \hookrightarrow \mathrm{GL}(n, \mathbb{F})$ be a representation of the finite group G over the field $\mathbb{F}$. If $n \geq 3$ then*

$$\text{hom-codim}(\mathbb{F}[V]^G) \geq 3.$$

So rings of invariants in three variables are Cohen-Macaulay.

PROOF: By Theorems 5.5.2 and 5.6.7 we may suppose that $\mathbb{F}$ has characteristic p and $G = P$ is a finite p-group. Then $V^P \neq \{0\}$. If $\dim_{\mathbb{F}}(V^P) = n-1$ then every nontrivial element of P is a transvection and $\mathbb{F}[V]^P$ is a polynomial algebra by Theorem 6.2.6 and $\text{hom-codim}(\mathbb{F}[V]^P) = n \geq 3$. So we may suppose that $\dim_{\mathbb{F}}(V^P) \leq n-2$. Let $W = V/V^P$ with the induced linear action of P. Then W has dimension at least two, so by Proposition 5.5.3 $\mathbb{F}[W]^P$ has homological codimension also at least two. Let $f_1, f_2 \in \mathbb{F}[W]^P$ be a regular sequence. The natural epimorphism $V \longrightarrow W$ induces monomorphisms $\mathbb{F}[W] \hookrightarrow \mathbb{F}[V]$ and $\mathbb{F}[W]^P \hookrightarrow \mathbb{F}[V]^P$. By Proposition 5.1.1 f_1, f_2 form a regular sequence in $\mathbb{F}[W]$, and since $\mathbb{F}[V]$ is a free $\mathbb{F}[W]$-module, also in $\mathbb{F}[V]$. So again by Proposition 5.1.1 they form a regular sequence in $\mathbb{F}[V]^P$.

Let $y, z \in W^*$ be a linearly independent. Identify W^* with the annihilator of V^P in V^*, and extend y, z to a linearly independent 3-tuple x, y, z for V^*. We claim that the sequence $f_1, f_2, c_{\text{top}}(x)$ of length 3 is a regular sequence in $\mathbb{F}[V]^P$. To simplify notation set $f_3 = c_{\text{top}}(x)$.

To show that $f_1, f_2, f_3 \in \mathbb{F}[V]^P$ is a regular sequence we must show that whenever

❶ $$h_1 f_1 + h_2 f_2 + h_3 f_3 = 0$$

for forms $h_1, h_2, h_3 \in \mathbb{F}[V]^P$ then h_3 belongs to the ideal $(f_1, f_2) \subset \mathbb{F}[V]^P$.

If $F \in \mathbb{F}[V]$ is a form, we may regard it as a polynomial in x with coefficients which are forms in the other variables, viz.,

$$F = F_0 x^d + F_1 x^{d-1} + \cdots + F_d$$

where $F_0, \ldots, F_d \in \mathbb{F}[W]$ and $F_0 \neq 0$. We call d the x-degree of F and denote it by $\deg_x(F)$. Observe that the P-action preserves the x-degree.

Regarding h_1, h_2 and f_3 as polynomials in x with coefficients which are forms in y, z, and the remaining elements of a basis for V^* we may perform polynomial division to obtain

❷ $$h_i = q_i f_3 + r_i, \quad \deg_x(r_i) < \deg_x(f_3) \quad i = 1, 2.$$

Recall from Section 2.4 the twisted differential operator $\partial_g = 1 - g$. Since the P action preserves the x-degree of forms we have

❸ $$\deg_x(\partial_g(r_i)) \leq \deg_x(r_i) < \deg_x(f_3) \quad i = 1, 2.$$

If we apply ∂_g to ❷ we obtain

$$0 = \partial_g(h_i) = \partial_g(q_i) f_3 + \partial_g(r_i) \quad i = 1, 2$$

since h_1, h_2 and f_3 are P-invariant. If $\partial_g(q_i) \neq 0$ this equation would say that f_3 divides r_i contradicting ❸. Hence $\partial_g(q_i) = 0$ and this entails that $\partial_g(r_i) = 0$ also. Since this holds for all $g \in P$ the forms q_i and r_i are P-invariant for $i = 1, 2$. Set

$$H = r_1 f_1 + r_2 f_2 \in \mathbb{F}[V]^P.$$

Substituting ❷ into ❶ then yields

❹ $$0 = (q_1 f_1 + q_2 f_2 + h_3) f_3 + H.$$

Recall that $f_1, f_2 \in \mathbb{F}[W]$ so $\deg_x(f_1) = 0 = \deg_x(f_2)$ and hence

❺ $$\deg_x(H) = \max\{\deg_x(r_1), \deg_x(r_2)\} < \deg_x(f_3).$$

Unless

$$q_1 f_1 + q_2 f_2 + h_3 = 0$$

equation ❹ would imply f_3 divides H contradicting ❺. Therefore

$$h_3 = -(q_1 f_1 + q_2 f_2)$$

showing that h_3 belongs to the ideal generated by f_1 and f_2 in $\mathbb{F}[V]^P$. □

5.7 The Gorenstein and Other Homological Properties

Many properties of rings (in the widest sense of the word) often have a homological characterization. For example, the theorems of D. Hilbert and J.-P. Serre (Theorems 5.2.1 and 5.3.3) characterize polynomial algebras in the graded case, and a theorem of E. F. Assmus ([19] Theorem 2.7) characterizes complete intersections. In this section we will examine one other of these properties and comment on several others.

We begin with a short discussion of Gorenstein rings, which are a special class of Cohen-Macaulay rings. For a more extensive account see, e.g., [47] Chapter 3, [400], [403] and of course the classic [27]. We confine ourselves to the graded connected case. To define Gorenstein algebras in ring theoretic terms, recall that an ideal I in a commutative graded connected algebra is called **irreducible** if whenever $I = I' \cap I''$ for ideals I', I'', then either $I = I'$ or $I = I''$. If A is Noetherian, a **parameter ideal** for A is an ideal generated by a system of parameters for A.

DEFINITION: *A commutative graded connected Noetherian algebra over a field is called* **Gorenstein** *if it is Cohen-Macaulay and every parameter ideal is irreducible.*

PROPOSITION 5.7.1: *A commutative graded connected Noetherian Cohen-Macaulay algebra A is Gorenstein if and only if for every parameter ideal $I \subset A$ the quotient algebra $H = A/I$ satisfies Poincaré duality.*

PROOF: Suppose that A is Gorenstein and that $I \subset A$ is a parameter ideal. There is no loss in generality in assuming that $I \neq \overline{A}$, so $H \neq \mathbb{F}$. Since $I \subset A$ is irreducible, $(0) \subset H$ is irreducible. Let J be the intersection of all the nonzero ideals of H. Since (0) is irreducible in H, it follows that $J \neq (0)$, and is the unique nonzero minimal ideal of H. The totalization of H is finite dimensional because $I \subset A$ is a parameter ideal. Let d be the largest integer such that $H_d \neq 0$. If $\dim_{\mathbb{F}}(H_d) \neq 1$ then H would have more than one minimal ideal. For if $x, y \in H_d$ are linearly independent, then $(x), (y) \subset H$ are distinct nonzero minimal ideals in H. Let $[H] \in H_d$ be a basis vector. Then $J = ([H])$. If $w \neq 0 \in H$ then the ideal (w) contains J since J is the unique minimal ideal in H. This says that there is an element $u \in H$ such that $uw = [H]$, and hence H satisfies Poincaré duality with fundamental class $[H]$.

Conversely, if H satisfies Poincaré duality with fundamental class $[H]$, then the principal ideal generated by $[H]$ is the unique nonzero minimal ideal of H. Therefore $(0) \subset H$ is irreducible in H which implies that $I \subset A$ is irreducible in A. □

The homological characterization of Gorenstein algebras depends on the following remarkable result of J.-P. Serre ([344] Proposition IV.5 or [372] Theorem 6.6.2) which shows that the graded module $\mathrm{Hom}_A(\mathbb{F},\ A/(a_1, \ldots, a_r))_*$ is up to a shift in the grading [6] *independent* of the choice of the elements $a_1, \ldots, a_r \in A$ so long as they form a regular sequence.

THEOREM 5.7.2 (J. -P. Serre): *If* $a_1, \ldots, a_r \in A$ *is a regular sequence then* $\mathrm{Ext}^i_A(\mathbb{F},\ A) = 0$ *for* $i < r$ *and*

$$\mathrm{Ext}^r_A(\mathbb{F},\ A) \cong \mathrm{Ext}^{r-1}_A(\mathbb{F},\ A/(a_1)) \cong \cdots \cong \mathrm{Ext}^0_A(\mathbb{F}, A/(a_1, \ldots, a_r)).$$

(We use $\mathrm{Ext}^0_A(-,\ -)$ *as an alternate notation for* $\mathrm{Hom}_A(-,\ -)_*$*.)* □

Theorem 5.7.2 shows that for a parameter ideal I in a Cohen-Macaulay algebra A the graded vector space $\mathrm{Hom}_A(\mathbb{F}, A/I)$ is up to a shift of grading independent of I. It is isomorphic to the **socle**, socle(A/I), of A/I, i.e., to the set of elements in A/I annihilated by the maximal ideal.

PROPOSITION 5.7.3: *A commutative graded connected Noetherian Cohen-Macaulay algebra* A *of Krull dimension* d *is Gorenstein if and only if* $\mathrm{Tot}(\mathrm{Ext}^d_A(\mathbb{F}, A))$ *is one-dimensional.*

PROOF: If A is Gorenstein then for any parameter ideal $I \subset A$ it follows that $\mathrm{Ext}^d_A(\mathbb{F}, A)$ and $\mathrm{Hom}_A(\mathbb{F}, A/I)_*$ are isomorphic as graded vector spaces up to a shift of grading by Theorem 5.7.2. By Proposition 5.7.1 the only elements in $H = A/I$ with annihilator ideal $\overline{A}$ are multiples of the fundamental class $[H]$ so $\mathrm{Tot}(\mathrm{Hom}_A(\mathbb{F}, H)_*)$ is one-dimensional. The argument is reversible, so the result follows. □

It follows from Proposition 5.7.3 and the remarks preceding it that although Gorenstein algebras are defined in terms of a property of all parameter ideals, it suffices for a single such ideal to have this property. In other words, we have the following characterization of Gorenstein algebras.

COROLLARY 5.7.4: *Let* A *be a Noetherian commutative graded connected Cohen-Macaulay algebra of Krull dimension* d *and* $I \subset A$ *a parameter ideal. Then the following conditions are equivalent:*

(i) A *is Gorenstein.*
(ii) I *is irreducible.*
(iii) A/I *is a Poincaré duality algebra.* □

The Gorenstein property of rings of invariants was investigated by K. Watanabe in [443] and [444]. We treat these theorems using a result of R.P. Stanley,

[6] That is, the module of graded A-module homomorphisms, where $\mathbb{F}$ is regarded as a graded module concentrated in degree zero. Shifting the grading means suspending or desuspending.

whose proof would take us a bit too far afield for a book ostensibly on invariant theory. The motivation for this result is a consequence of Proposition 5.7.1 and Theorem 5.5.1 (4). Namely, if A is Gorenstein, then its Poincaré series has the form $P(A, t) = P(\mathbf{N}, t) \cdot P(A//\mathbf{N}, t)$, where $\mathbf{N} \subseteq A$ is a Noether normalization of A. The quotient algebra $A//\mathbf{N}$ satisfies Poincaré duality, so $P(A//\mathbf{N}, t)$ is a **palindromic polynomial**, i.e.

$$t^d P(A//\mathbf{N}, \frac{1}{t}) = P(A//\mathbf{N}, t),$$

where d is the degree of the fundamental class of $A//\mathbf{N}$. Given the special nature of $P(\mathbf{N}, t)$ one sees that $P(A, t)$ satisfies a functional equation equation

$$P(A, \frac{1}{t}) = (-1)^n t^s P(A, t),$$

for some integer s. R. P. Stanley's theorem ([400] Theorem 4.4) says that for integral domains the converse holds.

THEOREM 5.7.5 (R. P. Stanley): *Over a field, a commutative graded connected Noetherian Cohen-Macaulay integral domain A of Krull dimension n is Gorenstein if and only if its Poincaré series satisfies a functional equation*

$$P(A, \frac{1}{t}) = (-1)^n t^s P(A, t)$$

for some $s \in \mathbb{Z}$. □

R. P. Stanley actually only states this result for $\mathbb{F} = \mathbb{C}$, however his proof works for any ground field.

PROPOSITION 5.7.6 (K. Watanabe): *Let $\rho : G \hookrightarrow \mathrm{GL}(n, \mathbb{F})$ be a representation of a finite group G over the field $\mathbb{F}$. Let $|G|$ be invertible in $\mathbb{F}$. Then*

(i) *If $\rho(G) \subseteq \mathrm{SL}(n, \mathbb{F})$, then $\mathbb{F}[V]^G$ is Gorenstein.*

(ii) *If $\rho(G)$ contains no pseudoreflections and $\mathbb{F}[V]^G$ is Gorenstein, then $\rho(G) \subseteq \mathrm{SL}(n, \mathbb{F})$.*

PROOF: If $\rho(G) \subseteq \mathrm{SL}(n, \mathbb{F})$ then by Theorem 3.1.6 $P(\mathbb{F}[V]^G, t)$ satisfies the functional equation

$$P\left(\mathbb{F}[V]^G, \frac{1}{t}\right) = (-1)^n t^n P(\mathbb{F}[V]^G, t),$$

and *(i)* follows from Theorem 5.7.5.

To prove *(ii)* we note from the functional equation, T. Molien's theorem (Theorem 3.1.3), and Theorem 3.1.4 that

$$\sum_{g \in G} \frac{1}{\det(1 - g\,t)} = \sum_{g \in G} \frac{\det(g)}{\det(1 - g\,t)}.$$

If we put $t = 0$ into this identity we obtain $|G| = \sum_{g \in G} \det(g)$. Since each $\det(g)$ is a root of unity we must have $\det(g) = 1$. □

We refer to [400] R. P. Stanley for further characterizations of Gorenstein algebras via their Poincaré series. For a treatment of the Gorenstein property of rings of invariants using the canonical module (see e.g., [47] Chapter 3) see [30], Chapter 4, especially Sections 4.5 and 4.6.

$\mathbb{F}[V]^G$ is a polynomial algebra
$\Rightarrow \mathbb{F}[V]^G$ is a hypersurface
$\Rightarrow \mathbb{F}[V]^G$ is a complete intersection
$\Rightarrow \mathbb{F}[V]^G$ is a Gorenstein ring
$\Rightarrow \mathbb{F}[V]^G$ is a Cohen-Macaulay ring

We have looked at just one possible refinement of the Cohen-Macaulay property in this section. In R. P. Stanley's lovely survey article [401] he introduces a hierarchy of conditions on an algebra that refine the Cohen-Macaulay property. These may be tabulated as a series of implications as pictured. The classification of nonmodular groups, or better put representations, such that the ring of invariants is a complete intersection, is rather long and complicated. We refer to the original sources for details: [147], [188], [282], [330], [445], and the references there. The following necessary condition that a ring of invariants be a complete intersection is an analog of J.-P. Serre's condition[7] that it be a polynomial algebra [346].

PROPOSITION 5.7.7 (R. L. Gordeev, V. Kac-K. Watanabe): *For a representation $\rho : G \hookrightarrow \mathrm{GL}(n, \mathbb{F})$ of a finite group G, if $\mathbb{F}[V]^G$ is a complete intersection, then $\rho(G)$ is generated by bireflections, i.e., elements g such that the rank of $g - \mathbf{I}$ is at most 2.* □

This result was reworked in [196]: On the one hand it is sharpened by relaxing the assumption to $\mathbb{F}[V]^G$ is Cohen-Macaulay, and on the other hand weakend in that the conclusion holds only for p-groups.

Finally, nonmodular representations having a hypersurface as their ring of invariants were classified by H. Nakajima [278] in terms of linear characters associated to certain subgroups. He extended this way results obtained by R. P. Stanley in [399].

We will come back to the hierarchy of ring theoretic properties in Section 10.6 and 10.7. in connection with stabilizer subgroups. If $\rho : G \hookrightarrow \mathrm{GL}(n, \mathbb{F})$ is a representation of a finite group and $U \leq V = \mathbb{F}^n$ is a linear subspace, then the **pointwise satbilizer of U in G** is the subgroup G_U of G of elements that act trivially on U. Many homological properties of $\mathbb{F}[V]^G$ are inherited by $\mathbb{F}[V]^{G_U}$, loc.cit.

[7] If $\rho : G \hookrightarrow GL(n, \mathbb{F})$ is a representation of a finite group over the field $\mathbb{F}$ and $\mathbb{F}[V]^G$ is a polynomial algebra, then G is generated by pseudoreflections.

5.8 Examples

We collect some examples in this section to illustrate the Cohen-Macaulay property, or lack of it, in rings of invariants in the modular case.

EXAMPLE 1 (M.-J. Bertin [36]): Let $\rho : \mathbb{Z}/4 \hookrightarrow \mathrm{GL}(4, \mathbb{F})$ be the regular representation of $\mathbb{Z}/4$ over the field $\mathbb{F} = \mathbb{F}_2$ with two elements. (See Example 3 in Section 1.1 and Example 1 in Section 2.4.) The ring of invariants of this representation is not Cohen-Macaulay, [97]. To show this we reason as follows (see [292]).

Since the regular representation ρ is a permutation representation, the ring of invariants $\mathbb{F}[x_1, x_2, x_3, x_4]^{\mathbb{Z}/4}$ contains the four elementary symmetric polynomials

$$\begin{aligned} e_1 &= x_1 + x_2 + x_3 + x_4, \\ e_2 &= x_1x_2 + x_1x_3 + x_1x_4 + x_2x_3 + x_2x_4 + x_3x_4, \\ e_3 &= x_1x_2x_3 + x_1x_2x_4 + x_1x_3x_4 + x_2x_3x_4, \\ e_4 &= x_1x_2x_3x_4. \end{aligned}$$

Direct computation yields another quadratic invariant, viz.,

$$q = \mathrm{Tr}^{\mathbb{Z}/4}(x_1x_2) = x_1x_2 + x_2x_3 + x_3x_4 + x_4x_1.$$

The four polynomials $e_1, e_2 + q, e_3, e_4 \in \mathbb{F}[x_1, x_2, x_3, x_4]^{\mathbb{Z}/4}$ form a system of parameters, since they do so in $\mathbb{F}[x_1, x_2, x_3, x_4]$. If $\mathbb{F}[x_1, x_2, x_3, x_4]^{\mathbb{Z}/4}$ were Cohen-Macaulay, then these four polynomials would have to be a regular sequence in $\mathbb{F}[x_1, x_2, x_3, x_4]^{\mathbb{Z}/4}$. However, if we set $k = \mathrm{Tr}^{\mathbb{Z}/4}(x_1^2x_2)$ and $\ell = \mathrm{Tr}^{\mathbb{Z}/4}(x_1^3x_2)$ then there is the relation

$$e_1(e_2q + e_1k + e_1e_3 + \ell) + (e_2 + q)k + e_2e_3 = 0.$$

This shows that e_3 is a zero divisor in $\mathbb{F}[V]^{\mathbb{Z}/4}/(e_1, e_2 + q)$, because $e_2 \notin (e_1, e_2 + q)$, and therefore the ring of invariants $\mathbb{F}[V]^{\mathbb{Z}/4}$ is not Cohen-Macaulay.

As we have pointed out several times (see, e.g., Section 2.3, Example 2, and Section 3.2, Example 1) vector invariants serve often as negative examples, e.g., that some nice property fails to hold after the sum of enough copies of a fixed representation has been formed. This was done to show that Emmy Noether's bound for the degrees of algebra generators can fail in the modular case. The Cohen-Macaulay property is also in this class as the following result shows.

PROPOSITION 5.8.1 (H. E. A. Campbell-A. Geramita-I. P. Hughes-R. J. Shank-D. L. Wehlau): *Let $\rho : P \hookrightarrow \mathrm{GL}(n, \mathbb{F})$ be a representation of a finite p-group over the field $\mathbb{F}$ of characteristic p. Denote by $\bigoplus_m \rho$ the m-fold direct sum of ρ with itself. Then for $m \geq 3$ the vector invariants $\mathbb{F}[\bigoplus_m V]^P$ are not Cohen-Macaulay.*

PROOF: Without loss of generality we suppose that $\rho(P) \subseteq \mathrm{Uni}(n, \mathbb{F})$. So the matrices in $\rho(P)$ acting on the standard basis $z_1, \ldots, z_n$ for the linear forms are all strictly upper triangular as shown. Again, without loss of generality we can assume that for at least one element in P the matrix entry $a_{n-1,n}$ is not zero. Then $z_n \in \mathbb{F}[z_1, \ldots, z_n]^P$, and the orbit of z_{n-1} consists of elements of the form $z_{n-1} + \lambda z_n$ with $\lambda \in \mathbb{F}$.

$$\begin{bmatrix} 1 & a_{1,2} & \cdots & a_{1,n} \\ \vdots & 1 & & \vdots \\ 0 & 0 & \ddots & a_{n-1,n} \\ \cdots & \cdots & 0 & 1 \end{bmatrix}$$

Denote by $z_{i,j}$, $j = 1, \ldots, n$, the corresponding basis for the jth summand in $\underset{m}{\oplus} V$ for $i = 1, \ldots, m$. Then $z_{i,n} \in \mathbb{F}[\underset{m}{\oplus} V]^P$. Direct computation shows that

$$f_{k,l} = z_{k,n}z_{l,n-1} - z_{l,n}z_{k,n-1} \in \mathbb{F}[\underset{m}{\oplus} V]^P, \quad k, l = 1, \ldots, m.$$

Next, note that we have the relation

$$z_{1,n}f_{2,3} - z_{2,n}f_{1,3} + z_{3,n}f_{1,2} = 0.$$

This says that $z_{3,n}f_{1,2} \in (z_{1,n}, z_{2,n}) \subset \mathbb{F}[\underset{m}{\oplus} V]^P$.

If $\mathbb{F}[\underset{m}{\oplus} V]^P$ were Cohen-Macaulay, then the linear forms $z_{1,n}, \ldots, z_{m,n} \in \mathbb{F}[\underset{m}{\oplus} V]^P$ could be completed to a system of parameters for $\mathbb{F}[\underset{m}{\oplus} V]^P$, which would have to be a regular sequence, by Macaulay's theorem, and therefore $f_{1,2} \in (z_{1,n}, z_{2,n})$. This, in turn, implies that there exist invariants $h, k \in \mathbb{F}[\underset{m}{\oplus} V]^P$ such that

$$z_{1,n}z_{2,n-1} - z_{2,n}z_{1,n-1} = f_{1,2} = hz_{1,n} + kz_{2,n},$$

from which we obtain by rearranging terms

$$(z_{2,n-1} - h)z_{1,n} = (z_{1,n-1} + k)z_{2,n}.$$

The elements $z_{1,n}, z_{2,n} \in \mathbb{F}[\underset{m}{\oplus} V]^P$, being linear, are also prime by Corollary 1.7.4, so this equation implies that $z_{1,n} \mid k + z_{1,n-1}$ and $z_{2,n} \mid h + z_{2,n-1}$. For degree reasons we then must have $z_{1,n} = \alpha k + \alpha z_{1,n-1}$ and $z_{2,n} = -\beta h + \beta z_{2,n-1}$ for suitable $\alpha, \beta \in \mathbb{F}^\times$. From the first of these equations we conclude

$$z_{1,n-1} = \frac{1}{\alpha}(z_{1,n} - \alpha h),$$

which says that $z_{1,n-1} \in \mathbb{F}[\underset{m}{\oplus} V]^P$.

But by hypothesis, P contains an element g such that the action of g on the linear forms in V^* is by a matrix as shown above with $a_{n-1,n} \neq 0$, and for this g, $gz_{1,n-1} = z_{1,n-1} + a_{n-1,n}z_{1,n} \neq z_{1,n-1}$, which is a contradiction. Hence $\mathbb{F}[\underset{m}{\oplus} V]^P$ is not Cohen-Macaulay. □

We close this chapter with an example of a permutation representation for which many, *but not all*, of the properties of the ring of invariants are independent of the characteristic. It is taken from [297], Example 21.5.1.

EXAMPLE 2: The dihedral group of order 10 has a five-dimensional permutation representation $\rho : D_{10} \hookrightarrow \mathrm{GL}(5, \mathbb{F})$ afforded by the matrices

$$\mathbf{D} = \begin{bmatrix} 0 & 1 & 0 & 0 & 0 \\ 0 & 0 & 1 & 0 & 0 \\ 0 & 0 & 0 & 1 & 0 \\ 0 & 0 & 0 & 0 & 1 \\ 1 & 0 & 0 & 0 & 0 \end{bmatrix} \quad \text{and} \quad \mathbf{S} = \begin{bmatrix} 0 & 1 & 0 & 0 & 0 \\ 1 & 0 & 0 & 0 & 0 \\ 0 & 0 & 0 & 0 & 1 \\ 0 & 0 & 0 & 1 & 0 \\ 0 & 0 & 1 & 0 & 0 \end{bmatrix}.$$

The Poincaré series can be calculated with Molien's theorem, Theorem 3.1.3, and Proposition 3.2.2. We obtain

$$P(\mathbb{F}[x_1, \ldots, x_5]^{D_{10}}, t) = \frac{1 + t^2 + t^3 + 2t^4 + 2t^5 + 2t^6 + t^7 + t^8 + t^{10}}{(1-t^3)(1-t^4)(1-t^5)}.$$

The ring of invariants is generated by orbit sums of special monomials, thus by invariants of degree less than or equal to 10 by M. Göbel's bound, Theorem 3.4.2, and Corollary 3.4.3. Note that in this case M. Göbel's bound coincides with Emmy Noether's bound, i.e., Emmy Noether's bound holds *no matter what the ground field is*.

If the characteristic of $\mathbb{F}$ is zero or coprime to 10, then the ring of invariants is Cohen-Macaulay, and even Gorenstein by K. Watanabe's criterion, Proposition 5.7.6.

If the characteristic of $\mathbb{F}$ is 2, then a 2-Sylow subgroup is generated by $\mathbf{S}$. Its ring of invariants is a hypersurface

$$\mathbb{F}[x_1, \ldots, x_5]^{\mathbb{Z}/2} = \mathbb{F}[x_1 + x_2, x_1x_2, x_3 + x_5, x_3x_5, x_4]/(r),$$

where the relation is given by

$$\begin{aligned} r = (x_1x_2)(x_3 + x_5)^2 + (x_3x_5)(x_1 + x_2)^2 + (x_1x_3 + x_2x_5)^2 \\ + (x_1x_3 + x_2x_5)(x_3 + x_5)(x_1 + x_2), \end{aligned}$$

cf. [288]. In particular, $\mathbb{F}[x_1, \ldots, x_5]^{\mathbb{Z}/2}$ is Cohen-Macaulay. Therefore, by Corollary 5.6.8, the ring of invariants $\mathbb{F}[x_1, \ldots, x_5]^{D_{10}}$ is Cohen-Macaulay also in characteristic 2.

Finally, we come to characteristic 5. The 5-Sylow subgroup is generated by $\mathbf{D}$, i.e., we need to compute the ring of invariants of the regular representation of the cyclic group of order 5 in the modular case. This is not known! (See, however, [124].) But we do know that the depth of $\mathbb{F}[x_1, \ldots, x_5]^{\mathbb{Z}/5}$ is precisely 3, [105], [375], [386], or Proposition 6.5.7. The only thing we can derive from Theorem 5.6.7 is that in characteristic 5 the ring of invariants of the dihedral group has depth at least 3.

Chapter 6
Modular Invariant Theory

MODULAR invariant theory[1] is concerned with the invariants of finite groups over fields of nonzero characteristic, which may well divide the order of the group. Many of the most interesting problems in the invariant theory of finite groups occur in the modular case: Indeed, some of them make sense only in this case. We can only scratch the surface of what is a fast-growing body of literature and a very active research area.

In this chapter we will examine a number of aspects of modular invariant theory, some quite old, such as the Dickson algebra (see Section 6.1), and others quite new, such as transvection groups (see Section 6.2) and the transfer variety (see Section 6.4). If $\rho : G \hookrightarrow \mathrm{GL}(n, \mathbb{F})$ is a representation of a finite group whose order is divisible by the characteristic of $\mathbb{F}$, then the averaging operator derived from the transfer (see Section 2.2) is no longer defined, and the transfer homomorphism $\mathrm{Tr}^G : \mathbb{F}[V] \longrightarrow \mathbb{F}[V]^G$ itself is no longer surjective in positive degrees. Indeed, the transfer ideal $\mathrm{Im}(\mathrm{Tr}^G) \subset \mathbb{F}[V]^G$ has height at most $n-1$ by Theorem 2.4.5. The affine variety defined by the extended ideal $(\mathrm{Im}(\mathrm{Tr}^G))^e \subset \mathbb{F}[V]$ turns out to have a particularly simple description; it is the union of the fixed-point sets of the elements of order p in G as we show in Section 6.4.

If in addition to having positive characteristic the ground field is the finite field $\mathbb{F}_q$, then $\mathrm{GL}(n, \mathbb{F}_q)$ is a finite group, and the invariants of this group in its tautological representation were computed by L. E. Dickson in [85]. The answer is decidedly nice, viz., $\mathbb{F}_q[V]^{\mathrm{GL}(n,\mathbb{F}_q)}$ is a polynomial algebra $\mathbb{F}_q[V]^{\mathrm{GL}(n,\mathbb{F}_q)} = \mathbb{F}_q[\mathbf{d}_{n,0}, \ldots, \mathbf{d}_{n,n-1}] = \mathbf{D}(n)$ (see Theorem 6.1.4). The gen-

[1] This is not the usage of the term in the classical literature, so some care should be taken when consulting papers or books written before ca. 1940.

erators $\mathbf{d}_{n,0}, \ldots, \mathbf{d}_{n,n-1}$ belong to $\mathbb{F}_q[V]^G$ for any $\rho : G \hookrightarrow \mathrm{GL}(n, \mathbb{F}_q)$ and are a universal system of parameters for rings of invariants of degree n over the finite field $\mathbb{F}_q$. This makes possible many interesting problems related to the structure of $\mathbb{F}_q[V]^G$ as a $\mathbf{D}(n)$-module. We will examine a few of these in Section 6.5.

The Dickson algebra has also played an important role in other branches of mathematics, particularly algebraic topology, and we will encounter it again when we study Steenrod operations in later chapters, which provide another tool, one originating in algebraic topology, for studying invariants over finite fields.

As a last topic for this chapter we choose to examine some results for p-groups in characteristic p. This we do in Section 6.3.

6.1 The Dickson Algebra

Let $p \in \mathbb{N}$ be a fixed prime, $q = p^\nu$ a fixed positive power of p, and $\mathbb{F}_q$ the Galois field with q elements. If $n \in \mathbb{N}$, then the general linear group $\mathrm{GL}(n, \mathbb{F}_q)$ acts on $V = \mathbb{F}_q^n$, and we define the **Dickson algebra** $\mathbf{D}(n)$ to be the ring of invariants $\mathbb{F}_q[V]^{\mathrm{GL}(n, \mathbb{F}_q)}$. (See also Section 5.6.) This algebra of invariants was originally computed by L. E. Dickson in [85]. Since then many different proofs have appeared, e.g., [389], [456], and [372] Chapter 8. The proof that we offer here is based on one of R. Steinberg [408]. It is particularly short as befits a proof benefitting from hindsight. The following two lemmas are the essential steps in the proof.

LEMMA 6.1.1: *Let $q = p^\nu$ be a positive prime power, $\mathbb{F}_q$ the Galois field with q elements, and $V = \mathbb{F}_q^n$ the n-dimensional vector space over $\mathbb{F}_q$. Set*

$$\Phi_n(X) = \prod_{z \in V^*} (X + z) \in \mathbb{F}_q[V][X].$$

Then $\Phi_n(X)$ is a q-polynomial, in the sense that (see, e.g., [382])

$$\Phi_n(X) = \sum_{i=0}^{n} \mathbf{d}_{n,i} X^{q^i};$$

in other words, the only powers of X occurring in $\Phi_n(X)$ are the qth powers $q^0, q^1, \ldots, q^n$.

PROOF: By induction on n. For $n = 1$ choose a nonzero linear form $z \in V^*$. Each element of V^* is an $\mathbb{F}_q$-multiple of z so

$$\Phi_1(X) = \prod_{\lambda \in \mathbb{F}_q} (X + \lambda z) = X^q - z^{q-1} X$$

is a q-polynomial. If we assume that the result has been established for $n - 1$, then, since raising to the qth power is an $\mathbb{F}_q$-linear map, and $\lambda^q = \lambda$

for any $\lambda \in \mathbb{F}_q$, we have the identity

$$\Phi_{n-1}(\alpha X' + \beta X'') = \alpha\Phi_{n-1}(X') + \beta\Phi_{n-1}(X''), \qquad \alpha, \beta \in \mathbb{F}_q.$$

Let $z_1, \ldots, z_n \in V^*$ be a basis, and set $W^* = \mathrm{Span}_{\mathbb{F}_q}\{z_1, \ldots, z_{n-1}\}$. Then

$$\begin{aligned}\Phi_n(X) &= \prod_{z \in V^*}(X+z) = \prod_{\substack{w\in W^* \\ \lambda\in\mathbb{F}_q}}(X+w+\lambda z_n) = \prod_{\lambda\in\mathbb{F}_q}\prod_{w\in W^*}(X+w+\lambda z_n) \\ &= \prod_{\lambda\in\mathbb{F}_q}\Phi_{n-1}(X+\lambda z_n) = \prod_{\lambda\in\mathbb{F}_q}(\Phi_{n-1}(X)+\lambda\Phi_{n-1}(z_n)) \\ &= \Phi_{n-1}(X)^q - \Phi_{n-1}(z_n)^{q-1}\Phi_{n-1}(X),\end{aligned}$$

completing the induction. □

The coefficients of the polynomial $\Phi(X) \in \mathbb{F}_q[V][X]$ are called the **Dickson polynomials**. Notice that $\deg(\mathbf{d}_{n,i}) = q^n - q^i$ for $i = 0, \ldots, n$, and that $\mathbf{d}_{n,n} = 1$.

LEMMA 6.1.2 (R. Steinberg): *Let $q = p^\nu$ be a positive prime power, $\mathbb{F}_q$ the Galois field with q elements, and $V = \mathbb{F}_q^n$ the n-dimensional vector space over $\mathbb{F}_q$. Choose a basis $z_1, \ldots, z_n$ for the space V^* of linear forms, and let j be an integer between 1 and n. Then z_j is a root of a monic polynomial of degree $q^n - q^{j-1}$ with coefficients in $z_1, \ldots, z_{j-1}, \mathbf{d}_{n,j-1}, \ldots, \mathbf{d}_{n,n-1}$.*

PROOF: Every linear form is a root of $\Phi_n(X)$ and every linear form in $z_1, \ldots, z_{j-1}$ is a root of $\Phi_{j-1}(X)$. Therefore $\Phi_{j-1}(X)$ divides $\Phi_n(X)$. Consider the quotient $\Psi_j(X) = \Phi_n(X)/\Phi_{j-1}(X)$. This is a monic polynomial of degree $q^n - q^{j-1}$ and has z_j as a root. The coefficients of Φ_{j-1} belong to $\mathbb{F}_q[z_1, \ldots, z_{j-1}]$, and the coefficients of $\Phi_n(X)$ are $\mathbf{d}_{n,0}, \ldots, \mathbf{d}_{n,n}$. Since $\deg(\mathbf{d}_{n,i}) = q^n - q^i > q^n - q^{j-1} = \deg(\Psi_j(X))$ for $i = 0, \ldots, j$, the Dickson polynomials $\mathbf{d}_{n,0}, \ldots, \mathbf{d}_{n,j}$ cannot figure in the coefficients, which therefore must be polynomials in $z_1, \ldots, z_{j-1}, \mathbf{d}_{n,j-1}, \ldots, \mathbf{d}_{n,n-1}$. □

PROPOSITION 6.1.3: *Let $q = p^\nu$ be a positive prime power, $\mathbb{F}_q$ the Galois field with q elements, and $V = \mathbb{F}_q^n$ the n-dimensional vector space over $\mathbb{F}_q$. For $j = 1, \ldots, n$ the polynomials $z_1, \ldots, z_{j-1}, \mathbf{d}_{n,j-1}, \ldots, \mathbf{d}_{n,n-1} \in \mathbb{F}_q[V]$ are a system of parameters.*

PROOF: From R. Steinberg's lemma (Lemma 6.1.2), we conclude that the subalgebra $\mathbb{F}_q\langle z_1, \ldots, z_j, \mathbf{d}_{n,j}, \ldots, \mathbf{d}_{n,n-1}\rangle$ of $\mathbb{F}_q[V]$ generated by the polynomials $z_1, \ldots, z_j, \mathbf{d}_{n,j}, \ldots, \mathbf{d}_{n,n-1}$ is a finite extension of the subalgebra $\mathbb{F}_q\langle z_1, \ldots, z_{j-1}, \mathbf{d}_{n,j-1}, \ldots, \mathbf{d}_{n,n-1}\rangle$ of $\mathbb{F}_q[V]$ generated by the forms $z_1, \ldots, z_{j-1}, \mathbf{d}_{n,j-1}, \ldots, \mathbf{d}_{n,n-1}$. This gives us a nested sequence of finite extensions

$$\begin{aligned}\mathbb{F}_q\langle \mathbf{d}_{n,0}, \ldots, \mathbf{d}_{n,n-1}\rangle \subset \cdots \subset \mathbb{F}_q\langle z_1, \ldots, z_{j-1}, \mathbf{d}_{n,j-1}, \ldots, \mathbf{d}_{n,n-1}\rangle \\ \subset \mathbb{F}_q\langle z_1, \ldots, z_j, \mathbf{d}_{n,j}, \ldots, \mathbf{d}_{n,n-1}\rangle \subset \cdots \\ \subset \mathbb{F}_q\langle z_1, \ldots, z_n\rangle = \mathbb{F}_q[z_1, \ldots, z_n],\end{aligned}$$

and the result follows. □

THEOREM 6.1.4 (L. E. Dickson): *Let $q = p^\nu$ be a positive prime power, $\mathbb{F}_q$ the Galois field with q elements. Then the Dickson algebra is a polynomial algebra generated by the Dickson polynomials, i.e., $\mathbf{D}(n) = \mathbb{F}_q[\mathbf{d}_{n,0}, \ldots, \mathbf{d}_{n,n-1}]$.*

PROOF: The Dickson polynomials $\mathbf{d}_{n,0}, \ldots, \mathbf{d}_{n,n-1}$ in $\mathbb{F}_q[V]$ form a system of parameters by Proposition 6.1.3. From their definition they also lie in $\mathbf{D}(n)$. Since

$$\prod_{i=0}^{n-1} \deg(\mathbf{d}_{n,i}) = \prod (q^n - q^i) = |\mathrm{GL}(n, \mathbb{F}_q)|,$$

the result follows from Proposition 4.5.5. □

Combining Propositions 5.1.1 and 6.1.3 we obtain the following result.

COROLLARY 6.1.5: *Let $q = p^\nu$ be a positive power of the prime integer p. If $\rho : G \hookrightarrow \mathrm{GL}(n, \mathbb{F})$ is a representation of a finite group over the Galois field $\mathbb{F}_q$, and $n \geq 2$, then any two distinct Dickson polynomials $\mathbf{d}_{n,i}$, $\mathbf{d}_{n,j}$ are a regular sequence in $\mathbb{F}_q[V]^G$.* □

This result does not extend for larger n to more than two Dickson polynomials without attention to which of the Dickson polynomials are chosen (see Section 8.5).

In [408] R. Steinberg considers the nested sequence of subgroups $G(0) < \cdots < G(n)$ of $\mathrm{GL}(n, \mathbb{F}_q)$ defined by the requirement that $g \in G(r)$ if and only if the first r rows of the matrix of g agree with the first r rows of the identity matrix $\mathbf{I} \in \mathrm{GL}(n, \mathbb{F}_q)$. The group $G(r)$ is just the semidirect product $\mathrm{Mat}_{n-r,r} \rtimes \mathrm{GL}(n-r, \mathbb{F}_q)$, where $\mathrm{GL}(n-r, \mathbb{F}_q)$ acts on the additive group of $(n-r) \times r$ matrices $\mathrm{Mat}_{n-r,r}$ by left multiplication. The order of this group is $|G(r)| = q^{r(n-r)} |\mathrm{GL}(n-r, \mathbb{F}_q)|$. The polynomials $z_1, \ldots, z_r, \mathbf{d}_{n,r}, \ldots, \mathbf{d}_{n,n-1}$ belong to $\mathbb{F}_q[V]^{G(r)}$, and the product of their degrees is $(q^n - q^r) \cdots (q^n - q^{n-1}) = q^{r(n-r)} |\mathrm{GL}(n-r, \mathbb{F}_q)|$ also. So by Proposition 6.1.3 and Proposition 4.5.5, we obtain R. Steinberg's generalization of L. E. Dickson's theorem:

THEOREM 6.1.6 (R. Steinberg): *Let $q = p^\nu$ be a positive prime power, $\mathbb{F}_q$ the Galois field with q elements, r an integer from 1 to n, and $G(r)$ the subgroup of $\mathrm{GL}(n, \mathbb{F}_q)$ consisting of matrices g that agree with the* *identity matrix* $\mathbf{I} \in \mathrm{GL}(n, \mathbb{F}_q)$ *in the first* r *rows. Then*

$$\mathbb{F}_q[V]^{G(r)} = \mathbb{F}_q[z_1, \ldots, z_r, \mathbf{d}_{n,r}, \ldots, \mathbf{d}_{n,n-1}]. \quad \square$$

The definition of the Dickson polynomials is not very practical for computations. The following formula for them is, by contrast, quite useful.

THEOREM 6.1.7 (R. E. Stong, T. Tamagawa): *Let* $n \in \mathbb{N}$, p *a prime, and* q *a power of* p. *Then the Dickson polynomials are given by the formulae*

$$\mathbf{d}_{n,i} = (-1)^{n-i} \sum_{\substack{W^* \le V^* \\ \dim(W^*)=i}} \prod_{z \notin W^*} z.$$

PROOF: The Poincaré series of $\mathbf{D}(n)$ is

$$P(\mathbf{D}(n), t) = \prod_{i=0}^{n-1} \frac{1}{1 - t^{q^n - q^i}},$$

and since $(q^n - q^i) + (q^n - q^j) > q^n - 1$, it follows that

$$P(\mathbf{D}(n), t) = t^{q^n - q^{n-1}} + t^{q^n - q^{n-2}} + \cdots + t^{q^n - 1} + \cdots.$$

Hence the homogeneous component of $\mathbf{D}(n)$ of degree $q^n - q^i$ is 1-dimensional. It therefore suffices to show that the forms defined by the above formulae are nonzero, since they clearly belong to $\mathbf{D}(n)$. Denote these temporarily by $\widetilde{\mathbf{d}}_{n,i}$, $i = 0, \ldots, n-1$. We show that they are nonzero by induction on n. The form $\widetilde{\mathbf{d}}_{n,0}$ is the product of all the nonzero elements of V^*, so is always nonzero, which establishes the result for $n = 1$. For $n > 1$ it remains only to show that $\widetilde{\mathbf{d}}_{n,1}, \ldots, \widetilde{\mathbf{d}}_{n,n-1}$ are nonzero. To this end let $W < V$ be a codimension-one subspace. The inclusion induces a restriction map $\lambda : \mathbf{D}(n) \longrightarrow \mathbf{D}(n-1)$, and one readily checks[2] from the above formulae that

$$\lambda(\widetilde{\mathbf{d}}_{n,i}) = \begin{cases} \widetilde{\mathbf{d}}^{\,q}_{n-1,i-1}, & 1 \le i \le n-1, \\ 0, & i = 0, \end{cases}$$

and so by induction the result follows. $\square$

Note that the Dickson polynomial $\mathbf{d}_{n,0}$ of maximal degree, the **top Dickson polynomial,** is the product of all the nonzero linear forms in V^*.

The invariants of the special linear groups $\mathrm{SL}(n, \mathbb{F}_q)$ can easily be derived from what we have already done. Notice that the top Dickson polynomial $\mathbf{d}_{n,0}$ is the Euler class of the $\mathrm{GL}(n, \mathbb{F}_q)$-orbit $V^* \setminus \{0\}$. This set, i.e., the

[2] Compare this with the fractal property of the Dickson algebra in Section 9.2 and Appendix A.3 in [290].

nonzero linear forms, are also an $\mathrm{SL}(n, \mathbb{F}_q)$-orbit, and the $\mathrm{SL}(n, \mathbb{F}_q)$-Euler class of this orbit, $\mathbf{L}$, is the product of one linear form from each line in V^*. This is the Euler class of the configuration of all hyperplanes in V regarded as an $\mathrm{SL}(n, \mathbb{F}_q)$-invariant configuration of hyperplanes (see Section 4.4).

THEOREM 6.1.8 (L. E. Dickson): *Let $q = p^\nu$ be a positive prime power, $\mathbb{F}_q$ the Galois field with q elements. If $n \in \mathbb{N}$ and $V = \mathbb{F}_q^n$, then*

$$\mathbb{F}_q[V]^{\mathrm{SL}(n,\mathbb{F}_q)} = \mathbb{F}_q[\mathbf{L}, \mathbf{d}_{n,1}, \ldots, \mathbf{d}_{n,n-1}].$$

PROOF: Since $\mathbf{d}_{n,0}, \ldots, \mathbf{d}_{n,n-1} \in \mathbb{F}_q[V]$ are a system of parameters, by Proposition 6.1.3, so are $\mathbf{L}, \mathbf{d}_{n,1}, \ldots, \mathbf{d}_{n,n-1} \in \mathbb{F}_q[V]$. Since the polynomials $\mathbf{L}, \mathbf{d}_{n,1}, \ldots, \mathbf{d}_{n,n-1}$ belong to $\mathbb{F}_q[V]^{\mathrm{SL}(n,\mathbb{F}_q)}$ and

$$\deg(\mathbf{L}) \cdot \prod_{n=1}^{n-1} \deg(\mathbf{d}_{n,i}) = \frac{q^n - 1}{q - 1} \cdot \prod_{i=1}^{n-1} (q^n - q^i) = |\mathrm{SL}(n, \mathbb{F}_q)|,$$

the result follows from Proposition 4.5.5. □

6.2 Transvection Groups

In this section we will examine a special class of groups that appear only in the modular case, the finite transvection groups. We begin with a definition.

DEFINITION: *An element $\mathbf{T} \in \mathrm{GL}(n, \mathbb{F})$ is called a* **transvection** *if* $\ker(\mathbf{I} - \mathbf{T}) \subset V$ *has codimension* 1 *and* $\mathrm{Im}(\mathbf{I} - \mathbf{T}) \subseteq \ker(\mathbf{I} - \mathbf{T})$. *The hyperplane $H = \ker(\mathbf{I} - \mathbf{T})$ is called the* **hyperplane** *of* $\mathbf{T}$, *and the line* $\mathrm{Im}(\mathbf{I} - \mathbf{T})$ *the* **direction** *of* $\mathbf{T}$. *A nonzero vector $x \in \mathrm{Im}(\mathbf{I} - \mathbf{T})$ is called a* **transvector** *of* $\mathbf{T}$.

Note that if the kernel of a linear transformation is a hyperplane, then the dimension formula of elementary linear algebra (see, e.g., [374] Theorem 8.4.2) implies that the image is a line. Transvections of finite order are precisely the nondiagonalizable pseudoreflections; see Section 7.1.

If $\mathbf{T} \in \mathrm{GL}(n, \mathbb{F})$ is a transvection, and $x \in \mathrm{Im}(\mathbf{I} - \mathbf{T})$ is a fixed transvector for $\mathbf{T}$, then

$$\mathbf{T}(v) = v + \varphi(v) \cdot x \qquad \forall\, v \in V,$$

where $\varphi(v) \in \mathbb{F}$. The linearity of $\mathbf{T}$ entails the linearity of φ, so $\varphi : V \longrightarrow \mathbb{F}$ is a linear functional. If $H = \ker(\mathbf{I} - \mathbf{T})$ is the hyperplane of $\mathbf{T}$, then for $u \in H$ we have $\mathbf{T}(u) = u$, so $\varphi(u) \cdot x = 0$, and since $x \neq 0$, that $\varphi(u) = 0$, so $H \subseteq \ker(\varphi)$. On the other hand, $\mathbf{T} \neq \mathbf{I}$, since $\mathbf{I} - \mathbf{T}$ is not zero, so $\varphi \neq 0$ and hence $H = \ker(\varphi)$. For fixed x this linear functional is unique, since a transvector is nonzero, so $(\mathbf{I} - \mathbf{T})(v) \in \mathrm{Span}_{\mathbb{F}}\{x\}$ means that $(\mathbf{I} - \mathbf{T})(v) = \varphi(v) \cdot x$ for a unique $\varphi(v) \in \mathbb{F}$.

On the other hand, if ψ is a fixed linear functional on V with kernel H, the hyperplane of the transvection $\mathbf{T}$, then $\psi = a \cdot \varphi$ for some nonzero $a \in \mathbb{F}$

and hence we may write

$$\mathsf{T}(v) = v + \psi(v) \cdot y,$$

where $y = a \cdot x \in \mathrm{Im}(\mathsf{I} - \mathsf{T})$ is also a transvector for T.

Before going further it will be convenient to introduce some notation.

NOTATION: *Let $\mathbb{F}$ be a field, $V = \mathbb{F}^n$, φ a nonzero linear functional on V, and $x \in V$ a vector with $\varphi(x) = 0$. Define the linear transformation*

$$\mathsf{t}(\varphi, x) : V \longrightarrow V$$

by $\mathsf{t}(\varphi, x)(v) = v + \varphi(v) \cdot x$ for all $v \in V$. (Note that $\mathsf{t}(\varphi, 0) = \mathsf{I}$.)

The preceding discussion can be summarized in the following lemma:

LEMMA 6.2.1: *Let $\mathbb{F}$ be a field, $V = \mathbb{F}^n$, and $\mathsf{T} \in GL(n, \mathbb{F})$ a transvection with hyperplane H and direction L. Then,*

(i) if φ is a linear functional with kernel H, there is a unique transvector $x \in L$ such that $\mathsf{T} = \mathsf{t}(\varphi, x)$, and

(ii) if $x \in L$ is a transvector for T, there is a unique linear functional φ with kernel H such that $\mathsf{T} = \mathsf{t}(\varphi, x)$. □

We collect some elementary properties of transvections, and the construction $\mathsf{t}(\varphi, x)$ for pairs consisting of a linear functional φ and a vector $x \in \ker(\varphi)$.

LEMMA 6.2.2: *Let $\mathbb{F}$ be a field, $V = \mathbb{F}^n$, φ a nonzero linear functional on V, and $x', x'' \in V$ nonzero vectors with $\varphi(x') = 0 = \varphi(x'')$. Then*

$$\mathsf{t}(\varphi, x'') \cdot \mathsf{t}(\varphi, x') = \mathsf{t}(\varphi, (x' + x'')) = \mathsf{t}(\varphi, x') \cdot \mathsf{t}(\varphi, x'').$$

In particular, $\mathsf{t}(\varphi, x)^{-1} = \mathsf{t}(\varphi, -x)$, and $\mathsf{t}(\varphi, x)$ is a transvection with hyperplane $H = \ker(\varphi)$ and transvector x.

PROOF: The formula follows from the simple computation

$$\begin{aligned}
\mathsf{t}(\varphi, (x' + x'')) &= v + \varphi(v) \cdot (x' + x'') = v + \varphi(v) \cdot x' + \varphi(v) \cdot x'' \\
&= v + \varphi(v + \varphi(v) \cdot x'') \cdot x' + \varphi(v) \cdot x'' \\
&= v + \varphi(v) \cdot x'' + \varphi(v + \varphi(v) \cdot x'') \cdot x' \\
&= \mathsf{t}(\varphi, x') \cdot \mathsf{t}(\varphi, x'')
\end{aligned}$$

and the commutativity of vector addition. The rest is then immediate. □

LEMMA 6.2.3: *Let $\mathbb{F}$ be a field, $V = \mathbb{F}^n$, φ', φ'' nonzero linear functionals on V, and $x', x'' \in V$ nonzero vectors with $\varphi'(x') = 0 = \varphi''(x'')$. Then $\mathsf{t}(\varphi', x') = \mathsf{t}(\varphi'', x'')$ if and only if there is a nonzero scalar $a \in \mathbb{F}^\times$ such that*

$$\varphi'' = a \cdot \varphi' \text{ and } x'' = a^{-1} x'.$$

PROOF: Suppose $\mathrm{t}(\varphi', x') = \mathrm{t}(\varphi'', x'')$. Then $\varphi'(v) \cdot x' = \varphi''(v) \cdot x''$ for all $v \in V$. Let $u \in \ker(\varphi')$. Then $\varphi'(u) \cdot x' = 0$ so $\varphi''(u) \cdot x'' = 0$ also. Since $x'' \neq 0$, this means that $\varphi''(u) = 0$, so $\ker(\varphi') \subseteq \ker(\varphi'')$, and since both kernels are hyperplanes, this means that they are equal. Hence there is a nonzero $a \in \mathbb{F}$ such that $\varphi'' = a \cdot \varphi'$. If we choose $u \in V$ with $\varphi'(u) = 1$, then we obtain

$$x' = \varphi'(u) \cdot x' = \varphi''(u) \cdot x'' = a \cdot x''.$$

The converse is clear from the definitions and the result follows. □

In a similar manner one establishes the following result:

LEMMA 6.2.4: *Let $\mathbb{F}$ be a field, $V = \mathbb{F}^n$, φ a nonzero linear functional on V, and $x \in V$ a nonzero vector with $\varphi(x) = 0$. Then for any nonzero $a \in \mathbb{F}$, $\mathrm{t}(a\varphi, x) = \mathrm{t}(\varphi, ax)$.* □

NOTATION: *Let $\mathbb{F}$ be a field, $V = \mathbb{F}^n$, and $H \subset V$ a hyperplane. Denote by $\mathcal{T}(H)$ the set of all transvections with hyperplane H together with $\mathbf{I}$. Note that Lemma 6.2.2 shows that $\mathcal{T}(H)$ forms a group.*

LEMMA 6.2.5: *Let $\mathbb{F}$ be a field, $V = \mathbb{F}^n$, $H \subset V$ a hyperplane, and φ a fixed nonzero linear functional on V with kernel H. Then the map $\tau : H \longrightarrow \mathcal{T}(H)$ defined by $\tau(x) = \mathrm{t}(\varphi, x)$ is an isomorphism of groups. In particular, if $\mathbb{F}$ is finite, then $\mathcal{T}(H)$ is an elementary abelian p-subgroup of $\mathrm{SL}(n, \mathbb{F})$. If $|\mathbb{F}| = q = p^\nu$, then $|\mathcal{T}(H)| = q^{n-1}$.*

PROOF: The map τ is bijective by Lemma 6.2.1 and a homomorphism of groups by Lemma 6.2.2. H is an elementary abelian p-group of order q^{n-1}. If $\mathbb{F}$ is a finite field, then $\mathbb{F}^\times$ contains no elements of order p, so elements of order some power of p must have determinant 1 so belong to $\mathrm{SL}(n, \mathbb{F})$. □

REMARK: If the ground field $\mathbb{F}$ has characteristic p then every element apart from $\mathbf{I}$ in $\mathcal{T}(H)$ has order p so is a pseudoreflection. If the ground field has characteristic zero then the elements of $\mathcal{T}(H)$ apart from $\mathbf{I}$ all have infinite order, so are not pseudoreflections, although they fix a hyperplane. Put another way, a transvection is a pseudoreflection only if the ground field has nonzero characteristic.

The invariants of the groups $\mathcal{T}(H)$ over finite fields were first considered by P. S. Landweber and R. E. Stong in connection with their study of the depth of rings of invariants in [224]. For this reason we call the groups $\mathcal{T}(H)$ over a finite field **Landweber-Stong groups**. Here is what they found.

THEOREM 6.2.6 (P. S. Landweber-R. E. Stong): *Let $\mathbb{F}_q$ be a finite field with $q = p^\nu$ elements and $H \subset V = \mathbb{F}_q^n$ a hyperplane. Let $z_1, \ldots, z_n \in V^*$ be a basis with $H = \ker(z_n)$. Then*

$$\mathbb{F}_q[V]^{\mathcal{T}(H)} = \mathbb{F}_q[c_q(z_1), \ldots, c_q(z_{n-1}), z_n].$$

PROOF: If $x_1, \ldots, x_n$ is the dual basis to $z_1, \ldots, z_n$, then $H = \mathrm{Span}_{\mathbb{F}_q}\{x_1, \ldots, x_{n-1}\}$, and the matrices with respect to this basis in $\mathcal{T}(H)$ are of the form

$$\begin{bmatrix} 1 & 0 & 0 & \cdots & a_1 \\ 0 & 1 & 0 & \cdots & a_2 \\ \vdots & \ddots & \ddots & \cdots & \vdots \\ 0 & 0 & \cdots \ddots & 1 & a_{n-1} \\ 0 & 0 & \cdots & 0 & 1 \end{bmatrix}, \qquad a_1, \ldots, a_{n-1} \in \mathbb{F}_q.$$

Therefore, $z_n \in \mathbb{F}_q[V]^{\mathcal{T}(H)}$. For $1 \leq i \leq n-1$ the top Chern class of the orbit of z_i is

$$c_q(z_i) = \prod_{a \in \mathbb{F}_q} (z_i + az_n) = z_i^q + \text{lower terms}.$$

Hence $c_q(z_1), \ldots, c_q(z_{n-1}), z_n \in \mathbb{F}_q[V]^{\mathcal{T}(H)}$ are a system of parameters, and the result follows from Proposition 4.5.5. □

Interestingly enough, years earlier H. Nakajima had studied the dual representations in [273]. We call these groups **Nakajima groups**. Here is what he found.

THEOREM 6.2.7 (H. Nakajima): *Let $\mathbb{F}_q$ be a finite field with $q = p^\nu$ elements and $H \subset V = \mathbb{F}_q^n$ a hyperplane. Let $z_1, \ldots, z_n \in V^*$ be a basis with $H = \ker(z_n)$. Let $\mathcal{T}(H)^{\mathrm{tr}} < \mathrm{SL}(n, \mathbb{F}_q)$ be the subgroup obtained by transposing the matrices in $\mathcal{T}(H)$. Then*

$$\mathbb{F}_q[V]^{\mathcal{T}(H)^{\mathrm{tr}}} = \mathbb{F}_q[z_1, \ldots, z_{n-1}, c_{q^{n-1}}(z_n)].$$

PROOF: The matrices in $\mathcal{T}(H)^{\mathrm{tr}}$ are

$$\begin{bmatrix} 1 & 0 & 0 & \cdots & 0 \\ 0 & 1 & 0 & \cdots & 0 \\ \vdots & \ddots & \ddots & \cdots & \vdots \\ 0 & 0 & \cdots 0 & 1 & 0 \\ a_1 & a_2 & \cdots & a_{n-1} & 1 \end{bmatrix}, \qquad a_1, \ldots, a_{n-1} \in \mathbb{F}_q,$$

so $z_1, \ldots, z_{n-1} \in \mathbb{F}_q[V]^{\mathcal{T}(H)^{\mathrm{tr}}}$, and the top Chern class of the orbit of z_n is

$$c_{q^{n-1}}(z_n) = \prod_{a_1, \ldots, a_{n-1} \in \mathbb{F}_q} (a_1 z_1 + \cdots + a_{n-1} z_{n-1} + z_n) = z_n^{q^{n-1}} + \text{lower terms}.$$

Again the proof is completed by observing that $z_1, \ldots, z_{n-1}, c_{q^{n-1}}(z_n) \in \mathbb{F}_q[V]^{\mathcal{T}(H)^{\mathrm{tr}}}$ are a system of parameters and applying Proposition 4.5.5. □

If $H \subset V = \mathbb{F}^n$ is a hyperplane in the n-dimensional vector space V over the field $\mathbb{F}$ and $\varphi : V \longrightarrow \mathbb{F}$ is a linear form with kernel H, then for any subspace

$W \subseteq H$ we may restrict the isomorphism $\tau : H \longrightarrow \mathcal{T}(H)$ to W, and the image is then a subgroup $\mathcal{T}(W) < \mathcal{T}(H)$ consisting of transvections with hyperplane H. It is not hard to see that the analogues of the theorems of P. S. Landweber and R. E. Stong, Theorem 6.2.6, and H. Nakajima, Theorem 6.2.7, hold for these groups also. For more information on the invariant theory of these transvection groups see [372] Section 8.2 and [273].

So far we have considered only subgroups of $\mathrm{GL}(n, \mathbb{F})$ all of whose nonidentity elements are transvections. We could consider more generally subgroups of $\mathrm{GL}(n, \mathbb{F})$ generated by transvections. Such a group we call a **transvection group**. In [271] H. Nakajima proves the following result about irreducible representations generated by transvections.

PROPOSITION 6.2.8 (H. Nakajima): *Let $\rho : G \hookrightarrow \mathrm{GL}(n, \mathbb{F})$ be an irreducible representation of G. Assume that the characteristic of $\mathbb{F}$ is odd, the degree n is at least 3, and that $\rho(G)$ is generated by transvections. Then $\mathbb{F}[V]^G$ is a polynomial algebra if and only if G is conjugate in $\mathrm{GL}(n, \mathbb{F})$ to $\mathrm{SL}(n, \mathbb{F})$.*

From this it follows that not every **transvection group** has an invariant algebra that is a polynomial algebra. Even worse in [273] H. Nakajima presents examples of transvection groups with non-Cohen–Macaulay invariants.

Finally, there is another way to generalize the results by P. S. Landweber–R. E. Stong and H. Nakajima. Instead of restricting the hyperplane H to a subspace W, one takes an **arrangement of hyperplanes** $H_1, \dots, H_l$ and considers the invariants of stabilizer and hyperplanewise stabilizer subgroups associated to this arrangement; see [313] for a general treatment of arrangements of hyperplanes, and [299] for a discussion of the invariants of groups associated to such arrangements.

6.3 p-Groups in Characteristic p

In this section we will sample what is known about the invariants of finite p-groups in characteristic p. Among the many problems involving invariants of finite p-groups in characteristic p we choose to begin our discussion of their invariants with some results concerning the cyclic group $\mathbb{Z}/p$ from [378], [288], and [292]. We will apply these results in Section 6.5.

Let $\rho : \mathbb{Z}/p \hookrightarrow \mathrm{GL}(n, \mathbb{F})$ be a representation of $\mathbb{Z}/p$ over the field $\mathbb{F}$ of characteristic p implemented by the matrix $\mathbf{P}$. By a suitable change of basis, $\mathbf{P}$ can be put into Jordan canonical form, viz.,

$$\mathbf{P} = \begin{bmatrix} \mathbf{J}_1 & 0 & 0 & \cdots & 0 \\ 0 & \mathbf{J}_2 & 0 & \cdots & 0 \\ \vdots & 0 & \ddots & \cdots & 0 \\ 0 & \cdots & \cdots & 0 & \mathbf{J}_k \end{bmatrix},$$

where each Jordan block $\mathbf{J}_i$, $i = 1, \ldots, k$, is of the form

$$\mathbf{J}_i = \begin{bmatrix} 1 & 1 & 0 & \cdots & 0 \\ 0 & 1 & 1 & \ddots & 0 \\ \mathbf{0} & \ddots & \ddots & \ddots & \mathbf{0} \\ \vdots & \vdots & \ddots & 1 & 1 \\ 0 & \cdots & \cdots & 0 & 1 \end{bmatrix} \in \mathrm{GL}(n_i, \mathbb{F})$$

for some integer n_i, with $1 \leq n_i \leq p$ and $n_1 + \cdots + n_k = n$. This shows, among other things, that the representation ρ is defined over the prime field $\mathbb{F}_p$, and so there is no loss in generality to restrict the ground field to be $\mathbb{F}_p$. If there is only one block and it has size 2, then the invariants $\mathbb{F}_p[x, y]^{\mathbb{Z}/p}$ are rather easy to compute.

EXAMPLE 1 ($\mathbb{Z}/p < \mathrm{GL}(2, \mathbb{F}_p)$): The matrices in $\mathbb{Z}/p$ are

$$\begin{bmatrix} 1 & a \\ 0 & 1 \end{bmatrix} \in \mathrm{GL}(2, \mathbb{F}_p), \quad a \in \mathbb{F}_p,$$

so we see that $y \in \mathbb{F}[x, y]^{\mathbb{Z}/p}$, and the top Chern class of the orbit of x is $c_p(x) = x(x^{p-1} - y^{p-1})$. These two invariants form a system of parameters for the ring of invariants, and the product of their degrees is p, so by Proposition 4.5.5 the ring of invariants $\mathbb{F}[x, y]^{\mathbb{Z}/p} = \mathbb{F}[c_p(x), y]$ is a polynomial algebra generated by these forms.

On the basis of this example it might seem reasonable to try to compute the invariants of a single Jordan block by induction on the rank of the block. It turns out that this is misleading; the smaller Jordan blocks have proven very difficult to handle (see, e.g., [15], [124], [288], and [347]), and it is wiser to work from the top down (see, e.g., [292], [295], [375], [378] and [386]), i.e., to start with the Jordan block of size p. This Jordan block implements the regular representation of $\mathbb{Z}/p$, so is a permutation representation, and it is this that we will exploit. The following result[3] is in fact more general. It is valid for any permutation representation of a cyclic p-group, if we replace the equality in *(i)* by the inequality $\leq$.

THEOREM 6.3.1: *Let p be a prime, $\mathbb{F}$ a field of characteristic p, and X a finite $\mathbb{Z}/p$-set. Denote by $X/\mathbb{Z}/p$ the set of $\mathbb{Z}/p$-orbits in X. Then,*

(i) the ideal $\mathrm{Im}(\mathrm{Tr}^{\mathbb{Z}/p})$ *is prime,*

(ii) $\mathfrak{ht}(\mathrm{Im}(\mathrm{Tr}^{\mathbb{Z}/p})) = |X| - |X/\mathbb{Z}/p|$, *and*

(iii) $\mathbb{F}[X]^{\mathbb{Z}/p}/\mathrm{Im}(\mathrm{Tr}^{\mathbb{Z}/p}) \cong \mathbb{F}[c_{\mathrm{top}}(x) \mid x \in X\,]$.

PROOF: By Proposition 3.2.2 $\mathbb{F}[X]^{\mathbb{Z}/p}$ has an additive basis consisting of orbit sums $\mathfrak{S}_{x^K}$ of monomials $x^K \in \mathbb{F}[X]$. For any monomial $x^K \in \mathbb{F}[X]$ the orbit of x^K consists of x^K alone, or p distinct monomials. If the

[3] In this result, note that *(i)* is also a consequence of Theorem 6.4.7.

orbit consists of x^K alone, then x^K is fixed by $\mathbb{Z}/p$, and $\mathrm{Tr}^{\mathbb{Z}/p}(x^K) = 0$, so $\mathrm{Im}(\mathrm{Tr}^{\mathbb{Z}/p})$ consists precisely of the orbit sums of the monomials with free orbits.

Let $S \subset \mathbb{F}[X]$ be the subalgebra generated by the top Chern classes $c_{\mathrm{top}}(x)$ for $x \in X$. Note that if the element $x \in X$ has a free orbit, then the top Chern class has degree p; otherwise, the degree is 1. The top Chern class $c_{\mathrm{top}}(x)$ of an element $x \in X$ is a monomial, and the orbit of this monomial consists of $c_{\mathrm{top}}(x)$ alone. Moreover, if x^K is a monomial with only one element in its orbit, then it is a product of such top orbit Chern classes, so $\mathfrak{S}_{x^K} = x^K$ lies in S.

Hence S together with $\mathrm{Im}(\mathrm{Tr}^{\mathbb{Z}/p})$ generates $\mathbb{F}[X]^{\mathbb{Z}/p}$ as an algebra, and $S \cap \mathrm{Im}(\mathrm{Tr}^{\mathbb{Z}/p}) = (0)$. Therefore, the composition

$$S \hookrightarrow \mathbb{F}[X]^{\mathbb{Z}/p} \longrightarrow \mathbb{F}[X]^{\mathbb{Z}/p}/\mathrm{Im}(\mathrm{Tr}^{\mathbb{Z}/p})$$

is an isomorphism. Since $S = \mathbb{F}[c_{\mathrm{top}}(x) \mid x \in X]$, it is a polynomial algebra, so an integral domain of Krull dimension equal to the number of $\mathbb{Z}/p$-orbits in X. Therefore, $\mathrm{Im}(\mathrm{Tr}^{\mathbb{Z}/p}) \subset \mathbb{F}[X]^{\mathbb{Z}/p}$ is a prime ideal of height $|X| - |X/\mathbb{Z}/p|$. □

If one examines the proof of Theorem 6.3.1, then it becomes clear that for permutation groups of $\mathbb{Z}/p$ the image of the transfer must be generated by the orbit sums of the free orbits. The analogue of this for a general modular representation is based on the following proposition, which seems well known, but for which we have no precise reference.

PROPOSITION 6.3.2: *Let G be a finite group and M an indecomposable G-module over the field $\mathbb{F}$. Then $\mathrm{Tr}^G : M \longrightarrow M$ is nonzero if and only if M is isomorphic to $\mathbf{P}_{\mathbb{F}}$, the projective cover of the trivial G-module $\mathbb{F}$.*

PROOF: Choose a finite-dimensional projective module P and an epimorphism $\varphi : P \longrightarrow M$. Let $P = P_1 \oplus \cdots \oplus P_r$ be a decomposition of P into indecomposable G-modules. Since each P_i is a direct summand of the projective module P, it is itself projective.

Choose $x \in M$ such that $\mathrm{Tr}^G(x) \neq 0$ and let $u \in P$ be a preimage of x, i.e., $\varphi(u) = x$. Write $u = u_1 + \cdots + u_r$ with $u_i \in P_i$ for $i = 1, \ldots, r$. Then

$$0 \neq \mathrm{Tr}^G(x) = \mathrm{Tr}^G(\varphi(u)) = \varphi(\mathrm{Tr}^G(u_1)) + \cdots + \varphi(\mathrm{Tr}^G(u_r))$$

certainly implies that $\mathrm{Tr}^G(u_s) \neq 0$ for some $1 \leq s \leq r$. Choose such an index s. Then $\mathrm{Span}_{\mathbb{F}}\{\mathrm{Tr}^G(u_s)\} \subseteq P_s$ is a 1-dimensional trivial G-submodule. Since P_s is indecomposable, its socle must be simple ([16] Corollary 5 in Section 6, or [345] §14.3 Corollaire 1), and since $\mathrm{Span}_{\mathbb{F}}\{\mathrm{Tr}^G(u_s)\} \subseteq \mathrm{soc}(P_s)$, it follows that we must have equality, i.e., $\mathrm{Span}_{\mathbb{F}}\{\mathrm{Tr}^G(u_s)\} = \mathrm{soc}(P_s)$. So P_s is isomorphic to $\mathbf{P}_{\mathbb{F}}$.

To complete this part of the proof we show that the restriction of φ to P_s induces an isomorphism $\varphi_s : P_s \xrightarrow{\cong} M$. First note that φ_s is injective, for if $\ker(\varphi_s) \neq 0$, then $\mathrm{soc}(P_s) \subseteq \ker(\varphi_s)$, which contradicts the fact that $0 \neq \varphi(u_s)$ and $u_s \in \mathrm{soc}(P_s)$. Hence φ_s is injective.

Since P_s is projective, it is also injective by Theorem 4 in Section 6 of [16]. So $\varphi(P_s) \subseteq M$ is a direct summand. M, however, is indecomposable, so contains no nontrivial proper direct summands, and since $\varphi(P_s) \neq (0)$, it follows that $\varphi(P_s) = M$. Hence φ_s is also surjective and so an isomorphism.

Any finite-dimensional G-module is a direct sum of indecomposable G-modules. From what we have just shown, if $\mathrm{Tr}^G : \mathbf{P}_{\mathbb{F}} \circlearrowleft$ is zero, then $\mathrm{Tr}^G : M \circlearrowleft$ is likewise trivial for any finite-dimensional G-module. So to finish the proof it suffices to exhibit a finite-dimensional G-module M where $\mathrm{Tr}^G : M \circlearrowleft$ is nonzero. If $M = \mathbb{F}(G)$ is the left regular representation of G with the elements of G as basis, then $\mathrm{Tr}^G(1) = \sum_{g \in G} g \neq 0 \in M$. □

COROLLARY 6.3.3: *Let G be a finite group and M a finite-dimensional G-module over the field $\mathbb{F}$. Then $\dim(\mathrm{Im}(\mathrm{Tr}^G))$ is the multiplicity of $\mathbf{P}_{\mathbb{F}}$, the projective cover of the trivial G-module $\mathbb{F}$, as an indecomposable factor of M.*

PROOF: Write $M = M_1 \oplus \cdots \oplus M_k$, where $M_1, \ldots, M_k$ are indecomposable G-modules. Then

$$\mathrm{Im}(\mathrm{Tr}^G : M \longrightarrow M) = \mathrm{Im}(\mathrm{Tr}^G : M_1 \longrightarrow M_1) \oplus \cdots \oplus \mathrm{Im}(\mathrm{Tr}^G : M_k \longrightarrow M_k).$$

If we apply Proposition 6.3.2 to each indecomposable factor, we obtain the desired conclusion. □

Here is an application of this to invariant theory; see also [257].

COROLLARY 6.3.4: *Let $\rho : G \hookrightarrow \mathrm{GL}(n, \mathbb{F})$ be a representation of a finite group over the field $\mathbb{F}$. Then the Poincaré series of $\mathrm{Im}(\mathrm{Tr}^G)$ is given by*

$$P(\mathrm{Im}(\mathrm{Tr}^G), t) = \sum_{k=0}^{\infty} \mu(\mathbf{P}_{\mathbb{F}}, k) t^k,$$

where $\mu(\mathbf{P}_{\mathbb{F}}, k)$ is the multiplicity of the projective cover $\mathbf{P}_{\mathbb{F}}$ of the trivial module as an indecomposable factor of $\mathbb{F}[V]_k$. □

It is known, [59], that $\mathrm{Im}\left(\mathrm{Tr}^{\mathrm{GL}(n, \mathbb{F}_q)}\right)$ is the principal ideal generated by $\mathbf{d}_{n,0}^{n-1}$. We therefore have the following consequence for the first occurrence problem (see, e.g., [440] for a discussion of first occurrence problems of the general linear and symmetric groups) of the projective cover $\mathbf{P}_{\mathbb{F}_q}(\mathrm{GL}(n, \mathbb{F}_q))$ of the trivial 1-dimensional $\mathrm{GL}(n, \mathbb{F}_q)$-module as an indecomposable factor in $\mathbb{F}_q[V]$.

COROLLARY 6.3.5: *The first occurrence of* $\mathbf{P}_{\mathbb{F}_q}(\mathrm{GL}(n, \mathbb{F}_q))$, *the projective cover of the trivial* 1*-dimensional* $\mathrm{GL}(n, \mathbb{F}_q)$*-module as an indecomposable factor in* $\mathbb{F}_q[V]$, *is in degree* $(q^n - 1)^{n-1}$. □

After this digression we return to our study of the invariants of finite p-groups in characteristic p.

Another central theme in invariant theory has been the study of when the invariants of a representation $\rho : G \hookrightarrow \mathrm{GL}(n, \mathbb{F})$ are a polynomial algebra. For the case of finite p-groups this question was studied intensively by H. Nakajima in a series of papers [270]-[281]. Here is one of his most interesting results. It is taken from [277].

THEOREM 6.3.6 (H. Nakajima): *Let* $\rho : P \hookrightarrow \mathrm{GL}(n, \mathbb{F})$ *be a representation of a finite* p*-group* P *over the field* $\mathbb{F}$ *of characteristic* p. *Suppose that* P *contains subgroups* $P_1, \ldots, P_n$ *for which there is a basis* $z_1, \ldots, z_n \in V^*$ *such that for* $i = 1, \ldots, n$

(i) *the* P*-orbit* $[z_i]$ *of* z_i *and the* P_i*-orbit of* z_i *are the same,*
(ii) P_i *acts freely on the orbit of* z_i,
(iii) $[z_i] \subseteq \mathrm{Span}_{\mathbb{F}} \{z_1, \ldots, z_i\}$, *and finally, that*
(iv) $|P_1| \cdots |P_n| = |P|$.

Then $\mathbb{F}[V]^P$ *is a polynomial algebra generated by the top Chern classes of the orbits* $[z_1], \ldots, [z_n]$.

PROOF: Let

$$V^* = V_n^* \supset V_{n-1}^* \supset \cdots \supset V_0^* = \{0\}$$

be the complete flag defined by the basis $z_1, \ldots, z_n$, i.e., we define the vector subspaces $V_i^* = \mathrm{Span}\{z_1, \ldots, z_i\}$, for $i = 1, \ldots, n$. Then these are P-invariant subspaces fulfilling the hypotheses of Theorem 4.5.6 and the result follows. □

For $\mathbb{F} = \mathbb{F}_p$ the prime field, H. Nakajima shows that the converse holds [277]: If $\mathbb{F}_p[V]^P$ is a polynomial algebra, then the representation of P satisfies conditions *(i)-(iv)* and the generators of the ring of invariants are precisely the top Chern classes of the orbits of $[z_1], \ldots, [z_n]$. When $\mathbb{F}$ is not the prime field this is no longer the case: There is an example due to R. E. Stong, [388], that we explain next. We are grateful to R. E. Stong for his permission to include this example here.

EXAMPLE 2 (R. E. Stong): Fix a prime integer, $p \in \mathbb{N}$, and choose elements $\alpha, \beta \in \mathbb{F}_{p^3}$ in the Galois field with p^3 elements such that $1, \alpha, \beta \in \mathbb{F}_{p^3} \cong \mathbb{F}_p^3$ are a basis for $\mathbb{F}_{p^3}$ over $\mathbb{F}_p$ as an $\mathbb{F}_p$-vector space, when $\mathbb{F}_p$ is identified with the prime subfield in $\mathbb{F}_{p^3}$. The matrices

$$\mathrm{T}_1 = \begin{bmatrix} 1 & 0 & 0 \\ 1 & 1 & 0 \\ 0 & 0 & 1 \end{bmatrix}, \quad \mathrm{T}_2 = \begin{bmatrix} 1 & 0 & 0 \\ 0 & 1 & 0 \\ 1 & 0 & 1 \end{bmatrix}, \quad \mathrm{T}_3 = \begin{bmatrix} 1 & 0 & 0 \\ \alpha & 1 & 0 \\ \beta & 0 & 1 \end{bmatrix} \in \mathrm{GL}(3, \mathbb{F}_{p^3})$$

define a representation $\rho : E \hookrightarrow \mathrm{GL}(3, \mathbb{F}_{p^3})$ of the elementary abelian p-group $E = \mathbb{Z}/p \oplus \mathbb{Z}/p \oplus \mathbb{Z}/p$ on $V = \mathbb{F}_{p^3}^3$. Let $z_1, z_2, z_3 \in V^*$ be the canonical basis for the dual vector space. The following formulae will prove useful in the sequel:

$$\begin{array}{lll} \mathrm{T}_1(z_1) = z_1, & \mathrm{T}_1(z_2) = z_1 + z_2, & \mathrm{T}_1(z_3) = z_3, \\ \mathrm{T}_2(z_1) = z_1, & \mathrm{T}_2(z_2) = z_2, & \mathrm{T}_2(z_3) = z_1 + z_3, \\ \mathrm{T}_3(z_1) = z_1, & \mathrm{T}_3(z_2) = \alpha z_1 + z_2, & \mathrm{T}_3(z_3) = \beta z_1 + z_3, \end{array}$$

and

$$\mathrm{T}_1^a\mathrm{T}_2^b\mathrm{T}_3^c(z_1) = z_1, \quad \mathrm{T}_1^a\mathrm{T}_2^b\mathrm{T}_3^c(z_2) = (a + c\alpha)z_1 + z_2, \quad \mathrm{T}_1^a\mathrm{T}_2^b\mathrm{T}_3^c(z_3) = (b + c\beta)z_1 + z_3.$$

Let $E' < E$ be the subgroup of E generated by T_1 and T_2. The orbits of the linear forms z_1, z_2, and z_3 under E' are

$$\begin{aligned} [z_1]' &= \{z_1\}, \\ [z_2]' &= \{z_2 + \lambda z_1 \mid \lambda \in \mathbb{F}_p\}, \\ [z_3]' &= \{z_3 + \mu z_1 \mid \mu \in \mathbb{F}_p\}, \end{aligned}$$

and therefore by the theorem of H. Nakajima and R. E. Stong (Proposition 4.5.6) we obtain

$$\mathbb{F}_{p^3}[z_1, z_2, z_3]^{E'} = \mathbb{F}_{p^3}[z_1, \Phi, \Psi],$$

where

$$\Phi = z_2^p - z_2 z_1^{p-1} \text{ and } \Psi = z_3^p - z_3 z_1^{p-1}$$

are the top Chern classes of the E'-orbits of z_2 and z_3, respectively.

The group $E'' = E/E'$ acts on $\mathbb{F}_{p^3}[z_1, z_2, z_3]^{E'}$ via the matrix T_3, and one obtains

$$\begin{aligned} \mathrm{T}_3(z_1) &= z_1, \\ \mathrm{T}_3(\Phi) &= (\alpha^p - \alpha)z_1^p + \Phi, \\ \mathrm{T}_3(\Psi) &= (\beta^p - \beta)z_1^p + \Psi. \end{aligned}$$

Note that neither α nor β belongs to $\mathbb{F}_p$, and hence $(\alpha^p - \alpha) \neq 0 \neq (\beta^p - \beta)$. From these formulae it follows that the polynomial

$$\Lambda = (\beta^p - \beta)\Phi - (\alpha^p - \alpha)\Psi \neq 0$$

is invariant under T_3, the orbit of Φ under T_3 is

$$\left\{\Phi + \gamma(\alpha^p - \alpha)z_1^p \mid \gamma \in \mathbb{F}_p\right\},$$

and hence the top Chern class of this orbit, viz.,

$$\Gamma = \prod_{\gamma \in \mathbb{F}_p} (\Phi + \gamma(\alpha^p - \alpha)z_1^p) = \Phi^p - (\alpha^p - \alpha)^{p-1}\Phi z_1^{p(p-1)},$$

is also a nonzero T_3-invariant polynomial. Therefore, since

$$\mathbb{F}_{p^3}[z_1, z_2, z_3]^E = \left(\mathbb{F}_{p^3}[z_1, z_2, z_3]^{E'}\right)^{E''},$$

[372] Proposition 1.5.1, it follows that $z_1, \Lambda, \Gamma \in \mathbb{F}_{p^3}[z_1, z_2, z_3]^E$. The polynomials Φ and Ψ satisfy

$$\Phi \equiv z_2^p \bmod (z_1) \text{ and } \Psi \equiv z_3^p \bmod (z_1),$$

so $\Lambda \equiv (\beta^p - \beta)z_2^p - (\alpha^p - \alpha)z_3^p \bmod (z_1)$, and $\Gamma \equiv z_2^{p^2} \bmod (z_1)$. Therefore,

$$\frac{\mathbb{F}_{p^3}[z_1, z_2, z_3]}{(z_1, \Gamma, \Lambda^p)} \cong \frac{\mathbb{F}_{p^3}[z_2, z_3]}{(z_2^{p^2}, z_3^{p^2})}$$

is totally finite, and hence by the Noether normalization theorem (Theorem A.3.1) the polynomials $z_1, \Lambda^p, \Gamma \in \mathbb{F}_{p^3}[z_1, z_2, z_3]^E$ form a homogeneous system of parameters. This, in turn, implies that the elements $z_1, \Lambda, \Gamma \in \mathbb{F}_{p^3}[z_1, z_2, z_3]$ form a homogeneous system of parameters, and therefore, by Proposition 4.5.5, we have

$$\mathbb{F}_{p^3}[z_1, z_2, z_3]^E = \mathbb{F}_{p^3}[z_1, \Lambda, \Gamma],$$

where

$$\deg(z_1) = 1, \qquad \deg(\Lambda) = p, \qquad \deg(\Gamma) = p^2.$$

R. E. Stong next shows that Λ cannot be the top Chern class of an orbit of E on V^* as follows. Let $z \in V^*$ and denote by $E_z < E$ the isotropy group of z. The orbit of z contains $|E : E_z|$ elements, so if we can show that $|E_z| = 1, p$, or p^3, it will follow that no E-orbit contains p elements, so no polynomial of degree p such as Λ can be the top Chern class of an E-orbit.

Let $0 \neq z \in V^*$ and write

$$z = rz_1 + sz_2 + tz_3.$$

If $s = 0 = t$, then $E_z = E_{z_1} = E$ has order p^3, so we may suppose that $(s, t) \neq (0, 0)$. If $\mathrm{T} = \mathrm{T}_1^a\mathrm{T}_2^b\mathrm{T}_3^c \in E$, then using the formulae at the start of this discussion we obtain

$$\begin{aligned} \mathrm{T}(z) &= r\mathrm{T}_1^a\mathrm{T}_2^b\mathrm{T}_3^c(z_1) + s\mathrm{T}_1^a\mathrm{T}_2^b\mathrm{T}_3^c(z_2) + t\mathrm{T}_1^a\mathrm{T}_2^b\mathrm{T}_3^c(z_3) \\ &= z + ((a + c\alpha)s + (b + c\beta)t)\, z_1. \end{aligned}$$

So $\mathrm{T} \in E_z$ if and only if

$$(a + c\alpha)s + (b + c\beta)t = 0.$$

If $\mathrm{T}' = \mathrm{T}_1^{a'}\mathrm{T}_2^{b'}\mathrm{T}_3^{c'}$, $\mathrm{T}'' = \mathrm{T}_1^{a''}\mathrm{T}_2^{b''}\mathrm{T}_3^{c''} \in E_z$, then

$$\begin{aligned} (a' + c'\alpha)s + (b' + c'\beta)t &= 0, \\ (a'' + c''\alpha)s + (b'' + c''\beta)t &= 0. \end{aligned}$$

Since $(s, t) \neq (0, 0)$, we may assume without loss of generality that $t \neq 0$. If we multiply the first equation by $(a'' + c''\alpha)$, the second equation by $(a' + c'\alpha)$, and then subtract the second equation from the first equation, we get

$$0 = (a''b' - a'b'') + (b'c'' - b''c')\alpha + (a''c' - a'c'')\beta.$$

Since $1, \alpha, \beta \in \mathbb{F}_{p^3} = \mathbb{F}_p^3$ are linearly independent over $\mathbb{F}_p$, we conclude that the determinants of any of the 2×2 minors of the matrix

$$\begin{bmatrix} a' & b' & c' \\ a'' & b'' & c'' \end{bmatrix}$$

are zero. Hence this matrix must have rank 1, so $\mathbf{T}'$ and $\mathbf{T}''$ are linearly dependent in $E = \mathbb{F}_p^3$. Therefore, $|E_z| = 1$ or p, and hence $|E : E_z| = 1, p^2$, or p^3 as claimed.

To recapitulate: $E = \mathbb{Z}/p \oplus \mathbb{Z}/p \oplus \mathbb{Z}/p$ acts on $V = \mathbb{F}_{p^3}^3$ with ring of invariants

$$\mathbb{F}_{p^3}[V]^P = \mathbb{F}_{p^3}[z_1, \Lambda, \Gamma]$$

where $\deg(z_1) = 1$, $\deg(\Lambda) = p$, and $\deg(\Gamma) = p^2$. No orbit of E on V^* contains p elements, and hence there is no way to choose Λ to be the top Chern class of an orbit of E on V^*. Therefore, Nakajima's theorem becomes false over a general Galois field, nor is it clear what to expect should replace it.

We have already commented several times that the vector invariants of the tautological representation τ of $\Sigma_2 = \mathbb{Z}/2$ over a field $\mathbb{F}$ of characteristic 2 serve as examples of a negative nature. In particular, if $\mathbb{Z}/2$ acts on $\mathbb{F}_2\begin{bmatrix} x_1, \ldots, x_k \\ y_1, \ldots, y_k \end{bmatrix}$ by simultaneous interchange of $x_1, \ldots, x_k$ with $y_1, \ldots, y_k$, then

$$x_1 \cdots x_k + y_1 \cdots y_k \in \mathbb{F}_2\begin{bmatrix} x_1, \ldots, x_k \\ y_1, \ldots, y_k \end{bmatrix}^{\mathbb{Z}/2}$$

is indecomposable, so $\beta\left(\mathbb{F}_2\begin{bmatrix} x_1, \ldots, x_k \\ y_1, \ldots, y_k \end{bmatrix}^{\mathbb{Z}/2}\right) \geq k$. What increases the interest in, and the importance of, these examples is the remarkable fact that the Hilbert ideal $\mathfrak{h}(\mathbb{Z}/2) \subset \mathbb{F}_2\begin{bmatrix} x_1, \ldots, x_k \\ y_1, \ldots, y_k \end{bmatrix}$ (see Section 2.3) is in fact generated by linear and quadratic forms, i.e., forms of degree less than or equal to $2 = |\mathbb{Z}/2|$. Here is how this goes.

PROPOSITION 6.3.7: *Let τ_k be the permutation representation of $\mathbb{Z}/2$ acting on $\mathbb{F}_2\begin{bmatrix} x_1, \ldots, x_k \\ y_1, \ldots, y_k \end{bmatrix}$ by simultaneous interchange of $x_1, \ldots, x_n$ with $y_1, \ldots, y_n$. Then the Hilbert ideal $\mathfrak{h}(\tau_k) \subset \mathbb{F}_2\begin{bmatrix} x_1, \ldots, x_k \\ y_1, \ldots, y_k \end{bmatrix}$ is generated by linear and quadratic forms.*

PROOF: The form $x_i + y_i$ is invariant, so belongs to $\mathfrak{h}(\tau_k)$ for $i = 1, \ldots, k$. Since the ground field has characteristic 2 this implies $x_i \equiv y_i$ mod $\mathfrak{h}(\tau_k)$ for $i = 1, \ldots, k$.

Since τ_k is a permutation representation, the orbit sums of monomials generate the ring of invariants $\mathbb{F}_2\begin{bmatrix} x_1, \ldots, x_k \\ y_1, \ldots, y_k \end{bmatrix}^{\mathbb{Z}/2}$. Hence the Hilbert ideal is generated by the orbit sums of the monomials of strictly positive degree. The orbit sum of a monomial $x^C y^C$ is $x^C y^C$ again which is a product of invariant quadratic forms, viz., $x^C y^C = \prod_{i=1}^{k}(x_i y_i)^{c_i}$.

For a monomial $x^A y^B$ with $A \neq B$ the orbit sum is $x^A y^B + x^B y^A$. If such a monomial has degree at least 2 we need to show that its orbit sum is not needed to generate the Hilbert ideal. Without loss of generality we may suppose that $a_1 \neq 0$. Let $A' = (a_1 - 1, a_2, \ldots, a_k)$. Then

$$\begin{aligned} x^A y^B + x^B y^A &= x_1 x^{A'} y^B + x^B y_1 y^{A'} \\ &\underset{\mathfrak{h}(\tau_k)}{\cong} y_1 x^{A'} y^B + x^B y_1 y^{A'} = y_1(x^{A'} y^B + x^B y^{A'}). \end{aligned}$$

The binomial $x^{A'} y^B + x^B y^{A'}$ is the orbit sum of the monomial $x^{A'} y^B$ so in $\mathfrak{h}(\tau_k)$. Since we assumed $\deg(x^A y^B) \geq 2$ the degree of $x^{A'} y^B + x^B y^{A'}$ is strictly positive. This shows that $x^A y^B + x^B y^A$ is not needed to generate the Hilbert ideal $\mathfrak{h}(\tau_k)$. □

A similar argument can be applied to the vector invariants of permutation representations of the cyclic group $\mathbb{Z}/p$ where p is an odd prime. As for the case $p = 2$ the Hilbert ideal is generated by forms of degree at most $p = |\mathbb{Z}/p|$.

6.4 The Transfer Variety

If $\rho : G \hookrightarrow \mathrm{GL}(n, \mathbb{F})$ is a representation of a finite group and $\mathrm{Im}(\mathrm{Tr}^G) \subseteq \mathbb{F}[V]^G$ the transfer ideal, then we extend $\mathrm{Im}(\mathrm{Tr}^G)$ to an ideal $(\mathrm{Im}(\mathrm{Tr}^G))^e$ in $\mathbb{F}[V]$ by taking the ideal in $\mathbb{F}[V]$ that is generated by the elements in $\mathrm{Im}(\mathrm{Tr}^G)$. This extended ideal defines an affine algebraic set, the transfer variety, which we study in this section. The material of this section is from the unpublished manuscripts [215] and [216]. We are grateful to K. Kuhnigk for her permission to use it here. Other papers treating the transfer variety are, e.g., [55] and [119]. For the basic facts and notation in connection with the transfer we refer to Sections 2.2 and 2.4.

DEFINITION: *Let* $\rho : G \hookrightarrow \mathrm{GL}(n, \mathbb{F})$ *be a representation of a finite group over the field* $\mathbb{F}$. *The* **transfer variety**, *denoted by* $\mathfrak{X}_G \subseteq V$, *is defined by*

$$\mathfrak{X}_G = \left\{ x \in V \mid \mathrm{Tr}^G(f)(x) = 0 \ \forall f \in \mathrm{Tot}(\mathbb{F}[V]) \right\}.$$

N.b. Since $\mathfrak{X}_G$ is an ***affine*** variety, we must use all polynomial functions to define it, and not just homogeneous ones.

Recall that $I_g \subseteq \mathbb{F}[V]$ denotes the ideal generated by $\partial_g(V^*) = (1-g)(V^*)$, where, of course, V^* is the dual space of $V = \mathbb{F}^n$. See Section 2.4 for the definition of ∂_g and its basic properties.

LEMMA 6.4.1: *Let* $g \in \mathrm{GL}(n, \mathbb{F})$ *and* $\mathfrak{V}(I_g) \subseteq V$ *the affine variety defined by the ideal* $I_g \subset \mathrm{Tot}(\mathbb{F}[V])$. *Then* $\mathfrak{v}(I_g) = V^g$, *where* $V^g \subseteq V$ *is the fixed-point set of the element* g *acting on* V.

PROOF: An element $x \in V$ belongs to $\mathfrak{v}(I_g)$ if and only if $\partial_g \ell(x) = 0$ for every linear form ℓ. If $x \in V^g$, then

$$\partial_g \ell(x) = \ell(x) - \ell(g^{-1}x) = \ell(x) - \ell(x) = 0,$$

so $x \in \mathfrak{v}(I_g)$. Conversely, since

$$\partial_g \ell(x) = \ell(x) - \ell(g^{-1}x),$$

it follows that $x \in \mathfrak{v}(I_g)$ if and only if $\ell(x) = \ell(g^{-1}x)$ for every linear form ℓ. Since the linear forms separate the points in V, it follows that $x = g^{-1}x$, and hence x belongs to V^g. □

The transfer variety is defined by the radical of the ideal in $\mathbb{F}[V]$ generated by $\mathrm{Im}(\mathrm{Tr}^G)$, which by Feshbach's transfer theorem (Theorem 2.4.5) is contained in the intersection of the ideals I_g, where g ranges over the elements of order p in G. Passing to varieties turns the inclusion around and the intersection into a union, so we get the following:

COROLLARY 6.4.2: *Let* $\rho : G \hookrightarrow \mathrm{GL}(n, \mathbb{F})$ *be a representation of a finite group over the field* $\mathbb{F}$. *Then*

$$\bigcup_{|g|=p} V^g \subseteq \mathfrak{X}_G,$$

where the union runs over all the elements of order p *in* G. □

We next show that the inclusion of Corollary 6.4.2 is in fact an equality.

LEMMA 6.4.3: *Let* $B \subseteq V = \mathbb{F}^n$ *be a finite subset and* $b_0 \in B$. *Then there exists a polynomial function* $h \in \mathrm{Tot}(\mathbb{F}[V])$ *such that*

$$h(b_0) = 1, \text{ and } h(b) = 0, \ \forall\, b \in B, b \neq b_0.$$

PROOF: It is enough to consider the case of a pair $x, y \in V$ of distinct points, and prove the existence of a polynomial function $h_{x,y} \in \mathrm{Tot}(\mathbb{F}[V])$ with the property

$$h_{x,y}(x) = 1, \qquad h_{x,y}(y) = 0.$$

For given this, the general case is solved by setting

$$h = \prod_{\substack{b \in B \\ b \neq b_0}} h_{b_0,b}.$$

So, suppose $x \neq y \in V = \mathbb{F}^n$. Write $x = (x_1, \ldots, x_n)$, $y = (y_1, \ldots, y_n)$, and choose i between 1 and n with $x_i \neq y_i$. The function $h_{x,y} : V \longrightarrow \mathbb{F}$ defined by

$$h_{x,y}(a_1, \ldots, a_n) = \frac{a_i - y_i}{x_i - y_i}$$

is linear, though not homogeneous, and has the desired property. □

LEMMA 6.4.4: *Let $\rho : G \hookrightarrow \mathrm{GL}(n, \mathbb{F})$ be a representation of a finite group G over the field $\mathbb{F}$, $x \in V$, and $G_x \le G$ the isotropy group of x. Then for any $f \in \mathbb{F}[V]^G$ of positive degree we have*

$$\mathrm{Tr}^G(f)(x) = |G_x|\,\mathrm{Tr}^G_{G_x}(f)(x).$$

PROOF: Choose elements $g_1, \ldots, g_m \in G$ that are simultaneously a left and right transversal for G_x in G. This is always possible by König's lemma, [466] pp. 12–13. Then $g_1^{-1}, \ldots, g_m^{-1}$ is also a left transversal for G_x in G, and

$$\begin{aligned}
\mathrm{Tr}^G(f)(x) &= \sum_{g \in G} gf(x) = \sum_{i=1}^{m} \sum_{h \in G_x} hg_i f(x) = \sum_{i=1}^{m} \sum_{h \in G_x} f((hg_i)^{-1}x) \\
&= \sum_{i=1}^{m} \sum_{h \in G_x} f(g_i^{-1}hx) = \sum_{i=1}^{m} (\underbrace{f(g_i^{-1}x) + \cdots + f(g_i^{-1}x)}_{|G_x|}) \\
&= \sum_{i=1}^{m} |G_x| f(g_i^{-1}x) = |G_x| \sum_{i=1}^{m} g_i f(x) = |G_x|\,\mathrm{Tr}^G_{G_x}(f)(x)
\end{aligned}$$

by the definition of the relative transfer. □

PROPOSITION 6.4.5: *Let $\rho : G \hookrightarrow \mathrm{GL}(n, \mathbb{F})$ be a representation of a finite group G over the field $\mathbb{F}$ of characteristic p. Then a point $x \in V$ belongs to the transfer variety $\mathfrak{X}_G$ if and only if p divides $|G_x|$.*

PROOF: The conditions of Lemma 6.4.4 are fulfilled, so if p divides the order of the isotropy group G_x, then,

$$\mathrm{Tr}^G(f)(x) = |G_x| \sum_{i=1}^{m} g_i f(x) = |G_x|\,\mathrm{Tr}^G_{G_x}(f)(x) = 0,$$

and hence $x \in \mathfrak{X}_G$.

On the other hand, suppose p does not divide $|G_x|$. The orbit $B \subset V$ of x is a finite set, so, by Lemma 6.4.3, there is a polynomial function h on V such that

$$h(y) = 0 \ \forall\, y \in B, y \neq x, \text{ and } h(x) = 1.$$

Then, by Lemma 6.4.4, and the fact that $g_1(x), \ldots, g_m(x)$ are pairwise distinct, with $g_i(x) = x$ only if $g_i \in G_x$, we get

$$\begin{aligned}
\mathrm{Tr}^G(h)(x) &= |G_x| \sum_{i=1}^{m} g_i h(x) = |G_x| \sum_{i=1}^{m} h(g_i^{-1}x) = |G_x|\,\mathrm{Tr}^G_{G_x}(h)(x) \\
&= |G_x| \Big(\sum_{b \in B} h(b) \Big) = |G_x| h(x) = |G_x| \neq 0 \in \mathbb{F}
\end{aligned}$$

and hence $x \notin \mathfrak{X}_G$. □

COROLLARY 6.4.6: *Let $\varrho : G \hookrightarrow \mathrm{GL}(n, \mathbb{F})$ be a representation of a finite group over the field $\mathbb{F}$ of characteristic p. Then*

$$\mathfrak{X}_G = \bigcup_{\substack{g \in G \\ |g| = p}} V^g,$$

in other words, the transfer variety is the union of the fixed-point sets of the elements in G of order p.

PROOF: By Proposition 6.4.5 any $x \in \mathfrak{X}_G$ is fixed by some element $g \in G$ with $|g| = p$, and hence $\mathfrak{X}_G \subseteq \bigcup_{|g|=p} V^g$. On the other hand, by Corollary 6.4.2, $\bigcup_{|g|=p} V^g \subseteq \mathfrak{X}_G$. □

By combining Proposition 6.4.5 with our previous work we arrive at M. Feshbach's main result for the modular transfer in invariant theory, [114]. For another treatment of M. Feshbach's result see [349].

THEOREM 6.4.7 (M. Feshbach): *Let $\varrho : G \hookrightarrow \mathrm{GL}(n, \mathbb{F})$ be a representation of a finite group over the field $\mathbb{F}$ of characteristic p. Then*

$$\sqrt{\mathrm{Im}(\mathrm{Tr}^G)} = \Big(\bigcap_{\substack{|g| = p \\ g \in G}} I_g \Big) \cap \mathbb{F}[V]^G,$$

and the ideals $\mathfrak{p}_g = I_g \cap \mathbb{F}[V]^G$, $|g| = p$, and $g \in G$ are prime. Hence $\mathit{ht}(\mathrm{Im}(\mathrm{Tr}^G)) = \mathit{ht}\Big(\sqrt{\mathrm{Im}(\mathrm{Tr}^G)}\Big) = n - \max\{\dim_{\mathbb{F}}(V^g) \mid g \in G \text{ and } |g| = p\} < n$.

PROOF: Since the hypotheses and conclusion are not altered by passage to a field extension we may suppose that $\mathbb{F}$ is algebraically closed, where Hilbert's Nullstellensatz ([25] Corollary 1.1.12) holds. By Corollary 6.4.6 $\mathfrak{X}_G = \bigcup_{\substack{|g|=p \\ g \in G}} V^g$. Passing back to ideals leads to the equality

$$\sqrt{(\mathrm{Im}(\mathrm{Tr}^G))^{\mathrm{e}}} = \bigcap_{\substack{|g| = p \\ g \in G}} I_g$$

as ideals in $\mathbb{F}[V]$, where $\Big(\mathrm{Im}(\mathrm{Tr}^G)\Big)^{\mathrm{e}}$ is the ideal in $\mathbb{F}[V]$ generated by $\mathrm{Im}(\mathrm{Tr}^G)$. The ideals I_g are prime, since they are generated by linear forms. So their intersection with $\mathbb{F}[V]^G$ are also prime, and the result follows. □

COROLLARY 6.4.8: *Let $\varrho : G \hookrightarrow \mathrm{GL}(n, \mathbb{F})$ be a representation of a finite group G over the field $\mathbb{F}$ of characteristic p. Then $\sqrt{\mathrm{Im}(\mathrm{Tr}^G)}$ is determined by the cyclic subgroups of order p in G. In particular,*

$$\sqrt{\mathrm{Im}(\mathrm{Tr}^G)} = \bigcap_{P = \mathrm{Syl}_p(G)} \sqrt{\mathrm{Im}(\mathrm{Tr}^P)} = \bigcap_{|g| = p} I_g,$$

where the intersection runs over the p-Sylow subgroups of G, respectively one element from each subgroup of order p in G. □

For representations defined over finite fields M. Feshbach's transfer (Theorem 6.4.7) and the Stong-Tamagawa formulae (Theorem 6.1.7) can be used to show that certain Dickson polynomials must be present in the radical of the transfer ideal, while certain others must be absent. Here is a simple result of this type. We will return to this topic in Section 9.3 where we will sharpen this using other methods.

COROLLARY 6.4.9: *Let $\rho : G \longrightarrow \mathrm{GL}(n, \mathbb{F})$ be a representation of the finite group G over the Galois field $\mathbb{F}$ of characteristic p and suppose that $\mathcal{ht}(\mathrm{Im}(\mathrm{Tr}^G)) = h$. Then $\mathbf{d}_{n,h} \notin \sqrt{\mathrm{Im}(\mathrm{Tr}^G)}$.*

PROOF: By Feshbach's transfer theorem there is an element $g \in G$ of order p with $\dim_{\mathbb{F}}(V^g) = n - h$. Let $U \subset V^*$ be the annihilator of V^g. Let $0 \neq x \in V^g \subseteq \mathfrak{X}_{\mathrm{Im}(\mathrm{Tr}^G)}$. If we can show $\mathbf{d}_{n,h}(x) \neq 0$ then by Hilbert's Nullstellensatz ([25] Corollary 1.1.12) $\mathbf{d}_{n,h} \notin \sqrt{\mathrm{Im}(\mathrm{Tr}^G)}$. By the Tamagawa-Stong formula we have

$$\mathbf{d}_{n,h}(x) = \sum_{\dim_{\mathbb{F}}(W)=h} \prod_{z \notin W} z(x) = \prod_{z \notin U} z(x) + \sum_{\substack{\dim_{\mathbb{F}}(W)=h \\ W \neq U}} \prod_{z \notin W} z(x).$$

If $W \subset V^*$ is an h-dimensional subspace different from U, then there is a nonzero linear form u in U but not in W, so

$$\prod_{z \notin W} z(x) = u(x) \cdot \prod_{\substack{z \notin W \\ z \neq u}} z(x) = 0.$$

Hence

$$\mathbf{d}_{n,h}(x) = \prod_{z \notin U} z(x) \neq 0. \quad \square$$

Here is an elementary example to illustrate some of the ideas discussed in this section.

EXAMPLE 1: If $\rho : G \hookrightarrow \mathrm{GL}(2, \mathbb{F}_q)$, $q = p^\nu$, $p \in \mathbb{N}$ a prime, is a representation of degree 2, then the elements of G of order p each fix a line in $V = \mathbb{F}_q^2$, and no more. So, if $p \mid |G|$, the transfer variety is a union of lines. If for each such line L we choose a linear form ℓ_L with kernel L, then

$$\sqrt{\mathrm{Im}(\mathrm{Tr}^G)} = \prod_{L \in \mathcal{L}} \ell_L = L_G,$$

where $\mathcal{L}$ denotes the set $\{V^g \mid |g| = p\}$. The generator L_G of this principal ideal is irreducible if and only if G acts transitively on the set $\mathcal{L}$. (See Example 1 in Section 9.3 for a detailed explanation of why this is so.)

PROPOSITION 6.4.10: *Let $\mathbb{F}$ be a field of characteristic p, $s \in \mathbb{N}$ an integer, and $\rho : \mathbb{Z}/p^s \hookrightarrow \mathrm{GL}(n, \mathbb{F})$ a representation of a cyclic group of order p^s over $\mathbb{F}$. Then $\sqrt{\mathrm{Im}(\mathrm{Tr}^{\mathbb{Z}/p^s})} \subset \mathbb{F}[V]^{\mathbb{Z}/p^s}$ is a prime ideal.*

PROOF: The group $\mathbb{Z}/p^s$ contains a unique subgroup of order p, and hence by Corollary 6.4.8 $\sqrt{\mathrm{Im}(\mathrm{Tr}^{\mathbb{Z}/p^s})} = \mathfrak{p}_g$, where $g \in \mathbb{Z}/p^s$ is any element of order p, and the result follows. □

EXAMPLE 2: If p is an odd prime, then the only p-groups with a unique subgroup of order p are the cyclic groups $\mathbb{Z}/p^s$, $s \in \mathbb{N}$. On the other hand (see, e.g., [156] Theorem 12.5.2), the prime 2 is a bit special in this context, since the generalized quaternion group Q_n of order 2^n contains a unique involution. Hence for any representation $\rho : Q_n \hookrightarrow \mathrm{GL}(n, \mathbb{F})$ of Q_n over a field of characteristic 2 we have that $\sqrt{\mathrm{Im}(\mathrm{Tr}^{Q_n})}$ is a prime ideal in $\mathbb{F}[V]^{Q_n}$.

6.5 The Koszul Complex and Invariant Theory

If we want to study the homological codimension of a ring of invariants, or any commutative graded algebra over the field $\mathbb{F}$, then one way to approach the problem is first to find candidates $h_1, \ldots, h_k \in H$ for a regular sequence, and then to *test* whether indeed they are such. If $H = \mathbb{F}[V]^G$ is the ring of invariants of a representation $\rho : G \hookrightarrow \mathrm{GL}(n, \mathbb{F})$ of a finite group, then candidates for a regular sequence are often easy to find. For example. we could choose any of the following.

(1) If $(V^*)^G \neq \{0\}$, then a basis for the fixed-point set.
(2) The top Chern classes of a Dade basis (when one exists; see Section 4.3) will give us a system of parameters for $\mathbb{F}[V]^G$, so taking these, or some subset of them, would again give us some test candidates.
(3) In the case the ground field $\mathbb{F}$ is finite, the Dickson polynomials (see Section 6.1) would also give us some reasonable test invariants.

In each of these cases we would certainly obtain a regular sequence in $\mathbb{F}[V]$, but do they form a regular sequence in $\mathbb{F}[V]^G$? Perhaps the best tool for checking whether this is the case is the Koszul complex (see Section 5.1). So, suppose that $\rho : G \hookrightarrow \mathrm{GL}(n, \mathbb{F})$ is a representation of a finite group G over the field $\mathbb{F}$ and that $h_1, \ldots, h_k \in \mathbb{F}[V]^G$ are invariant polynomials that form a regular sequence in $\mathbb{F}[V]$. What we would like to do is to show that the Koszul complex (see Theorem 5.1.3 for why)

$$\mathcal{L} = \mathbb{F}[V]^G \otimes E(s^{-1}h_1, \ldots, s^{-1}h_k),$$
$$\partial(s^{-1}h_i) = h_i \text{ for } i = 1, \ldots, k$$

is acyclic. Let $(\mathcal{K}, \partial)$ denote the Koszul complex

$$\mathcal{K} = \mathbb{F}[V] \otimes E(s^{-1}h_1, \ldots, s^{-1}h_k),$$
$$\partial(s^{-1}h_i) = h_i \text{ for } i = 1, \ldots, k$$

with the action of G extended from $\mathbb{F}[V]$ to $\mathcal{K}$ by by letting G act trivially on the elements of $E(s^{-1}h_1, \ldots, s^{-1}h_k)$. If we do this, then, as complexes, $\mathcal{K}^G \cong \mathcal{L}$. As is well known, taking fixed points and taking homology rarely commute. There is, however, a standard way to deal with such a situation, which goes back to [73] and is often called the composite functor theorem [151]. Here is how this works in our case.

We denote by $\mathbb{F}(G)$ the group algebra of G over $\mathbb{F}$ and let $\mathcal{P}(G) \longrightarrow \mathbb{F}$ denote a projective resolution of $\mathbb{F}$ regarded as an $\mathbb{F}(G)$-module via the augmentation homomorphism $\varepsilon : \mathbb{F}(G) \longrightarrow \mathbb{F}$, such as for example the bar construction $\mathcal{B}(G)$ of G over $\mathbb{F}$ (see, e.g., [46] or [73]). Introduce the double complex[4]

$$\mathcal{C} = \mathrm{Hom}_{\mathbb{F}(G)}(\mathcal{P}(G), \ \mathcal{K}).$$

A bit of care is needed with the gradings to turn this into an acceptable double complex, i.e., one satisfying the standard grading conventions (see, e.g., [232]). The differential d coming from the projective resolution $\mathcal{P}(G)$ appears in the contravariant variable of $\mathrm{Hom}_{\mathbb{F}(G)}(-, \ -)$ and hence the differential d^* induced by d on $\mathcal{C}$ ***increases*** the grading in $\mathcal{C}$ coming from $\mathcal{P}(G)$, i.e., $d^* : \mathcal{C}^{s,t} \longrightarrow \mathcal{C}^{s+1,t}$. Therefore, we must grade the Koszul complex $\mathcal{K}$ compatibly, i.e., so that ∂ (which appears in the covariant variable of $\mathrm{Hom}_{\mathbb{F}(G)}(-, \ -)$) also raises the resolution degree (this time in $\mathcal{K}$) by 1, and the induced differential on $\mathcal{C}$ satisfies $\partial : \mathcal{C}^{s,t} \longrightarrow \mathcal{C}^{s,t+1}$. This is the standard grading of the Koszul complex à la Eilenberg-Moore [357]. In other words, we bigrade $\mathcal{K}$ by giving the elements of $\mathbb{F}[V]$ resolution degree 0 (and internal degree $\deg(f)$) and the elements $s^{-1}h_i$ the resolution degree -1 (and internal degree $\deg(h_i)$) for $i = 1, \ldots, k$. Doing so, $(\mathcal{C}, \ d^*, \ \partial)$ becomes a double complex (of graded modules!), the grading on $\mathcal{C}^{s,t}$ coming from the grading of $\mathbb{F}[V]$).

Since $(\mathcal{C}, \ d^*, \ \partial)$ is a double complex we can totalize it, i.e., we can form the associated graded complex $\left\{\mathrm{Tot}(\mathcal{C})^m = \bigoplus_{s+t=m} \mathcal{C}^{s,t} \mid m \in \mathbb{Z}\right\}$ with differential $d^* + \partial$. We denote the cohomology of this complex by $H^*(G; (\mathcal{K}, \ \partial))$ and refer to it laxly as the **cohomology of G with coefficients in the Koszul complex** $(\mathcal{K}, \ \partial)$.

Associated to this double complex are two spectral sequences, [232] Chapter XI, which are independent of the choice of the resolution $\mathcal{P}(G)$. The spectral sequence where we first apply the differential of the Koszul complex, and then that of the $\mathbb{F}(G)$-resolution is trivial. To see this note that the augmentation map

$$\eta : (\mathcal{K}, \ \partial) \longrightarrow \left(\frac{\mathbb{F}[V]}{(h_1, \ldots, h_k)}, \ 0\right)$$

[4] See also [46] Chapter VII, Section 5, for a similar construction in connection with the cohomology of semidirect products.

is an equivalence of complexes, and hence so is the induced map of double complexes

$$\eta : (\mathcal{C}, d^*, \partial) \longrightarrow \mathrm{Hom}_{\mathbb{F}(G)}\Big(\mathcal{P}(G), \Big(\frac{\mathbb{F}[V]}{(h_1, \ldots, h_k)}, 0\Big)\Big).$$

Therefore, we obtain the following:

PROPOSITION 6.5.1: *With the preceding hypotheses and notation the augmentation induces an isomorphism*

$$H^*(G; (\mathcal{K}, \partial)) \longrightarrow H^*\Big(G; \frac{\mathbb{F}[V]}{(h_1, \ldots, h_k)}\Big). \quad \square$$

The spectral sequence where we first apply the differential of the projective resolution (e.g., bar construction differential) and then the differential of the Koszul complex becomes a technical tool for studying the depth of rings of invariants. If we denote this spectral sequence by $\{E_r, d_r\}$, then as a consequence of Proposition 6.5.1 ($\Rightarrow$ denotes **converges to**)

$$E_r \Rightarrow H^*\Big(G; \frac{\mathbb{F}[V]}{(h_1, \ldots, h_k)}\Big).$$

Since G acts trivially on $E(s^{-1}h_1, \ldots, s^{-1}h_k)$, the term E_1 of the spectral sequence takes the form

$$E_1 = H^*(G; \mathbb{F}[V]) \otimes E(s^{-1}h_1, \ldots, s^{-1}h_k).$$

Therefore, the term E_2 may be identified with the cohomology of yet another Koszul complex, namely for the elements

$$h_1, \ldots, h_k \in H^0(G; \mathbb{F}[V]) = \mathbb{F}[V]^G \subseteq H^*(G; \mathbb{F}[V]).$$

Therefore, recalling that $h_1, \ldots, h_k$ are a regular sequence in $\mathbb{F}[V]$, we conclude they are algebraically independent (see e.g. [372] Proposition 6.2.1). So from the discussion in Section 5.1 we see

$$E_2^{s,t} = \mathrm{Tor}^s_{\mathbb{F}[h_1, \ldots, h_k]}(H^t(G; \mathbb{F}[V]), \mathbb{F}),$$

where $H^*(G; \mathbb{F}[V])$ is regarded as an $\mathbb{F}[h_1, \ldots, h_k]$-module via the inclusion

$$\mathbb{F}[h_1, \ldots, h_k] \subseteq \mathbb{F}[V]^G = H^0(G; \mathbb{F}[V]) \subseteq H^*(G; \mathbb{F}[V])$$

and the product in group cohomology. To summarize, we have proved the following:

PROPOSITION 6.5.2: *With the preceding hypotheses and notation there is a convergent second-quadrant spectral sequence* $\{E_r, d_r\}$ *with*

$$E_r \Rightarrow H^*\Big(G; \frac{\mathbb{F}[V]}{(h_1, \ldots, h_k)}\Big),$$

$$E_2^{s,t} = \mathrm{Tor}^s_{\mathbb{F}[h_1, \ldots, h_k]}(H^t(G; \mathbb{F}[V]), \mathbb{F}).$$

PROOF: We need only remark that convergence is a consequence of the fact that $E_2^{s,t} = 0$ if $s < -k$. □

This spectral sequence is a precursor of A. Grothendieck's local cohomology spectral sequence [152]. We are going to use it to rederive the main result of [105]. A spectral sequence may seem like a formidable tool, and it is: It tends to break a problem down into such small parts that, either the solution is immediate, or it is apparent exactly what it is that is missing and needs to be proved. We illustrate this with a high-tech proof of a result P. S. Landweber and R. E. Stong [224] whose original proof in [224] is entirely elementary, consisting of an intricate use of unique factorization. (See also Proposition 5.1.1.)

PROPOSITION 6.5.3: *Suppose that $\rho : G \hookrightarrow \mathrm{GL}(n, \mathbb{F})$, $n \geq 2$, is a representation of a finite group G over the field $\mathbb{F}$. If $h_1, h_2 \in \mathbb{F}[V]^G$ are a regular sequence in $\mathbb{F}[V]$, then they are a regular sequence in $\mathbb{F}[V]^G$ also.*

PROOF: The polynomial algebra $\mathbb{F}[h_1, h_2]$ has global dimension 2, so the functors $\mathrm{Tor}^s_{\mathbb{F}[h_1, h_2]}(-, -)$ are identically zero for $s < -2$. Figure 6.5.1 represents E_2. It shows that there is no way that a nonzero differential can either arrive at or leave from $E_2^{-1,0}$. The elements of $E_2^{-1,0}$ have negative total degree. $H^*\left(G; \frac{\mathbb{F}[V]}{(h_1, h_2)}\right)$ is zero in negative degrees, and hence $E_\infty^{s,t}$ is also zero in negative degrees. Therefore,

$$0 = E_2^{-1,0} = \mathrm{Tor}^{-1}_{\mathbb{F}[h_1, h_2]}(H^0(G; \mathbb{F}[V]), \mathbb{F}) = \mathrm{Tor}^{-1}_{\mathbb{F}[h_1, h_2]}(\mathbb{F}[V]^G, \mathbb{F}),$$

and the result follows from Proposition A.1.5. □

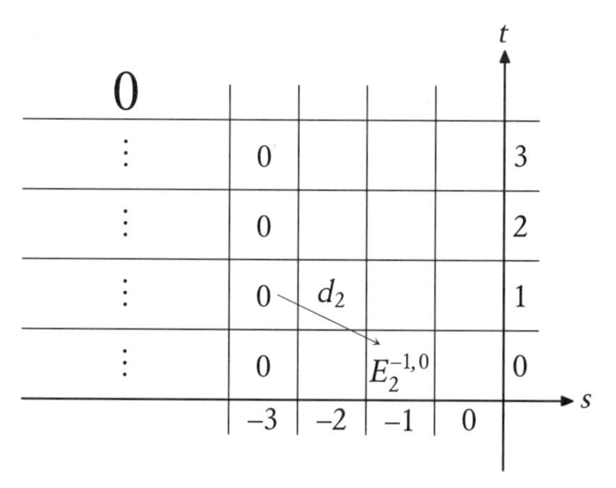

FIGURE 6.5.1: E_2 in the dimension-2 case

The proof of Proposition 6.5.3 serves as a paradigm for future applications. Its essential input was the presence of well placed zeros in the E_2-term which were consequences of the hypotheses. In general we will need to prove that such zeros occur. We set out to show how this can be done in the case of the cyclic group $\mathbb{Z}/q$ and its representations in characteristic p, where $q = p^\nu$ is a power of the characteristic. We will concentrate on the case where $\mathbb{F} = \mathbb{F}_q$ is a finite field, so the Dickson polynomials $\mathbf{d}_{n,n-1}, \ldots, \mathbf{d}_{n,0}$ are a universal system of parameters for rings of invariants. This is no loss of generality since $\mathbb{F}_q$ is a splitting field for $\mathbb{Z}/q$.

If the ground field is finite and the Dickson polynomials are used as a system of parameters, then the spectral sequence of Proposition 6.5.2 takes the form

$$E_r \Rightarrow H^*\left(G; \mathbb{F}[V]_{\mathrm{GL}(n,\mathbb{F})}\right),$$
$$E_2^{s,t} = \mathrm{Tor}^s_{\mathbf{D}^*(n)}(H^t(G; \mathbb{F}[V]),\ \mathbb{F}),$$

where $\mathbb{F}[V]_{\mathrm{GL}(n,\mathbb{F})}$ is the ring of coinvariants of the group $\mathrm{GL}(n, \mathbb{F})$. Notice that $H^*\left(G; \mathbb{F}[V]_{\mathrm{GL}(n,\mathbb{F})}\right)$ is zero for $* < 0$, and therefore there are no elements of negative total degree in E_∞. So in general the E_∞-term looks as pictured in Figure 6.5.2: This is the fact that will be exploited in computations. The terms on the border of the vanishing area, i.e., where $s + t = 0$ (referred to as the **vanishing line**) are connected with the **stable invariants** introduced in [187] (see Section 1.2) and studied in [285].

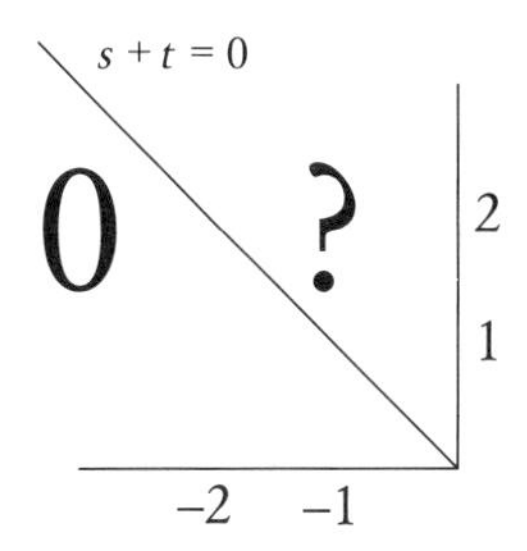

FIGURE 6.5.2: The vanishing area $s + t < 0$ for E_∞.

To work with this spectral sequence we need to obtain information about $H^*(G; \mathbb{F}[V])$ and how the Dickson algebra acts on it. So the first step is to compute these modules. The cyclic group $\mathbb{Z}/q$ and its representations in characteristic p, where $q = p^\nu$ is a power of the characteristic, provide a useful, nontrivial, collection of examples to start with. As a reference for the cohomology of the cyclic group see one of the books dealing with group cohomology, such as, [4], [29], [68], or [109].

We begin by reviewing how to compute $H^*(\mathbb{Z}/q; A)$ where A is a commutative graded connected algebra over the field $\mathbb{F}$ of characteristic p and $\alpha : \mathbb{Z}/q \hookrightarrow \mathrm{Aut}(A)$ a faithful representation of $\mathbb{Z}/q$ on A by grading preserving automorphisms. The standard resolution of $\mathbb{F}$ as a $\mathbb{Z}/q$-module allows us to compute the cohomology modules $H^*(\mathbb{Z}/q; A)$ from the periodic cocomplex

$$0 \longrightarrow A \xrightarrow{\partial} A \xrightarrow{\mathrm{Tr}^{\mathbb{Z}/q}} A \xrightarrow{\partial} \cdots$$

where $\gamma \in \mathbb{Z}/q$ is a generator, $\partial = 1 - \gamma$, and $\mathrm{Tr}^{\mathbb{Z}/q} = \partial^{q-1}$ is the transfer homomorphism. Note that $A^{\mathbb{Z}/q} = \ker(\partial)$. From this cocomplex one sees that

$$H^0(\mathbb{Z}/q;A) = A^{\mathbb{Z}/q}$$
$$H^1(\mathbb{Z}/q;A) = \ker(\mathrm{Tr}^{\mathbb{Z}/q})/\mathrm{Im}(\partial)$$
$$H^2(\mathbb{Z}/q;A) = A^{\mathbb{Z}/q}/\mathrm{Im}(\mathrm{Tr}^{\mathbb{Z}/q})$$

and that $H^{i+2}(\mathbb{Z}/q;A) = H^i(\mathbb{Z}/q;A)$ for all $i > 0$. We write $H^{\mathrm{od}}(\mathbb{Z}/q;A)$ for $H^i(\mathbb{Z}/q;A)$ for some fixed, but arbitrary, odd integer i. Likewise $H^{\mathrm{ev}}(\mathbb{Z}/q;A)$ denotes $H^j(\mathbb{Z}/q;A)$ for some fixed strictly positive, but otherwise arbitrary, even integer j. This is justified by periodicity. It may come as a bit of a surprise that $H^{\mathrm{od}}(\mathbb{Z}/q;A)$ and $H^{\mathrm{ev}}(\mathbb{Z}/q;A)$ are isomorphic as graded vector spaces. This is a special case of Herbrand's lemma (see, e.g., [226] Theorem 5.2 or [342] VIII Proposition 8) and follows from the two exact sequences

$$0 \longrightarrow A^{\mathbb{Z}/q} \longrightarrow A \xrightarrow{\quad\partial\quad} A \longrightarrow \mathrm{coker}(\partial) \longrightarrow 0$$

$$0 \longrightarrow H^{\mathrm{od}}(\mathbb{Z}/q;A) \longrightarrow \mathrm{coker}(\partial) \xrightarrow{\quad\mathrm{Tr}^{\mathbb{Z}/p}\quad} A^{\mathbb{Z}/q} \longrightarrow H^{\mathrm{ev}}(\mathbb{Z}/q;A) \longrightarrow 0$$

by taking Euler characteristics. In fact more is true: We are going to describe an isomorphism[5]

$$[su^*] \cup - : H^{\mathrm{ev}}(\mathbb{Z}/q;A) \longrightarrow H^{\mathrm{od}}(\mathbb{Z}/q;A)$$

of $A^{\mathbb{Z}/q}$-modules.

Recall that the group ring $\mathbb{F}(\mathbb{Z}/q)$ of $\mathbb{Z}/q$ over $\mathbb{F}$ is isomorphic to

$$\frac{\mathbb{F}[\gamma]}{(\gamma^q - 1)} = \frac{\mathbb{F}[\gamma]}{(\gamma - 1)^q}$$

since $\mathbb{F}$ has characteristic p. So substituting u for $(\gamma - 1)$ we see that

$$\mathbb{F}(\mathbb{Z}/q) = \frac{\mathbb{F}[u]}{(u^q)},$$

which is a local ring. A (co-)multiplicative resolution of $\mathbb{F}$ as $\mathbb{Z}/q$-module is given by the modified Koszul complex, or Tate complex [424]

$$\mathcal{T}_* = \mathbb{F}(\mathbb{Z}/q) \otimes E[su] \otimes \Gamma(\tau u),$$

where su is the suspension of u, τu the transpotence, $E[su]$ denotes the exterior algebra generated by su, and $\Gamma(\tau u)$ denotes the divided polynomial

[5] The idea to construct this isomorphism as a map from $H^2(\mathbb{Z}/q;A)$ to $H^3(\mathbb{Z}/q;A)$ by cupping with an appropriate class in $H^1(\mathbb{Z}/q;A)$ is due to F. R. Cohen and S. Gitler.

algebra generated by τu. The homological degrees of su and τu are 1 and 2 respectively. The differential d in the resolution is given by

$$d(su) = u$$
$$d(\tau u) = u^{q-1} su.$$

We may therefore compute the cohomology $H^*(\mathbb{Z}/q;A)$ from the cocomplex

$$\begin{aligned}\mathcal{T}^*(A) &= \mathrm{Hom}_{\mathbb{F}(\mathbb{Z}/q)}(\mathbb{F}(\mathbb{Z}/q) \otimes E[su] \otimes \mathbb{F}[\tau u], A)\\ &\cong \mathrm{Hom}_{\mathbb{F}}(E[su] \otimes \Gamma[\tau u], A) \cong \mathrm{Hom}_{\mathbb{F}}(E[su] \otimes \Gamma[\tau u], \mathbb{F}) \otimes A\\ &\cong E[su^*] \otimes \mathbb{F}[\tau u^*] \otimes A,\end{aligned}$$

where su^* and τu^* are dual to su and τu respectively, and $E[su^*]$ and $\mathbb{F}[\tau u^*]$ are the dual Hopf algebras to $E[su]$ and $\Gamma(\tau u)$ respectively.

The coboundary map

$$d^* : \mathcal{T}^1(A) \longrightarrow \mathcal{T}^2(A)$$

may be computed as follows: $\mathcal{T}^1(A) = su^* \otimes A$ and $\mathcal{T}^2(A) = \tau u^* \otimes A$. So

$$d^*(su^*)(\tau u \otimes a) = su^*(su \otimes \mathrm{Tr}^{\mathbb{Z}/q}(a))$$

for any $a \in A$. By definition of su^* it is zero on an element of the form $su \otimes a$ whenever $\deg(a) > 0$. On the other hand $\mathrm{Tr}^{\mathbb{Z}/q}(a) = 0$ for any $a \in A_0$ since A is connected, and $\mathrm{Tr}^{\mathbb{Z}/q}(1) = 1 + \gamma(1) + \cdots \gamma^{q-1}(1) = q = 0$, as $\mathbb{F}$ has characteristic p and $q = p^\nu$. Therefore su^* is a cocycle and defines an element $[su^*]$ in $H^1(\mathbb{Z}/q;A)$. In fact it defines the class[6] corresponding to $1 \in H^{\mathrm{od}}(\mathbb{Z}/q;A) = \ker(\mathrm{Tr}^{\mathbb{Z}/q})/\mathrm{Im}(\partial)$ in the standard resolution. We let

$$[su^*] \cup - : H^{\mathrm{ev}}(\mathbb{Z}/q;A) \longrightarrow H^{\mathrm{od}}(\mathbb{Z}/q;A)$$

be the map defined by cupping with $[su^*]$. This is a map of $A^{\mathbb{Z}/q} = H^0(\mathbb{Z}/q;A)$-modules.

LEMMA 6.5.4: *Let $p \in \mathbb{N}$ a prime integer, $q = p^\nu$, A be a graded connected algebra of finite type over the field $\mathbb{F}$ of characteristic p and $\alpha : \mathbb{Z}/q \hookrightarrow \mathrm{Aut}(A)$ a faithful representation of $\mathbb{Z}/q$ on A by grading preserving automorphisms. Then $[su^*] \cup - : H^{\mathrm{ev}}(\mathbb{Z}/q;A) \longrightarrow H^{\mathrm{od}}(\mathbb{Z}/q;A)$ is an isomorphism of $A^{\mathbb{Z}/q}$-modules.*

PROOF: Cupping with su^* at the cochain level in $\mathcal{T}^*(A)$ defines an isomorphism from $\mathcal{T}^{\mathrm{ev}}(A) = \mathcal{T}^{2t}(A)$ to $\mathcal{T}^{2t+1}(A) = \mathcal{T}^{\mathrm{od}}(A)$ for any positive integer t, and since su^* is a cocycle, commutes with the coboundary map d^*. Let $w \in \mathcal{T}^{\mathrm{od}}(A)$ be a cocycle. Choose $z \in \mathcal{T}^{\mathrm{ev}}(A)$ with $su^* \cup (z) = w$. Then

$$su^* \cup (d^*(z)) = d^*(su^* \cup (z)) = d^*(w) = 0$$

[6] Since $\mathbb{F}$ has characteristic p and $\mathbb{Z}/q$ has order q we have $\mathrm{Tr}^{\mathbb{Z}/q}(1) = 0 \in A$.

since $su^* \cup -$ commutes with d^*. As $su^* \cup -$ is an isomorphism from $\mathcal{J}^{\mathrm{ev}}(A)$ to $\mathcal{J}^{\mathrm{od}}(A)$ we conclude that z is also a cocycle. Hence the map

$$[su^*] \cup - : H^{\mathrm{ev}}(\mathbb{Z}/q;A) \longrightarrow H^{\mathrm{od}}(\mathbb{Z}/q;A)$$

is an epimorphism. In addition, we know from ❻ that $H^{\mathrm{ev}}(\mathbb{Z}/q;A)$ and $H^{\mathrm{od}}(\mathbb{Z}/q;A)$ are graded $\mathbb{F}$-vector spaces of finite type with the same Poincaré series, so $[su^*] \cup -$ must be an isomorphism. □

The main result of G. Ellingsrud and T. Skjelbred in [105] is a formula for the depth of the ring of invariants $\mathbb{F}[V]^{\mathbb{Z}/q}$ when $\mathbb{F}$ is a field of characteristic p, $q = p^{\nu}$, and $\rho : \mathbb{Z}/q \hookrightarrow \mathrm{GL}(n, \mathbb{F})$. The following lemma provides us with a lower bound. A portion of the argument used in its proof is adapted from [55] Proposition 17 and was used in the proof of Proposition 5.6.10.

LEMMA 6.5.5: *Let $\rho : \mathbb{Z}/q \hookrightarrow \mathrm{GL}(n, \mathbb{F})$ be a representation of the finite group $\mathbb{Z}/q$, p an odd prime, $q = p^{\nu}$, over the field $\mathbb{F}$ of characteristic p. If $n \geq 3$ then* $\text{hom-codim}(\mathbb{F}[V]^{\mathbb{Z}/q}) \geq 2 + \dim_{\mathbb{F}}(V^{\mathbb{Z}/q})$.

PROOF: Let $V = V_{d_1} \oplus V_{d_2} \oplus \cdots \oplus V_{d_k}$ be a decomposition of V into indecomposable $\mathbb{Z}/q$-modules. Each indecomposable factor has a 1-dimensional fixed point set, so $\dim_{\mathbb{F}}(V^{\mathbb{Z}/q}) = k$. For $i = 1, \ldots, k$ let $x_1(i), x_2(i), \ldots, x_{d_i}(i)$ be a Jordan basis for the ith factor. Specifically, the generator γ is mapped to the matrix $\mathbf{J}_{d_i}$ as shown in Figure 6.5.3. Let the corresponding dual basis for the space of linear forms $V_{d_i}^*$ on the i-th factor be $z_1(i), z_2(i), \ldots, z_{d_i}(i)$. With these notations we have

$$\mathbf{J}_{d_i} = \begin{bmatrix} 1 & 1 & & & \\ 0 & 1 & 1 & & \\ \mathbf{0} & \ddots & \ddots & \ddots & \mathbf{0} \\ & & \ddots & 1 & 1 \\ & & & 0 & 1 \end{bmatrix} \in \mathrm{GL}(d_i, \mathbb{F})$$

FIGURE 6.5.3: A Jordan Block

$$V_{d_i}^{\mathbb{Z}/q} = \bigcap_{j=2}^{d_i} \ker(z_j(i))$$

and $z_1(i)$ is nonzero on the fixed point set $V_{d_i}^{\mathbb{Z}/q}$. Since trivial factors may be split off we may suppose that $d_i \geq 2$ for $i = 1, \ldots, k$. Hence $k \leq \left[\frac{n}{2}\right]$ where $[r]$ denotes the integral part of the rational number r.

Let $W_{d_i}^*$ denote the linear span in $V_{d_i}^*$ of $z_2(i), z_3(i), \ldots, z_{d_i}(i)$ and set $W = W_{d_1} \oplus \cdots \oplus W_{d_i}$. Since $n \geq 3$, it follows $\dim_{\mathbb{F}}(W) = n - k \geq 2$, so the ring of invariants

$$\mathbb{F}[W]^{\mathbb{Z}/q} = \mathbb{F}\left[z_j(i) \mid {}_{i=1,\ldots,k}^{2 \leq j \leq d_j}\right]^{\mathbb{Z}/q}$$

has homological codimension at least 2: This follows from Proposition

6.1.5. Choose a pair of forms $f_1, f_2 \in \mathbb{F}[W]^{\mathbb{Z}/q}$ that are a regular sequence. By Proposition 5.1.1 they are a regular sequence in $\mathbb{F}[W]$ and hence also in $\mathbb{F}[V]$. Since $\mathbb{F}[W]^{\mathbb{Z}/q} \subset \mathbb{F}[V]^{\mathbb{Z}/q}$ it follows from Proposition 5.1.1 that $f_1, f_2 \in \mathbb{F}[V]^{\mathbb{Z}/q}$ are also a regular sequence. They are homogeneous polynomials in the forms $z_2(i), z_3(i), \dots, z_{d_i}(i)$.

We claim that the sequence of length $2+k$

$$f_1, f_2, c_q(z_1(1)), \dots, c_q(z_1(k)) \in \mathbb{F}[V]^{\mathbb{Z}/q}$$

is a regular sequence in $\mathbb{F}[V]^{\mathbb{Z}/q}$. To simplify notation set

$$f_{j+2} = c_q(z_1(j)) \quad \text{for } j = 1, \dots, k,$$

and let us suppose inductively that we have shown $f_1, \dots, f_m$ are a regular sequence in $\mathbb{F}[V]^{\mathbb{Z}/q}$. Since f_1, f_2 are a regular sequence in $\mathbb{F}[V]^{\mathbb{Z}/q}$ the induction starts, and we may suppose that $m \geq 2$.

To show that $f_1, \dots, f_m, f_{m+1} \in \mathbb{F}[V]^{\mathbb{Z}/q}$ is a regular sequence we must show that whenever

$$❶ \qquad h_1 f_1 + h_2 f_2 + \cdots + h_m f_m + h_{m+1} f_{m+1} = 0$$

for forms $h_1, \dots, h_m, h_{m+1} \in \mathbb{F}[V]^{\mathbb{Z}/q}$ then h_{m+1} belongs to the ideal $(f_1, \dots, f_m) \subset \mathbb{F}[V]^{\mathbb{Z}/q}$.

If $F \in \mathbb{F}[V]$ is a form, we may regard it as a polynomial in $z_{m+1}(1)$ with coefficients which are forms in the other variables, viz.,

$$F = F_0 z_{m+1}(1)^d + F_1 z_{m+1}(1)^{d-1} + \cdots + F_d$$

where

$$F_0, \dots, F_d \in \mathbb{F}\left[z_i(j) \mid \begin{matrix} i = 1, \dots, d_j \\ j = 1, \dots, k \\ (i, j) \neq (m+1, 1) \end{matrix} \right]$$

and $F_0 \neq 0$. We call d the $z_{m+1}(1)$-degree of F and denote it by $\deg_{m+1}(F)$. Observe that the $\mathbb{Z}/q$-action preserves the $z_{m+1}(1)$-degree.

Regarding $h_1, \dots, h_m, f_{m+1}$ as polynomials in $z_{m+1}(1)$ with coefficients which are forms in the other variables we may perform polynomial division to obtain

$$❷ \qquad h_i = q_i f_{m+1} + r_i, \quad \deg_{m+1}(r_i) < \deg_{m+1}(f_{m+1})$$

for $i = 1, \dots, m$. Since the $\mathbb{Z}/q$-action preserves the $z_{m+1}(1)$-degree of forms we have

$$❸ \qquad \deg_{m+1}(\partial(r_i)) \leq \deg_{m+1}(r_i) < \deg_{m+1}(f_{m+1})$$

for $i = 1, \dots, m$, where $\partial = 1 - \gamma$ and $\gamma \in \mathbb{Z}/q$ is a generator. If we apply ∂ to ❷ we obtain for $1 \leq i \leq m$

$$0 = \partial(h_i) = \partial(q_i) f_{m+1} + \partial(r_i)$$

since h_i and f_{m+1} are $\mathbb{Z}/q$-invariant. If $\partial(q_i) \neq 0$ this equation would say that f_{m+1} divides r_i contradicting ❸. Hence $\partial(q_i) = 0$ and this entails that $\partial(r_i) = 0$ also. Therefore the forms q_i and r_i are invariant for $i = 1, \ldots, m$. Set

$$H = r_1 f_1 + r_2 f_2 + \cdots + r_m f_m \in \mathbb{F}[V]^{\mathbb{Z}/q}.$$

Substituting ❷ into ❶ then yields

$$❹ \qquad 0 = (q_1 f_1 + q_2 f_2 + \cdots + q_m f_m + h_{m+1}) f_{m+1} + H.$$

Note that

$$\deg_{m+1}(f_1) = \deg_{m+1}(f_2) = \cdots = \deg_{m+1}(f_m) = 0$$

so

$$❺ \qquad \deg_{m+1}(H) = \max\left\{\deg_{m+1}(r_i) \mid 1 \leq i \leq m\right\} < \deg_{m+1}(f_{m+1}).$$

Unless

$$q_1 f_1 + q_2 f_2 + \cdots + q_m f_m + h_{m+1} = 0$$

equation ❹ would imply f_{m+1} divides H contradicting ❺. Therefore

$$h_{m+1} = -(q_1 f_1 + q_2 f_2 + \cdots + q_m f_m)$$

as required. This completes the induction step and hence the proof. □

The following result is implicit in [105]. We reach ahead in the proof to use a result from Section 9.2 (Proposition 9.2.9).

PROPOSITION 6.5.6: *Let p be an odd prime, $\mathbb{F}$ a field of characteristic p and $\rho : \mathbb{Z}/q \hookrightarrow \mathrm{GL}(n, \mathbb{F})$ a representation of $\mathbb{Z}/q$, $q = p^{\nu}$, and $n \geq 3$. If $n \geq 2 + \dim_{\mathbb{F}}(V^{\mathbb{Z}/p})$, where $\mathbb{Z}/p$ is the unique subgroup of order p in $\mathbb{Z}/q$ then*

$$\text{hom-codim}(\mathbb{F}[V]^{\mathbb{Z}/q}) = 2 + \text{hom-codim}\left(\frac{\mathbb{F}[V]^{\mathbb{Z}/q}}{\mathrm{Im}(\mathrm{Tr}^{\mathbb{Z}/q})}\right).$$

PROOF: By the Auslander-Buchsbaum equality [47] Theorem 1.3.3 it is equivalent to prove

$$\text{hom-dim}_{\mathbf{D}(n)}(\mathbb{F}[V]^{\mathbb{Z}/q}) + 2 = \text{hom-dim}_{\mathbf{D}(n)}\left(\frac{\mathbb{F}[V]^{\mathbb{Z}/q}}{\mathrm{Im}(\mathrm{Tr}^{\mathbb{Z}/q})}\right).$$

To this end consider the Koszul spectral sequence with

$$E_r \Rightarrow H^*(\mathbb{Z}/q, \mathbb{F}[V]_{\mathrm{GL}(n,\mathbb{F})})$$
$$E_2^{s,t} = \mathrm{Tor}^s_{\mathbf{D}(n)}(\mathbb{F}, H^t(\mathbb{Z}/q; \mathbb{F}[V])).$$

Let $\text{hom-dim}_{\mathbf{D}(n)}\left(\frac{\mathbb{F}[V]^{\mathbb{Z}/q}}{\mathrm{Im}(\mathrm{Tr}^{\mathbb{Z}/q})}\right) = d$. Then $\text{hom-dim}_{\mathbf{D}(n)}(H^t(\mathbb{Z}/q, \mathbb{F}[V])) = d$ for all $t > 0$ by Lemma 6.5.4. Recall by Feshbach's transfer theorem (Theorem 6.4.7) that

$$\mathscr{h}t(\mathrm{Im}(\mathrm{Tr}^{\mathbb{Z}/q})) = n - \dim_{\mathbb{F}}(V^{\mathbb{Z}/p}).$$

Hence $\mathit{ht}(\mathrm{Im}(\mathrm{Tr}^{\mathbb{Z}/q})) \geq 2$. By Proposition 9.2.9 it follows that $\mathbf{d}^m_{n,0}, \mathbf{d}^m_{n,1}$ annihilate $\mathbb{F}[V]^{\mathbb{Z}/q} \big/ \mathrm{Im}(\mathrm{Tr}^{\mathbb{Z}/q})$ for all m sufficiently large. Therefore, for $m >> 0$,

$$\mathrm{Tor}^{-2}_{\mathbb{F}[\mathbf{d}^m_{n,0}, \mathbf{d}^m_{n,1}, \ldots, \mathbf{d}^m_{n,n-1}]}(\mathbb{F}, \mathbb{F}[V]^{\mathbb{Z}/q}) \big/ \mathrm{Im}(\mathrm{Tr}^{\mathbb{Z}/q}) \neq 0,$$

and by flat base change we conclude

$$d = \text{hom-dim}_{\mathbf{D}(n)}\left(\mathbb{F}[V]^{\mathbb{Z}/q} \big/ \mathrm{Im}(\mathrm{Tr}^{\mathbb{Z}/q})\right) = \text{hom-dim}\left(H^t(\mathbb{Z}/q; \mathbb{F}[V])\right) \geq 2$$

for all $t > 0$. So the E_2-term of the Koszul spectral sequence is as depicted in Diagram 6.5.4. The terms marked ⊞ are nonzero: The indicated differential is the only possible one originating or terminating at $E_2^{-d,1}$. Since $d \geq 2$, $-d + 1 < 0$ and $E_2^{-d,1}$ is of negative total degree. As E_∞ is zero in negative total degrees the indicated differential must be nonzero. Therefore

$$E_2^{0,-d+2} = \mathrm{Tor}_{\mathbf{D}(n)}(\mathbb{F}, \mathbb{F}[V]^{\mathbb{Z}/q}) \neq 0.$$

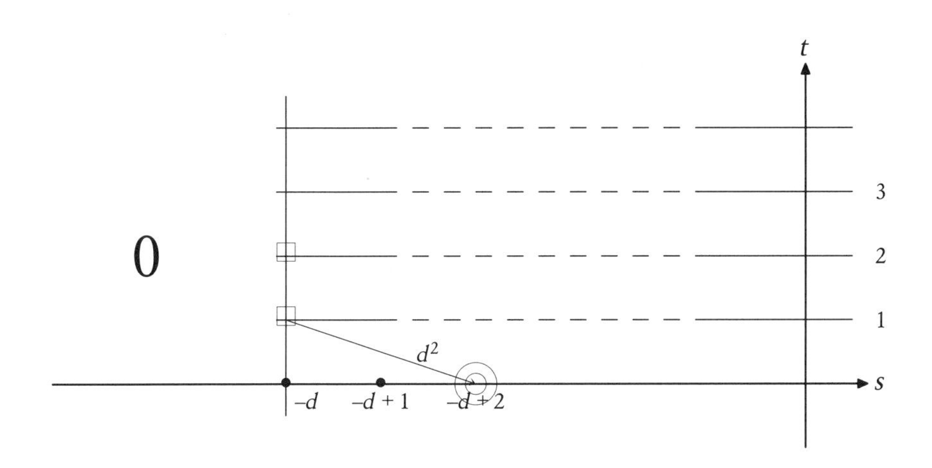

DIAGRAM 6.5.4: $E_2^{s,t}$

The terms marked • in the diagram are also of negative total degree, and no nonzero differential can either originate or terminate at them, so they too must be zero. This means

$$E_2^{0,-d+1} = \mathrm{Tor}_{\mathbf{D}(n)}(\mathbb{F}, \mathbb{F}[V]^{\mathbb{Z}/q}) = 0.$$

This implies that $\text{hom-dim}(\mathbb{F}[V]^{\mathbb{Z}/q}) = d - 2$ as was to be shown. □

THEOREM 6.5.7 (G. Ellingsrud-T. Skjelbred [105]): *Suppose that $\mathbb{F}$ is a field of characteristic p, and $\rho : \mathbb{Z}/p \hookrightarrow \mathrm{GL}(n, \mathbb{F})$ is a representation of the* *cyclic group $\mathbb{Z}/p$ of order p and degree n. Then*

$$\text{hom-codim}(\mathbb{F}[V]^{\mathbb{Z}/p}) = \min\left\{n, 2 + \dim_{\mathbb{F}}(V^{\mathbb{Z}/p})\right\}.$$

PROOF: By Proposition 5.1.1 may suppose that $n \geq 3$. We know from Lemma 6.5.5

$$\text{hom-codim}(\mathbb{F}[V]^{\mathbb{Z}/p}) \geq 2 + \dim_{\mathbb{F}}(V^{\mathbb{Z}/p})$$

so it will suffice to show

$$\text{hom-codim}(\mathbb{F}[V]^{\mathbb{Z}/p}) \leq 2 + \dim_{\mathbb{F}}(V^{\mathbb{Z}/p}).$$

We also have from Proposition 6.5.6 the equality

$$\text{hom-codim}(\mathbb{F}[V]^{\mathbb{Z}/p}) = 2 + \text{hom-codim}\left(\frac{\mathbb{F}[V]^{\mathbb{Z}/p}}{\text{Im}(\text{Tr}^{\mathbb{Z}/p})}\right),$$

so we need to verify the inequality

$$\text{hom-codim}\left(\frac{\mathbb{F}[V]^{\mathbb{Z}/p}}{\text{Im}(\text{Tr}^{\mathbb{Z}/p})}\right) \leq \dim_{\mathbb{F}}(V^{\mathbb{Z}/p}).$$

Since Feshbach's transfer theorem (Theorem 6.4.7) implies $\mathit{ht}(\text{Im}(\text{Tr}^{\mathbb{Z}/p})) = n - \dim(V^{\mathbb{Z}/p})$ it follows that

$$\dim\left(\frac{\mathbb{F}[V]^{\mathbb{Z}/p}}{\text{Im}(\text{Tr}^{\mathbb{Z}/p})}\right) = \dim(V^{\mathbb{Z}/p}),$$

and since the homological codimension is always less than or equal to the Krull dimension the result follows. □

The preceding result remains true if we replace the cyclic group of order p, $\mathbb{Z}/p$, by a cyclic group $\mathbb{Z}/p^s$ of order p^s; see [105]. For further applications of the Koszul spectral sequence the reader is referred to [375] and [386]. For an elementary proof of the theorem of G. Ellingsrud and T. Skjelbred, as well as several extensions of it, see [55].

Chapter 7
Special Classes of Invariants

VERY often the rings of invariants that occur in practice are for special classes of groups, representations or both. For example a finite classical group in its defining representation, a pseudoreflection representation, or a low degree representation of a solvable group. It would be fair to say that the first of these is not yet well understood. It is the invariant theory of the latter classes that we choose to present in this chapter. There is a very large body of literature dealing with these matters, so we have had to pick and choose what we presented.

We will start with a discussion of pseudoreflection representations, which have a rich history going back to the origins of invariant theory in the nineteenth century. We will present a mixture of classical and modern results involving these representations. The theorem of G. C. Shephard and J. A. Todd on the invariants of complex pseudoreflection groups culminated a century of work on these groups and their classification. The basic result has been generalized to the nonmodular case, and given a classification-free proof [70], which we will present in Section 7.1. An offshoot of this proof has been the more recent study of the rings of coinvariants of pseudoreflection representations, in both the modular and nonmodular cases, and we will sample some of the more recent results in this area in Section 7.2.

The solvable groups and their invariants are the next topic we investigate. A useful tool here is a relative version of Noether's bound, which allows us to relate $\beta(G)$ to $\beta(H)$ and $|G{:}H|$, where $H \leq G$ is a subgroup of G. Based on [376] and [379], we present an algorithm to compute the invariants of a solvable group in the nonmodular situation. This, in turn, leads to various refinements of Noether's bound for these groups.

Representations of groups in low degree present a very restricted class of problems, because from the point of view of invariant theory it is the image of the representation, i.e., a subgroup of $\mathrm{GL}(n, \mathbb{F})$, that matters. For small n, the finite subgroups of $\mathrm{GL}(n, \mathbb{Z})$ have been classified up to conjugation, and for a finite field $\mathbb{F}_q$ L. E. Dickson made extensive lists of the subgroups of $\mathrm{GL}(n, \mathbb{F}_q)$ at the start of the twentieth century. By availing ourselves of these lists, we can compute the invariants of basically all groups in degree 2. In the process we display how the tools we have developed so far, as well as some that we will develop in the succeeding chapters, can be brought to bear on invariant-theoretic computations. We will discuss the computations of invariants in degree two in Sections 7.4 and 7.5. Although it is often misleading to generalize the patterns seen in these computations, because, for example, these rings of invariants must always be Cohen–Macaulay (Proposition 5.6.9), such patterns can serve to guide future research. What seems appropriate to us will be mentioned towards the end of the chapter.

We certainly do not intend to imply that these are the only choices possible for this chapter: They are, however, the ones that we are most familiar with. Other classes of groups worthy of discussion at this point are the vector invariants of $\mathbb{Z}/p$ in characteristic p ([8], [9], [10], [15], [54], [325], [326], and [327]) for which so many problems remain unanswered.

7.1 Pseudoreflections and Pseudoreflection Groups

The study of which representations $\rho : G \hookrightarrow \mathrm{GL}(n, \mathbb{F})$ of which groups G over which fields $\mathbb{F}$ lead to polynomial rings of invariants $\mathbb{F}[V]^G$ is one of the more beautiful chapters of invariant theory. As usual, there is a nonmodular and a modular case. This section will skim the surface of the nonmodular case, where there is a relatively complete picture, as well as a classification theorem. As is often the case in invariant theory, the modular case is somewhat more subtle and obscure.

Recall that a linear automorphism $s : V \longrightarrow V$ in a finite-dimensional vector space V over the field $\mathbb{F}$ is called a **pseudoreflection** if it is not the identity, has finite order, and leaves a codimension-1 subspace, denoted by H_s, or V^s, and called the **hyperplane** of s, pointwise fixed. The subspace $\mathrm{Im}(1-s)$ of V is therefore of dimension 1. It is called the **direction** of s. Compare with the terminology of transvections used in Section 6.2: This is no accident, since transvections of finite order are pseudoreflections (if $\mathbb{F}$ has nonzero characteristic p, transvections are always of order p by Lemma 6.2.5). It is not hard to see that a pseudoreflection $s : V \longrightarrow V$ acts on the dual space $s : V^* \longrightarrow V^*$ also as a pseudoreflection, since the characteristic polynomial of a matrix and its transpose are the same.

In the nonmodular case, the representations $\rho : G \hookrightarrow \mathrm{GL}(n, \mathbb{F})$ for which $\mathbb{F}[V]^G$ is a polynomial algebra are precisely those where $\rho(G)$ is generated by pseudoreflections. Such a representation is called a **pseudoreflection representation**. Somewhat less precisely, one calls G a **pseudoreflection group** if such a representation exists. The imprecision lies in the fact that the same group may have more than one pseudoreflection representation, even over the complex field $\mathbb{C}$, as well as over other fields.

H. S. M. Coxeter, in [79], based on his classification of the real reflection groups, i.e., Coxeter groups, had observed that the invariants of such a real Coxeter representation are always a polynomial algebra. Later, G. C. Shephard and J. A. Todd [352] observed the same phenomenon for $\mathbb{F} = \mathbb{C}$ based on a list of all examples of finite complex pseudoreflection groups. Table 7.1.1 shows the classification as it appeared in [352] and some additional information extracted from [73].

In reading the table, the last column lists those primes that do not divide the order of G, and for which the character field of G lies in $\mathbb{Q}_{\hat{p}}$, the p-adic rational numbers. This information is important for many topological applications beginning with [73]. See also [372] Chapter 10 and the references listed there. The next to last column gives the character field of the pseudoreflection representation: ζ_m denotes a primitive mth root of unity. In general, a complex representation of a finite group is not necessarily equivalent to a representation whose matrix entries are in the field generated by the character values. The Schur index measures the extent to which this fails. However, for a group generated by pseudoreflections, the Schur index is always 1, see, e.g., [34], [73], or [372] Theorem 7.3.1. Hence for those primes listed as *good primes* in the table there is a mod p reduction of the representation, which is again generated by pseudoreflections. Just because a prime is not good does not mean that it is bad; see Example 3 in Section 7.4 of [372] for more information.

C. Chevalley, in [70], provided a proof of H. S. M. Coxeter's observation from first principles, and J.-P. Serre, in [346] (see also [39] Chapitre 5, §5, N° 2 Théorème 1), noticed that C. Chevalley's proof was adequate to handle the nonmodular case. We confine our discussion of pseudoreflection groups to a version of this proof, since a complete discussion of pseudoreflection groups and their invariants would require a book in itself, but see [372] Chapters 7 and 10 and the many references listed there. We choose to use the condition of Corollary 5.3.4 to verify that $\mathbb{F}[V]^G$ is a polynomial algebra. By Proposition A.1.5 this is equivalent to showing that $\mathbb{F}[V]$ is a flat $\mathbb{F}[V]^G$-module, i.e., that $\mathrm{Tor}_1^{\mathbb{F}[V]^G}(\mathbb{F}, \mathbb{F}[V]) = 0$. We divide the work into three lemmas.

LEMMA 7.1.1: *Let* $\rho : G \hookrightarrow \mathrm{GL}(n, \mathbb{F})$ *be a representation of a finite group* G *over the field* $\mathbb{F}$. *Then the map*

$$\mathrm{Tor}_1^{\mathbb{F}[V]^G}(\mathbb{F}, \mathrm{Tr}^G) : \mathrm{Tor}_1^{\mathbb{F}[V]^G}(\mathbb{F}, \mathbb{F}[V]) \longrightarrow \mathrm{Tor}_1^{\mathbb{F}[V]^G}(\mathbb{F}, \mathbb{F}[V])$$

is zero.

	DIM	$\lvert G \rvert$	DEGREES	χ – FIELD	GOOD PRIMES
1	n	$(n+1)!$	$2,3,\ldots,n+1$	$\mathbb{Q}$	$>n+1$
2a	n	$qm^{n-1}n!$ $(q\mid m$ and $m>1)$	$m,2m,\ldots,(n-1)m,qn$	$\mathbb{Q}(\zeta_m)$	$1 \pmod{m}, >n$
2b	2	$2m \quad (m>2)$	$2,m$	$\mathbb{Q}(\zeta_m+\zeta_m^{-1})$	$\pm 1 \pmod{m}$
3	1	m	m	$\mathbb{Q}(\zeta_m)$	$1 \pmod{m}$
4	2	$24=2^3 3$	$4,6$	$\mathbb{Q}(\zeta_3)$	$1 \pmod{3}$
5	2	$72=2^3 3^2$	$6,12$	$\mathbb{Q}(\zeta_3)$	$1 \pmod{3}$
6	2	$48=2^4 3$	$4,12$	$\mathbb{Q}(\zeta_{12})$	$1 \pmod{12}$
7	2	$144=2^4 3^2$	$12,12$	$\mathbb{Q}(\zeta_{12})$	$1 \pmod{12}$
8	2	$96=2^5 3$	$8,12$	$\mathbb{Q}(i)$	$1 \pmod{4}$
9	2	$192=2^6 3$	$8,24$	$\mathbb{Q}(\zeta_8)$	$1 \pmod{8}$
10	2	$288=2^5 3^2$	$12,24$	$\mathbb{Q}(\zeta_{12})$	$1 \pmod{12}$
11	2	$576=2^6 3^2$	$24,24$	$\mathbb{Q}(\zeta_{24})$	$1 \pmod{24}$
12	2	$48=2^4 3$	$6,8$	$\mathbb{Q}(\sqrt{-2})$	$1,3 \pmod{8}, \neq 3$
13	2	$96=2^5 3$	$8,12$	$\mathbb{Q}(\zeta_8)$	$1 \pmod{8}$
14	2	$144=2^4 3^2$	$6,24$	$\mathbb{Q}(\zeta_3,\sqrt{-2})$	$1,19 \pmod{24}$
15	2	$288=2^5 3^2$	$12,24$	$\mathbb{Q}(\zeta_{24})$	$1 \pmod{24}$
16	2	$600=2^3 3\cdot 5^2$	$20,30$	$\mathbb{Q}(\zeta_5)$	$1 \pmod{5}$
17	2	$1200=2^4 3\cdot 5^2$	$20,60$	$\mathbb{Q}(\zeta_{20})$	$1 \pmod{20}$
18	2	$1800=2^3 3^2 5^2$	$30,60$	$\mathbb{Q}(\zeta_{15})$	$1 \pmod{15}$
19	2	$3600=2^4 3^2 5^2$	$60,60$	$\mathbb{Q}(\zeta_{60})$	$1 \pmod{60}$
20	2	$360=2^3 3^2 5$	$12,30$	$\mathbb{Q}(\zeta_3,\sqrt{5})$	$1,4 \pmod{15}$
21	2	$720=2^4 3^2 5$	$12,60$	$\mathbb{Q}(\zeta_{12},\sqrt{5})$	$1,49 \pmod{60}$
22	2	$240=2^4 3\cdot 5$	$12,20$	$\mathbb{Q}(i,\sqrt{5})$	$1,9 \pmod{20}$
23	3	$120=2^3 3\cdot 5$	$2,6,10$	$\mathbb{Q}(\sqrt{5})$	$1,4 \pmod{5}$
24	3	$336=2^4 3\cdot 7$	$4,6,14$	$\mathbb{Q}(\sqrt{-7})$	$1,2,4 \pmod{7}$
25	3	$648=2^3 3^4$	$6,9,12$	$\mathbb{Q}(\zeta_3)$	$1 \pmod{3}$
26	3	$1296=2^4 3^4$	$6,12,18$	$\mathbb{Q}(\zeta_3)$	$1 \pmod{3}$
27	3	$2160=2^4 3^3\cdot 5$	$6,12,30$	$\mathbb{Q}(\zeta_3,\sqrt{5})$	$1,4 \pmod{15}$
28	4	$1152=2^7 3^2$	$2,6,8,12$	$\mathbb{Q}$	$\neq 2$ or 3
29	4	$7680=2^9 3\cdot 5$	$4,8,12,20$	$\mathbb{Q}(i)$	$1 \pmod{4}, \neq 5$
30	4	$2^6 3^2 5^2$	$2,12,20,30$	$\mathbb{Q}(\sqrt{5})$	$1,4 \pmod{5}$
31	4	$2^{10} 3^2 5$	$8,12,20,24$	$\mathbb{Q}(i)$	$1 \pmod{4}, \neq 5$
32	4	$2^7 3^5 5$	$12,18,24,30$	$\mathbb{Q}(\zeta_3)$	$1 \pmod{3}$
33	5	$2^7 3^4 5$	$4,6,10,12,18$	$\mathbb{Q}(\zeta_3)$	$1 \pmod{3}$
34	6	$2^9 3^7 5\cdot 7$	$6,12,18,24,30,42$	$\mathbb{Q}(\zeta_3)$	$1 \pmod{3}, \neq 7$
35	6	$2^7 3^4 5$	$2,5,6,8,9,12$	$\mathbb{Q}$	$\neq 2,3$ or 5
36	7	$2^{10} 3^4 5\cdot 7$	$2,6,8,10,12,14,18$	$\mathbb{Q}$	$\neq 2,3,5$ or 7
37	8	$2^{14} 3^5 5^2 7$	$2,8,12,14,18,20,24,30$	$\mathbb{Q}$	$\neq 2,3,5$ or 7

TABLE 7.1.1: The finite irreducible complex pseudoreflection groups

PROOF: This is immediate from the commutative diagram

$$\begin{array}{ccc} \mathrm{Tor}_1^{\mathbb{F}[V]^G}(\mathbb{F}, \mathbb{F}[V]^G) & \xrightarrow{\mathrm{Tor}_1^{\mathbb{F}[V]^G}(\mathbb{F},\mathrm{Tr}^G)} & \mathrm{Tor}_1^{\mathbb{F}[V]^G}(\mathbb{F}, \mathbb{F}[V]^G) = 0 \\ & {\scriptstyle \mathrm{Tor}_1^{\mathbb{F}[V]^G}(\mathbb{F},\mathrm{Tr}^G)} \searrow \quad \swarrow {\scriptstyle \mathrm{Tor}^{\mathbb{F}[V]^G}(\mathbb{F},i)} & \\ & \mathrm{Tor}_1^{\mathbb{F}[V]^G}(\mathbb{F}, \mathbb{F}[V]) & \end{array}$$

where $i : \mathbb{F}[V]^G \hookrightarrow \mathbb{F}[V]$ is the inclusion. □

LEMMA 7.1.2: *Let $\varrho : G \hookrightarrow \mathrm{GL}(n, \mathbb{F})$ be a representation of a finite group G over the field $\mathbb{F}$. If $|G|$ is invertible in $\mathbb{F}$, then*

$$\mathrm{Tor}_1^{\mathbb{F}[V]^G}(\mathbb{F}, \mathbb{F}[V])^G = 0.$$

PROOF: Since $|G| \in \mathbb{F}^\times$, the transfer homomorphism

$$\mathrm{Tr}^G = \mathrm{Tor}_1^{\mathbb{F}[V]^G}(\mathbb{F}, \mathrm{Tr}^G) : \mathrm{Tor}_1^{\mathbb{F}[V]^G}(\mathbb{F}, \mathbb{F}[V]) \longrightarrow \left(\mathrm{Tor}_1^{\mathbb{F}[V]^G}(\mathbb{F}, \mathbb{F}[V])\right)^G$$

is a surjective map of $\mathbb{F}[V]^G$-modules, and the result follows from Lemma 7.1.1. □

We need some preliminaries before stating and proving the last lemma needed for the proof of the main theorem.

Let $s \in \mathrm{GL}(n, \mathbb{F})$ be a pseudoreflection and $\ell_s \in \mathbb{F}[V]$ a linear form with $\ker(\ell_s) = H_s = \{v \in V \mid sv = v\}$. The linear form ℓ_s depends on s up to a nonzero scalar. The action of the pseudoreflection s on the space of linear forms V^* has direction ℓ_s, i.e., $\mathrm{Im}(1 - s) = \mathrm{Span}_{\mathbb{F}}\{\ell_s\} \subseteq V^*$. Therefore for any linear form $z \in V^*$ we have $s(z) \equiv z \bmod (\ell_s)$. Since $\mathbb{F}[V]$ is generated by linear forms it follows that $sf \equiv f \bmod (\ell_s)$ for any $f \in \mathbb{F}[V]$. The algebra $\mathbb{F}[V]$ is a unique factorization domain, and ℓ_s is a prime element, so

$$(s - 1)(f) = \ell_s \cdot \Delta_s(f)$$

for a unique $\Delta_s(f) \in \mathbb{F}[V]$ (see, e.g., [285], or [372] Section 7.1 and the references there). If f has degree k, then $\Delta_s(f)$ has degree $k-1$, and $\Delta_s(f)=0$ if and only if $sf = f$. For $a \in \mathbb{F}[V]$ of degree 0 set $\Delta_s(a) = 0$.

The operator $\Delta_s : \mathbb{F}[V] \longrightarrow \mathbb{F}[V]$ is linear and satisfies the following twisted derivation formula:

$$\Delta_s(fh) = \Delta_s(f) \cdot h + s(f) \cdot \Delta_s(h)$$

(see, e.g., [82]). Hence inductively one obtains

$$\Delta_s(f^k) = \Delta_s(f) \left(f^{k-1} + f^{k-2} \cdot s(f) + \cdots + s(f)^{k-1}\right)$$

for any $k \in \mathbb{N}$.

LEMMA 7.1.3: *Let* $\rho : G \hookrightarrow \mathrm{GL}(n, \mathbb{F})$ *be a representation of a finite group* G *over the field* $\mathbb{F}$. *If* $\rho(G)$ *is generated by pseudoreflections then* $\mathrm{Tor}_1^{\mathbb{F}[V]^G}(\mathbb{F}, \mathbb{F}[V])^G = 0$ *if and only if* $\mathrm{Tor}_1^{\mathbb{F}[V]^G}(\mathbb{F}, \mathbb{F}[V]) = 0$.

PROOF: If $\mathrm{Tor}_1^{\mathbb{F}[V]^G}(\mathbb{F}, \mathbb{F}[V]) = 0$, then passing to invariants cannot change this, so $\mathrm{Tor}_1^{\mathbb{F}[V]^G}(\mathbb{F}, \mathbb{F}[V])^G = 0$ also. To prove the converse, suppose $\mathrm{Tor}_1^{\mathbb{F}[V]^G}(\mathbb{F}, \mathbb{F}[V]) \neq 0$. Let $s \in G$ be a pseudoreflection. Then the operator $\Delta_s : \mathbb{F}[V] \longrightarrow \mathbb{F}[V]$ is an $\mathbb{F}[V]^G$-module homomorphism, so it induces a map

$$\mathrm{Tor}_1^{\mathbb{F}[V]^G}(\mathbb{F}, \Delta_s) : \mathrm{Tor}_1^{\mathbb{F}[V]^G}(\mathbb{F}, \mathbb{F}[V]) \longrightarrow \mathrm{Tor}_1^{\mathbb{F}[V]^G}(\mathbb{F}, \mathbb{F}[V])$$

lowering the grading by 1. If $u \in \mathrm{Tor}_1^{\mathbb{F}[V]^G}(\mathbb{F}, \mathbb{F}[V])$ is a nonzero element of minimal degree, then $\Delta_s(u) = 0$. Let $\mu_{\ell_s} : \mathbb{F}[V] \longrightarrow \mathbb{F}[V]$ be the $\mathbb{F}[V]^G$-module homomorphism induced by multiplication by ℓ_s. Then

$$(s - 1) = \mu_{\ell_s} \cdot \Delta_s \in \mathrm{End}(\mathbb{F}[V]),$$

from which it follows that in $\mathrm{End}(\mathrm{Tor}_1^{\mathbb{F}[V]^G}(\mathbb{F}, \mathbb{F}[V]))$ we have

$$\mathrm{Tor}_1^{\mathbb{F}[V]^G}(\mathbb{F}, s - 1) = \mathrm{Tor}_1^{\mathbb{F}[V]^G}(\mathbb{F}, \mu_{\ell_s}) \cdot \mathrm{Tor}_1^{\mathbb{F}[V]^G}(\mathbb{F}, \Delta_s).$$

Applying this formula to u yields $\mathrm{Tor}_1^{\mathbb{F}[V]^G}(\mathbb{F}, s - 1)(u) = 0$, since for degree reasons we have $\mathrm{Tor}_1^{\mathbb{F}[V]^G}(\mathbb{F}, \Delta_s)(u) = 0$. In other words, $s(u) = u$. Since G is generated by pseudoreflections, it follows that u is G-invariant. So $\mathrm{Tor}_1^{\mathbb{F}[V]^G}(\mathbb{F}, \mathbb{F}[V])^G$ contains the nonzero element u as was to be shown. □

THEOREM 7.1.4 (G. C. Shephard–J. A.Todd, C. Chevalley): *Let* V *be a finite-dimensional vector space over the field* $\mathbb{F}$ *and* $\rho : G \hookrightarrow \mathrm{GL}(V)$ *a representation of a finite group* G. *Assume that* $|G|$ *is relatively prime to the characteristic of* $\mathbb{F}$. *Then the following are equivalent:*

(i) G *is generated by pseudoreflections.*

(ii) $\mathbb{F}[V]^G$ *is a polynomial algebra.*

PROOF: Suppose that ρ is a pseudoreflection representation. Since $|G| \in \mathbb{F}^\times$, it follows from Lemma 7.1.2 that $\mathrm{Tor}_1^{\mathbb{F}[V]^G}(\mathbb{F}, \mathbb{F}[V])^G = 0$. Since $\rho(G)$ is generated by pseudoreflections, we can apply Lemma 7.1.3 and obtain $\mathrm{Tor}_1^{\mathbb{F}[V]^G}(\mathbb{F}, \mathbb{F}[V]) = 0$. Therefore, $\mathbb{F}[V]$ is a flat $\mathbb{F}[V]^G$-module, and hence $\mathbb{F}[V]^G$ is a polynomial algebra by Corollary 5.3.4.

Suppose, on the other hand, that $\mathbb{F}[V]^G$ is a polynomial algebra, with polynomial generators $f_1, \ldots, f_n$ of degrees $d_1 \leq \cdots \leq d_n$. Let $H \leq G$ be the subgroup of G generated by $s(G)$, the set of pseudoreflections in G. By what has already been shown $\mathbb{F}[V]^H$ is a polynomial algebra. Choose generators $h_1, \ldots, h_n$ for $\mathbb{F}[V]^H$, and let their degrees be $e_1 \leq \cdots \leq e_n$. Of course, $f_i \in \mathbb{F}[h_1, \ldots, h_n]$ for $i = 1, \ldots, n$. By Corollary 3.1.5 we have

$$|G| = d_1 \cdots d_n, \quad |s(G)| = \sum_{i=1}^{n}(d_i - 1),$$
$$|H| = e_1 \cdots e_n, \quad |s(H)| = \sum_{i=1}^{n}(e_i - 1).$$

We claim that $d_i \geq e_i$ for $i = 1, \ldots, n$. For $i = 1$ this is clear. Assume that $e_i \leq d_i$ for $i = 1, \ldots, m$ and consider f_{m+1}. If $d_{m+1} < e_{m+1}$, then f_{m+1} must be a polynomial in $h_1, \ldots, h_m$. Since $d_i \leq d_{m+1} < e_{m+1}$ for $i = 1, \ldots, m$, $f_1, \ldots, f_m$ are also polynomials in $h_1, \ldots, h_m$, and hence $f_1, \ldots, f_{m+1} \in \mathbb{F}[h_1, \ldots, h_m]$ would be algebraically independent, which is impossible. Therefore, $d_i \geq e_i$ for $i = 1, \ldots, n$ as claimed.

Since G and H have the same pseudoreflections, we obtain from Corollary 3.1.5 that

$$\sum_{i=1}^{n}(d_i - 1) = \sum_{i=1}^{n}(e_i - 1),$$

and it follows that $d_i = e_i$ for $i = 1, \ldots, n$. But then $|G| = d_1 \cdots d_n = e_1 \cdots e_n = |H|$, so $H = G$ and G is generated by pseudoreflections. □

Example 1 (The Icosahedral Group I_6): The full group of symmetries of the icosahedron is group number 23 in Table 7.1.1. The group I_6 is generated by its rotation subgroup $\mathcal{I}$ (see Example 4 in Section 3.1) and the matrix $-\mathbf{I}$, where $\mathbf{I}$ is the identity matrix.

To compute the invariants of this group in its defining representation we recall that the icosahedron has six polar axes $v_0, \ldots, v_5$ consisting of the lines joining pairs of opposite vertices, and ten polar axes joining midpoints of opposite faces $f_0, \ldots, f_9$. If $\lambda_0, \ldots, \lambda_5$ are linear forms whose kernels are planes through the origin orthogonal to $v_0, \ldots, v_5$, then the action of $\mathcal{I}$ on the icosahedron permutes $\lambda_0, \ldots, \lambda_5$, and the action of $-\mathbf{I}$ partitions the set $\Lambda = \{\lambda_0, \ldots, \lambda_5\}$ into three orbits each consisting of a pair $\lambda, -\lambda$ with $\lambda \in \Lambda$. Hence $f = \prod_{i=0}^{5} \lambda_i$ is I_6-invariant. Likewise, if $\mu_0, \ldots, \mu_9$ are linear forms whose kernels are orthogonal to the polar axes $f_0, \ldots, f_9$, then $h = \prod_{i=0}^{9} \mu_i$ is also an I_6-invariant.

Since $\mathrm{I}_6 \subset \mathbb{O}(3, \mathbb{R})$, the norm $Q = x^2 + y^2 + z^2 \in \mathbb{F}[x, y, z]$ is also I_6-invariant. To see that the three invariant forms $Q, f, h \in \mathbb{R}[x, y, z]^{\mathrm{I}_6}$ form a system of parameters we need to consider the intersection $\mathfrak{V}$ of the complex varieties $\mathfrak{X}_{\mathbb{C}} = \{z \in \mathbb{C}^3 \mid f(z) = 0\}$, $\mathfrak{Y}_{\mathbb{C}} = \{z \in \mathbb{C}^3 \mid h(z) = 0\}$, and $\mathfrak{Z}_{\mathbb{C}} = \{z \in \mathbb{C}^3 \mid Q(z) = 0\}$. Since the polynomials are homogeneous, this intersection is a cone, i.e., if $u \in \mathfrak{V}$, then so is ζu for any complex number ζ. The corresponding real varieties $\mathfrak{X}_{\mathbb{R}} = \{z \in \mathbb{R}^3 \mid f(z) = 0\}$ and $\mathfrak{Y}_{\mathbb{R}} = \{z \in \mathbb{R}^3 \mid h(z) = 0\}$ are each a union of lines, and hence the intersection of $\mathfrak{X}_{\mathbb{C}}$ and $\mathfrak{Y}_{\mathbb{C}}$ is the union of the complexification of these lines. Therefore, if $u = (x, y, z) \in \mathfrak{X}_{\mathbb{C}} \cap \mathfrak{Y}_{\mathbb{C}}$, we can find a complex number ζ such that ζu has only real coordinates. However, the variety $\mathfrak{Z}_{\mathbb{C}}$ contains no real points apart from (0, 0, 0), and hence $\mathfrak{V} = \{0\}$ as required.

Since $Q, f, h \in \mathbb{R}[x, y, z]^{\mathrm{I}_6}$ are a system of parameters, and the product of their degrees is $2 \cdot 6 \cdot 10 = 120 = |\mathrm{I}_6|$, we can apply Proposition 4.5.5 to conclude that

$$\mathbb{R}[x, y, z]^{\mathrm{I}_6} = \mathbb{R}[Q, f, h].$$

The invariants of the rotation subgroup $\mathcal{I}$ can be derived from this with the aid of Proposition 5.5.8, confirming the computation in Section 3.1, Example 4. The missing form k is the product of the 15 linear forms whose kernels are orthogonal to the polar axes joining the midpoints of opposite edges.

In the modular case it may happen that a pseudoreflection representation $\rho : G \hookrightarrow \mathrm{GL}(n, \mathbb{F})$ does not have a ring of invariants that is a polynomial algebra. This occurs, for example, for the Weyl group $W(\mathbf{F}_4)$ in its natural representation in characteristic 3; see, e.g., [426] or [372] Section 7.4, Example 4. One thing that is known is that when $\mathbb{F}[V]^G$ is a polynomial algebra, then ρ is a pseudoreflection representation (see, e.g., [346], or [39] Chapitre V §6 Exercise 7 and 8), and moreover, for any nonzero subspace $U \subset V$ the pointwise stabilizer $G_U = \left\{g \in G \mid g\,|_U = \mathbf{1}\right\}$ is also generated by pseudoreflections (see Section 10.6). This latter criterion applies nicely to the Weyl group of type $W(\mathbf{F}_4)$.

EXAMPLE 2: The Weyl group of type $\mathbf{F}_4$ has order $1152 = 3^2 \cdot 2^7$, so its 3-Sylow subgroup has order 9. For a discussion of this group and its subgroups see [372] Section 7.4, Example 4. The group $W(\mathbf{F}_4)$ acts on the Dynkin diagram of its subgroup of type $\mathbf{D}_4$. One of the elements of the 3-Sylow subgroup induces the rotational symmetry of the diagram called the triality. The 3-Sylow subgroup fixes the middle root. In characteristic different from 3 this root spans the fixed-point set of the action of $W(\mathbf{F}_4)$ on the vector space $\mathbb{F}^4$. In characteristic 3 the sum of the four roots of $W(\mathbf{D}_4)$ are also fixed by the entire 3-Sylow subgroup, and not just the triality (this depends on the fact that $-2 \equiv 1 \bmod 3$), so $U = (\mathbb{F}_3^4)^{\mathrm{Syl}_3(W(\mathbf{F}_4))}$ is 2-dimensional. The pointwise stabilizer $W(\mathbf{F}_4)_U$ of this subspace is a proper nontrivial subgroup of $W(\mathbf{F}_4)$ and contains the 3-Sylow subgroup of $W(\mathbf{F}_4)$. But no proper subgroup of $W(\mathbf{F}_4)$ generated by pseudoreflections in characteristic 3 has a 3-Sylow subgroup of order 9. Hence $W(\mathbf{F}_4)_U$ is not a pseudoreflection group, so the invariants of $W(\mathbf{F}_4)$ are not a polynomial algebra in characteristic 3 even though $W(\mathbf{F}_4)$ is generated by pseudoreflections. This means that the nonmodular characteriza-

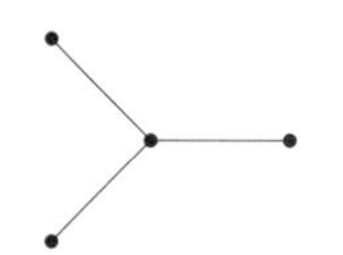

DYNKIN DIAGRAM 7.1.2: Type $\mathbf{D}_4$

tion of representations whose rings of invariants are polynomial algebras by G. C. Shephard and J. A. Todd, fails in the modular case.

For irreducible pseudoreflection representations $\rho : G \hookrightarrow \mathrm{GL}(n, \mathbb{F})$, G. Kemper and G. Malle, [199], have shown that the conditions

(i) G is generated by pseudoreflections, and
(ii) the pointwise stabilizers of proper nonzero subspaces have polynomial invariants,

are necessary and sufficient for $\mathbb{F}[V]^G$ to have polynomial invariants by examining all the cases. For finite p-groups P represented in characteristic p, H. Nakajima has given a sufficient condition, [277], that the ring of invariants be a polynomial algebra (see Theorem 6.3.6), which when the ground field is the prime field is also necessary. If the ground field is not the prime field, this can fail. There is an example over $\mathbb{F}_{p^3}$ due to R. E. Stong, [388], that has been discussed in Section 6.3. For a discussion of the more restricted family of transvection groups we refer to Section 6.2.

So, altogether we can say that in the nonmodular case pseudoreflection groups form the family with the nicest rings of invariants, namely polynomial rings. The discussions here and in Section 6.2 has shown that in the modular case the representations with polynomial invariant rings form a proper subset of all pseudoreflection representations. One might, however, hope that also in the modular case the rings of invariants of pseudoreflection groups are not "arbitrarily bad", i.e., one might hope that they are, e.g., at least Cohen–Macaulay. This is not true; in [270] and [273] H. Nakajima gives examples of pseudoreflection groups with non-Cohen–Macaulay rings of invariants. H. Nakajima has also shown ([276] Proposition 2.1) that, at least for abelian groups, the property that $\mathbb{F}[V]^G$ be a polynomial algebra is controlled by the p-Sylow subgroup in characteristic p.

PROPOSITION 7.1.5 (H. Nakajima): *Let $\rho : G \hookrightarrow \mathrm{GL}(n, \mathbb{F})$ be a pseudoreflection representation of an abelian group G. Then $\mathbb{F}[V]^G$ is a polynomial algebra if and only if $\mathbb{F}[V]^{\mathrm{Syl}_p(G)}$ is a polynomial algebra.*

PROOF: Since G is abelian it splits into a direct product of its p-Sylow subgroup $\mathrm{Syl}_p(G)$ and a subgroup $\mathrm{Syl}_{p'}(G)$ of order relatively prime to p. Moreover by passing to a suitable field extension we can split V into a direct sum $V_p \oplus V_{p'}$ of a representation V_p of $\mathrm{Syl}_p(G)$ and $V_{p'}$ of $\mathrm{Syl}_{p'}(G)$. The result is then immediate from the isomorphism $\mathbb{F}[V]^G \cong \mathbb{F}[V_p]^{\mathrm{Syl}_p(G)} \otimes \mathbb{F}[Vp']^{\mathrm{Syl}_{p'}(G)}$. □

This result is the first step in H. Nakajima's determination of all representations of abelian groups with rings of invariants that are polynomial algebras. For more details see [276].

7.2 Coinvariants of Pseudoreflection Groups

In this section $\mathbb{F}$ will be an arbitrary field, $\rho : G \hookrightarrow \mathrm{GL}(n, \mathbb{F})$ a representation of a finite group G over $\mathbb{F}$, and $\mathbb{F}[V]_G$ the ring of coinvariants. The Hilbert ideal $\mathfrak{h}(G)$ is stable under the operation of G on $\mathbb{F}[V]$, so the action of G passes to the quotient $\mathbb{F}[V]_G$. This gives us a finite-dimensional representation of G, which in the case of a pseudoreflection group in coprime characteristic was first computed by Chevalley [70]:

THEOREM 7.2.1 (C. Chevalley): *Let $G \hookrightarrow \mathrm{GL}(n, \mathbb{F})$ be a finite group generated by pseudoreflections whose order is prime to the characteristic of $\mathbb{F}$. Then $\mathrm{Tot}(\mathbb{F}[V]_G)$ is the regular representation of G.*

PROOF: By the theorem of Shephard-Todd (Theorem 7.1.4) $\mathbb{F}[V]^G$ is a polynomial algebra on $n = \dim_{\mathbb{F}}(V)$ generators, so by Macaulay's theorem (Theorem A.3.5) $\mathbb{F}[V]$ is a free module over $\mathbb{F}[V]^G$ and there is an isomorphism

$$\mathbb{F}[V] \cong \mathbb{F}[V]^G \otimes \mathbb{F}[V]_G$$

as graded $\mathbb{F}(G)$-modules, i.e., as graded G-representations. Choose elements $h_1, \ldots, h_d \in \mathbb{F}[V]$ that generate $\mathbb{F}[V]$ freely as an $\mathbb{F}[V]^G$-module and note that they project to a vector space basis $\overline{h}_1, \ldots, \overline{h}_d$ in $\mathbb{F}[V]_G$. Write

$$g \cdot h_i = \sum a_{i,j}(g) h_i,$$

where $(a_{i,j}(g))$ is a $d \times d$ matrix with entries in $\mathbb{F}[V]$. The extension of fields of fractions $\mathbb{F}(V) \mid \mathbb{F}(V)^G$ is Galois. So $\mathbb{F}(V)$ regarded as a G-representation over $\mathbb{F}(V)^G$ is the regular representation. Hence taking traces gives

$$\sum_{i=1}^{n} a_{i,i}(g) \mathrm{tr}(a_{i,j}(g)) = \chi_{\mathrm{reg}}(g),$$

where reg denotes the regular representation of G and χ_{reg} its character. The action of $g \in G$ on $\mathbb{F}[V]_G$ is represented by the matrix $(\varepsilon(a_{i,j}(g)))$, where $\varepsilon : \mathbb{F}[V] \longrightarrow \mathbb{F}$ is the augmentation. By homogeneity the entries $a_{i,i}(g)$ have degree zero, and hence

$$\mathrm{tr}(\varepsilon(a_{i,j}(g))) = \mathrm{tr}(a_{i,j}(g)) = \chi_{\mathrm{reg}},$$

and the result follows. (As reference for the use of character theory in the preceding proof see, for example, [346] §2.3 Corollary 3.) □

If $\mathbb{F}[V]^G = \mathbb{F}[f_1, \ldots, f_n]$ is a polynomial algebra, another basic property of the algebra $\mathbb{F}[V]_G$ of coinvariants is that it is a Poincaré duality algebra; see Theorem 5.4.1. In the nonmodular case this is a well-studied finite-dimensional representation of G. In particular, as just shown, it is isomorphic to the regular representation of G over $\mathbb{F}$, and a fundamental class is represented by the Jacobian determinant $\det[\partial f_i / \partial z_j]$, so is a $\det^{-1}$-relative invariant; see Corollary 5.4.2. Our first goal in this section is to prove that the last of these conclusions, viz., that the fundamental class of $\mathbb{F}[V]_G$ is a $\det^{-1}$-relative invariant, holds in general. The proof is characteristic free, and provides a new proof (see [384]) of this fact even in the nonmodular case.

THEOREM 7.2.2: *Let* $\rho : G \hookrightarrow \mathrm{GL}(n, \mathbb{F})$ *be a representation of a finite group* G *over the field* $\mathbb{F}$. *If* $\mathbb{F}[V]^G$ *is a polynomial algebra, then the fundamental class,* $[\mathbb{F}[V]_G]$, *of the ring of coinvariants is a* $\det^{-1}$*-relative invariant.*

PROOF: Choose homogeneous polynomials $f_1, \ldots, f_n \in \mathbb{F}[V]$ such that $\mathbb{F}[V]^G \cong \mathbb{F}[f_1, \ldots, f_n]$. We begin by recalling how to compute a fundamental class for $\mathbb{F}[V]_G$, or indeed, for $\mathbb{F}[z_1, \ldots, z_n]/(f_1, \ldots, f_n)$, whenever $f_1, \ldots, f_n \in \mathbb{F}[z_1, \ldots, z_n]$ is a system of parameters, and $z_1, \ldots, z_n$ is a basis for V^* (see the proof of Theorem 5.4.1). Here is how this goes: Write

$$f_i = \sum_{j=1}^{n} a_{i,j} z_j, \qquad i = 1, \ldots, n,$$

for homogeneous polynomials $a_{i,j} \in \mathbb{F}[z_1, \ldots, z_n]$, $i, j = 1, \ldots, n$. The determinant, $\det[a_{i,j}]$, regarded as an element of the quotient algebra $\mathbb{F}[z_1, \ldots, z_n]/(f_1, \ldots, f_n)$, is a fundamental class.

To begin the proof proper, we recall [39] Chapitre V, §6, Exercise 8, that since $\mathbb{F}[V]^G$ is a polynomial algebra, G (or better put $\rho(G)$) is generated by pseudoreflections. Therefore, to show that the fundamental class of $\mathbb{F}[V]_G$ is a $\det^{-1}$-relative invariant, it will suffice to show that

$$s([\mathbb{F}[V]_G]) = \det(s)^{-1} \cdot [\mathbb{F}[V]_G] \in \mathbb{F}[V]_G$$

for all pseudoreflections $s \in G$.

The pseudoreflections in G can be of two types: diagonalizable, or transvections. We consider these two cases separately.

CASE: $s \in s_\Delta(G)$, *i.e.,* s *is diagonalizable.*

We choose a basis $z_1, \ldots, z_n$ for V^* such that s is represented by the matrix

$$\begin{bmatrix} 1 & 0 & \cdots & 0 \\ 0 & \ddots & \cdots & 0 \\ \vdots & \cdots & 1 & 0 \\ 0 & \cdots & 0 & \lambda \end{bmatrix} \in \mathrm{GL}(n, \mathbb{F}), \quad \lambda \in \mathbb{F}^\times.$$

Then

$$(\maltese) \qquad s(z_i) = \begin{cases} z_i & \text{for } i = 1, \ldots, n-1, \\ \lambda z_n & \text{for } i = n. \end{cases}$$

Choose $a_{i,j} \in \mathbb{F}[V]$ such that

$$(\Delta) \qquad f_i = \sum_{i=1}^{n} a_{i,j} z_j, \qquad i = 1, \ldots, n.$$

This way of writing f_i is not unique, and we choose to write it in this way so that none of the polynomials $a_{i,j}$ for $j \neq n$ is divisible by z_n. This is possible, because if

$$a_{i,j} = \sum_{E=(e_1,\ldots,e_{n-1},0)} \alpha_E z^E + \sum_{K=(k_1,\ldots,k_n)\,|\,k_n \neq 0} \beta_K z^K,$$

then the desired form can be achieved by using the equalities

$$\beta_K z^K z_j = \beta_K z^{K+\Delta_j-\Delta_n} z_n, \qquad j = 1, \ldots, n-1,$$

where Δ_k is the exponent sequence with a 1 in the kth position and 0's elsewhere.

Having done this, we note that it means that $a_{i,j} \in \mathbb{F}[z_1, \ldots, z_{n-1}]$ for $j = 1, \ldots, n-1$ and $i = 1, \ldots, n$, and hence from (✠) we get

$$\binom{\Delta}{\Delta} \qquad s(a_{i,j}) = a_{i,j} \quad \text{for } j = 1, \ldots, n-1, \text{ and } i = 1, \ldots, n.$$

If we apply s to (Δ) and use $\binom{\Delta}{\Delta}$, we get

$$f_i = sf_i = \sum_{j=1}^{n} s(a_{i,j}) s(z_j) = \sum_{j=1}^{n-1} a_{i,j} z_j + \lambda s(a_{i,n}) z_n,$$

since f_i is invariant for $i = 1, \ldots, n$. Equating this representation for f_i with (Δ) and simplifying gives

$$a_{i,n} z_n = \lambda s(a_{i,n}) z_n \quad \text{for } i = 1, \ldots, n,$$

from which we conclude that

$$s(a_{i,n}) = \lambda^{-1} a_{i,n} \quad \text{for } i = 1, \ldots, n.$$

Therefore,

$$\det[s(a_{i,j})] = \det \begin{bmatrix} a_{1,1} & \cdots & a_{1,n-1} & \lambda^{-1} a_{1,n} \\ \vdots & \vdots & \vdots & \vdots \\ a_{n,1} & \cdots & a_{n,n-1} & \lambda^{-1} a_{n,n} \end{bmatrix} = \lambda^{-1} \det[a_{i,j}]$$

by an elementary property of determinants, [374] page 240. Since $\det(s) = \lambda$, and we may choose $[\mathbb{F}[V]_G] = \det[a_{i,j}]$, we have shown as claimed that $s([\mathbb{F}[V]_G]) = \det(s)^{-1} [\mathbb{F}[V]_G]$.

CASE: $t \in s_{\Delta}(G)$, *i.e., t is a transvection.*

We choose a basis $z_1, \ldots, z_n$ for V^* such that the matrix of t is

$$\begin{bmatrix} 1 & 0 & \cdots & 0 \\ 0 & \ddots & \ddots & \vdots \\ \vdots & & \ddots & 1 \\ 0 & \cdots & 0 & 1 \end{bmatrix} \in \mathrm{GL}(n, \mathbb{F}),$$

in other words,

$$(\maltese) \qquad t(z_i) = \begin{cases} z_i & \text{for } i = 1, \ldots, n-1, \\ z_n + z_{n-1} & \text{for } i = n. \end{cases}$$

This time we choose to write

$$(\triangle) \qquad f_i = \sum_{j=1}^{n} a_{i,j} z_j \quad \text{for } i = 1, \ldots, n$$

in such a way that $a_{i,j} \in \mathbb{F}[z_1, \ldots, z_{n-1}]$ for $j = 1, \ldots, n-2, n$ and $i = 1, \ldots, n$. As noted in the previous case this sort of rewriting is always possible. If we apply t to ($\triangle$), use ($\maltese$), and that f_i is invariant for $i = 1, \ldots, n$, we obtain the following sequence of equalities:

$$\begin{aligned} f_i = t(f_i) &= \sum_{j=1}^{n} t(a_{i,j}) t(z_j) = \sum_{j=1}^{n-2} a_{i,j} z_j + t(a_{i,n-1}) z_{n-1} + a_{i,n}(z_n + z_{n-1}) \\ &= \sum_{j=1}^{n-2} a_{i,j} z_j + \big(t(a_{i,n-1}) + a_{i,n}\big) z_{n-1} + a_{i,n} z_n. \end{aligned}$$

Equating this with ($\triangle$) gives after further simplification

$$a_{i,n-1} z_{n-1} = \big(t(a_{i,n-1}) + a_{i,n}\big) z_{n-1},$$

from which we conclude that

$$\binom{\triangle}{\triangle} \qquad t(a_{i,n-1}) = a_{i,n-1} - a_{i,n} \quad \text{for } i = 1, \ldots, n.$$

Hence, from ($\triangle$) and $\binom{\triangle}{\triangle}$ we obtain

$$\det[t(a_{i,j})] = \det \begin{bmatrix} a_{1,1} & \cdots & a_{1,n-1} - a_{1,n} & a_{1,n} \\ \vdots & \vdots & \vdots & \vdots \\ a_{n,1} & \cdots & a_{n,n-1} - a_{n,n} & a_{n,n} \end{bmatrix} = \det[a_{i,j}]$$

by the same elementary property of determinants ([374] page 240) as in the previous case. Since we may choose to represent $[\mathbb{F}[V]_G]$ by $\det[a_{i,j}]$ and $\det(t) = 1$, we have shown that

$$t([\mathbb{F}[V]_G]) = \det(t)^{-1} [\mathbb{F}[V]_G]$$

as was claimed.

Since G is generated by pseudoreflections, it follows that $g([\mathbb{F}[V]_G]) = \det(g)^{-1}\,[\mathbb{F}[V]_G]$ for any $g \in G$. $\square$

Elementary examples, e.g., representations of p-groups in characteristic p with rings of invariants that are polynomial algebras, show that the Jacobian determinant may very well not represent a fundamental class for $\mathbb{F}[V]_G$. One can't help wondering, though, when it does. A glance at the example of the coinvariants of Σ_3 in its tautological representation examined in Section 1.2, Example 1, shows that the answer cannot be a simple dichotomy between modular and nonmodular cases. The answer appears to be a bit more subtle. To deal with this problem we need to introduce some notation, and review some material from [285] and [372] Section 7.1.

Let $s \in \mathrm{GL}(n, \mathbb{F})$ be a pseudoreflection. The reflecting hyperplane $V^s \subsetneq V$ of s will be denoted by H_s. The set of linear forms in V^* with kernel H_s is 1-dimensional, and ℓ_s will denote an arbitrary, but fixed, linear form with $\ker(\ell_s) = H_s$. Since s fixes a hyperplane pointwise, 1 is an eigenvalue of s of multiplicity at least $n-1$, so the characteristic polynomial of s splits into linear factors. The remaining eigenvalue, which could also be 1, is denoted by λ_s. Note that $\det(s) = \lambda_s$. Associated with s is a twisted differential operator Δ_s, introduced by C. Chevalley in [70] (see Section 7.1) with the properties

$$\Delta_s(f) = (sf - f)/\ell_s,$$
$$\Delta_s(fh) = \Delta_s(f)h + s(f)\Delta_s(h)$$

for any $f \in \mathbb{F}[V]$. With these notations one has

$$\Delta_s(\ell_s) = \lambda_s - 1 \quad \text{and} \quad s(\ell_s) = \lambda_s \cdot \ell_s.$$

From this one obtains by induction

$$\Delta_s(\ell_s^k) = (1 + \lambda_s + \cdots + \lambda_s^{k-1})(\lambda_s - 1)\,\ell_s^{k-1} = (\lambda_s^k - 1)\,\ell_s^{k-1},$$

and therefore it follows that

$$\Delta_s^k(\ell_s^k) = (\lambda_s^k - 1)\cdots(\lambda_s - 1) \in \mathbb{F}.$$

In particular, if $\lambda_s \in \mathbb{F}^\times$ has order $|s|$, then

$$(\bigstar) \qquad \Delta_s^{|s|-1}(\ell_s^{|s|-1}) = (\lambda_s^{|s|-1} - 1)\cdots(\lambda_s - 1) \neq 0 \in \mathbb{F}.$$

If $\rho : G \hookrightarrow \mathrm{GL}(n, \mathbb{F})$ is a representation of a finite group G over the field $\mathbb{F}$, as previously we denote by $s(G) \subset G$ the set of pseudoreflections in G, i.e., the set of $s \in G$ such that $\rho(s)$ is a pseudoreflection. As before we denote by $s_\Delta(G)$ the set of diagonalizable pseudoreflections in $s(G)$, i.e., those $s \in s(G)$ with $\lambda_s \neq 1$, and by $s_{\not\Delta}(G)$ the set of nondiagonalizable pseudoreflections, these are the transvections. The subgroup generated by the transvections forms a normal subgroup of G, which lies in the kernel of the homomorphism $\det : G \longrightarrow \mathbb{F}^\times$.

Likewise, $\mathcal{H}(G) = \{H_s \mid s \in s(G)\}$ denotes the set of reflecting hyperplanes of the pseudoreflections in G, with $\mathcal{H}_\Delta(G)$ the subset of $\mathcal{H}(G)$ where s is in $s_\Delta(G)$ and $\mathcal{H}_{\not\Delta}(G)$ the subset of $\mathcal{H}(G)$ where s is in $s_{\not\Delta}(G)$. If $H \in \mathcal{H}(G)$, we denote by ℓ_H a linear form with kernel H. The set of all pseudoreflections in G with reflecting hyperplane H is a semidirect product of an elementary abelian p-group, the subgroup of transvections with reflecting hyperplane H together with 1, and a cyclic group of order a divisor of $p^\alpha - 1$ for some $\alpha \in \mathbb{N}_0$. A generator for this group will be denoted by s_H. Note that $s_H \in s_\Delta(G)$. See, e.g., [372] Section 8.2.

After this review of [285] we can establish the facts that we need to determine precisely when the Jacobian determinant will represent a fundamental class for the ring of coinvariants.

We introduce a form $L_G \in \mathbb{F}[V]$ by the formula

$$L_G = \prod_{H \in \mathcal{H}_\Delta(G)} \ell_{s_H}^{|s_H|-1}.$$

The configuration of hyperplanes $\mathcal{H}_{\Delta(G)} \subset V = \mathbb{F}^n$ is the affine variety defined by L_G .

LEMMA 7.2.3: *Let $\rho : G \hookrightarrow \mathrm{GL}(n, \mathbb{F})$ be a representation of a finite group G over the field $\mathbb{F}$. Then L_G divides any* $\det^{-1}$*-relative invariant.*

PROOF: The type of argument given by R. P. Stanley in [399] Theorem 3.1 is valid more generally, see e.g., [158]. Specifically, let $f \in \mathbb{F}[V]^G_{\det^{-1}}$. Since the linear forms $\ell_{H'}$, $\ell_{H''}$ are relatively prime for $H' \neq H'' \in \mathcal{H}_\Delta(G)$ it is enough to show that $\ell_H^{|s_H|-1}$ divides f for each $H \in \mathcal{H}_\Delta(G)$. So let $H \in \mathcal{H}_\Delta(G)$ and choose an eigenbasis $\ell_H = z_1, z_2, \dots, z_n$ for V^* so that the matrix of s_H is of the form

$$s_H = \begin{bmatrix} \lambda_H & & & \\ 0 & 1 & & \\ & 0 & \ddots & 0 \\ & & & 1 \end{bmatrix} \in \mathrm{GL}(n, \mathbb{F}).$$

Regard f as a polynomial in z_1 with coefficients in $z_2, \dots, z_n$, say

$$f = h_0 + z_1 h_1 + \cdots + h_k z^k \qquad h_0, \dots, h_k \in \mathbb{F}[z_2, \dots, z_n].$$

Then

$$sf = h_0 + \lambda z_1 h_1 + \cdots + \lambda^k z_1^k h_k,$$

and, since $sf = \lambda^{-1} f = \lambda^{|s_H|-1} f$ we obtain

$$sf = \lambda^{|s_H|-1} h_0 + \lambda^{|s_H|-1} z_1 h_1 + \cdots + \lambda^{|s_H|-1} z_1^k h_k = h_0 + \lambda z_1 h_1 + \cdots + \lambda^k z_1^k h_k = f,$$

which shows that $h_i = 0$ unless $i \equiv 0 \bmod (|s_H| - 1)$ and the result follows. □

Since the Jacobian is a $\det^{-1}$ invariant (see e.g., [372] page 21), Lemma 7.2.3 implies:

LEMMA 7.2.4: *Let $\rho : G \hookrightarrow \mathrm{GL}(n, \mathbb{F})$ be a representation of a finite group G over the field $\mathbb{F}$. Suppose that $\mathbb{F}[V]^G = \mathbb{F}[f_1, \ldots, f_n]$ is a polynomial algebra and $z_1, \ldots, z_n \in V^*$ Then L_G divides $J_G = \det[\partial f_i/\partial z_j]$.* □

LEMMA 7.2.5: *Let $s \in \mathrm{GL}(n, \mathbb{F})$ be a pseudoreflection and $\mathcal{H}$ a finite set of hyperplanes in V. Suppose*
(i) $H_s \notin \mathcal{H}$, and
(ii) if $H \in \mathcal{H}$, then $sH \in \mathcal{H}$.
For every $H \in \mathcal{H}$ choose a linear form $\ell_H \in \mathbb{F}[V]$ with kernel $\mathbb{H}$, and a positive integer $a_H \in \mathbb{N}$. S set $L = \prod_{H \in \mathcal{H}} \ell_H^{a_H}$. Then $\Delta_s(L) = 0$.

PROOF: Since s permutes the hyperplanes in $\mathcal{H}$, it also permutes the collection of linear forms $\{\ell_H \mid H \in \mathcal{H}\}$ up to some nonzero scalar multiples $\alpha(H)$. Therefore we find

$$s(L) = \prod_{H \in \mathcal{H}} s(\ell_H)^{a_H} = \prod_{H \in \mathcal{H}} (\alpha(H)\ell_H)^{a_H} = \Big(\prod_{H \in \mathcal{H}} \alpha(H)^{a_H}\Big) \cdot L.$$

If we set $\prod_{H \in \mathcal{H}} \alpha(H)^{a_H} = \alpha \in \mathbb{F}^\times$ we obtain

$$\ell_s \Delta_s(L) = (s-1)(L) = (\alpha - 1)L.$$

Since $H_s \notin \mathcal{H}$, the linear form ℓ_s is prime to ℓ_H for each $H \in \mathcal{H}$, and hence also to L. If $\alpha - 1$ were not 0, then this equation would imply, to the contrary, that ℓ_s divides L. Hence $\alpha - 1 = 0$, and since $\ell_s \neq 0$, we conclude that $\Delta_s(L) = 0$. □

LEMMA 7.2.6: *Let $\rho : G \hookrightarrow \mathrm{GL}(n, \mathbb{F})$ be a representation of a finite group over the field $\mathbb{F}$. Suppose that $\mathbb{F}[V]^G = \mathbb{F}[f_1, \ldots, f_n]$ is a polynomial algebra and $L_G = \prod_{H \in \mathcal{H}_\Delta(G)} \ell_{s_H}^{|s_H|-1}$. Then L_G is a $\det^{-1}$-relative invariant.*

PROOF: Since $\mathbb{F}[V]^G$ is a polynomial algebra, G is generated by pseudoreflections (see [346]). We rewrite the formula for L_G in the form $L_G = \prod_{H \in \mathcal{H}_\Delta(G)} \ell_{s_H}^{|s_H|-1}$. For $H \in \mathcal{H}_\Delta(G)$ we define L by the requirement that $L_G = (\ell_{s_H}^{|s_H|-1})L$. Then

$$\begin{aligned} s_H(L_G) &= s_H(\ell_{s_H}^{|s_H|-1}) s_H(L) = \lambda_{s_H}^{|s_H|-1} \ell_{s_H}^{|s_H|-1} L \quad \text{(by Lemma 7.2.5)} \\ &= \det(s_H)^{-1} \ell_{s_H}^{|s_H|-1} L = \det(s_H)^{-1} L_G, \end{aligned}$$

and the pseudoreflection s_H is represented by the matrix

$$\begin{bmatrix} 1 & 0 & \cdots & 0 \\ \mathbf{0} & \ddots & \cdots & \mathbf{0} \\ \vdots & \cdots & 1 & 0 \\ 0 & \cdots & 0 & \lambda_{s_H} \end{bmatrix} \in \mathrm{GL}(n, \mathbb{F}), \quad \lambda_{s_H} \in \mathbb{F}^\times.$$

On the other hand, if $t \in s_{\Delta}(G)$ is a transvection, then setting $H = V^t$ we have from Lemma 7.2.5

$$t(L_G) = t(\ell_{s_H}^{|s_H|-1})t(L) = \ell_{s_H} L,$$

since $t\ell_{s_H} = \ell_{s_H}$, because $\lambda_t = 1 = \det(t)$. The result follows because $s(G)$ generates G. □

The final ingredient that we need is J. Hartmann's generalization of the formula of D. J. Benson and W. W. Crawley-Bovey [31], [283]. Her formulation runs as follows ([158] Corollary 5):

PROPOSITION 7.2.7 (J. Hartmann): *Let $\rho : G \hookrightarrow \mathrm{GL}(n, \mathbb{F})$ be a representation of a finite group G over the field $\mathbb{F}$ such that $\mathbb{F}[V]^G = \mathbb{F}[f_1, \ldots, f_n]$. Then*

$$\sum_{i=1}^{n} (\deg(f_i) - 1) \geq \deg(L_G)$$

with equality if and only if G contains no transvections. □

THEOREM 7.2.8: *Let $\rho : G \hookrightarrow \mathrm{GL}(n, \mathbb{F})$ be a representation of a finite group over the field $\mathbb{F}$ of characteristic p, which may be 0. Suppose that $\mathbb{F}[V]^G = \mathbb{F}[f_1, \ldots, f_n]$ is a polynomial algebra. Then $\det[\partial f_i/\partial z_j]$ represents a fundamental class of $\mathbb{F}[V]_G$ if and only if G contains no transvections.*

PROOF: Suppose that G contains no transvections. Then, by [39] Chapitre V, §6, Exercise 8, $s(G) = s_{\Delta}(G)$ generates G, so G is generated by pseudoreflections of order relatively prime to the characteristic of $\mathbb{F}$. Therefore, by [285] Theorem 3.4, the ideal of stable invariants $\mathcal{I}_{\infty}(G)$, and the ideal of generalized invariants associated to the set $s_{\Delta}(G)$, $\mathcal{I}(s_{\Delta}(G))$, coincide. This has two consequences that we will make use in the remainder of the proof:

- The ideal $\mathcal{I}_{\infty}(G)$ is generated by a regular sequence of length n ([285] Theorem 2.9).
- A polynomial $f \in \mathbb{F}[V]$ of degree m belongs to the ideal $\mathcal{I}_{\infty}(G)$ if and only if for any m-tuple (with repetitions allowed) $s_1, \ldots, s_m \in s_{\Delta}(G)$ the value of the operator $\Delta_{s_1} \cdots \Delta_{s_m}$ on f is zero ([285] the definition of $\mathcal{I}(S)$ in Section 2 and Theorem 3.4). Note that $\Delta_{s_1} \cdots \Delta_{s_m}(f)$ has degree zero so is a scalar.

Let $\deg(f_i) = d_i$, for $i = 1, \ldots, n$, so J_G and a fundamental class for $\mathbb{F}[V]_G$ have the same degree, namely $\sum(d_i - 1)$. This is also the degree of L_G by Hartmann's result 7.2.7. Since L_G divides J_G by Lemma 7.2.4, and J_G is nonzero, it follows that they are nonzero multiples of each other. So it is sufficient to show that $L_G \neq 0$ in the algebra of coinvariants $\mathbb{F}[V]_G$. From

[285] we have

$$\mathfrak{h}(G) = \left(\overline{\mathbb{F}[V]^G}\right)^{\mathrm{e}} = (f_1, \ldots, f_n) = \mathcal{J}_1(G) \subseteq \cdots \subseteq \mathcal{J}_\infty(G) = \mathcal{J}(s_\Delta(G)),$$

where ideals $\mathcal{J}_m(G)$ are defined inductively (see [285] Section 1) by

$$\mathcal{J}_m(G) = \begin{cases} (0) & \text{for } m = 0 \\ (\{f \in \mathbb{F}[V] \mid gf - f \in J_{m-1}\}) & \text{for } m > 0. \end{cases}$$

So it is more than enough to show that $L_G \notin \mathcal{J}_\infty(G)$. For this we evaluate the operator $\prod_{H \in \mathcal{H}_\Delta(G)} \Delta_{s_H}^{|s_H|-1}$ on L_G. We find, with L as in the proof of Lemma 7.2.5, that, for $i \geq 1$

$$\Delta_{s_H}^i(L_G) = \Delta_{s_H}^i(\ell_H^{|s_H|-1} L) = \Delta_{s_H}^i(\ell_{s|H|-1})L$$

by Lemma 7.2.4 and the fact that Δ_{s_H} is a twisted. derivation. Thus by formula (★)

$$\Delta_{s_H}^{|s_H|-1}(L_G) = (\lambda_{s_H}^{-1} - 1)L$$

Hence

$$\Big(\prod_{H \in \mathcal{H}_\Delta(G)} \Delta_{s_H}^{|s_H|-1}\Big)(L_G) = \prod_{H \in \mathcal{H}_\Delta(G)} (\lambda_{s_H}^{-1} - 1)\,!,$$

and this scalar is nonzero. Therefore $L_G \notin \mathcal{J}_\infty(G)$ as required.

Conversely, if $s_\Delta(G) \neq \varnothing$ then the formula of J. Hartmann, loc.cit., gives

$$\deg(J_G) > \sum (d_i - 1) = \deg(L_G)$$

By Lemma 7.2.4 $J_G = L_G \cdot K$, and $\deg(K) = \deg(J_G) - \deg(L_G) > 0$. Since L_G is a $\det^{-1}$ relative invariant by Lemma 7.2.6, and J_G is also, it follows that K must be invariant. Then $J_G = L_G \cdot K$ says that $J_G \in (f_1, \ldots, f_n)$, so J_G is zero in $\mathbb{F}[V]_G$ and cannot represent a fundamental class. □

REMARK: Note that the proof shows that when G contains no pseudoreflections then $\det[\partial f_i / \partial z_j]$ and L_G are nonzero scalar multiples of each other so L_G also represents a fundamental class of $\mathbb{F}[V]_G$ if and only if G contains no transvections.

These results have been used by J. Hartmann in [158] to study the invariants of pseudoreflection representations $\rho : G \hookrightarrow \mathrm{GL}(n, \mathbb{F})$ on the tensor product of a polynomial algebra and an exterior algebra $H(V) = \mathbb{F}[V] \otimes E[V]$ of the representation space. She shows that when the invariants $\mathbb{F}[V]^G$ are a polynomial algebra $\mathbb{F}[f_1, \ldots, f_n]$ that $H(V)^G$ has the form $\mathbb{F}[f_1, \ldots, f_n] \otimes E[df_1, \ldots, df_n]$ if and only if the group $\rho(G)$ contains no transvections. This generalizes a result of L. Solomon [392] from the nonmodular case. She also obtains a generalization to the modular case of a result of R. P.

Stanley [399] to the effect that $\mathbb{F}[V]^G_{\det^{-1}}$ is a free module over $\mathbb{F}[V]^G$ of rank one when $\mathbb{F}[V]^G = \mathbb{F}[f_1, \dots, f_n]$ and G contains no transvections.

These results also have consequences for the transfer variety. Recall from Section 6.4 that the transfer ideal of $\rho : G \hookrightarrow GL(n, \mathbb{F})$ is the affine variety $\mathfrak{X}_G \subset V$ defined by the ideal $\mathrm{Im}(\mathrm{Tr}^G)^e \subset \mathbb{F}[V]$. It is the union of the fixed subspaces of the elements of order p in G, and has dimension at most $n-1$. The only way it can have dimension $n-1$ is if G contains transvections. Therefore we have the following corollary:

COROLLARY 7.2.9: *Let $\rho : G \hookrightarrow GL(n, \mathbb{F})$ be a representation of a finite group over the field $\mathbb{F}$ of characteristic p, which may be 0. Suppose that $\mathbb{F}[V]^G = \mathbb{F}[f_1, \dots, f_n]$ is a polynomial algebra. Then $\det[\partial f_i/\partial z_j]$ represents a fundamental class of $\mathbb{F}[V]_G$ if and only if the dimension of the transfer variety $\mathfrak{X}_G$ is at most $n-2$.* □

7.3 Solvable, Nilpotent and Alternating Groups

The nonmodular representations of the solvable groups were the first large family of groups for which Emmy Noether's bound could be proved in the general nonmodular case, [370], [379], and [376]. Indeed, the proof, based on an argument[1] due to Emmy Noether, [301], provides us with an algorithm to compute all algebra generators.

To place these results in a proper setting, we start off this section with a presentation of a relative version of Emmy Noether's bound, and B. Schmid's congruence semigroup lemma from her thesis [332]. After that we show how a sharper version of B. Schmid's results allows us to improve Noether's degree bound for various families of groups, e.g., for nilpotent groups. This material is taken from the Diplomarbeit of D. Krause, [211].

We require some terminology. Suppose that $C \subseteq A$ is a finite integral extension of commutative graded connected Noetherian algebras over a field. If $a \in A$, then a is a root of a monic polynomial of minimal degree $m_a(X) \in C[X]$, called the **minimal polynomial of a over C**. We set

$$\deg(A \mid C) = \max\{\deg(m_a(X)) \mid a \in A\}$$

and call it the **degree of A over C**. As usual, if there is no maximum, we set $\deg(A \mid C) = \infty$. Recall the following combinatorial lemma.

---

[1] In this paper Emmy Noether gives two complete proofs of the finiteness theorem over $\mathbb{C}$. Both proofs of the finiteness theorem have been reworked many times, the account of H. Weyl in [451] pp. 275-276 being perhaps the most famous, and those of J. Fogarty [122] and P. Fleischmann [120] (see Section 2.3) being perhaps the most successful. See also [26] for a reworking in a completely different context.

LEMMA 7.3.1: *Let V be a vector space over a field $\mathbb{F}$ and $u_1, \ldots, u_j \in \mathbb{F}[V]$. If $j! \neq 0 \in \mathbb{F}$, then the monomial $u_1 \cdots u_j$ is a linear combination of jth powers of sums of elements of $\{u_1, \ldots, u_j\}$.*

PROOF: This follows from the formula

$$(-1)^j j!\, u_1 \cdots u_j = \sum_{I \subseteq \{1,\ldots,j\}} (-1)^{|I|} \Big(\sum_{i \in I} u_i\Big)^j.$$

In this formula, I runs over all subsets of $\{1, \ldots, j\}$, and $|I|$ is the cardinality of I. □

Recall that $\beta(A)$ denotes the maximum degree of a generator in a minimal algebra generating set for an algebra A.

THEOREM 7.3.2: *Let $C \subseteq A$ be a finite integral extension of commutative graded connected Noetherian algebras over the field $\mathbb{F}$ of characteristic p. Suppose*

(i) $\deg(A \mid C)$ is finite,

(ii) $p > \deg(A \mid C)$, and

(iii) there is a C-module splitting $\pi : A \longrightarrow C$ to the inclusion $C \hookrightarrow A$.

Then $\beta(C) \leq \beta(A) \cdot \deg(A \mid C)$.

PROOF: Set $\beta(A) = m$ and $\deg(A \mid C) = d$. Let B be the subalgebra of C generated by elements of degree at most md. Our goal is to show that $B = C$. To this end introduce

$$N = \operatorname{Span}_{\mathbb{F}} \left\{ a \in A \mid \deg(a) \leq m \right\},$$

$$M = \operatorname{Span}_{\mathbb{F}} \left\{ a_1^{e_1} \cdots a_k^{e_k} \;\middle|\; \begin{array}{c} k, e_1, \ldots, e_k \in \mathbb{N},\ e_1 + \cdots + e_k < d \\ a_1, \ldots, a_k \in N \end{array} \right\}.$$

We are going to show that $B \cdot M = A$, i.e., that M generates A as a B-module. The reason why this is useful will appear in the penultimate step of the proof.

If $a \in A$, then the minimal polynomial of a has degree at most d. Hence we can find $b_1, \ldots, b_d \in C$ such that

$$(*) \qquad a^d = -(b_1 a^{d-1} + \cdots + b_d).$$

If $\deg(a) \leq m$, then

$$\deg(b_1) < \deg(b_2) < \cdots < \deg(b_d) \leq dm,$$

so $b_1, \ldots, b_d \in B$. The elements $a, a^2, \ldots, a^{d-1}$ belong to M, so $(*)$ shows that $a^d \in B \cdot M$ for any $a \in N$.

Next suppose that $a^E = a_1^{e_1} \cdots a_k^{e_k}$ with $a_1, \ldots, a_k \in N$ and $e_1 + \cdots + e_k = d$.

From Lemma 7.3.1 we obtain

$$(**)\qquad (-1)^d d!\,a^E = \sum_{I\subseteq\{1,\ldots,d\}} (-1)^{|I|}\Big(\sum_{i\in I} a_i\Big)^d = \sum_{I\subseteq\{1,\ldots,d\}} (-1)^{|I|} h_I^d,$$

where $h_I \in N$. Since $d! \neq 0 \in \mathbb{F}$, it follows that $a^E \in B\cdot M$.

Assume inductively that all monomials $a^E = a_1^{e_1}\cdots a_k^{e_k}$ with $k, e_1, \ldots, e_k \in \mathbb{N}$, $a_1, \ldots, a_k \in N$, and $e_1+\cdots+e_k \leq d+i$ belong to $B\cdot M$. Consider a monomial $a^E = a_1^{e_1}\cdots a_k^{e_k}$ with $k, e_1, \ldots, e_k \in \mathbb{N}$, $a_1, \ldots, a_k \in N$, and $e_1+\cdots+e_k = d+i+1$. Without loss of generality we may suppose $a^E = a^{E'}\cdot a_k$. By the induction hypothesis we have $a^{E'} \in B\cdot M$. Therefore we may choose $h_1, \ldots, h_l \in N$ and $d_1, \ldots, d_l \in \mathbb{N}$ with $d_1+\cdots+d_l < d$ and $c_D \in B$, so that

$$a^{E'} = \sum c_D h^D = \sum_{|D|<d-1} c_D h^D + \sum_{|D|=d-1} c_D h^D,$$

where $h^D = \prod_{i=1}^l h_i^{d_i}$. If $|D| < d-1$, then $h^D a_k \in M$ for degree reasons, and hence

$$\sum_{|D|<d-1} c_D h^D a_k \in B\cdot M.$$

If $|D| = d-1$, then by $(**)$ $h^D a_k \in B\cdot M$, and hence

$$\sum_{|D|=d-1} c_D h^D a_k \in B\cdot M.$$

Combining these inclusions gives

$$a^E = a^{E'}\cdot a_k = \sum c_D h^D a_k = \sum_{|D|<d-1} c_D h^D a_k + \sum_{|D|=d-1} c_D h^D a_k \in B\cdot M.$$

Therefore, by induction, any monomial $a^E = a_1^{e_1}\cdots a_k^{e_k}$, with $a_1, \ldots, a_k \in N$, belongs to $B\cdot M$. Since N generates A as an algebra, we have shown that $B\cdot M = A$ as required.

To complete the proof that $B = C$ we apply the projection π to A and obtain

$$C = \pi(A) = \pi(B\cdot M) = B,$$

since $\pi(M) \subset B$ and B is a subalgebra of C. □

THEOREM 7.3.3 (Weak Relative Noether Bound): *Let $\mathbb{F}$ be a field of characteristic p and $\varrho : G \hookrightarrow \mathrm{GL}(n, \mathbb{F})$ be a representation of a finite group G over $\mathbb{F}$. If $H \leq G$ is a subgroup*[2] *such that $p > |G:H|$, then*

$$\beta(\varrho) \leq \beta(\varrho|_H)\cdot |G:H|.$$

---

[2] We call this result a *weak version*, because we need to assume that the characteristic of the ground field is strictly larger than $|G:H|$, and not just coprime.

PROOF: Consider the inclusion $\mathbb{F}[V]^G \subseteq \mathbb{F}[V]^H$. This is a finite extension, and every $f \in \mathbb{F}[V]^H$ is a root of the polynomial

$$\Phi_f(X) = \prod_{g \in G/H} (X - gf),$$

where the product is taken over a set of coset representatives of H in G. The polynomial $\Phi_f(X)$ has degree $|G:H|$, and therefore we find that $\deg(\mathbb{F}[V]^H | \mathbb{F}[V]^G) \leq |G:H|$. Since p does not divide $|G:H|$, there is the averaging operator, derived from the relative transfer Tr^G_H,

$$\pi^G_H = \frac{1}{|G:H|} \sum_{g \in G/H} g = \frac{1}{|G:H|} \mathrm{Tr}^G_H : \mathbb{F}[V]^H \longrightarrow \mathbb{F}[V]^G,$$

which splits the inclusion $\mathbb{F}[V]^G \hookrightarrow \mathbb{F}[V]^H$. Therefore, the hypotheses of Theorem 7.3.2 are satisfied, and applying this theorem yields the desired conclusion. □

We next take up the investigation of the invariants of solvable groups. Problems involving finite solvable groups are particularly amenable to study by inductive methods over chains of subgroups and reduction to cyclic groups of prime order. So it shouldn't come as a surprise that we will start with cyclic groups. We need a combinatorial lemma of B. Schmid, Lemma 2.1 in [333], which we formulate as follows.

LEMMA 7.3.4 (Congruence Semigroup Lemma, B. Schmid): *Let $m, n \in \mathbb{N}$. Suppose we are given $a_1, \ldots, a_n \in \mathbb{N}$ satisfying $a_i \not\equiv 0 \bmod m$ and let $\mathbb{B} \subseteq \underbrace{\mathbb{N} \times \cdots \times \mathbb{N}}_{n}$ be the congruence semigroup defined by*

$$b = (b_1, \ldots, b_n) \in \mathbb{B} \text{ if and only if } a_1 b_1 + \cdots + a_n b_n \equiv 0 \bmod m.$$

Then $\mathbb{B}$ is generated as a semigroup by the elements $b \in \mathbb{B}$ satisfying $b_1 + \cdots + b_n \leq m$.

PROOF: We proceed by induction on n. The case $n = 1$ is immediate, since $\mathbb{B} = \{m, 2m, 3m, \ldots\}$ is generated by m. Let $b = (b_1, \ldots, b_n) \in \mathbb{B}$, and suppose $b_1 + \cdots + b_n > m$. If some $b_i = 0$, then b lies in a congruence semigroup in $\underbrace{\mathbb{N} \times \cdots \times \mathbb{N}}_{n-1}$ and we are done by induction. Otherwise, we proceed to show the existence of elements $b', b'' \in \mathbb{B}$ such that $0 \neq b', b'' \neq b$, and $b = b' + b''$. To this end write

$$a_1 b_1 + \cdots + a_n b_n \equiv 0 \bmod m$$

in the form

$$c_1 + \cdots + c_{|b|} = \underbrace{(a_1 + \cdots + a_1)}_{b_1} + \cdots + \underbrace{(a_n + \cdots + a_n)}_{b_n} \equiv 0 \bmod m.$$

By assumption the number of terms is $|b| = b_1 + \cdots + b_n > m$, and hence at least two of the partial sums

$$\sigma_k = \sum_{i=1}^{k} c_i \qquad 1 \leq k \leq |b|,$$

must be equal modulo m by the pigeonhole principle. The difference $\sigma_{k'} - \sigma_{k''}$, with $k' > k''$, of two such partial sums that are equal modulo m belongs to $\mathbb{B}$, is nonzero, and has the form

$$b' = a_{k''+1}b_{k''+1} + a_{k''+2}b_{k''+2} + \cdots + a_{k'-1}b_{k'-1} + a_{k'}b_{k'}.$$

Setting $b'' = b - b'$ we have the desired elements $b', b'' \in \mathbb{B}$ with $b = b' + b''$. □

From the congruence semigroup lemma it is easy to construct a generating set for the ring of invariants of nonmodular representations of cyclic groups. The idea is to diagonalize the generating matrix and convert the condition that a monomial be invariant to a congruence on the exponent sequence. In tis way we verify explicitly that Emmy Noether's bound holds.

LEMMA 7.3.5: *Let $m \in \mathbb{N}$ and $\rho : \mathbb{Z}/m \hookrightarrow \mathrm{GL}(n, \mathbb{F})$ be a faithful representation of the finite cyclic group $\mathbb{Z}/m$ over the field $\mathbb{F}$. If the characteristic of $\mathbb{F}$ does not divide m, then $\mathbb{F}[V]^{\mathbb{Z}/m}$ is generated by elements of degree at most m. Moreover, if the ground field is large enough, then we can choose monomials with exponents satisfying the congruence of Lemma 7.3.4 as algebra generators.*

PROOF: By flat base change we may suppose that $\mathbb{F}$ contains a primitive mth root of unity λ. The representation ρ is then implemented by a diagonal matrix

$$\mathrm{T} = \begin{bmatrix} \lambda^{a_1} & \cdots & 0 \\ 0 & \ddots & 0 \\ 0 & \cdots & \lambda^{a_n} \end{bmatrix} \in \mathrm{GL}(n, \mathbb{F}),$$

where $0 \leq a_i \leq m-1$ for $i = 1, \ldots, n$. By splitting off the fixed-point set of T we may also suppose that $a_i \neq 0$ for $i = 1, \ldots, n$. The action of T on a monomial $z^B = z_1^{b_1} \cdots z_n^{b_n}$ is given by

$$\mathrm{T}(z^B) = \lambda^{a_1 b_1 + \cdots + a_n b_n} z^B.$$

Hence $f \in \mathbb{F}[V]^{\mathbb{Z}/m}$ if and only if f is a sum of monomials satisfying

$$a_1 b_1 + \cdots + a_n b_n \equiv 0 \bmod m.$$

The set of monomials $\{z^B \mid a_1 b_1 + \cdots + a_n b_n \equiv 0 \bmod m\}$ regarded as a semigroup under multiplication is isomorphic to the subsemigroup $\mathbb{B}$ of $\underbrace{\mathbb{N} \times \cdots \times \mathbb{N}}_{n}$ consisting of those n-tuples $(b_1, \ldots, b_n)$ such that $a_1 b_1 + \cdots + a_n b_n \equiv 0 \bmod m$. The result then follows from the congruence semigroup lemma B. Schmid, Lemma 7.3.4. □

In [416] a table is constructed listing the number of algebra generators in every degree for the special case of the complex regular representation of $\mathbb{Z}/m$, $m \leq 10$. We copy this table:

↓ m \ Degree →	1	2	3	4	5	6	7	8	9	10	Total
1	1										1
2	1	1									2
3	1	1	2								4
4	1	2	2	2							7
5	1	2	4	4	4						15
6	1	3	6	6	2	2					20
7	1	3	8	12	12	6	6				48
8	1	4	10	18	16	8	4	4			65
9	1	4	14	26	32	18	12	6	6		119
10	1	5	16	36	48	32	12	8	4	4	166

TABLE 7.3.1

C. W. Strom also gives a complete description of the algebra generators for arbitrary m. We take the presentation from [297], Examples 21.4.1 and 21.4.2.

EXAMPLE 1 (C. W. Strom [416]): Consider the regular representation of $\mathbb{Z}/m$ over the complex numbers $\mathbb{C}$. Then the ring of invariants $\mathbb{C}[x_0, \ldots, x_{m-1}]^{\mathbb{Z}/m}$ is generated as an algebra by forms $y_0, y_1^{a_1}, \ldots, y_{m-1}^{a_{m-1}}$, where

$$a_1 + 2a_2 + \cdots + (m-1)a_{m-1} \equiv 0 \bmod m$$

and,

$$y_j = x_0 + \exp\left(\frac{2\pi \boldsymbol{i}}{m}\right)^j x_1 + \exp\left(\frac{2\pi \boldsymbol{i}}{m}\right)^{2j} x_2 + \cdots + \exp\left(\frac{2\pi \boldsymbol{i}}{m}\right)^{(m-1)j} x_{m-1}.$$

From this C. W. Strom derives Table 7.3.1 containing the number of $\mathbb{C}$-algebra generators in each degree for $m = 1, \ldots, 10$. The preceding Lemma 7.3.5 combined with repeated use of Corollary 2.3.5 leads to the following result, that was independently proved in [327] and [370].

PROPOSITION 7.3.6: *Let $\rho : G \hookrightarrow \mathrm{GL}(n, \mathbb{F})$ be a nonmodular representation of a solvable group G. Then $\beta(\rho) \leq |G|$.* □

We come to another way to refine Emmy Noether's bound, which is based on the weak relative Noether bound, Theorem 7.3.3. Recall from Chapter 4 the construction of fine orbit Chern classes, and in particular Theorem 4.2.4: If we can find a chain of subgroups

$$\{1\} = G_m < G_{m-1} < \cdots < G_1 < G_0 = G$$

such that for the index of two successive subgroups $e_i = |G_i : G_{i+1}|$ the characteristic p of the ground field $\mathbb{F}$ is

$$p > \max\{e_{m-1}, \ldots, e_0\},$$

then $\mathbb{F}[V]^G$ is generated as an algebra by the fine Chern classes associated to this chain and all its subchains. In the class of groups that admit a chain like this one finds the nonmodular alternating groups, [376] and [379], so we have the following result:

PROPOSITION 7.3.7: *Let $\rho : A_m \hookrightarrow \mathrm{GL}(n, \mathbb{F})$ be a nonmodular faithful representation of the alternating group in m letters. Then the ring of invariants is generated by fine orbit Chern classes.* □

B. Schmid proved[3] in Proposition 2.2 of [332] for noncyclic abelian groups the following sharper version of her congruence semigroup lemma (Lemma 7.3.4):

PROPOSITION 7.3.8 (B. Schmid): *Let $\rho : G \hookrightarrow \mathrm{GL}(n, \mathbb{F})$ be a faithful representation of a noncyclic abelian group G. If the characteristic p does not divide the group order, then $\beta(\rho) < |G|$.* □

Combining Theorem 2.3.3 and Corollary 2.3.5 with Proposition 7.3.8 opens the door to a series of improved (relative) Noether bounds for quite a number of families of groups. We summarize results from [211], Satz 3.1.3, Folgerung 3.1.4, and Folgerung 3.1.6.

PROPOSITION 7.3.9 (D. Krause): *Let $\rho : G \hookrightarrow \mathrm{GL}(n, \mathbb{F})$ be a faithful representation. Let H be a normal subgroup of G such that the quotient group G/H is noncyclic abelian.*

(1) *If the characteristic p does not divide the index $|G : H|$, then $\beta(\rho) \leq \beta(\rho|_H) \cdot (|G : H| - 1)$.*
(2) *Assume that p does not divide the group order $|G|$. If there is a normal series $G = G_0 \rhd G_1 \rhd \cdots \rhd G_{m-1} \rhd G_m = \{1\}$ for G with a noncyclic abelian factor G_i/G_{i+1} for some $i \in \{1, \ldots, m-1\}$, then*

$$\beta(\rho) \leq |G| - \frac{|G|}{|G_i : G_{i+1}|}.$$ □

Note that the preceding result applies in particular to solvable groups with a noncyclic factor.

A proper subfamily of the solvable groups is that of the nilpotent groups. For this family we find a better degree bound. Recall some notations (see, e.g., Chapter 6 in [331], Section 2.3 in [149], or Kapitel III in [179]): Denote by

$$Z(G) = \{g \in G \mid gh = hg \ \forall\, h \in G\}$$

[3] B. Schmid states and proves the following only for representations over the complex numbers. However, her proof generalizes to the nonmodular case.

the **center** of the group G. Set $Z^1(G) = Z(G)$ and define iteratively the **higher centers** $Z^j(G)$ by requiring

$$Z^{j+1}(G)/Z^j(G) = Z(G/Z^j(G)), \quad \forall\, j \geq 1.$$

This leads to an ascending chain of subgroups of G called the **upper** or **ascending central series,**

$$\{1\} = Z^0(G) \leq Z^1(G) \leq \cdots \leq Z^j(G) \leq \cdots \leq G.$$

DEFINITION: *A group G is called* **nilpotent** *if there exists an $s \in \mathbb{N}$ such that $Z^s(G) = G$. The smallest natural number s with this property is the* **class of nilpotency**, *denoted by* $\mathrm{nil}(G) = s$.

Since the ascending central series is a normal series with abelian quotients, we know that every nilpotent group is solvable. Moreover, if G is not abelian, then the last factor group $G/Z^{s-1}(G)$ is not cyclic ([156] Theorem 9.4.2). Therefore, we can sharpen Emmy Noether's bound for nilpotent groups in the following way (Theorem 3.2.5 in [211]).

COROLLARY 7.3.10 (D. Krause): *Let $\rho : G \hookrightarrow \mathrm{GL}(n, \mathbb{F})$ be a faithful nonmodular representation of a nonabelian nilpotent group of nilpotency class $s > 1$. Then $\beta(G) \leq |G| - Z^{s-1}(G) \leq |G| - 2^{s-1}$.*

PROOF: The first inequality follows from Corollary 2.3.5 together with Proposition 7.3.8 and Theorem 2.3.3:

$$\begin{aligned}\beta(G) \leq \beta(Z^{s-1}(G)) \cdot \beta(G/Z^{s-1}(G))\\ \leq |Z^{s-1}(G)| \cdot (|G/Z^{s-1}(G)| - 1)\\ = |G| - |Z^{s-1}(G)|.\end{aligned}$$

We come to the second inequality. We consider the ascending central series

$$\{1\} = Z^0(G) < Z^1(G) < \cdots < Z^s(G) = G.$$

Since all factor groups of this chain are nontrivial, we obtain

$$|Z^j(G)| \geq 2|Z^{j-1}(G)| \quad \forall\, 2 = 1, \ldots, s.$$

Hence, by induction, $|Z^{s-1}(G)| \geq 2^{s-1}$, and the second inequality follows. □

A special family of nilpotent groups is that of p-groups. Since the successive factor groups of their ascending central series have order at least p, we can improve the preceding result in this case to the following; see Theorem 3.3 [211].

COROLLARY 7.3.11 (D. Krause): *Let $\rho : P \hookrightarrow \mathrm{GL}(n, \mathbb{F})$ be a faithful nonmodular representation of a nonabelian p-group of class of nilpotency $s > 1$. Then $\beta(G) \leq |G| - p^{s-1}$.* □

Finally, instead of combining B. Schmid's result, Proposition 7.3.8, with the relative Noether bound, Corollary 2.3.5, we want to combine it with Theorem 7.3.3 to find sharper degree bounds. So, we consider a group G with a chain of subgroups (rather than a normal series)

$$\{1\} = G_m < G_{m-1} < \cdots < G_1 < G_0 = G$$

such that for the index of two successive subgroups, $e_i = |G_i : G_{i+1}|$,

$$p > \max\{e_{m-1}, \ldots, e_0\},$$

where p is the characteristic of the ground field.

As a first example we note that we can improve the bound for the alternating groups given in Proposition 7.3.7; see Satz 4.1.1 [211].

PROPOSITION 7.3.12 (D. Krause): *Let $\varrho : A_m \hookrightarrow \mathrm{GL}(n, \mathbb{F})$ be a faithful representation of the alternating group in m letters. If the characteristic p of the ground field does not divide $|A_m|$, then*

$$\beta(\varrho) \leq \begin{cases} |A_m| & \text{for } m = 2, 3, \\ \frac{3}{4}|A_m| & \text{for } m > 3. \end{cases}$$

PROOF: For $m = 2, 3$ there is nothing to prove; see Theorem 2.3.3. The alternating group A_4 is solvable with commutator series

$$A_4 > \mathbb{Z}/2 \times \mathbb{Z}/2 > \{1\}.$$

The Klein 4-group is noncyclic but abelian, so by an application of Proposition 7.3.9 we obtain the desired inequality. We proceed by induction on m, and consider the chain of subgroups

$$A_m > A_{m-1} > \cdots > A_4.$$

Since p does not divide the group order $|A_m|$, we find that $p > |A_i : A_{i-1}|$ for all $i = 4, \ldots, m$. Hence

$$\begin{aligned} \beta(\varrho) &\leq \beta(\varrho|_{A_{m-1}})|A_m : A_{m-1}| = \beta(\varrho|_{A_{m-1}})m \\ &\leq \beta(\varrho|_{A_{m-2}})m(m-1) \leq \cdots \leq \beta(\varrho|_{A_4})|A_m : A_4| \\ &\leq \frac{3}{4}|A_m|. \quad \square \end{aligned}$$

EXAMPLE 2 (D. Krause): A similar argument to the one used in Proposition 7.3.12 leads to the following degree bounds for nonmodular representations $\varrho : \Sigma_m \hookrightarrow \mathrm{GL}(n, \mathbb{F})$ of the symmetric group.

$$\beta(\varrho(\Sigma_m)) \leq \begin{cases} |\Sigma_m| & \text{for } m = 2, 3, \\ \frac{3}{4}|\Sigma_m| & \text{for } m > 3. \end{cases}$$

This method of finding a degree bound for algebra generators can be applied to any group where we are able to find an appropriate chain of subgroups. D. Krause illustrates this method in her Diplomarbeit, [211], also for the Coxeter groups. We summarize the results briefly, and refer to the Atlas of Finite Groups, [77], for notation.[4]

PROPOSITION 7.3.13 (D. Krause): *We consider nonmodular representations of the Coxeter groups of type A_n, B_n, D_n, F_4, G_2, H_n, I_3 and I_4 over the field $\mathbb{F}$. Then one has:*

(1) $$\beta_{\mathbb{F}}(W(A_n)) \leq \begin{cases} |W(A_2)| & \text{for } n = 2, \\ \frac{3}{4}|W(A_n)| & \text{otherwise.} \end{cases}$$

(2) $$\beta_{\mathbb{F}}(W(B_n)) \leq \begin{cases} \frac{3}{4}|W(B_n)| & \text{for } n = 2, 3, \\ \left(\frac{3}{4}\right)^{n/2+1}|W(B_n)| & \text{for even } n > 3, \\ \left(\frac{3}{4}\right)^{(n+1)/2}|W(B_n)| & \text{for odd } n > 3. \end{cases}$$

(3) $$\beta_{\mathbb{F}}(W(D_n)) \leq \begin{cases} \frac{3}{4}|W(D_n)| & \text{for } n = 2, 3, \\ \left(\frac{3}{4}\right)^{n/2}|W(D_n)| & \text{for even } n > 3, \\ \left(\frac{3}{4}\right)^{(n+1)/2}|W(D_n)| & \text{for odd } n > 3. \end{cases}$$

(4) $\beta_{\mathbb{F}}(W(F_4)) \leq \frac{9}{16}|W(F_4)| = 648.$

(5) $\beta_{\mathbb{F}}(W(G_2)) \leq \frac{3}{4}|W(G_2)| = 9.$

(6) $$\beta_{\mathbb{F}}(W(H_n)) \leq \begin{cases} |W(H_n)| & \text{for odd } n, \\ \frac{3}{4}|W(H_n)| & \text{for even } n. \end{cases}$$

(7) $\beta_{\mathbb{F}}(W(I_3)) \leq \frac{3}{4}|W(I_3)| = 90.$

(8) $\beta_{\mathbb{F}}(W(I_4)) \leq \frac{9}{16}|W(I_4)| = 8100.$

(9) $\beta_{\mathbb{F}}(W(E_n)) \leq \frac{45}{64}|W(E_n)|$ *if the characteristic is zero and* $n = 6, 7$ *or* 8, *or* $p > 23$ *and* $n = 6, 7$, *or* $p > 113$ *and* $n = 8$. □

7.4 GL(2, $\mathbb{F}_p$) and Some of Its Subgroups

In this section we deal with the 2-dimensional groups over a prime field $\mathbb{F}_p$. This choice is motivated by the following two observations. The dimension $n = 2$ is small enough to find (in most cases) an explicit description of the ring of invariants in question, giving us the opportunity to illustrate the techniques introduced so far with real-life examples. On the other hand, $n = 2$ is already large enough often to require very many different tools to find the answers.

---

[4] In particular, A_n is not the alternating group in the Proposition which follows but the Coxeter group with that diagram.

We refer to Chapter 13 and Chapter 22 in [297] for a more complete treatment, as well as many more references.

EXAMPLE 1 (L. E. Dickson [86]): We start with the general linear group GL(2, $\mathbb{F}_p$) itself. It has order $(p^2 - 1)(p^2 - p)$. Its invariant ring for the tautological representation is $\mathbf{D}(2) = \mathbb{F}_p[\mathbf{d}_{2,0}, \mathbf{d}_{2,1}]$, the Dickson algebra, as we have already seen in Theorem 6.1.4. The two Dickson polynomials are orbit Chern classes of the only nontrivial orbit $V^* \setminus 0$. See the discussion of Example 1 in Section 4.1.

EXAMPLE 2 (L. E. Dickson [86]): The special linear group SL(2, $\mathbb{F}_p$) consisting of matrices with determinant one has order $(p + 1)(p^2 - p) = p \cdot (p^2 - 1)$. Its invariants are $\mathbb{F}_p[\mathbf{L}, \mathbf{d}_{2,1}]$, where $\mathbf{L}$, the Euler class, is the product of linear forms one taken out of each 1-dimensional subspace in V^*, Example 1 in Section 4.4.

EXAMPLE 3 (L. E. Dickson [86]): The groups between the special and the general linear group are $\mathrm{SL}_k(2, \mathbb{F}_p)$, consisting of the matrices $\mathbf{M}$ such that $\det(\mathbf{M})^k = 1$. This group has order $k(p + 1)(p^2 - p) = kp \cdot (p^2 - 1)$. Using Corollary 4.5.4 we find that its invariant ring is $\mathbb{F}_p[\mathbf{L}^k, \mathbf{d}_{2,1}]$. Note that in the terminology of Section 4.4 the polynomial $\mathbf{L}$ is a pre-Euler class for $\mathrm{SL}_k(2, \mathbb{F}_p)$ with Euler class $\mathbf{L}^k$.

EXAMPLE 4: The groups $\mathrm{SL}_k(2, \mathbb{F}_p)$ of the preceding example are cyclic extensions of the special linear group afforded by a scalar matrix

$$\begin{bmatrix} \lambda & 0 \\ 0 & \lambda \end{bmatrix} \in \mathrm{GL}(2, \mathbb{F}_p)$$

generating a cyclic group $\mathbb{Z}/k$ of order k, where λ is a kth root of unity. The invariants of this cyclic group $\mathbb{Z}/k$ are given by

$$\mathbb{F}_p \left[x^k, x^{k-1}y, \ldots, xy^{k-1}, y^k\right] /I,$$

where the ideal I of relations is generated by

$$(x^{k-i}y^i)(x^{k-j}y^j) = (x^{k-l}y^l)(x^{k-m}y^m) \quad \forall\, i + j = l + m.$$

The generators for the ring of invariants are the polarized Chern classes of the two orbits of x and y. The ring is Cohen-Macaulay, by Theorem 5.5.2 or Proposition 5.6.9, but it is not Gorenstein, unless the characteristic is 2, by K. Watanabe's criterion (Proposition 5.7.6). In characteristic 3 the ring is a hypersurface. Note that this is the 2-fold vector representation of a nontrivial 1-dimensional representation of $\mathbb{Z}/k$.

EXAMPLE 5 (M.-J. Bertin [97]): We come to the 2-dimensional orthogonal groups. We make the assumption that p is an odd prime, because otherwise the orthogonal and symplectic groups coincide (and indeed, since we are in dimension 2, they are the same as the special linear group, [331] Chapter 8). We consider a nondegenerate quadratic form Q. Choosing a suitable basis we can assume that it is one of the following two canonical forms:

$$Q = \begin{cases} xy & \text{or} \\ x^2 - \lambda y^2 & \text{for } \lambda \in \mathbb{F}_p^\times \text{ a fixed nonsquare.} \end{cases}$$

This gives us two orthogonal groups

$$\mathbb{O}_+(2, \mathbb{F}_p), \qquad \mathbb{O}_-(2, \mathbb{F}_p),$$

where the first is the isotropy group of $Q_+ = xy$, while the second is the isotropy group of $Q_- = x^2 - \lambda y^2$. The orders of these groups are

$$|\, \mathbb{O}_+(2, \mathbb{F}_p) \,| = 2(p-1) \quad \text{and} \quad |\, \mathbb{O}_-(2, \mathbb{F}_p) \,| = 2(p+1),$$

as was calculated in [87], Theorem 172, or [233]. L. E. Dickson also gives a complete list of generators: The first group is generated by matrices of the form

$$\begin{bmatrix} a & 0 \\ 0 & a^{-1} \end{bmatrix} \quad \text{and} \quad \begin{bmatrix} 0 & 1 \\ 1 & 0 \end{bmatrix},$$

where $a \in \mathbb{F}_p^\times$. The ring of invariants turns out to be a polynomial ring

$$\mathbb{F}_p[x, y]^{\mathbb{O}_+(2, \mathbb{F}_p)} = \mathbb{F}_p[xy, x^{p-1} + y^{p-1}]$$

generated by the defining form $Q_+ = xy$ and the top orbit Chern class of the orbit of the linear form $x + y$. This is an application of Proposition 4.5.5. However, the original proof works a bit differently: see Section 2.D in [97]. For the investigation of the second group we need to consider two cases. If $p \equiv 3 \bmod 4$, then we can choose $\lambda = -1$ and hence $Q_- = x^2 + y^2$. It is generated by matrices of the form

$$\begin{bmatrix} -1 & 0 \\ 0 & 1 \end{bmatrix} \quad \text{and} \quad \begin{bmatrix} a & -b \\ b & a \end{bmatrix},$$

where $a^2 + b^2 = 1$ and $a, b \in \mathbb{F}_p$. If $p \equiv 1 \bmod 4$, then we can choose λ to be a generator of the cyclic group $\mathbb{F}_p^\times$, so the orthogonal group is generated by matrices of the form

$$\begin{bmatrix} -1 & 0 \\ 0 & 1 \end{bmatrix} \quad \text{and} \quad \begin{bmatrix} a & b\lambda^{-1} \\ b & a \end{bmatrix},$$

with $a^2 - \lambda^{-1} b^2 = 1$ and $a, b \in \mathbb{F}_p$. In both cases the ring of invariants turns out to be a polynomial ring generated by the form Q_- and an additional form $\mathcal{P}^1(Q_-)$,

$$\mathbb{F}_p[x, y]^{\mathbb{O}_-(2, \mathbb{F}_p)} = \begin{cases} \mathbb{F}_p[x^2 + y^2, x^{p+1} + y^{p+1}] & p \equiv 3 \bmod 4, \\ \mathbb{F}_p[x^2 - \lambda y^2, x^{p+1} - \lambda y^{p+1}] & p \equiv 1 \bmod 4. \end{cases}$$

Again, this is an application of Proposition 4.5.5, and once again, the original proof works a bit differently: see Section 2.D in [97]. The additional generator $\mathcal{P}^1(Q_-)$ is what is called the first reduced Steenrod power of Q_-:

What this is all about will be explained in Chapter 8. In any case, it is easy enough to verify by hand that the two polynomials are invariant.

EXAMPLE 6 (M.-J. Bertin [97]): If we restrict out attention to the special orthogonal groups, we again have two groups,

$\mathbb{SO}_+(2, \mathbb{F}_p)$ for $Q_+ = xy$ with order $p-1$, and

$\mathbb{SO}_-(2, \mathbb{F}_p)$ for $Q_- = x^2 - \lambda y^2$ with order $p+1$.

The first group is cyclic of order $p-1$, and consists of matrices of the form

$$\begin{bmatrix} a & 0 \\ 0 & a^{-1} \end{bmatrix}$$

for $a \in \mathbb{F}_p^\times$. The ring of invariants turns out to be a hypersurface (cf. Proposition 5.5.8)

$$\mathbb{F}_p[x, y]^{\mathbb{SO}_+(2,\mathbb{F}_p)} = \mathbb{F}_p[xy, x^{p-1}, y^{p-1}]/((xy)^{p-1} - x^{p-1}y^{p-1}).$$

The generators are the defining form $Q_+ = xy$ and the top orbit Chern classes of the orbits of x and y. This result appears in [97], Section 2.D. The ring is visibly not a unique factorization domain, since

$$(xy)^{p-1} = x^{p-1}y^{p-1}$$

gives two different factorizations of the same polynomial.

The second group is cyclic of order $p+1$. The ring of invariants turns out also to be a hypersurface (see Proposition 5.5.8) generated by the quadratic form Q_-, its first Steenrod power[5] $\mathcal{P}^1(Q_-)$, and the Euler class $\mathbf{L}=x^py-xy^p$,

$$\mathbb{F}_p[x, y]^{\mathbb{SO}_-(2,\mathbb{F}_p)} = \mathbb{F}_p[x^2 + y^2, x^{p+1} + y^{p+1}, x^py - xy^p]/(r),$$

where the relation r is given by

$$r = (x^{q+1} + y^{p+1})^2 - (x^2 + y^2)^{p+1} + (x^py - xy^p)^2.$$

This is from Section 2.D in [97]. As above we see that this is not a unique factorization domain, viz.,

$$(x^{p+1} + y^{p+1})^4 = \left((x^2 + y^2)^{p+1} - (x^py - xy^p)^2\right)^2.$$

EXAMPLE 7: A p-Sylow subgroup Uni(2, $\mathbb{F}_p$) of GL(2, $\mathbb{F}_p$) has order p and may be chosen to be the set of all matrices

$$\begin{bmatrix} 1 & a \\ 0 & 1 \end{bmatrix}, \quad a \in \mathbb{F}_p.$$

Its invariant ring is generated by two top orbit Chern classes , viz.,

$$\mathbb{F}_p[c_{\text{top}}(x), c_{\text{top}}(y)] = \mathbb{F}_p[x^p - xy^{p-1}, y].$$

[5] See Chapter 8 for a discussion of Steenrod operations.

See Example 2 in Section 4.5. Note that the p-Sylow subgroups are Landweber-Stong (or dually Nakajima) groups (see Theorem 6.2.6 and Theorem 6.2.7 in Section 6.2).

EXAMPLE 8: The Borel (or parabolic) subgroup of $\mathrm{GL}(2, \mathbb{F}_p)$ consists of all upper triangular matrices. Hence it contains the unimodular group as a normal subgroup. Indeed,

$$\mathrm{Up}(2, \mathbb{F}_p) = \mathrm{Uni}(2, \mathbb{F}_p) \rtimes (\mathbb{Z}/(p-1) \times \mathbb{Z}/(p-1)).$$

$\mathrm{Up}(2, \mathbb{F}_p)$ has order $|\ \mathrm{Up}(2, \mathbb{F}_p)\ | = p(p-1)^2$. The ring of invariants is

$$\begin{aligned}\mathbb{F}_p[x, y]^{\mathrm{Up}(2,\mathbb{F}_p)} &= \left(\mathbb{F}_p[x, y]^{\mathrm{Syl}_p}\right)^{\mathbb{Z}/(p-1)\times\mathbb{Z}/(p-1)} \\ &= \mathbb{F}_p[x^p - xy^{p-1}, y]^{\mathbb{Z}/(p-1)\times\mathbb{Z}/(p-1)} = \mathbb{F}_p[c_{\mathrm{top}}(x),\ c_{\mathrm{top}}(y)],\end{aligned}$$

a polynomial ring generated by the top orbit Chern classes of the basis elements x, y.

These are all the subgroups in $\mathrm{GL}(2, \mathbb{F}_p)$ that can be summarized under the title *classical groups and relatives*. We turn next to other types of subgroups. We follow L. E. Dickson's classification of subgroups in $\mathrm{SL}(2, \mathbb{F}_p)$, [86] and [84], and start with nonmodular cyclic subgroups.

EXAMPLE 9: Consider the representation

$$\rho : \mathbb{Z}/k \hookrightarrow \mathrm{SL}(2, \mathbb{F}_p)$$

afforded by the matrix

$$\begin{bmatrix} \lambda & 0 \\ 0 & \lambda^{-1} \end{bmatrix},$$

where $\lambda \in \mathbb{F}_p^\times$ is a kth root of unity and $k \mid (p-1)$. The ring of invariants is

$$\mathbb{F}_p[x, y]^{\mathbb{Z}/k} = \mathbb{F}_p[x^k, y^k, xy]/\left((xy)^k - x^k y^k\right).$$

The two invariants of degree k are the top orbit Chern classes of x, resp. y, while xy is (up to a coefficient) the polarized orbit Chern class of type (1, 1) of these two orbits. Note that this is a family of subgroups of $\mathbb{SO}_+(2, \mathbb{F}_p)$, see Example 6 above, so the generator xy can also be explained in this way. The ring of invariants is a complete intersection, but not a unique factorization domain, since $x^k y^k$ has the two factorizations

$$x^k y^k = (xy)^k.$$

EXAMPLE 10: We come to the quaternion group Q_8. It has a two-dimensional representation over $\mathbb{F}_3$ afforded by the matrices

$$\mathbf{A} = \begin{bmatrix} 0 & 1 \\ -1 & 0 \end{bmatrix} \quad \text{and} \quad \mathbf{B} = \begin{bmatrix} 1 & 1 \\ 1 & -1 \end{bmatrix}.$$

Its ring of invariants is

$$\mathbb{F}_3[x, y]^{Q_8} = \mathbb{F}_3[\mathbf{L}, \mathbf{d}_{2,1}, (x^2+y^2)^2] \Big/ \left(((x^2+y^2)^2)^3 - \mathbf{d}_{2,1}^2 + 2(x^2+y^2)^2\mathbf{L}^2\right),$$

where $\mathbf{L} = xy(x+y)(x-y)$ denotes the Euler class, so $\mathbf{L}^2 = \mathbf{d}_{2,0}$. Note that $(x^2+y^2)^2$ is *not* an orbit Chern class, but a *fine* orbit Chern class, cf. Example 3 in Section 4.1, as well as Examples 2 and 3 in Section 4.2. The ring is a hypersurface. It is not factorial, because

$$\mathbf{d}_{2,1}^2 = (x^2+y^2)^2\left(((x^2+y^2)^2)^2 + 2\mathbf{L}^2)\right).$$

This is the first group in a whole family of groups of order $p^{2m} - 1$ in GL(2, $\mathbb{F}_p$), called **di-cyclic groups** $G_{p^{2m}-1}$ by L. E. Dickson, [84]. These are defined as follows. Let p be an odd prime, f an odd integer, and $m \in \mathbb{Z}$. Choose an element $\lambda \in \mathbb{F}_p$ that is not a square, and set

$$\mu = \left(a^2 - \lambda b^2\right)^{\frac{k(p^m-1)}{2}}.$$

Then the group $G_{p^{2m}-1}$ consists of all matrices of the form

$$\begin{bmatrix} a & \mu\lambda b \\ b & \mu a \end{bmatrix}.$$

For $p = 3, m = 1$ and $\mu = -1$ we obtain the quaternion group just encountered. For the general case the invariants are unknown.

EXAMPLE 11 (L. E. Dickson [86]): Consider the representation afforded by the matrices

$$\mathbf{T} = \begin{bmatrix} -1 & 0 \\ 0 & -1 \end{bmatrix} \quad \text{and} \quad \mathbf{E} = \begin{bmatrix} 0 & 1 \\ -1 & 0 \end{bmatrix} \in \mathrm{GL}(2, \mathbb{F}_p).$$

This is a faithful representation of the cyclic group of order 4 as long as the characteristic is not 2. Its ring of invariants is

$$\mathbb{F}_p[x, y]^{\mathbb{Z}/4} = \mathbb{F}_p[x^2+y^2, x^2y^2].$$

The first invariant, x^2+y^2, is the second orbit Chern class of the orbit of x, while x^2y^2 is $c_4(x)$. This is the first group in a family of generalized quaternion groups[6] of order $4e$ (see [73] page 253).

Choose $e \in \mathbb{Z}$ such that $2e \mid p-1$, and a primitive $2e$th root of unity $\lambda \in \mathbb{F}_p$. Then the two matrices

$$\mathbf{T}_\lambda = \begin{bmatrix} \lambda & 0 \\ 0 & \lambda^{-1} \end{bmatrix}, \quad \mathbf{E} = \begin{bmatrix} 0 & 1 \\ -1 & 0 \end{bmatrix}$$

generate a generalized quaternion group Q_{4e} of order $4e$ in SL(2, $\mathbb{F}_p$), and we computed their invariants for $\mathbb{F} = \mathbb{C}$ in Example 1 in Section 5.5. No

[6] These are called ***di-cyclic*** groups in [86] and much of L. E. Dickson's other papers.

structural changes take place in passing from the complex invariants to the $\mathbb{F}_p$-invariants: See the discussion preceding Proposition 3.1.2. The invariants form a hypersurface generated by the polynomials

$$f = (xy)^2,$$
$$g = x^{2e} + y^{2e}, \quad \text{and}$$
$$h = (xy)\left(x^{2e} - y^{2e}\right),$$

subject to the relation

$$h^2 = fg^2 - 4f^{e+1}.$$

See Proposition 5.5.8. The first polynomial $f = (xy)^2$ is the Euler class of the orbit of x, and xy is the pre-Euler class. The second polynomial is the sum of the top Chern classes of x and y under the subgroup generated by $\mathbf{T}$, i.e., g is a polarization of Chern classes, so a fine Chern class. The last generator h is also a fine Chern class: It is a polarization of the top Chern classes x^{2e}, y^{2e}, and the polarized Chern class xy under the action of $\mathbf{T}$.

EXAMPLE 12 (L. Smith–R. E. Stong [387]): Consider the representation of the dihedral group of order $2m$

$$\rho : D_{2m} \hookrightarrow \mathrm{GL}(2, \mathbb{F}_p)$$

afforded by the matrices

$$\mathbf{D} = \begin{bmatrix} \omega & 0 \\ 0 & \omega^{-1} \end{bmatrix} \quad \text{and} \quad \mathbf{S} = \begin{bmatrix} 0 & 1 \\ 1 & 0 \end{bmatrix},$$

where $\omega \in \mathbb{F}_p$ is an mth root of unity, so $m \mid (p-1)$. Its ring of invariants is

$$\mathbb{F}_p[x, y]^{D_{2m}} = \mathbb{F}_p[x^m + y^m, xy].$$

The first invariant, $x^m + y^m$, is the mth orbit Chern class of the orbit of $x + y$, while xy is $c_2(x + y)$. Note that this is a family of subgroups in the first orthogonal group $\mathbb{O}_+(2, \mathbb{F}_p)$.

We come to modular dihedral groups.

EXAMPLE 13 (C. W. Wilkerson): Consider the representation of the dihedral group of order $2p$

$$\rho : D_{2p} \hookrightarrow \mathrm{GL}(2, \mathbb{F}_p)$$

afforded by the matrices

$$\mathbf{D} = \begin{bmatrix} 1 & 1 \\ 0 & 1 \end{bmatrix} \quad \text{and} \quad \mathbf{S} = \begin{bmatrix} -1 & 0 \\ 0 & 1 \end{bmatrix}.$$

Its ring of invariants is

$$\mathbb{F}_p[x, y]^{D_{2p}} = \mathbb{F}_p[y, (xy^{p-1} - x^p)^2].$$

See Example 1, Section 5.6 in [372]. The generators are the top orbit Chern classes of the basis elements x and y. This is one group of the family D_{kp}, where $k \mid p-1$, of modular dihedral groups. They are generated by the p-Sylow subgroup consisting of the matrices

$$\begin{bmatrix} 1 & b \\ 0 & 1 \end{bmatrix}, \quad b \in \mathbb{F}_p,$$

and the cyclic group of order k generated by the diagonal matrix

$$\begin{bmatrix} a & 0 \\ 0 & 1 \end{bmatrix},$$

where a is a kth root of unity. All these groups are easily seen to have polynomial invariants generated by the top orbit Chern classes, $c_{\text{top}}(x) = (x^p - xy^{p-1})^k$ and $c_{\text{top}}(y) = y$, of the orbits of x and y.

EXAMPLE 14 (C. W. Wilkerson): Consider the dual of the previous representation of the dihedral group of order $2p$, viz.,

$$\varrho^{\text{tr}} : D_{2p} \hookrightarrow \text{GL}(2, \mathbb{F}_p),$$

afforded by the matrices

$$\mathbf{D} = \begin{bmatrix} 1 & 0 \\ 1 & 1 \end{bmatrix} \quad \text{and} \quad \mathbf{S} = \begin{bmatrix} -1 & 0 \\ 0 & 1 \end{bmatrix},$$

over a field of characteristic p. Its ring of invariants is

$$\mathbb{F}_p[x, y]^{D_{2p}} = \mathbb{F}_p[x^2, y^p - yx^{p-1}].$$

See Example 1, Section 5.6 in [372]. The generators are the top orbit Chern classes of the basis elements. This example shows that Proposition 3.1.4 is not valid in the modular situation, to wit, the number of pseudoreflections of D_{2p} is $2p-1$, which is strictly larger than $(2-1)+(p-1)$. Note also that this is a Nakajima group; cf. Section 6.2. As is the dual Landweber-Stong group above, we note that this is one group of the family D_{kp}, where $k \mid p-1$, of modular dihedral groups. They are generated by the p-Sylow subgroup consisting of the matrices

$$\begin{bmatrix} 1 & 0 \\ b & 1 \end{bmatrix}, \quad b \in \mathbb{F}_p,$$

and the cyclic group of order k generated by the diagonal matrix

$$\begin{bmatrix} a & 0 \\ 0 & 1 \end{bmatrix},$$

where a is a kth root of unity. The respective rings of invariants are generated by the top orbit Chern classes

$$c_{\text{top}}(x) = x^k \quad \text{and} \quad c_{\text{top}}(y) = y^p - yx^{p-1}$$

of the orbits of x and y.

EXAMPLE 15: We define a representation of the dihedral group of order $2(p+1)$

$$\rho : D_{2(p+1)} \hookrightarrow \mathrm{GL}(2, \mathbb{F}_p)$$

as follows. Let $-\lambda \in \mathbb{F}_p$ be a nonquadratic residue, so that $x^2+\lambda \in \mathbb{F}_p[x]$ is an irreducible polynomial and

$$\mathbb{F}_{p^2} \cong \mathbb{F}_p[x]/(x^2+\lambda).$$

Choose an element $\zeta = a + bx \in \mathbb{F}_{p^2}^{\times}$ of order $p+1$, for some $a, b \in \mathbb{F}_p$. Then the matrix

$$\mathbf{M} = \begin{bmatrix} a & -\lambda b \\ b & a \end{bmatrix} \in \mathrm{GL}(2, \mathbb{F}_p)$$

is just left multiplication with ζ in $\mathbb{F}_{p^2} = \mathrm{Span}_{\mathbb{F}_p}\{1, x\{$. It has order $(p+1)/2$ and determinant ± 1. Choose a and b such that the determinant is $+1$. To simplify the discussion further assume that $p \equiv 3 \bmod 4$ (i.e., -1 is not a quadratic residue mod p), and choose $\lambda = +1$. Then we have the element

$$\mathbf{M} = \begin{bmatrix} a & -b \\ b & a \end{bmatrix} \in \mathrm{GL}(2, \mathbb{F}_p)$$

of order $p+1$. Take, in addition, the element

$$\mathbf{N} = \begin{bmatrix} -1 & 0 \\ 0 & 1 \end{bmatrix} \in \mathrm{GL}(2, \mathbb{F}_p)$$

of order 2 and observe that $\mathbf{NMN} = \mathbf{M}^{-1}$, so our dihedral group is generated by $\mathbf{M}$ and $\mathbf{N}$. Its ring of invariants is

$$\mathbb{F}_p[x, y]^{D_{2(p+1)}} = \mathbb{F}_p\left[x^2+y^2, x^{p+1}+y^{p+1}\right];$$

see Example 6, Section 5.6 and Example 2, Section 10.4 in [372]. This representation coincides with the defining representation of the orthogonal group $\mathbb{O}_{-}(2, \mathbb{F}_p)$ with quadratic form x^2+y^2.

REMARK: This is not an exhaustive list of subgroups in $\mathrm{GL}(2, \mathbb{F}_p)$. In particular, L. E. Dickson lists, among others, three series of groups: tetrahedral, octahedral, and icosahedral groups related to the classical space groups with these names, see [86].

What kind of conclusions can we draw from these computations? Here is one such: In all the examples $\rho : G \hookrightarrow \mathrm{GL}(2, \mathbb{F}_p)$ one finds that $\beta(\rho) \le |G|$. We have no a priori explanation for why this should be true for the many modular examples among those we have presented. Another is that the worst example from the structural point of view is Example 4, which is the only occurrence of vector invariants in dimension 2. We also note that for the invariants of the irreducible representations there is always an invariant form of degree $4 = 2 \cdot 2$ or less. Is any generalization to higher dimensions possible? See, e.g., [425].

7.5 Integer Representations of Finite Groups

Let $\rho : G \hookrightarrow \mathrm{GL}(n, \mathbb{Z})$ be a faithful representation of a finite group over the integers. We have the following somewhat surprising result:

LEMMA 7.5.1: *Let $G \hookrightarrow \mathrm{GL}(n, \mathbb{Z})$ be a faithful representation of a finite group G and p an odd prime. Then the composition*

$$G \hookrightarrow \mathrm{GL}(n, \mathbb{Z}) \xrightarrow{r_p} \mathrm{GL}(n, \mathbb{Z}/p) = \mathrm{GL}(n, \mathbb{F}_p)$$

is a monomorphism, where r_p is induced by reduction mod p.

PROOF: Suppose that $\mathbf{I} \neq \mathbf{T} \in \mathrm{GL}(n, \mathbb{Z})$ has finite order k and $\mathbf{T} \equiv \mathbf{I}$ mod p. Write $\mathbf{T} = \mathbf{I} + p^e\mathbf{A}$, $\mathbf{A} \not\equiv \mathbf{0} \bmod p$. Then

$$\mathbf{0} = \mathbf{T}^k - \mathbf{I} = (\mathbf{I} + p^e\mathbf{A})^k - \mathbf{I} = kp^e\mathbf{A} + \binom{k}{2}p^{2e}\mathbf{A}^2 + \cdots,$$

which implies (after factoring out p^e)

$$\mathbf{0} = k\mathbf{A} + \binom{k}{2}p^e\mathbf{A}^2 + \cdots \equiv k\mathbf{A} \quad \bmod p,$$

and hence $k \equiv 0 \bmod p$. Write $k = pk'$. Replacing $\mathbf{T}$ by $\mathbf{T}^{k'}$ we may suppose that $\mathbf{T}$ has order p, i.e., that $k = p$. Then

$$\mathbf{I} = \mathbf{T}^p = (\mathbf{I} + p^e\mathbf{A})^p = \mathbf{I}, + \sum_{j=1}^{p} \binom{p}{j} p^{je}\mathbf{A}^j$$

which yields

$$(*) \qquad p^{e+1}\mathbf{A} = -\sum_{j=2}^{p} \binom{p}{j} p^{je}\mathbf{A}^j.$$

For $j = 2, 3, \ldots, p$ we have $ej \geq e + 1$ with equality if and only if $e = 1$ and $j = 2$. For p odd, however, $\binom{p}{2} \equiv 0 \bmod p$. So the right-hand side of $(*)$ is divisible by p^{e+2}. Hence the left-hand side must also by divisible by p^{e+2} so $\mathbf{A} \equiv 0 \bmod p$, which is a contradiction, and the result follows. □

This result gives rigid constraints on the orders of the finite groups that can occur as subgroups of $\mathrm{GL}(n, \mathbb{Z})$. In particular, their orders must divide the order of the general linear group over $\mathbb{F}_3$, which is

$$(3^n - 1)(3^n - 3) \cdots (3^n - 3^{n-1}).$$

Moreover, H. Minkowski proves in [254] and [255] that the order of a finite subgroup of $\mathrm{GL}(n, \mathbb{Z})$ cannot exceed the greatest common divisor of

$$2^{n^2} |\mathrm{GL}(n, \mathbb{F}_2)| = 2^{n^2}(2^n - 1) \cdots (2^n - 2^{n-1})$$

and

$$|\mathrm{GL}(n, \mathbb{F}_p)| = (p^n - 1) \cdots (p^n - p^{n-1})$$

for every odd prime p.

We consider the smallest possible nontrivial case, $n=2$. Combining the two conditions yields a complete list of possible group orders, viz., 1, 2, 3, 4, 6, 8, and 12. We find in [432] the complete list of finite subgroups of $\mathrm{GL}(2, \mathbb{Z})$ up to conjugation. Here it is:

(1) $G \cong \mathbb{Z}/2$ afforded by $\begin{bmatrix} -1 & 0 \\ 0 & -1 \end{bmatrix}$.

(2) $G \cong \mathbb{Z}/2$ afforded by $\begin{bmatrix} 0 & 1 \\ 1 & 0 \end{bmatrix}$.

(3) $G \cong \mathbb{Z}/2 \times \mathbb{Z}/2$ afforded by $\begin{bmatrix} -1 & 0 \\ 0 & -1 \end{bmatrix}$ and $\begin{bmatrix} 0 & 1 \\ 1 & 0 \end{bmatrix}$.

(4) $G \cong \mathbb{Z}/2 \times \mathbb{Z}/2$ afforded by $\begin{bmatrix} 1 & 0 \\ 0 & -1 \end{bmatrix}$ and $\begin{bmatrix} -1 & 0 \\ 0 & 1 \end{bmatrix}$.

(5) $G \cong \mathbb{Z}/3$ afforded by $\begin{bmatrix} 0 & -1 \\ 1 & -1 \end{bmatrix}$.

(6) $G \cong \mathbb{Z}/4$ afforded by $\begin{bmatrix} 0 & -1 \\ 1 & 0 \end{bmatrix}$.

(7) $G \cong \mathbb{Z}/6$ afforded by $\begin{bmatrix} 1 & -1 \\ 1 & 0 \end{bmatrix}$.

(8) $G \cong \Sigma_3$ afforded by $\begin{bmatrix} 0 & -1 \\ 1 & -1 \end{bmatrix}$ and $\begin{bmatrix} 0 & 1 \\ 1 & 0 \end{bmatrix}$.

(9) $G \cong \Sigma_3$ afforded by $\begin{bmatrix} 0 & -1 \\ 1 & -1 \end{bmatrix}$ and $\begin{bmatrix} 0 & -1 \\ -1 & 0 \end{bmatrix}$.

(10) $G \cong \mathbb{Z}/4 \rtimes \mathbb{Z}/2$ afforded by $\begin{bmatrix} 0 & 1 \\ -1 & 0 \end{bmatrix}$ and $\begin{bmatrix} 0 & 1 \\ 1 & 0 \end{bmatrix}$.

(11) $G \cong \mathbb{Z}/6 \rtimes \mathbb{Z}/2$ afforded by $\begin{bmatrix} 1 & -1 \\ 1 & 0 \end{bmatrix}$ and $\begin{bmatrix} 0 & 1 \\ 1 & 0 \end{bmatrix}$.

We reduce these representation over $\mathbb{Z}$ modulo a prime p. By the preceding result we still have a faithful representation as long as p is odd. We take things in order.

EXAMPLE 1 (the first representation of $\mathbb{Z}/2$): If we reduce this representation modulo 2, we get the identity matrix. Reducing modulo an odd prime gives for the invariants

$$\mathbb{F}_p[x, y]^{\mathbb{Z}/2} = \mathbb{F}_p[x^2, y^2, xy]/(x^2y^2 - (xy)^2),$$

since they form a subgroup of the scalar matrices; cf. Example 4 in Section 7.4.

EXAMPLE 2 (the second representation of $\mathbb{Z}/2$): If we reduce this representation modulo 2, we get the identity matrix, and nothing is to be done. Reducing modulo an odd prime gives the tautological representation of $\Sigma_2 = \mathbb{Z}/2$ and $\mathbb{F}_p[x, y]^{\mathbb{Z}/2} = \mathbb{F}_p[x+y, xy]$.

EXAMPLE 3 (the first representation of $\mathbb{Z}/2\times\mathbb{Z}/2$): If we reduce this representation modulo 2, we get Σ_2 in its tautological representation. Reducing modulo an odd prime gives

$$\mathbb{F}_p[x,y]^{\mathbb{Z}/2\times\mathbb{Z}/2}=\mathbb{F}_p[x^2+y^2,xy].$$

This is a special case of Example 12 of Section 7.4 with $m=2$.

EXAMPLE 4 (the second representation of $\mathbb{Z}/2\times\mathbb{Z}/2$): If we reduce this representation modulo 2, we get the trivial group. Reducing modulo an odd prime p gives

$$\mathbb{F}_p[x,y]^{\mathbb{Z}/2\times\mathbb{Z}/2}=\mathbb{F}_p[x^2,y^2].$$

The generators are the top orbit Chern classes of x, resp. y.

EXAMPLE 5 (the representation of $\mathbb{Z}/3$): This is the first example where even reducing modulo 2 gives a faithful representation. So we start with an investigation of the representations modulo $p\neq 3$, i.e., the nonmodular reductions. Note that this example comes from the regular representation of $\mathbb{Z}/3$ by dividing out the invariant subspace spanned by the sum of the group elements. We get (cf. Example 1 in Section 2.4 of [372])

$$\mathbb{F}_p[x,y]^{\mathbb{Z}/3}=\mathbb{F}_p[f,h,k]/(r),$$

where

$$\begin{aligned} f &= x^2-xy+y^2=-c_2(x),\\ h &= x^2y-xy^2=c_{\mathrm{top}}(x),\\ k &= x^3-3xy^2+y^3=\frac{1}{3}\mathrm{Tr}^{\mathbb{Z}/3}(x^2y),\\ r &= f^3-k^2+3hk-9h^2. \end{aligned}$$

Since we are in the nonmodular case, Noether's bound holds, and the transfer is surjective. So, calculating the image of the transfer up to degree 3 tells us that this is indeed a complete set of algebra generators. By Proposition 5.5.8 this ring of invariants is the hypersurface, as given, because the invariants f and k form a system of parameters and the product of their degrees is twice the group order (see Proposition 5.5.8). It is a unique factorization domain if $3\nmid p-1$ by Proposition 1.7.3.

Next, if we reduce modulo 3, our representation is conjugate to the 3-Sylow subgroup afforded by

$$\begin{bmatrix}1&1\\0&1\end{bmatrix},$$

and these invariants are a special case of Example 7 in Section 7.4.

EXAMPLE 6 (the representation of $\mathbb{Z}/4$): If we reduce this representation modulo 2, we get the tautological representation of Σ_2. Reducing modulo an odd prime p gives[7]

$$\mathbb{F}_p[x,y]^{\mathbb{Z}/4} = \mathbb{F}_p[\varphi_1, \varphi_2, \varphi_3]/(\varrho),$$

where

$$\begin{aligned}
\varphi_1 &= x^2 + y^2 = -c_2(x),\\
\varphi_2 &= x^2y^2 = c_{\text{top}}(x),\\
\varphi_3 &= x^3y - xy^3 = c_{\text{top}}(x+2y),\\
\varrho &= \varphi_3^2 - \varphi_1^2\varphi_2 + 4\varphi_2^2.
\end{aligned}$$

See Example 1 in Section 4.2 and Example 9 in Section 7.4. Note that $x \cdot y$ is the pre-Euler class of the orbit of x. Since φ_1 and φ_2 form a system of parameters, we can also describe the invariants by

$$\mathbb{F}_p[x,y]^{\mathbb{Z}/4} = \mathbb{F}_p[\varphi_1, \varphi_2] \oplus \mathbb{F}_p[\varphi_1, \varphi_2]\varphi_3.$$

The ring is a hypersurface, but not a unique factorization domain, because

$$\varphi_3^2 = \varphi_2(\varphi_1^2 - 4\varphi_2),$$

showing that φ_3^2 has two factorizations by nonassociates.

EXAMPLE 7 (the representation of $\mathbb{Z}/6$): If we reduce this representation modulo 2, we get the representation of $\mathbb{Z}/3$ above. Reducing modulo an odd prime $p > 3$ gives a complete intersection $\mathbb{F}_p[x,y]^{\mathbb{Z}/6} = \mathbb{F}_p[f, h^2, k^2, hk]/(r_1, r_2)$, where

$$\begin{aligned}
f &= x^2 - xy + y^2 = -\frac{1}{2}c_2(x),\\
h &= xy(x-y) \quad (\text{where } h^2 = c_{\text{top}}(x)),\\
k &= x^3 - 3xy^2 + y^3,\\
r_1 &= r = f^3 - k^2 + 3hk - 9h^2,\\
r_2 &= h^2k^2 - (hk)^2
\end{aligned}$$

(see Example 9 in Section 7.4). This can also be seen in the following way. Since $\mathbb{Z}/3$ is normal in $\mathbb{Z}/6$, we have by Example 7 above, that

$$\mathbb{F}_p[x,y]^{\mathbb{Z}/6} = \left(\mathbb{F}_p[x,y]^{\mathbb{Z}/3}\right)^{\mathbb{Z}/2} = \left(\mathbb{F}_p[f,h,k]/(r_1)\right)^{\mathbb{Z}/2}.$$

$\mathbb{Z}/2$ acts on the ring of invariants of $\mathbb{Z}/3$ by fixing f and changing the sign of h and k, which gives the result. The ring is a complete intersection, and visibly not a unique factorization domain, since $h^2k^2 = (hk)^2$.

---

[7] This example is treated at various places in [372]: To show how to construct invariants by orbit Chern classes in Example 2 of Section 3.2, to illustrate the method of polarization in Example 1 of Section 3.3, to illustrate Noether normalization in Example 1 in Section 5.3, and as a concrete example of our Proposition 5.5.8 in Example 3 of Section 5.5.

If we reduce modulo $p = 3$, we get

$$\mathbb{F}_3[x, y]^{\mathbb{Z}/6} = \mathbb{F}_3[(x + y)^2, h^2, h(x + y)]/(r),$$

where h is as above, and $r = (x + y)^2h^2 - (h(x + y))^2$. The result can be established by again using the normal subgroup $\mathbb{Z}/3 \lhd \mathbb{Z}/6$ to give

$$\mathbb{F}_3[x, y]^{\mathbb{Z}/6} = \left(\mathbb{F}_3[x, y]^{\mathbb{Z}/3}\right)^{\mathbb{Z}/2} = (\mathbb{F}_3[x + y, h])^{\mathbb{Z}/2},$$

where $\mathbb{Z}/2$ acts on $(x + y)$ and h by changing the sign, whence the result. The ring is a hypersurface, and not a unique factorization domain, because of the nonunique factorization of $(x + y)^2h^2$, viz.,

$$(x + y)^2h^2 = (h(x + y))^2.$$

EXAMPLE 8 (the first representation of Σ_3): If we reduce this representation modulo 2, we get the tautological representation of $GL(2, \mathbb{F}_2)$ into $GL(2, \mathbb{F}_2)$. Hence the ring of invariants is the Dickson algebra

$$\mathbb{F}_2[x, y]^{\Sigma_3} = \mathbb{F}_2[x^2 + xy + y^2,\ x^2y + xy^2],$$

where the generators are the Dickson polynomials. If we reduce modulo an odd prime $p \neq 3$, we have

$$\mathbb{F}_p[x, y]^{\Sigma_3} = \mathbb{F}_p[f,\ 2k - 3h]/(r_3),$$

where f, h, and k are as in Example 7. For the reduction modulo $p = 3$ we have

$$\mathbb{F}_3[x, y]^{\Sigma_3} = \mathbb{F}_3[x + y, h^2],$$

where h is as in Example 7. This may be verified with Proposition 4.5.5. This is a modular reduction of a Coxeter representation at a good prime, where the characteristic-zero and characteristic-p invariants are both polynomial algebras, but on generators of *different* degrees.

EXAMPLE 9 (the second representation of Σ_3): If this representation is reduced modulo 2, we get the the same representation as in Example 8. If we reduce modulo an odd prime $p \neq 3$, we have

$$\mathbb{F}_p[x, y]^{\Sigma_3} = \mathbb{F}_p[f, h],$$

where f and h are as in Example 7. This may be verified using Proposition 4.5.5. If we reduce modulo $p = 3$, we would have the same ring

$$\mathbb{F}_3[x, y]^{\Sigma_3} = \mathbb{F}_3[(x + y)^2, h],$$

so it has the same properties as above, but the transfer is no longer surjective, namely,

$$\mathrm{Im}\left(\mathrm{Tr}^{\Sigma_3}\right) = \left((x + y)^2\right).$$

This can be proven by calculating a basis for $\mathbb{F}_3[x, y]$ as a module over the ring of invariants and then taking the transfer of the module generators.

Notice that when $p=3$ this is a modular reduction of a Coxeter representation at a good prime, where the characteristic-zero and characteristic-p invariants are both polynomial algebras with generators of the same degrees.

EXAMPLE 10 (the representation of $\mathbb{Z}/4 \rtimes \mathbb{Z}/2$): If we reduce this representation modulo 2, we get the tautological representation of Σ_2.

If we reduce modulo an odd prime p, we have

$$\mathbb{F}_p[x, y]^{\mathbb{Z}/4\rtimes\mathbb{Z}/2} = \mathbb{F}_p[\varphi_1, \varphi_2],$$

where $\varphi_1 = x^2 + y^2$ and $\varphi_2 = (xy)^2$ are as in Example 6. Note that this is just the group of signed permutations, where an element of order 4 is given by, e.g., $x \longmapsto -y$ and $y \longmapsto x$.

EXAMPLE 11 (the representation of $\mathbb{Z}/6 \rtimes \mathbb{Z}/2$): If we reduce this representation modulo 2, we get a group of order 6, namely the tautological representation of $GL(2, \mathbb{F}_2)$, so the invariants are the Dickson algebra $\mathbf{D}(2)$ over $\mathbb{F}_2$.

If we reduce mod an odd prime $p \neq 3$, we have

$$\mathbb{F}_p[x, y]^{\mathbb{Z}/6\rtimes\mathbb{Z}/2} = \mathbb{F}_p\left[x^2 - xy + y^2, (xy(x-y))^2\right],$$

which may be verified by using Proposition 4.5.5. Note that in the notation of Example 7 above

$$f = x^2 - xy + y^2 \quad \text{and} \quad h^2 = (xy(x-y))^2.$$

If we reduce mod $p = 3$, we have formally the same ring, viz.,

$$\mathbb{F}_3[x, y]^{\mathbb{Z}/6\rtimes\mathbb{Z}/2} = \mathbb{F}_3[x^2 - xy + y^2, (xy(x-y))^2],$$

which may be verified as above. The only difference with the nonmodular situation is the transfer, which in this case turns out to be a principal ideal given by

$$\operatorname{Im}\left(\operatorname{Tr}^{\mathbb{Z}/6\rtimes\mathbb{Z}/2}\right) = ((x+y)^2).$$

As above, this can be proven by calculating a basis for $\mathbb{F}[x, y]$ as a module over the ring of invariants and then taking the transfer of the module generators.

In the next case, i.e., $n = 3$, we find a complete list of finite subgroups of $GL(3, \mathbb{Z})$ in [421]: There are 74 of them, of orders dividing 48. A complete treatment of their invariants appears in Chapter 23 of [297].

This completes our discussion of examples of degree 2. Among the thoughts provoked by these computations is the observation that these rings of invariants are complete intersections or polynomial algebras. Is this because the groups and the representations are small, or that they cannot contain scalar matrices $\lambda\mathbf{I} \in GL(2, \mathbb{Z})$, or just accidental?

Chapter 8
The Steenrod Algebra and Invariant Theory

RINGS of invariants over Galois fields support an additional structure derived from the Frobenius homomorphism. The operation of raising linear forms in $\mathbb{F}_q[V]$ to the qth power preserves invariants, and can be used to define the Steenrod algebra of a Galois field. Although originally conceived in the 1940s as a refinement of the cup product in cohomology, we prefer to view Steenrod operations as a mechanism (in this connection see [132]) for bringing to the fore information otherwise left hidden behind the Frobenius homomorphism. In the hope of making the Steenrod algebra familiar to as large an audience as possible we provide a complete account of its basic structure as a Hopf algebra.

If $\rho : G \hookrightarrow \mathrm{GL}(n, \mathbb{F}_q)$ is a representation of a finite group over the Galois field $\mathbb{F}_q$, then ignoring the grading, raising the linear forms to the qth power defines an algebra endomorphism of $\mathbb{F}_q[V]$, the Frobenius homomorphism,[1] that commutes with the G-action, so restricts to $\mathbb{F}_q[V]^G$. Let us denote this induced endomorphism by $\boldsymbol{P} : \mathbb{F}_q[V] \circlearrowleft$. It imposes on the algebra of invariants $\mathbb{F}_q[V]^G$ an additional structure, one not present on every subalgebra of $\mathbb{F}_q[V]$. We have chosen to organize the information contained in $\boldsymbol{P}$ as in algebraic topology, namely, we write informally

$$\boldsymbol{P}(f) = \sum_{i=0}^{\infty} \mathcal{P}^i(f),$$

where $\mathcal{P}^i(f) \in \mathbb{F}[V]^G$ are the homogeneous components of $\boldsymbol{P}(f)$: n.b., $\boldsymbol{P}$ plays havoc with the grading, and writing $\boldsymbol{P}(f)$ as a sum of homogeneous components recovers the grading. We will formalize this more precisely

[1] This is the endomorphism of $\mathbb{F}_q[V]$ induced from the classical Frobenius homomorphism $\Phi : V \longrightarrow V$ defined by raising the coordinates of a vector to the qth power. See, e.g., [343], where this viewpoint is adopted.

in Section 8.1. The operators $\mathcal{P}^i$, $i \geq 0$, can be assembled into an algebra, the Steenrod algebra $\mathscr{P}^*$, and in this way $\mathbb{F}_q[V]^G$ becomes what is called an unstable algebra over the Steenrod algebra. What this means, and the consequences that follow from it, are the subject of the remaining chapters.

The history of mathematics has often shown that it is better to develop key ideas in as wide a context as possible, rather than the narrow field of their original discovery. This was the case, for example, with D. Hilbert's original discoveries in invariant theory. As Emmy Noether realized, it is better to adopt a more general viewpoint than invariant theory for these ideas so as not to miss any hidden gems. In the case of Hilbert's basis theorem, this meant turning a theorem into a definition, i.e., introducing Noetherian rings, and developing their theory from first principles. We believe that the Steenrod algebra is a similar case, and we therefore choose to develop the theory not just for rings of invariants, or as a topologist might, for the mod p cohomology of topological spaces, but in a more general context: that of unstable algebras over the Steenrod algebra. This chapter may seem at times somewhat far afield, and the results achieved certainly have a broader range of applicability than actually needed for invariant theory, but the price paid well justifies the effort expended.

8.1 The Steenrod Operations

Throughout this section we deal with Galois fields $\mathbb{F}_q$ of characteristic p containing $q = p^\nu$ elements. Denote by $\mathbb{F}_q[V][[\xi]]$ the power series ring over $\mathbb{F}_q[V]$ in an additional variable ξ, and set $\deg(\xi) = 1 - q$. Define an $\mathbb{F}_q$-algebra homomorphism of degree[2] zero

$$P(\xi) : \mathbb{F}_q[V] \longrightarrow \mathbb{F}_q[V][[\xi]],$$

by requiring

$$P(\xi)(\ell) = \ell + \ell^q \xi \in \mathbb{F}_q[V][[\xi]] \quad \forall \text{ linear forms } \ell \in V^*.$$

In this way, we get, for an arbitrary polynomial $f \in \mathbb{F}_q[V]$, by separating out homogeneous components,

$$(\star) \qquad P(\xi)(f) = \begin{cases} \sum_{i=0}^{\infty} \mathcal{P}^i(f)\xi^i & \text{if } p \text{ is odd,} \\ \sum_{i=0}^{\infty} \mathrm{Sq}^i(f)\xi^i & \text{if } p = 2, \end{cases}$$

which defines $\mathcal{P}^i$, resp. Sq^i, as $\mathbb{F}_q$-linear maps

$$\mathcal{P}^i, \mathrm{Sq}^i : \mathbb{F}_q[V] \longrightarrow \mathbb{F}_q[V].$$

[2] Note carefully we ignore the usual topological sign conventions, since graded commutation rules play no role here, cf. Appendix A.1.

These maps are functorial in V. The operations $\mathcal{P}^i$, respectively Sq^i, are called **Steenrod reduced power operations**, respectively **Steenrod squaring operations**, or collectively, **Steenrod operations**. The two different symbols for p even, resp. odd, come from algebraic topology: using the grading conventions necessary for the investigation of cohomology operations on the cohomology of topological spaces (and that is what the Steenrod operations are) makes this unavoidable. However, we restrict our attention to commutative polynomial rings and can therefore ignore the difference; so with the indulgence of topologists we set $\mathrm{Sq}^i = \mathcal{P}^i$ for all $i \in \mathbb{N}_0$.

The sums appearing in (★) are actually finite, so $\mathrm{Im}(\boldsymbol{P}(\xi))$ actually lies in $\mathbb{F}_q[V][\xi]$. In fact, $\boldsymbol{P}(\xi)(f)$ is a *polynomial* in ξ of degree $\deg(f)$ with leading coefficient f^q. This means that the Steenrod operations acting on $\mathbb{F}_q[V]$ satisfy the **unstability condition**

$$\mathcal{P}^i(f) = \begin{cases} f^q & \text{if } i = \deg(f), \\ 0 & \text{if } i > \deg(f), \end{cases} \quad \forall\, f \in \mathbb{F}_q[V].$$

Note that these conditions express both a triviality condition, viz., $\mathcal{P}^i(f) = 0$ for all $i > \deg(f)$, and a nontriviality condition, viz., $\mathcal{P}^{\deg(f)}(f) = f^q$. It is the interplay of these two requirements that seems to endow the unstability condition with the power to yield unexpected consequences.

Next, observe that the multiplicativity of the operator $\boldsymbol{P}(\xi)$ leads to the formulae

$$\mathcal{P}^k(f'f'') = \sum_{i+j=k} \mathcal{P}^i(f')\mathcal{P}^j(f''), \quad \forall\, f', f'' \in \mathbb{F}_q[V].$$

These are called the **Cartan formulae** for the Steenrod operations. (N.b. In field theory, a family of operators satisfying these formulae is called a **higher-order differential operator**. See, e.g., [449].)

EXAMPLE 1: As a simple example of how one can compute with these operations consider the quadratic form

$$Q = x^2 + xy + y^2 \in \mathbb{F}_2[x, y].$$

Using linearity, the Cartan formula, and unstability[3] we obtain:

$$\begin{aligned} \mathrm{Sq}^1(Q) &= \mathrm{Sq}^1(x^2) + \mathrm{Sq}^1(xy) + \mathrm{Sq}^1(y^2) \\ &= 2x\mathrm{Sq}^1(x) + \mathrm{Sq}^1(x)\cdot y + x \cdot \mathrm{Sq}^1(y) + 2y\mathrm{Sq}^1(y) \\ &= 0 + x^2y + xy^2 + 0 = x^2y + xy^2, \\ \mathrm{Sq}^2(Q) &= Q^2 = x^4 + x^2y^2 + y^4, \\ \mathrm{Sq}^i(Q) &= 0 \text{ for } i > 2. \end{aligned}$$

[3] Note carefully that Sq^1 is a derivation. It is, indeed, the element of minimal degree in an infinite family of derivations present in the Steenrod algebra, see Corollary 8.3.4.

Since the Steenrod operations are natural with respect to linear transformations between vector spaces, they induce endomorphisms of the functor

$$\mathbb{F}_q[-] : \mathcal{V}ect_{\mathbb{F}_q} \longrightarrow \mathcal{A}lg_{\mathbb{F}_q}$$

from $\mathbb{F}_q$-vector spaces to commutative graded $\mathbb{F}_q$-algebras. They therefore commute with the action of $\mathrm{GL}(V)$ on $\mathbb{F}_q[V]$. If $G \hookrightarrow \mathrm{GL}(n, \mathbb{F}_q)$ is a faithful representation of a finite group G, then the Steenrod operations restrict to the ring of invariants $\mathbb{F}_q[V]^G$, i.e., map invariant forms to invariant forms. Therefore, they can be used to produce new invariants from old ones. This is a new feature of invariant theory over finite fields as opposed to arbitrary fields (but do see in this connection [458]). Here is an example to illustrate this. It is based on a result and the methods of [377].

EXAMPLE 2 (S. D. Cohen [76]): Let $\mathbb{F}_q$ be the Galois field with q elements of odd characteristic p. Up to a change of variables there is a unique nondegenerate quadratic form of degree 3 (see, e.g., [87] §199). Without loss of generality we may suppose it to be $Q = y^2 - xz \in \mathbb{F}_q[x, y, z]$. The orthogonal group $\mathbb{O}(3, \mathbb{F}_q)$ of this form has a tautological representation $\mathbb{O}(3, \mathbb{F}_q) \subset \mathrm{GL}(3, \mathbb{F}_q)$, and the ring of invariants $\mathbb{F}_q[x, y, z]^{\mathbb{O}(3,\mathbb{F}_q)}$ was computed by S. D. Cohen [76]. We show how Steenrod operations and some algebraic geometry can help us to make this computation.

By definition $Q \in \mathbb{F}[x, y, z]^{\mathbb{O}(3,\mathbb{F}_q)}$. If we apply the first Steenrod operation to this form, we receive the new invariant form of degree $q + 1$, viz.,

$$\mathcal{P}^1(Q) = 2y^{q+1} - x^q z - xz^q \in \mathbb{F}_q[x, y, z]^{\mathbb{O}(3,\mathbb{F}_q)}.$$

To complete the computation, let $\mathfrak{X}_Q \subset \mathbb{PF}(2)$ be the projective variety defined by the vanishing of the quadratic form Q. The Euler class E associated to the configuration of linear forms $\mathcal{L}$ defining the set of external lines to $\mathfrak{X}_Q$ (see [170] Section 8.2 and [377]) is an $\mathbb{O}(3, \mathbb{F}_q)$-invariant form of degree $q(q-1)$. The three forms $Q, \mathcal{P}^1(Q), \mathsf{E} \in \mathbb{F}_q[x, y, z]^{\mathbb{O}(3,\mathbb{F}_q)}$ turn out to be a system of parameters. Since the product of their degrees is $2 \cdot (q+1) \cdot q(q-1) = 2q(q^2 - 1) = |\mathbb{O}(3, \mathbb{F}_q)|$, it follows from Proposition 4.5.5 that

$$\mathbb{F}_q[x, y, z]^{\mathbb{O}(3,\mathbb{F}_q)} = \mathbb{F}_q[Q, \mathcal{P}^{\Delta_1}(Q), \mathsf{E}].$$

The special orthogonal group $\mathbb{SO}(3, \mathbb{F}_q)$ is of index two in $\mathbb{O}(3, \mathbb{F}_q)$ and the ring of invariants $\mathbb{F}_q[x, y, z]^{\mathbb{SO}(3,\mathbb{F}_q)}$ is Cohen–Macaulay by Proposition 5.6.10. The three forms $Q, \mathcal{P}^1(Q), \mathsf{E}$ belong to $\mathbb{F}_q[x, y, z]^{\mathbb{SO}(3,\mathbb{F}_q)}$, are a system of parameters, and the product of their degrees is $2|\mathbb{SO}(3, \mathbb{F}_q)|$. Therefore by Proposition 5.5.8

$$\mathbb{F}_q[x, y, z]^{\mathbb{SO}(3,\mathbb{F}_q)} = \mathbb{F}_q[x, y, z]^{\mathbb{O}(3,\mathbb{F}_q)} \oplus \mathbb{F}_q[x, y, z]^{\mathbb{O}(3,\mathbb{F}_q)} \cdot h$$

where $h \in \mathbb{F}_q[x, y, z]^{\mathbb{SO}(3,\mathbb{F}_q)}$ is an element of minimal degree that is not in $\mathbb{F}_q[x, y, z]^{\mathbb{O}(3,\mathbb{F}_q)}$. The pre-Euler $\mathbf{e}$ class of the configuration $\mathcal{L}$ of lines has degree q, is a det-relative invariant, and hence an $\mathbb{SO}(3, \mathbb{F}_q)$-invariant. It is an element of minimal degree in $\mathbb{F}_q[x, y, z]^{\mathbb{SO}(3,\mathbb{F}_q)}$ not in $\mathbb{F}_q[x, y, z]^{\mathbb{O}(3,\mathbb{F}_q)}$ and satisfies a quadratic equation over $\mathbb{F}_q[x, y, z]^{\mathbb{O}(3,\mathbb{F}_q)}$. Hence $\mathbb{F}_q[x, y, z]^{\mathbb{SO}(3,\mathbb{F}_q)}$ is a hypersurface.

8.2 The Steenrod Algebra

The Steenrod operations can be collected together to form an algebra, in fact a Hopf algebra, over the Galois field $\mathbb{F}_q$.

DEFINITION: The **Steenrod algebra** $\mathscr{P}^*$ is the $\mathbb{F}_q$-subalgebra of the endomorphism algebra of the functor $\mathbb{F}_q[-]$, generated by the operators $\mathcal{P}^0 = 1, \mathcal{P}^1, \mathcal{P}^2, \ldots$.

The Steenrod algebra is by no means freely generated by the Steenrod reduced powers. For example, when $p = 2$ it is easy to check that $\mathrm{Sq}^1\mathrm{Sq}^1 = 0$ by verifying this is the case for monomials $z^E = z_1^{e_1}, \ldots, z_n^{e_n}$: to do so one needs the formula, valid for any linear form, $\mathrm{Sq}^1(z^k) = kz^{k+1}$, which follows by induction from the Cartan formula. In fact every element in the Steenrod algebra is nilpotent: but the index of nilpotence is known only in a few cases, see e.g. [260], [261], [262], [437], [438], and lastly [459] for a resumé of what is known.

In this section we present one traditional algebraic picture of the Steenrod algebra by deriving the Adem-Wu relations, which are a complete set of relations between the generators. These relations are usually written as follows:

$$\mathcal{P}^i\mathcal{P}^j = \sum_{k=0}^{[i/q]}(-1)^{i-qk}\binom{(q-1)(j-k)-1}{i-qk}\mathcal{P}^{i+j-k}\mathcal{P}^k \quad \forall\, i, j \geq 0, i < qj.$$

Note for any Galois field $\mathbb{F}_q$ the coefficients are still elements in the prime subfield $\mathbb{F}_p$ of $\mathbb{F}_q$. They were originally conjectured by Wu Wen-Tsün based on his study of the mod p cohomology of Grassmann manifolds [461] and proved by J. Adem in [4], H. Cartan in [65], and for $p = 2$ by J.-P. Serre in [341].

By its very definition as a subalgebra of the endomorphisms of the functor

$$\mathbb{F}_q[-] : \mathcal{V}ect_{\mathbb{F}_q} \longrightarrow \mathcal{A}lg_{\mathbb{F}_q}$$

any relations between the Steenrod operations can be deduced from the action of $\mathscr{P}^*$ on $\mathbb{F}_q[V]$ provided only that the dimension of V is large enough, and this is the strategy used by H. Cartan in [65], and for $p = 2$ by J.-P. Serre in [341]. Their proofs are greatly simplified by the introduction of the **Bullett-Macdonald identity**, which provides us with a tightly wrapped description of the relations among the Steenrod operations, [48]. To describe this identity, as in [48] define $\boldsymbol{P}(\eta)$ to be the ring homomorphism

$$\boldsymbol{P}(\eta) : \mathbb{F}_q[V][\xi] \longrightarrow \mathbb{F}_q[V][\xi, \eta]$$

where, for $f_0, \ldots, f_m \in \mathbb{F}_q[V]$

$$\boldsymbol{P}(\eta)(f_0 + f_1\xi + \cdots + f_m\xi^m) = \boldsymbol{P}(\xi)(f_0) + \boldsymbol{P}(\xi)(f_1)\eta + \cdots + \boldsymbol{P}(\xi)(f_m)\eta^m$$

Next, introduce the variables u, t and s which are related to each other by $u = (1-t)^{q-1} = 1 + t + \cdots + t^{q-1}$ and $s = tu$. Then the Bullett-Macdonald identity is

$$\boldsymbol{P}(\eta)\big|_{\eta=s} \circ \boldsymbol{P}(\xi)\big|_{\xi=1} = \boldsymbol{P}(\eta)\big|_{\eta=u} \circ \boldsymbol{P}(\xi)\big|_{\xi=t^q} : \mathbb{F}_q[V] \longrightarrow \mathbb{F}_q[V][s].$$

Since $\boldsymbol{P}(\xi)$ and $\boldsymbol{P}(\eta)$ are ring homomorphisms, it is enough to check this equation for the basis elements of V^*. Here is how this goes: Let $z \in V^*$. Then

$$\begin{aligned}\boldsymbol{P}(\eta) \circ \boldsymbol{P}(\xi)(z) - \boldsymbol{P}(\eta)(z + z^q\xi) &= \boldsymbol{P}(\eta)(z) + \boldsymbol{P}(\eta)(z^q)\xi \\ &= z + z^q\eta + (z + z^q\eta)^q\xi = z + z^q\xi + z^q\eta + z^{q^2}\eta^q\xi \in \mathbb{F}_q[V][\xi, \eta].\end{aligned}$$

So if we substitute $\eta = s$ and $\xi = 1$ we obtain

$$(\text{∴}) \qquad \boldsymbol{P}(\eta)\big|_{\eta=s} \circ \boldsymbol{P}(\xi)\big|_{\xi=1}(z) = z + z^q + z^q s + z^{q^2}s^q,$$

and if we substitute $\eta = u$ and $\xi = t^q$ then the result is

$$(\maltese) \qquad \boldsymbol{P}(\eta)\big|_{\eta=u} \circ \boldsymbol{P}(\xi)\big|_{\xi=t^q}(z) = z + z^q t^q + z^q u + z^{q^2}u^q t^q.$$

The relations imposed on s, t and u imply

$$\begin{aligned}u + t^q &= (1-t)^{q-1} + t^q = (1 + t + t^2 + \cdots + t^{q-1}) + t^q \\ &= 1 + t(1 + t + \cdots + t^{q-1}) = 1 + s\end{aligned}$$

so indeed ($\maltese$) and (∴) are equal.

To derive the Adem-Wu relations from the Bullett-Macdonald identity we provide details for the residue computation[4] sketched in [48]. First of all, direct calculation gives:

$$\boldsymbol{P}(s)\boldsymbol{P}(1) = \boldsymbol{P}(\eta)\big|_{\eta=s} \circ \boldsymbol{P}(\xi)\big|_{\xi=1} = \sum_{a,k} s^a \mathcal{P}^a \mathcal{P}^k$$

$$\boldsymbol{P}(u)\boldsymbol{P}(t^q) = \boldsymbol{P}(\eta)\big|_{\eta=u} \circ \boldsymbol{P}(\xi)\big|_{\xi=t^q} = \sum_{a,b,j} u^{a+b-j} t^{qj} \mathcal{P}^{a+b-j}\mathcal{P}^j$$

which the Bullett-Macdonald identity says are equal. Recall from complex analysis that

$$\frac{1}{2\pi \boldsymbol{i}} \oint_\gamma z^m dz = \begin{cases} 1 & m = -1 \\ 0 & \text{otherwise,} \end{cases}$$

[4] The following discussion is based on conversations with E.H. Brown Jr.

where γ is a small circle around $0 \in \mathbb{C}$. Therefore we obtain

$$\begin{aligned}\sum_k \mathcal{P}^a \mathcal{P}^k &= \frac{1}{2\pi \mathbf{i}} \oint_\gamma \frac{\boldsymbol{P}(s)\boldsymbol{P}(1)}{s^{a+1}} ds \\ &= \frac{1}{2\pi \mathbf{i}} \oint_\gamma \frac{\boldsymbol{P}(u)\boldsymbol{P}(t^q)}{s^{a+1}} ds \\ &= \frac{1}{2\pi \mathbf{i}} \sum_{a,\, b,\, j} \oint_\gamma \frac{u^{a+b-j} t^{qj}}{s^{a+1}} ds \mathcal{P}^{a+b-j} \mathcal{P}^j .\end{aligned}$$

The formula $s = t(1-t)^{q-1}$ gives $ds = (1-t)^{q-2}(1-qt)dt$, so substituting gives

$$\begin{aligned}\frac{u^{a+b-j}t^{qj}}{s^{a+1}} ds &= \frac{(1-t)^{(q-1)(a+b-j)} t^{qj} (1-t)^{q-2}(1-qt)}{\left[t(1-t)^{q-1}\right]^{a+1}} dt \\ &= (1-t)^{(b-j-1)(q-1)+(q-2)} t^{qj-a-1}(1-qt)dt \\ &= (1-t)^{((b-j)(q-1)-1)} t^{qj-a-1}(1-qt)dt \\ &= \left[\sum_k (-1)^k \binom{(b-j)(q-1)-1}{k} t^k\right] t^{qj-a-1}(1-qt)dt \\ &= \sum_k (-1)^k \binom{(b-j)(q-1)-1}{k} \left[t^{k+qj-a-1} - qt^{k+qj-a}\right] dt.\end{aligned}$$

Therefore

$$\mathcal{P}^a \mathcal{P}^b = \sum_j \left[\frac{1}{2\pi \mathbf{i}} \oint_\gamma \frac{u^{a+b-j}t^{qj}}{s^{a+1}} ds\right] \mathcal{P}^{a+b-j} \mathcal{P}^j$$

$$= \sum_j \frac{1}{2\pi \mathbf{i}} \oint \sum_k (-1)^k \binom{(b-j)(q-1)-1}{k} \left[t^{k+qj-a-1} - qt^{k+qj-a}\right] dt \mathcal{P}^{a+b-j} \mathcal{P}^j .$$

Only the terms where

$$k + qj - a - 1 = -1 \quad (k = a - qj)$$

or

$$k + qj - a = -1 \quad (k = a - qj - 1)$$

contribute anything to the sum, so

$$\mathcal{P}^a \mathcal{P}^b = \sum_j \left[(-1)^{a-qj)} \binom{(b-j)(q-1)-1}{a-qj} + (-1)^{a-qj-1} q \binom{(b-j)(q-1)-1}{a-qj-1}\right] \mathcal{P}^{a+b-j} \mathcal{P}^j$$

and since

$$\binom{(b-j)(q-1)-1}{a-qj} - q\binom{(b-j)(q-1)-1}{a-qj-1} \equiv \binom{(b-j)(q-1)-1}{a-qj} \bmod p$$

we conclude

$$\mathcal{P}^a \mathcal{P}^b = \sum_j (-1)^{a-qj} \binom{(b-j)(q-1)-1}{a-qj} \mathcal{P}^{a+b-j} \mathcal{P}^j$$

establishing the Adem-Wu relations.

Thus there is a surjective map from the free associative algebra with 1 generated by the Steenrod operations modulo the ideal $\mathfrak{w}$ generated by the Adem-Wu relations

$$\mathcal{P}^a \mathcal{P}^b - \sum_j (-1)^{a-qj} \binom{(b-j)(q-1)-1}{a-qj} \mathcal{P}^{a+b-j} \mathcal{P}^j \quad a, b \in \mathbb{N} \text{ and } a < qb$$

onto the Steenrod algebra. To show that this map is an isomorphism we extend some theorems of H. Cartan, [65], J.-P. Serre, [341], and Wu Wen-Tsün, [461] from the case of the prime field to arbitrary Galois fields. We also rearrange their proofs so that they do not make any direct use of topology.

An **index sequence** is a sequence $I = (i_1, i_2, \ldots, i_k, \ldots)$ of nonnegative integers, almost all of which are zero. If I is an index sequence we denote by $\mathcal{P}^I \in \mathscr{P}^*$ the monomial $\mathcal{P}^{i_1} \cdot \mathcal{P}^{i_2} \cdots \mathcal{P}^{i_k} \cdots$ in the Steenrod operations $\mathcal{P}^i$, with the convention that trailing 1s are ignored. The degree of the element $\mathcal{P}^I$ is $(q-1)(j_1 + j_2 + \cdots + j_k + \cdots)$. These iterations of Steenrod operations are called **basic monomials**. An index sequence I is called **admissible** if $i_s \geq q i_{s+1}$ for $s \geq 1$. We call k the **length** of I if $i_k \neq 0$ but $i_s = 0$ for $s > k$. Write $\ell(I)$ for the length of I. It is often convenient to treat an index sequence as a finite sequence of nonnegative integers by truncating it to $\ell(I)$ entries.

A basic monomial is defined to be **admissible** if the corresponding index sequence is admissible. Note that if a monomial is admissible then no Adem-Wu relation can be applied to it. The strategy of H. Cartan and J.-P. Serre to show that the Adem-Wu relations are a complete set of defining relations for the Steenrod algebra is to prove that the admissible monomials are an $\mathbb{F}_q$-basis for $\mathscr{P}^*$. Denote by $\mathcal{B}^*$ the free, graded, associative algebra generated by the symbols $\mathcal{P}^k$ modulo the ideal $\mathfrak{w}$ obtained from the Adem-Wu relations described above. We have a surjective map $\mathcal{B}^* \longrightarrow \mathscr{P}^*$, and with his notation our goal is to prove:

THEOREM 8.2.1: *The admissible monomials span $\mathcal{B}^*$ as an $\mathbb{F}_q$-vector space. The images of the admissible monomials in the Steenrod algebra are linearly independent.*

PROOF: We begin by showing that the admissible monomials span $\mathcal{B}^*$. For an index sequence $I = (i_1, i_2, \ldots, i_k)$, we define the **moment** of I, denoted by $m(I)$, by $m(I) = \sum_{s=1}^{k} s \cdot i_s$. We first show that an inadmissible monomial is a linear combination of monomials of smaller moment. Granted this it follows by induction over the moment that the admissible monomials span $\mathcal{B}^*$.

Suppose that $\mathcal{P}^I$ is an inadmissible monomial. Then there is a smallest s such that $i_s < qi_{s+1}$, i.e.,

$$\mathcal{P}^I = \mathcal{Q}' \mathcal{P}^{i_s} \mathcal{P}^{i_{s+1}} \mathcal{Q}''$$

where $\mathcal{Q}'$, $\mathcal{Q}''$ are basic monomials, and $\mathcal{Q}'$ is admissible. It is therefore possible to apply an Adem-Wu relation to $\mathcal{P}^I$ to obtain

$$\mathcal{P}^I = \sum_j a_j \mathcal{Q}' \mathcal{P}^{i_s + i_{s+1} - j} \mathcal{P}^j \mathcal{Q}''$$

for certain coefficients $a_j \in \mathbb{F}_q$. The terms on the right hand side all have smaller moment than $\mathcal{P}^I$, and so, by induction on s we may express $\mathcal{P}^I$ as a sum of admissible monomials.

We next show that the admissible monomials are linearly independent as elements of the Steenrod algebra $\mathscr{P}^*$. For this we need a formula of Wu Wen-Tsün.

Denote by $e_i(x_1, \ldots, x_n)$ the i-th elementary symmetric polynomial in $x_1, \ldots, x_n$, so $e_n = x_1 x_2 \cdots x_n \in \mathbb{F}_q[x_1, \ldots, x_n]$. Then

$$\begin{aligned} \boldsymbol{P}(\xi)(e_n) &= \boldsymbol{P}(\xi)\Big(\prod_{i=1}^n x_i\Big) = \prod_{i=1}^n \boldsymbol{P}(\xi)(x_i) \\ &= \prod_{i=1}^n (x_i + x_i^q \xi) = \prod_{i=1}^n x_i \cdot \prod_{i=1}^n (1 + x_i^{q-1} \xi) \\ &= e_n(x_1, \ldots, x_n) \cdot \left(\sum_{i=1}^n e_i(x_1^{q-1}, \ldots, x_n^{q-1}) \xi^i \right). \end{aligned}$$

So we have obtained the formulae

$$\mathcal{P}^i(e_n) = e_n \cdot e_i(x_1^{q-1}, \ldots, x_n^{q-1}) \qquad i = 1, \ldots, n.$$

We claim that the monomials

$$\left\{ \mathcal{P}^I \mid \mathcal{P}^I \text{ admissible and } \deg(\mathcal{P}^I) \leq (q-1)n \right\}$$

are linearly independent in $\mathscr{P}^*$. To see this note that in case $\deg(\mathcal{P}^I) \leq (q-1)n$, then $\ell(I) \leq n$. Hence each entry in I is at most n (so the following formula makes sense), and

$$\mathcal{P}^I(e_n) = e_n \cdot \prod_{j=1}^s e_{i_j}(x_1^{q-1}, \ldots, x_n^{q-1}) + \ldots$$

where the remaining terms are lower in the lexicographic ordering on monomials. So, if $\ell(I) \leq n$, then $e_n \cdot \prod_{j=1}^s e_{i_j}(x_1^{q-1}, \ldots, x_n^{q-1})$ is the largest monomial in $\mathcal{P}^I(e_n)$ in the lexicographic order. Thus

$$\left\{\mathcal{P}^I(e_n) \mid \mathcal{P}^I \text{ admissible and } \deg(\mathcal{P}^I) \le (q-1)n\right\},$$

have distinct largest monomials, so are linearly independent.

By letting $n \longrightarrow \infty$ we obtain the assertion, completing the proof. □

Reformulated for the Steenrod algebra, the preceding result says:

THEOREM 8.2.2: *The Steenrod algebra $\mathscr{P}^*$ is isomorphic to the free associative $\mathbb{F}_q$-algebra generated by the reduced power operations $1 = \mathcal{P}^0$, $\mathcal{P}^1, \mathcal{P}^2, \ldots$ modulo the ideal $\mathfrak{w}$ generated by the Adem-Wu relations.* □

COROLLARY 8.2.3: *The admissible monomials are an $\mathbb{F}_q$-basis for the Steenrod algebra $\mathscr{P}^*$.* □

There are many other additive bases for the Steenrod algebra that find use in practice, such as the Milnor basis [252] that we will encounter in the next section. For alternative ways to define the Steenrod algebra see for example [441], [213], [37], [459] and of course the original source [405].

8.3 The Hopf Algebra Structure of the Steenrod Algebra

Our goal in this section is to complete the traditional picture of the Steenrod algebra by proving that $\mathscr{P}^*$ is a Hopf algebra[5] and extending Milnor's Hopf algebra [252] structure theorems from the prime field $\mathbb{F}_p$ to an arbitrary Galois field $\mathbb{F}_q$. It should be emphasized that this requires no new ideas, only a careful reworking of Milnor's proofs avoiding reference to algebraic topology and cohomology operations, and cautiously replacing p by q where appropriate.

PROPOSITION 8.3.1: *Let p be a prime integer, $q = p^\nu$ a power of p, and $\mathbb{F}_q$ the Galois field with q elements. Then the Steenrod algebra of $\mathbb{F}_q$ is a cocommutative Hopf algebra over $\mathbb{F}_q$ with respect to the coproduct*

$$\nabla : \mathscr{P}^* \longrightarrow \mathscr{P}^* \otimes \mathscr{P}^*$$

defined by the formulae

$$\nabla(\mathcal{P}^k) = \sum_{i+j=k} \mathcal{P}^i \otimes \mathcal{P}^j, \quad k = 1, 2, \ldots.$$

[5] One quick way to do this is to write down as comultiplication map

$$\nabla(\mathcal{P}^k) = \sum_{i+j=k} \mathcal{P}^i \otimes \mathcal{P}^j, \quad k = 1, 2, \ldots,$$

and verify that it is compatible with the Bullett-Macdonald identity, and hence also with the Adem-Wu relations.

PROOF: Consider the functor $V \rightsquigarrow \mathbb{F}_q[V] \otimes \mathbb{F}_q[V]$ that assigns to a finite dimensional vector space V over $\mathbb{F}_q$ the commutative graded algebra $\mathbb{F}_q[V] \otimes \mathbb{F}_q[V]$ over $\mathbb{F}_q$. There is a natural map of algebras

$$\mathscr{P}^* \otimes \mathscr{P}^* \longrightarrow \mathrm{End}(V \rightsquigarrow \mathbb{F}_q[V] \otimes \mathbb{F}_q[V])$$

given by the tensor product of endomorphisms. Since there is an isomorphism $\mathbb{F}_q[V] \otimes \mathbb{F}_q[V] \cong \mathbb{F}_q[V \oplus V]$, that is natural in V, the functor $\mathrm{End}(V \rightsquigarrow \mathbb{F}_q[V] \otimes \mathbb{F}_q[V])$ is a subfunctor of the functor $\mathrm{End}(V \rightsquigarrow \mathbb{F}_q[V])$ that assigns to a finite dimensional vector space V over $\mathbb{F}_q$ the polynomial algebra $\mathbb{F}_q[V]$. Hence restriction defines a map of algebras

$$\mathscr{P}^* \longrightarrow \mathrm{End}(V \rightsquigarrow \mathbb{F}_q[V] \otimes \mathbb{F}_q[V])$$

and we obtain a diagram of algebra homomorphisms

$$\begin{array}{ccc} & & \mathscr{P}^* \otimes \mathscr{P}^* \\ & \nearrow & \downarrow \tau \\ \mathscr{P}^* & \xrightarrow{\rho} & \mathrm{End}(V \rightsquigarrow \mathbb{F}_q[V] \otimes \mathbb{F}_q[V]) \end{array}$$

What we need to show is that $\mathrm{Im}(\rho) \subseteq \mathrm{Im}(\tau)$, for since τ is monic $\nabla = \tau^{-1}\rho$ would define the desired coproduct. Since $\mathcal{P}^k$ for $k = 1, 2, \ldots,$ generate $\mathscr{P}^*$ it is enough to check that $\rho(\mathcal{P}^k) \in \mathrm{Im}(\tau)$ for $k = 1, 2, \ldots, .$ But this is immediate from the Cartan formula. Since ∇ is a map of algebras the Hopf condition is satisfied, so $\mathscr{P}^*$ is a Hopf algebra. □

If J is an admissible index sequence then

$$e(J) = \sum_{s=1}^{\infty}(j_s - qj_{s+1})$$

is called the **excess** of J and of the corresponding basic monomial $\mathcal{P}^J$. For example, the sequences

$$M_k = (q^{k-1}, \ldots, q, 1), \ \ k = 1, 2, \ldots$$

are all the admissible sequences of excess zero. Note that

$$\deg(\mathcal{P}^{M_k}) = \sum_{j=1}^{k} q^{k-j}(q-1) = q^k - 1, \ \ \text{for } k = 1, 2, \ldots.$$

It is not hard to show that the excess $e(J)$ is the smallest integer e such that $\mathcal{P}^J(z_1 \cdots z_e) \neq 0 \in \mathbb{F}_q[z_1, \ldots, z_e]$. This observation allows us to extend the definition of the excess of any element of $\mathscr{P}^*$.

Recall by Corollary 8.2.3 that the admissible monomials are an $\mathbb{F}_q$-vector space basis for $\mathscr{P}^*$.

Let $\mathscr{P}_*(\mathbb{F}_q)$ denote the Hopf algebra dual to the Steenrod algebra $\mathscr{P}^*$. We define $\xi_k \in \mathscr{P}_*(\mathbb{F}_q)$ to be dual to the monomial $\mathcal{P}^{M_k} = \mathcal{P}^{q^{k-1}} \cdots \mathcal{P}^q \cdot \mathcal{P}^1$ with respect to the basis of admissible monomials for $\mathscr{P}^*$. This means that we have:

$$<\wp^J \mid \xi_k> = \begin{cases} 1 & \text{if } J = M_k, \\ 0 & \text{otherwise,} \end{cases}$$

where we have written $<\wp \mid \xi>$ for the value of an element $\wp \in \mathscr{P}^*$ on an element $\xi \in \mathscr{P}_*(\mathbb{F}_q)$. Note that $\deg(\xi_k) = q^k - 1$ for $k = 1, \ldots,$.

If $I = (i_1, i_2, \ldots, i_k, \ldots)$ is an index sequence we call ℓ the **length** of I, denoted by $\ell(I)$, if $i_k = 0$ for $k > \ell$, but $i_\ell \neq 0$. We associate to an index sequence $I = (i_1, i_2, \ldots, i_k, \ldots)$ the element $\xi^I = \xi_1^{i_1} \cdot \xi_2^{i_2} \cdots \xi_\ell^{i_\ell} \in \mathscr{P}_*(\mathbb{F}_q)$, where $\ell = \ell(I)$. Note that

$$\deg(\xi^I) = \sum_{s=1}^{\ell(I)} i_s(q^s - 1).$$

To an index sequence $I = (i_1, i_2, \ldots, i_k, \ldots)$ we also associate an admissible sequence $J(I) = (j_1, j_2, \ldots, j_k, \ldots)$ defined by

$$j_1 = \sum_{s=1}^{\infty} i_s q^{s-1}, \quad j_2 = \sum_{s=2}^{\infty} i_s q^{s-2}, \ldots, \quad j_k = \sum_{s=k}^{\infty} i_s q^{s-k}, \ldots.$$

It is easy to verify that as I runs over all index sequences that $J(I)$ runs over all admissible sequences. Finally, note that $\deg(\wp^{J(I)}) = \deg(\xi^I)$ for any index sequence I.

The crucial observation used by Milnor to prove the structure theorem of $\mathscr{P}_*(\mathbb{F}_q)$ is that the pairing of the admissible monomial basis for $\mathscr{P}^*$ against the monomials in the ξ_k is upper triangular in a suitable sense. To formulate this precisely we order the index sequences lexicographically from the right, so for example $(1, 2, 0, \ldots) \prec (0, 0, 1, \ldots)$.

LEMMA 8.3.2 (J. W. Milnor): *With the preceding notations we have that the inner product matrix $<\wp^{J(I)} \mid \xi^K>$ is upper triangular with 1s on the diagonal, i.e.,*

$$<\wp^{J(I)} \mid \xi^K> = \begin{cases} 1 & \text{if } I = K, \\ 0 & \text{if } I < K. \end{cases}$$

PROOF: Let the length of K be ℓ and define $K' = (k_1, k_2, \ldots, k_{\ell-1})$, so

$$\xi^K = \xi^{K'} \cdot \xi_\ell \in \mathscr{P}_*(\mathbb{F}_q).$$

If ∇ denotes the coproduct in $\mathscr{P}^*$, then we have the formula

$$(\divideontimes) \quad <\wp^{J(I)} \mid \xi^K> = <\wp^{J(I)} \mid \xi^{K'} \cdot \xi_\ell> = <\nabla(\wp^{J(I)}) \mid \xi^{K'} \otimes \xi_\ell>$$

If $J(I) = (j_1, j_2, \ldots, j_k, \ldots)$ then one easily checks that

$$\nabla(\wp^{J(I)}) = \sum_{J'+J''=J(I)} \wp^{J'} \otimes \wp^{J''}.$$

Substituting this into (⁘) gives

$$(\maltese) \qquad <\mathcal{P}^{J(I)} \mid \xi^K> = \sum_{J'+J''=J(I)} <\mathcal{P}^{J'} \mid \xi^{K'}> \cdot <\mathcal{P}^{J''} \mid \xi_\ell>.$$

By the definition of ξ_ℓ we have

$$<\mathcal{P}^{J''} \mid \xi_\ell> = \begin{cases} 1 & \text{if } J'' = M_\ell, \\ 0 & \text{otherwise.} \end{cases}$$

If $J'' = M_\ell$ then unraveling the definitions shows that $J' = J(I')$, for a suitable I', so if K and I have the same length ℓ, we have shown

$$<\mathcal{P}^{J(I)} \mid \xi^K> = <\mathcal{P}^{J(I')} \mid \xi^{K'}>,$$

and hence it follows from induction over the degree that

$$<\mathcal{P}^{J(I)} \mid \xi^K> = \begin{cases} 1 & \text{if } I = K, \\ 0 & \text{if } I < K. \end{cases}$$

If, on the other hand, $\ell(I) < \ell$ then all the terms

$$<\mathcal{P}^{J''} \mid \xi_\ell>$$

in the sum ($\maltese$) are zero and hence that $<\mathcal{P}^{J(I)} \mid \xi^K> = 0$ as required. □

THEOREM 8.3.3 (J. W. Milnor): *Let p be a prime integer, $q = p^\nu$ a power of p, and $\mathbb{F}_q$ the Galois field with q elements. Let $\mathcal{P}_*(\mathbb{F}_q)$ denote the dual Hopf algebra to the Steenrod algebra of the Galois field $\mathbb{F}_q$. Then, as an algebra*

$$\mathcal{P}_* \cong \mathbb{F}_q[\xi_1, \ldots, \xi_k, \ldots],$$

where $\deg(\xi_k) = q^k - 1$ for $k \in \mathbb{N}$. The coproduct is given by the formula

$$\nabla_*(\xi_k) = \sum_{i+j=k} \xi_i^{q^j} \otimes \xi_j, \quad k \geq 1.$$

PROOF: By Milnor's lemma (Lemma 8.3.2) the monomials $\{\xi^I\}$ where I ranges over all index sequences are linearly independent in $\mathcal{P}_*$. Hence $\mathbb{F}_q[\xi_1, \ldots, \xi_k, \ldots] \subseteq \mathcal{P}_*$. Both the algebras $\mathcal{P}_*$ and $\mathbb{F}_q[\xi_1, \ldots, \xi_k, \ldots]$ have the same Poincaré series, since $\deg(\mathcal{P}^{J(I)}) = \deg(\xi^I)$ for all index sequences I, and the admissible monomials $\mathcal{P}^{J(I)}$ are an $\mathbb{F}_q$-vector space basis for $\mathcal{P}^*$. So $\mathbb{F}_q[\xi_1, \ldots, \xi_k, \ldots] = \mathcal{P}_*$, and it remains to verify the formula for the coproduct.

To this end we use the test algebra $\mathbb{F}_q[u]$, the polynomial algebra on one generator, as in [252]. Note that for admissible sequences we have

$$(\bigstar) \qquad \mathcal{P}^J(u) = \begin{cases} u^{q^k} & \text{if } J = M_k, \\ 0 & \text{otherwise.} \end{cases}$$

Define the map

$$\lambda^* : \mathbb{F}_q[u] \longrightarrow \mathbb{F}_q[u] \otimes \mathscr{P}_*$$

by the formula

$$\lambda^*(u^i) = \sum \mathcal{P}^{J(I)}(u^i) \otimes \xi^I$$

where the sum is over all index sequences I. Note that in any given degree the sum is finite and that λ^* is a map of algebras. Moreover

$$(\lambda^* \otimes 1)\lambda^*(u) = (1 \otimes \nabla_*)\lambda^*(u),$$

i.e., the following diagram

☯

$$\begin{array}{ccc} \mathbb{F}_q[u] \otimes \mathscr{P}_* \otimes \mathscr{P}_* & \xleftarrow{1 \otimes \nabla_*} & \mathbb{F}_q[u] \otimes \mathscr{P}_* \\ \uparrow{\scriptstyle \lambda^* \otimes 1} & & \uparrow{\scriptstyle \lambda^*} \\ \mathbb{F}_q[u] \otimes \mathscr{P}_* & \xleftarrow{\lambda^*} & \mathbb{F}_q[u] \end{array}$$

is commutative.

From (★) it follows that

$$\lambda^*(u) = \sum u^{q^k} \otimes \xi_k$$

which when raised to the q^r-th power gives

$$\lambda^*(u^r) = \sum u^{q^{k+r}} \otimes \xi_k^{q^r},$$

and leads to the formula

$$(\lambda^* \otimes 1)(\lambda^*(u)) = (\lambda^* \otimes 1)\Big(\sum_k u^{q^k} \otimes \xi_k\Big) = \sum_r \sum_k u^{q^{k+r}} \otimes \xi_r^{q^k} \otimes \xi_k.$$

Whereas, the other way around the diagram ☯ leads to

$$(1 \otimes \nabla_*)(\lambda^*(u)) = \sum_j u^{q^j} \otimes \nabla_*(\xi_k),$$

and equating these two expressions leads to the asserted formula for the coproduct. □

The operations $\mathcal{P}^{p^i}$ for $i > 0$ are indecomposables in $\mathscr{P}^*$, so $\mathscr{P}^*$ is not generated by the operations $\mathcal{P}^{q^i}$ for $i \geq 0$; we need all the $\mathcal{P}^{p^i}$ for $i > 0$. This can be readily seen on hand from the dual Hopf algebra: Since $\mathbb{F}_q$ has characteristic p, the elements $\xi_1^{p^i}$ for $i \geq 0$ are all primitive, [253]. The following corollary also indicates that passing from the prime field $\mathbb{F}_p$ to a general Galois field $\mathbb{F}_q$ is not just a simple substitution of q for p.

COROLLARY 8.3.4: *Let p be a prime integer, $q = p^{\nu}$ a power of p, and $\mathbb{F}_q$ the Galois field with q elements. The indecomposable module $Q(\mathscr{P}^*)$ of the Steenrod algebra of $\mathbb{F}_q$ has a basis consisting of the elements $\mathcal{P}^{p^i}$ for $i \in \mathbb{N}_0$, and the primitive elements $P(\mathscr{P}^*)$ has a basis consisting of the elements $\mathcal{P}^{\Delta_k}$ for $k \in \mathbb{N}$, where, for $k \in \mathbb{N}$, $\mathcal{P}^{\Delta_k}$ is dual to ξ_k with respect to the monomial basis for $\mathscr{P}_*$.* □

8.4 The Inverse Invariant Theory Problem

In this section we indicate how the Steenrod algebra can be used to solve the inverse invariant theory problem (see Section 1.5, Problem 1) in the case where the ground field is a Galois field. In fact, we will consider the more general problem of determining when a commutative graded connected algebra H over a Galois field $\mathbb{F}_q$ is a ring of invariants. The solution to this problem is one of the first examples in which Steenrod algebra technology has been successfully applied to invariant theory. The essential results in this section go back to C.W. Wilkerson [454], [455] and J. F. Adams and C. W. Wilkerson [1], and were completed by the first author of this book in [290]. The proofs of the key theorems, Theorems 8.4.2 and 8.4.3, are far too long to be included here, and we refer the reader to the original sources.

With an eye toward characterizing rings of invariants, let us collect some essential properties shared by all rings of invariants. Let H be an $\mathbb{F}_q$-algebra. For H to be a ring of invariants we certainly need to assume that H is a commutative, graded, connected Noetherian algebra over $\mathbb{F}_q$. We will also need that there is an embedding $H \hookrightarrow \mathbb{F}_q[V]$, with $\mathbb{F}_q[V]$ integral over H; see Proposition 1.7.2. Since $\mathbb{F}_q$ is finite, so is $\mathrm{GL}(n, \mathbb{F}_q)$, so the full general linear group $\mathrm{GL}(n, \mathbb{F}_q)$ provides us with a set of universal invariants, namely the Dickson algebra $\mathbf{D}(n)$: i.e., H must contain $\mathbf{D}(n)$, where $n = \dim(H)$, in such a way that $\mathbf{D}(n) \hookrightarrow H$ is an integral extension. Finally, rings of invariants are integrally closed, i.e., in their field of fractions (see Theorem 1.7.1) so we also need to assume that H is integrally closed. Actually, this is enough to ensure that H is a ring of invariants, although assuming this much begs the question!

PROPOSITION 8.4.1: *Let H be a graded connected integrally closed commutative Noetherian algebra over $\mathbb{F}_q$. Suppose there are integral extensions*

$$\mathbf{D}(n) \hookrightarrow H \hookrightarrow \mathbb{F}_q[V],$$

such that the composition $\mathbf{D}(n) = \mathbb{F}_q[V]^{\mathrm{GL}(n,\mathbb{F})} \hookrightarrow H \hookrightarrow \mathbb{F}_q[V]$ is the natural inclusion. Then H is a ring of invariants.

PROOF: By what we have assumed we have the following situation:

$$\begin{array}{ccccc} \mathbf{D}(n) & \underset{\text{integral}}{\hookrightarrow} & H & \underset{\text{integral}}{\hookrightarrow} & \mathbb{F}_q[V] \\ \downarrow & & \downarrow & & \downarrow \\ \mathbb{F}\mathbb{F}(\mathbf{D}(n)) & \hookrightarrow & \mathbb{F}\mathbb{F}(H) & \hookrightarrow & \mathbb{F}_q(V), \end{array}$$

where $\mathbb{F}\mathbb{F}(-)$ denotes the field of fractions functor. The field extension $\mathbb{F}\mathbb{F}(\mathbf{D}(n)) \subseteq \mathbb{F}_q(V)$ is Galois with Galois group $\mathrm{GL}(n, \mathbb{F}_q)$. The fundamental theorem of Galois theory gives us a group $G \leq \mathrm{GL}(n, \mathbb{F}_q)$ such that

$$\mathbb{F}\mathbb{F}(H) = \mathbb{F}_q(V)^G \subseteq \mathbb{F}_q(V).$$

Hence we have an integral extension

$$H \hookrightarrow \mathbb{F}_q[V]^G$$

of integrally closed algebras with the same field of fractions $\mathbb{F}\mathbb{F}(H) = \mathbb{F}_q(V)^G$, so they are equal, i.e., $H = \mathbb{F}_q[V]^G$ as we claimed. □

So, in order to really solve the inverse invariant theory problem, we have to characterize those H's that have the following two properties: [6]

EMBEDDING PROPERTY: H admits an integral embedding into $\mathbb{F}_q[V]$.

IMBEDDING PROPERTY: H contains the natural copy of $\mathbf{D}(n) \subset \mathbb{F}[V]$, such that $\mathbf{D}(n) \hookrightarrow H$ is an integral extension.

An obvious necessary condition for H to embed in $\mathbb{F}_q[V]$, for some V, is that H contain no zero divisors. If it is going to be a ring of invariants it will admit an action of the Steenrod algebra satisfying the Cartan formulae and the unstability condition. Let us introduce some terminology to describe this.

DEFINITION: A graded connected commutative algebra H over $\mathbb{F}_q$ is called an **algebra over the Steenrod algebra** if it is a left $\mathscr{P}^*$-module satisfying the Cartan formulae. If, in addition, the unstability condition holds, then we say that H is an **unstable algebra over the Steenrod algebra**, or simply an **unstable algebra**.

EXAMPLE 1: Consider the polynomial algebra $\mathbb{F}_2[Q, T]$ over the field $\mathbb{F}_2$ with 2 elements, where the indeterminate Q has degree 2 and T has degree 3. If the Steenrod algebra were to act unstably on this algebra then the unstability condition would determine $\mathrm{Sq}^i(Q)$ and $\mathrm{Sq}^j(T)$ apart from $i = 1$ and $j = 1$ and 2. If we specify

$$(*) \qquad \mathrm{Sq}^1(Q) = T, \quad \mathrm{Sq}^1(T) = 0, \quad \mathrm{Sq}^2(T) = QT,$$

[6] Although **em**bedding and **im**bedding are alternative spellings of the same word, we will use the variation to distinguish the two properties that follow.

and demand that the Cartan formula hold, then using these formulae we can compute Sq^k on any monomial, and hence by linearity, on any polynomial in Q and T. For example

$$\mathrm{Sq}^1(QT) = \mathrm{Sq}^1(Q) \cdot T + Q \cdot \mathrm{Sq}^1(T) = T^2 + 0 = T^2,$$

and so on. If we denote by $\mathscr{A}$ the free associative algebra generated by the symbols Sq^k for $k \geq 1$ then we have specified a $\mathscr{A}$-module structure on $\mathbb{F}_2[Q, T]$. If we make $\mathscr{A}$ into a Hopf algebra by using the coproduct

$$\nabla(\mathrm{Sq}^k) = \sum_{k'+k''=k} \mathrm{Sq}^{k'} \otimes \mathrm{Sq}^{k''}$$

then $\mathbb{F}_2[Q, T]$ becomes an algebra over $\mathscr{A}$. In this formula we use the convention $\mathrm{Sq}^0 = 1 \in \mathscr{A}$. To verify that $\mathbb{F}_2[Q, T]$ is an algebra over the Steenrod algebra we need to verify that an element of the form

$$\mathrm{Sq}^a \mathrm{Sq}^b - \sum_{c=0}^{[\frac{a}{2}]} \binom{b-1-c}{a-2c} \mathrm{Sq}^{a+b-c} \mathrm{Sq}^c, \qquad 0 < a < 2b$$

acts trivially on $\mathbb{F}_2[Q, T]$. By use of the Cartan formulae this may be reduced to verifying such elements act trivially on the algebra generators Q and T (see Lemma 4.1 in [404]), and this is routine. Since the unstability condition holds for the algebra generators it follows from the Cartan formulae that it holds for any form $f \in \mathbb{F}_2[Q, T]$, so $\mathbb{F}_2[Q, T]$ is an unstable algebra over the Steenrod algebra.[7]

NOTATION: We denote by $\mathcal{K}$ the category whose objects are algebras over the Steenrod algebra $\mathscr{P}^*$ and whose morphisms are $\mathscr{P}^*$-algebra homomorphisms. We write $\mathcal{K}_{fg}$ for the full subcategory of finitely generated $\mathbb{F}_q$-algebras.

An algebra of invariants $\mathbb{F}_q[V]^G$ of a finite group over the Galois field $\mathbb{F}_q$ is a prototypical example of an unstable algebra over the Steenrod algebra. Of course the original example of such a structure is the cohomology algebra $H^*(X; \mathbb{F}_p)$ of a space with coefficients in the prime field $\mathbb{F}_p$ [405]. Being such an algebra imposes a very rigid structure on rings of invariants, since, as we have already remarked, the unstability condition is, on the one hand a triviality condition, viz., $\mathscr{P}^i(f) = 0$ if $i > \deg(f)$, and on the other hand a nontriviality condition, viz., $\mathscr{P}^i(f) = f^q$ if $i = \deg(f)$. This rigidity will allow us to deduce a number of quite remarkable results such as the following embedding theorem (see [1] Theorem 1.1 for the prime field and Corollary 6.1.5 in [290] for the general case).

---

[7] For those familiar with such things it is the mod 2 cohomology $H^*(B\mathbb{SO}(3); \mathbb{F}_2)$ of the classifying space $B\mathbb{SO}(3)$ of the real special orthogonal group $\mathbb{SO}(3)$ of $\mathbb{R}^3$.

THEOREM 8.4.2 (J. F. Adams–C. W. Wilkerson, M. D. Neusel, Embedding Theorem): *Let H be an unstable Noetherian integral domain over the Steenrod algebra with Krull dimension n. Then there exists an integral extension $H \hookrightarrow \mathbb{F}_q[V]$ in the category $\mathcal{K}_{fg}$.* $\square$

EXAMPLE 2: Consider the unstable algebra $\mathbb{F}_2[Q, T]$ of Example 1. The embedding theorem tells us we can find an injection of algebras over the Steenrod algebra $\alpha : \mathbb{F}_2[Q, T] \hookrightarrow \mathbb{F}_2[x, y]$ in the category $\mathcal{K}_{fg}$. One such is given by

$$\alpha(Q) = x^2 + xy + y^2, \qquad \alpha(T) = x^2y + xy^2,$$

which follows from a comparison with the formulae in Example 1 of Section 8.1. In fact, this is the only such embedding up to change of basis.

The imbedding property is a bit more subtle, and we need some further preparations before formally stating it. As we have already noted, the Steenrod algebra contains an infinite family of primitive [8] derivations, defined inductively by

$$\mathcal{P}^{\Delta_i} = \begin{cases} \mathcal{P}^1 & \text{if } i = 1, \\ \mathcal{P}^{\Delta_{i-1}}\mathcal{P}^{q^{i-1}} - \mathcal{P}^{q^{i-1}}\mathcal{P}^{\Delta_{i-1}} & \text{otherwise.} \end{cases}$$

To these we append the derivation of degree 0 defined by

$$\mathcal{P}^{\Delta_0}(h) = \deg(h)h, \quad \forall\, h \in H,$$

but which is *not* in $\mathscr{P}^*$.

By analogy with the classical terminology of inseparability we introduce the following definition:

DEFINITION: *An algebra H over the Steenrod algebra is called* $\mathscr{P}^*$-**inseparably closed** *if whenever $h \in H$ and*

$$\mathcal{P}^{\Delta_i}(h) = 0 \quad \forall\, i \geq 0,$$

there exists an element $\overline{h} \in H$ such that $(\overline{h})^p = h$.

In other words, if all the derivations $\mathcal{P}^{\Delta_i}$, $i \in \mathbb{N}_0$, vanish on an element in H, then that element must be a pth power. Note that the vanishing of $\mathcal{P}^{\Delta_0}$ is equivalent to assuming that the degree of the element is divisible by p.

Let's record the obvious definition of a $\mathscr{P}^*$-inseparable closure.

DEFINITION: *The* $\mathscr{P}^*$**-inseparable closure** *of H is a $\mathscr{P}^*$-inseparably closed algebra $\sqrt[\mathscr{P}^*]{H}$ containing H such that the following universal property holds: whenever we have a $\mathscr{P}^*$-inseparably closed algebra K containing H there exists an embedding $\varphi : \sqrt[\mathscr{P}^*]{H} \hookrightarrow K$ making the following diagram*

[8] Primitive in the sense of, a primitive element in a Hopf algebra.

commute:

$$\begin{array}{ccc} H & \subseteq & \sqrt[\mathscr{P}^*]{H} \\ \cap & \swarrow & \\ K & & \end{array}$$

The existence of the $\mathscr{P}^*$-inseparable closure for a given unstable algebra over the Steenrod algebra is a highly nontrivial fact; see Section 4.1 in [290], in particular Proposition 4.1.5, for an explicit construction. Equipped with this terminology we can cite [290], Theorem 8.1.5, which settles the imbedding problem, and [290] Theorem 7.3.3 which settles the uniqueness of such an imbedding (see also the Appendix of [41] and Appendix B of [245]).

THEOREM 8.4.3 (Imbedding Theorem): *Let H be a Noetherian unstable algebra over the Steenrod algebra. Then there exists an integral extension of unstable Noetherian algebras $\mathbf{D}(n)^{q^t} \hookrightarrow H$ in $\mathcal{K}_{fg}$ for some $t \in \mathbb{N}_0$. If H is $\mathscr{P}^*$-inseparably closed, then we have $t = 0$. For a given t, if an imbedding $\mathbf{D}(n)^{q^t} \hookrightarrow H$ exists, then it is unique.* □

Note carefully that the embedding theorem faithfully reflects the nonexistence of zero divisors, while the imbedding theorem tells us something about the existence of enough pth roots.

Combining Proposition 8.4.1, with Theorems 8.4.2 and 8.4.3 we obtain a complete solution to the inverse invariant theory problem; see [1] for prime fields and Theorem 7.1.1 in [290] for the general case.

THEOREM 8.4.4 (J. F. Adams-C. W. Wilkerson, M. D. Neusel, Galois Embedding Theorem): *Let H be an unstable algebra over the Steenrod algebra. Then H is a ring of invariants if and only if H is an integrally closed, $\mathscr{P}^*$-inseparably closed, Noetherian integral domain.*

In the case that the ring H in question happens to be a polynomial ring, then the condition of $\mathscr{P}^*$-inseparable closure can be replaced by the nonvanishing of the $\mathscr{P}^*$-generalized Jacobian determinant; see Theorem 1.2 in [384].

THEOREM 8.4.5: *Let $H = \mathbb{F}_q[h_1, \ldots, h_n]$ be an unstable polynomial algebra over the Steenrod algebra $\mathscr{P}^*$. Then H is the ring of invariants of a finite pseudoreflection group $G \hookrightarrow \mathrm{GL}(n, \mathbb{F}_q)$ if and only if*

$$\det\left(\mathscr{P}^{\Delta_i}(h_j)\right)_{\substack{i=0,\ldots,n-1\\ j=1,\ldots,n}} \neq 0. \quad \square$$

EXAMPLE 3: Consider again the polynomial algebra $\mathbb{F}_2[Q, T]$ of Example 2, where

$$\mathrm{Sq}^1(Q) = T, \quad \mathrm{Sq}^1(T) = 0, \quad \mathrm{Sq}^2(T) = QT.$$

The $\mathscr{P}^*$-generalized Jacobian matrix is

$$\begin{bmatrix} \mathrm{Sq}^{\Delta_0}(Q) & \mathrm{Sq}^{\Delta_0}(T) \\ \mathrm{Sq}^{\Delta_1}(Q) & \mathrm{Sq}^{\Delta_1}(T) \end{bmatrix} = \begin{bmatrix} 0 & T \\ T & 0 \end{bmatrix},$$

which has a nonzero determinant. Hence $\mathbb{F}_2[Q, T]$ embeds in $\mathbb{F}_2[x, y]$ as a ring of invariants. By the degree theorem (Theorem 4.5.3) the group in question must have order 6, since the Poincaré series of $\mathbb{F}_2[Q, T]$ is

$$\frac{1}{(1-t^2)(1-t^3)} = \frac{\frac{1}{6}}{1-t^2} + \cdots.$$

The full linear group $\mathrm{GL}(2, \mathbb{F}_2)$ itself has order 6, so we conclude that $\mathbb{F}_q[Q, T] = \mathbb{F}_q[x, y]^{\mathrm{GL}(2,\mathbb{F}_2)} = \mathbf{D}(2)$.

EXAMPLE 4: Consider the polynomial algebra $\mathbb{F}_2[u, w]$, where u has degree 2 and w has degree 4. If the Steenrod algebra acts on this polynomial algebra, then the unstability condition determines the action on u and w almost completely

$$\begin{aligned} &\mathrm{Sq}^0(u) = u, \ \mathrm{Sq}^0(w) = w, \\ &\mathrm{Sq}^1(u) = \mathrm{Sq}^1(w) = \mathrm{Sq}^3(w) = 0, \\ &\mathrm{Sq}^2(u) = u^2, \ \ \mathrm{Sq}^4(w) = w^4, \end{aligned}$$

where the middle equation is true for degree reasons and the last line follows from the unstability condition. The only operation left to consider is

$$\mathrm{Sq}^2(w) = \varepsilon uw, \qquad \varepsilon \in \mathbb{F}_2.$$

Either choice of ε is possible, and the result is an unstable algebra over the Steenrod algebra. Neither of these algebras can be a ring of invariants, since $\mathrm{Sq}^{\Delta_0}(u) = 0 = \mathrm{Sq}^{\Delta_0}(w)$, so the $\mathscr{P}^*$-generalized Jacobian matrix has a row of zeros. This is, of course, a special case of a more general phenomenon: *If all the generators of an unstable polynomial algebra over the Galois field* $\mathbb{F}_q$ *of characteristic* p *are divisible by* p, *then the algebra cannot be a ring of invariants.*

8.5 The Landweber–Stong Conjecture

One of the major advances in modular invariant theory made with the aid of the Steenrod algebra is the proof by D. Bourguiba and S. Zarati of the Landweber-Stong conjecture [41]. For a bit about the history of this conjecture see [96] and [224].

Recall that a ring of invariants $\mathbb{F}_q[V]^G$ over the Galois field $\mathbb{F}_q$ with $q = p^\nu$ elements supports two additional structures:

- it is a module (actually an algebra) over the Dickson algebra $\mathbf{D}(n)$, $n = \dim_{\mathbb{F}_q}(V)$, and

- it is an unstable algebra over the Steenrod algebra $\mathscr{P}^*$.

Moreover, these structures are related by the Cartan formulae. We summarize this by saying[9] $\mathbb{F}_q[V]^G$ is an **unstable $\mathbf{D}(n) \odot \mathscr{P}^*$-algebra**. Proposition 5.6.5 tells us that the codimension of $\mathbb{F}_q[V]^G$ as an algebra is equal to its codimension as a $\mathbf{D}(n)$-module. It is in this context that one can best understand the depth conjecture of P. S. Landweber and R. E. Stong: Namely, it should be viewed as a statement about the codimension of Noetherian unstable $\mathbf{D}(n) \odot \mathscr{P}^*$-modules.

CONJECTURE (P. S. Landweber-R. E. Stong): *If $\rho : G \hookrightarrow \mathrm{GL}(n, \mathbb{F}_q)$ is a representation of a finite group over a Galois field, then*

$$\text{hom-codim}(\mathbb{F}_q[V]^G) \geq k$$

if and only if $\mathbf{d}_{n,n-1}, \ldots, \mathbf{d}_{n,n-k} \in \mathbb{F}_q[V]^G$ is a regular sequence.

Note carefully, the order is important: It is the order of increasing degrees; cf. the discussion of the reverse Landweber-Stong conjecture at the end of this section, which is wrong. The conjecture is a consequence of the following more general result due to D. Bourguiba and S. Zarati [41]. We denote by $\mathrm{E}_R(M)$ the **injective hull** of an R-module M. We refer to [104] and [448] for the basics about injective modules and hulls.

THEOREM 8.5.1 (D. Bourguiba-S. Zarati): *If M is a Noetherian unstable $\mathbb{F}_q[V] \odot \mathscr{P}^*$-module, then*

$$\text{hom-codim}_{\mathbb{F}_q[V]}(M) \leq \text{hom-codim}_{\mathbb{F}_q[V]}(\mathrm{E}_{\mathbb{F}_q[V]\odot\mathscr{P}^*}(M)).$$

The proof of this theorem will occupy most of the rest of this section. We follow the path described in [373] to prove the depth conjecture, which means we need to solve Exercises A3.4 and A3.3 in Appendix 3 of [104] in this new context.

An essential ingredient in this approach is the existence, and structure, of injective hulls in the category of Noetherian unstable $\mathbf{D}(n) \odot \mathscr{P}^*$-modules. It is, however, easier, and for our purposes equivalent, to consider Noetherian unstable $\mathbb{F}_q[V] \odot \mathscr{P}^*$-modules and the structure of their injective hulls. The discussion that follows is of necessity brief, and we refer to the original sources [41], [227], [229], or the book [336] for the missing details.

An unstable $\mathbb{F}_q[V] \odot \mathscr{P}^*$-module M is a module over both $\mathbb{F}_q[V]$ and $\mathscr{P}^*$ that satisfies the **unstability condition for $\mathscr{P}^*$-modules**, namely $\mathcal{P}^k(x) = 0$ if $k > \deg(x)$, and the Cartan formula for the $\mathbb{F}_q[V]$-module structure,

---

[9] The **semitensor product** construction $A \odot H$ of a Hopf algebra H and an algebra A over H was introduced by W. S. Massey and F. P. Peterson in [240] to convert an algebra with such a mixed structure into an ordinary algebra over $A \odot H$. We do not make use of the actual construction, but find the notation to be convenient.

namely,

$$\mathcal{P}^k(f \cdot x) = \sum_{i+j=k} \mathcal{P}^i(f) \cdot \mathcal{P}^j(x), \qquad \forall f \in \mathbb{F}_q[V], x \in M.$$

Note that the first part of the unstability condition does not make sense for modules. The category of such modules is denoted by $\mathcal{U}_{\mathbb{F}_q[V]}$. For an unstable $\mathbb{F}_q[V] \odot \mathscr{P}^*$-module M we define functors

$$\Upsilon^k(M) = \mathrm{Hom}_{\mathbb{F}_q}(M_k, \mathbb{F}_q), \qquad k \in \mathbb{N}_0.$$

These are exact functors, and preserve direct sums, and therefore are representable (see, e.g., [336] Lemma 2.2.1). We denote by $\mathbf{J}_{\mathbb{F}_q[V]\odot\mathscr{P}^*}(k)$ the representing module,[10] so that

$$(\text{∻}) \qquad \mathrm{Hom}_{\mathbb{F}_q}(M, \Sigma^k\mathbb{F}_q) = \Upsilon^k(M) \cong \mathrm{Hom}_{\mathbb{F}_q[V]\odot\mathscr{P}^*}(M, \mathbf{J}_{\mathbb{F}_q[V]\odot\mathscr{P}^*}(k)),$$

where $\Sigma^k(-)$ denotes the k-fold suspension. The functors Υ^k are exact, so the modules $\mathbf{J}_{\mathbb{F}_q[V]\odot\mathscr{P}^*}(k)$ are injective objects in $\mathcal{U}_{\mathbb{F}_q[V]}$.

The module $\mathbf{J}_{\mathbb{F}_q[V]\odot\mathscr{P}^*}(k)$ is a sort of cofree injective module on one generator of degree k. It has the following direct description: Consider the vector space dual of $\mathbb{F}[V] \otimes \mathscr{P}^*$, viz., $\mathrm{Hom}^*_{\mathbb{F}_q}(\mathbb{F}[V] \otimes \mathscr{P}^*, \mathbb{F}_q)$. With the usual conventions, this is a negatively graded vector space over $\mathbb{F}_q$, where the elements of degree m are the linear maps from $(\mathbb{F}_q[V] \otimes \mathscr{P}^*)_{-m}$ to $\mathbb{F}_q$. This dual also carries an $\mathbb{F}_q[V] \odot \mathscr{P}^*$-module structure, given by the formula

$$(f \otimes P) \cdot \varphi(h \otimes Q) = \varphi(f \cdot P(h) \otimes PQ)$$

where $f, h \in \mathbb{F}_q[V]$, $P, Q \in \mathscr{P}^*$, and $\varphi : (\mathbb{F}_q[V] \otimes \mathscr{P}^*)_{-m} \longrightarrow \mathbb{F}_q$. The k-fold suspension is therefore zero in degrees greater than k. The module $\mathbf{J}_{\mathbb{F}_q[V]\odot\mathscr{P}^*}(k)$ is just the set of all unstable elements, i.e., the elements that satisfy the unstability condition, in the k-fold suspension of the $\mathscr{P}^*$-module $\mathrm{Hom}^*_{\mathbb{F}_q}(\mathbb{F}_q[V] \otimes \mathscr{P}^*, \mathbb{F}_q)$.

For any unstable $\mathbb{F}_q[V] \odot \mathscr{P}^*$-module M the natural map

$$M \longrightarrow \prod_{\varphi \in \Upsilon^k(M)} \mathbf{J}_{\mathbb{F}_q[V]\odot\mathscr{P}^*}(k) \cdot \varphi$$

given by the adjointness relation (∻) embeds M in an injective object of the category $\mathcal{U}_{\mathbb{F}_q[V]}$, so the category $\mathcal{U}_{\mathbb{F}_q[V]}$ has enough injectives, and each object M in $\mathcal{U}_{\mathbb{F}_q[V]}$ has an **injective hull** in $\mathcal{U}_{\mathbb{F}_q[V]}$, which we denote by $\mathbf{E}_{\mathbb{F}_q[V]\odot\mathscr{P}^*}(M)$.

For a Noetherian unstable $\mathbb{F}_q[V] \odot \mathscr{P}^*$-module the injective hull is a finite direct sum of certain basic injectives that we describe next. If $i_W : W \leq V$ is

[10] The modules $\mathbf{J}_{\mathbb{F}_q[V]\odot\mathscr{P}^*}(k)$ are the duals of the **Brown–Gitler modules**; see [336] Chapter 2.

a vector subspace, then the inclusion induces a map $i_W^* : \mathbb{F}_q[V] \longrightarrow \mathbb{F}_q[W]$ whose kernel is denoted by $\mathfrak{p}_W$. The quotient map $V \longrightarrow V/W$ induces an inclusion $\mathbb{F}_q[V/W] \subseteq \mathbb{F}_q[V]$, and $\mathfrak{p}_W$ is the ideal in $\mathbb{F}_q[V]$ generated by the linear forms in the image of this map. Via i_W^* we may regard $\Sigma^k (\mathbb{F}_q[W]) = \Sigma^k (\mathbb{F}_q[V]/\mathfrak{p}_W)$ as an object in $\mathcal{U}_{\mathbb{F}_q[V]}$, and there it has an injective hull, which we denote by $\mathsf{E}(V, W, k)$. The natural map

$$\Sigma^k (\mathbb{F}_q[W]) = \mathbb{F}_q[V] \otimes_{\mathbb{F}_q[V/W]} \Sigma^k(\mathbb{F}_q) \longrightarrow \mathbb{F}_q[V] \otimes_{\mathbb{F}_q[V/W]} \mathsf{J}_{\mathbb{F}_q[V/W]}(k)$$

induces an isomorphism, [229],

$$\mathbb{F}_q[V] \otimes_{\mathbb{F}_q[V/W]} \mathsf{J}_{\mathbb{F}_q[V/W]}(k) \longrightarrow \mathsf{E}(V, W, k).$$

The basic structure theorem that we need is the following; see [229]:

THEOREM 8.5.2 (J. Lannes-S. Zarati): *Let M be a Noetherian unstable $\mathbb{F}_q[V] \odot \mathscr{P}^*$-module. Then*

$$\mathsf{E}_{\mathbb{F}_q[V]\odot\mathscr{P}^*}(M) = \bigoplus \mathsf{E}(V, W, k)^{a_M(W,k)},$$

where

(i) W ranges over the subspaces $W \leq V$ (a finite set),
(ii) $k, a_M(W, k) \in \mathbb{N}_0$, and
(iii) for a given $W \leq V$ only finitely many $a_M(W, k)$ are nonzero. □

Before we can begin the proof proper of Theorem 8.5.1 we need to establish a number of basic facts about injective hulls in the category $\mathcal{U}_{\mathbb{F}_q[V]}$.

LEMMA 8.5.3: *Every element of $\mathsf{E}(V, W, k)$ is annihilated by some power of $\mathfrak{p}_W$.*

PROOF: By the Artin-Rees lemma, [104] Chapter 5, Lemma 5.1, there exists an integer $n \in \mathbb{N}$ such that $\forall\, m \in \mathbb{N}$

$$\begin{aligned}
&\left(\mathfrak{p}_W^{n+m} \cdot \mathsf{E}(V, W, k)\right) \cap \left(\Sigma^k (\mathbb{F}_q[W])\right)\\
&\quad = \mathfrak{p}_W^m \cdot \left(\left(\mathfrak{p}_W^n \cdot \mathsf{E}(V, W, k)\right) \cap \left(\Sigma^k (\mathbb{F}_q[W])\right)\right)\\
&\quad = \mathfrak{p}_W^{n+m} \mathsf{E}(V, W, k) \cap \mathfrak{p}_W^m \Sigma^k (\mathbb{F}_q[W])\\
&\quad \subseteq \mathsf{E}(V, W, k) \cap \{0\} = \{0\}.
\end{aligned}$$

The ideal $\mathfrak{p}_W$ is generated by linear forms and hence closed under the action of the Steenrod algebra: This may be verified directly or by glancing at the proof of Theorem 9.2.1. Therefore the subset $\mathfrak{p}_W^{n+m} \cdot \mathsf{E}(V, W, k) \subseteq \mathsf{E}(V, W, k)$ is an $\mathbb{F}_q[V] \odot \mathscr{P}^*$-submodule. The inclusion into the injective hull $\Sigma^k (\mathbb{F}_q[W]) \hookrightarrow \mathsf{E}(V, W, k)$ is an essential monomorphism in the category $\mathcal{U}_{\mathbb{F}_q[V]}$, so it follows that $\mathfrak{p}_W^n \cdot \mathsf{E}(V, W, k) = \{0\}$. In particular, we obtain $\mathfrak{p}_W^n \cdot \mathsf{E}(V, W, k) = \{0\}$, as required. □

The inclusion $M \hookrightarrow \mathsf{E}_{\mathbb{F}_q[V]\odot\mathscr{P}^*}(M)$ is an essential monomorphism, and

$$\mathsf{E}(V, W, k) \subseteq \mathsf{E}_{\mathbb{F}_q[V]\odot\mathscr{P}^*}(M)$$

for $a_M(W, k) \neq 0$. Therefore, $M \cap \mathsf{E}(V, W, k) \neq \{0\}$, and hence there is an element $0 \neq x \in M$ also annihilated by some power of $\mathfrak{p}_W$. Let m be the smallest integer such that $\mathfrak{p}_W^m \subseteq \mathrm{Ann}_{\mathbb{F}_q[V]}(x)$. If $m = 0$, set $y = x$; otherwise choose a nonzero element y in $\mathfrak{p}_W^{m-1} \cdot x$. This gives us in any case an element $0 \neq y \in M$ annihilated by $\mathfrak{p}_W$.

LEMMA 8.5.4: hom-dim$_{\mathbb{F}_q[V]}(M) \geq \mathscr{ht}(\mathfrak{p}_W)$.

PROOF: Let $r = \dim_{\mathbb{F}_q}(W)$ and $s = n - r = \dim_{\mathbb{F}_q}(V/W) = \mathscr{ht}(\mathfrak{p}_W)$. Choose a basis $x_1, \ldots, x_s$ for the linear forms in $\mathfrak{p}_W$ and adjoin $y_1, \ldots, y_r$ to them to obtain a basis for V^*, the linear forms in $\mathbb{F}_q[V]$. Then

$$\begin{aligned}\mathrm{Tor}_*^{\mathbb{F}_q[V]}(\mathbb{F}_q[W], M) &= \mathrm{Tor}_*^{\mathbb{F}_q[x_1,\ldots,x_s,y_1,\ldots,y_r]}(\mathbb{F}_q[y_1, \ldots, y_r], M)\\ &\cong \mathrm{Tor}_*^{\mathbb{F}_q[x_1,\ldots,x_s]}(\mathbb{F}_q, M).\end{aligned}$$

If we use the Koszul complex, cf. Section 5.1,

$$\begin{aligned}\mathscr{K} &= \mathbb{F}_q[x_1, \ldots, x_s] \otimes E(u_1, \ldots, u_s),\\ \partial(f \otimes 1) &= 0 \quad \forall f \in \mathbb{F}_q[x_1, \ldots, x_s],\\ \partial(1 \otimes u_i) &= x_i \otimes 1 \quad \text{for } i = 1, \ldots, s\end{aligned}$$

to compute this torsion product, we find that

$$0 \neq u_1 \cdots u_s \otimes y \in \mathscr{K} \otimes_{\mathbb{F}_q[x_1,\ldots,x_s]} M$$

is a nonzero cycle (since $x_i \cdot y = 0$ for $i = 1, \ldots, s$) and cannot be a boundary, since there are no chains of homological degree $s + 1$. Hence we conclude that

$$\mathrm{Tor}_s^{\mathbb{F}_q[V]}(\mathbb{F}_q[W], M) = \mathrm{Tor}_s^{\mathbb{F}_q[V]}(\mathbb{F}_q[y_1, \ldots, y_r], M) \neq 0$$

and therefore hom-dim$_{\mathbb{F}_q[V]}(M) \geq s = \mathscr{ht}(\mathfrak{p}_W)$ as claimed. □

LEMMA 8.5.5: hom-dim$_{\mathbb{F}_q[V]}(\mathsf{E}(V, W, k)) = \mathscr{ht}(\mathfrak{p}_W)$.

PROOF: We use the same notation as in the preceding Lemma 8.5.4. Recall that

$$\mathsf{E}(V, W, k) = \mathbb{F}_q[V] \otimes_{\mathbb{F}_q[V/W]} \mathsf{J}_{\mathbb{F}_q[V/W]}(k),$$

and $\mathsf{J}_{\mathbb{F}_q[V/W]}(k)$ itself is a totally finite graded vector space. If we employ the Koszul complex

$$\begin{aligned}\mathscr{L} &= E(u_1, \ldots, u_s, w_1, \ldots, w_r) \otimes \mathbb{F}_q[V],\\ \partial(1 \otimes f) &= 0 \quad \forall f \in \mathbb{F}_q[V],\\ \partial(u_i \otimes 1) &= 1 \otimes x_i \quad \text{for } i = 1, \ldots, s,\\ \partial(w_j \otimes 1) &= 1 \otimes y_j \quad \text{for } j = 1, \ldots, r\end{aligned}$$

to resolve $\mathbb{F}_q$ as an $\mathbb{F}_q[V]$-module, we obtain

$$\begin{aligned}\operatorname{Tor}_*^{\mathbb{F}_q[V]}(\mathbb{F}_q, \mathsf{E}(V, W, k)) &= H^*\left(\mathcal{L} \otimes_{\mathbb{F}_q[V]} \mathsf{E}(V, W, k)\right)\\ &= H^*\left(E(u_1, \dots, u_s, w_1, \dots, w_r) \otimes \mathbb{F}_q[V] \otimes_{\mathbb{F}_q[V/W]} \mathsf{J}_{\mathbb{F}_q[V/W]}(k)\right)\\ &= H^*\left(E(u_1, \dots, u_s) \otimes \left(E(w_1, \dots, w_r) \otimes \mathbb{F}_q[W]\right) \otimes \mathsf{J}_{\mathbb{F}_q[V/W]}(k)\right)\\ &= E(u_1, \dots, u_s) \otimes \mathsf{J}_{\mathbb{F}_q[V/W]}(k).\end{aligned}$$

This follows from the fact that the Koszul complex

$$E(w_1, \dots, w_r) \otimes \mathbb{F}_q[W],$$
$$\partial(1 \otimes f) = 0, \ \forall f \in \mathbb{F}_q[W], \quad \text{and} \quad \partial(w_j \otimes 1) = y_j \text{ for } j = 1, \dots, r$$

is acyclic, and the differential in the complex

$$E(u_1, \dots, u_s) \otimes \mathsf{J}_{\mathbb{F}_q[V/W]}(k)$$

is trivial, because $x_1, \dots, x_s \in \mathfrak{p}_W$ and $\mathfrak{p}_W$ annihilates $\mathsf{J}_{\mathbb{F}_q[V/W]}(k)$. Therefore,

$$\operatorname{Tor}_s^{\mathbb{F}_q[V]}(\mathbb{F}_q, \mathsf{E}(V, W, k)) \neq 0, \quad \text{but} \quad \operatorname{Tor}_{s+1}^{\mathbb{F}_q[V]}(\mathbb{F}_q, \mathsf{E}(V, W, k)) = 0,$$

and hence $\text{hom-dim}_{\mathbb{F}_q[V]}(\mathsf{E}(V, W, k)) = s$ as claimed. □

PROOF OF THEOREM 8.5.1: By the structure theorem for $\mathcal{U}_{\mathbb{F}_q[V]}$ injective hulls, Theorem 8.5.2,

$$\mathsf{E}_{\mathbb{F}_q[V] \odot \mathscr{P}^*}(M) = \bigoplus \mathsf{E}(V, W, k)^{a_M(W,k)},$$

and therefore we have the equality

$$\begin{aligned}&\text{hom-codim}_{\mathbb{F}_q[V]}(\mathsf{E}_{\mathbb{F}_q[V] \odot \mathscr{P}^*}(M))\\ &\quad = \min\left\{\text{hom-codim}_{\mathbb{F}_q[V] \odot \mathscr{P}^*}(\mathsf{E}(V, W, k)) \mid a_M(W, k) \neq 0\right\}.\end{aligned}$$

Fix once and for all a subspace $W \leq V$ such that $a_M(W, k) \neq 0$ and

$$\text{hom-codim}_{\mathbb{F}_q[V]}(\mathsf{E}_{\mathbb{F}_q[V] \odot \mathscr{P}^*}(M)) = \text{hom-codim}_{\mathbb{F}_q[V]}(\mathsf{E}(V, W, k)).$$

Combining Lemmas 8.5.4 and 8.5.5 we obtain

$$\text{hom-dim}_{\mathbb{F}_q[V]}(M) \geq \mathit{ht}(\mathfrak{p}_W) = \text{hom-dim}_{\mathbb{F}_q[V]}(\mathsf{E}(V, W, k)),$$

and therefore the theorem follows from the Auslander-Buchsbaum equality ([21], or [47], Theorem 1.3.3). □

We next investigate when the Dickson polynomials $\mathbf{d}_{n,n-1}, \dots, \mathbf{d}_{n,n-r} \in \mathbb{F}_q[V]$ are a regular sequence on one of the modules $\mathsf{E}(V, W, k)$.

LEMMA 8.5.6 (D. Bourguiba-S. Zarati): *Let $V = \mathbb{F}_q^n$ and $W \leq V$ be an r-dimensional vector subspace. If $N \neq 0$ is a totally finite $\mathbb{F}_q[V/W] \odot \mathscr{P}^*$-module, then $\mathbf{d}_{n,n-1}, \dots, \mathbf{d}_{n,n-r} \in \mathbb{F}_q[V]$ is a regular sequence on the module $\mathbb{F}_q[V] \otimes_{\mathbb{F}_q[V/W]} N$.*

PROOF: Suppose first that N is a trivial $\mathbb{F}_q[V/W]$-module. Then

$$\mathbb{F}_q[V] \otimes_{\mathbb{F}_q[V/W]} N \cong \mathbb{F}_q[W] \otimes_{\mathbb{F}_q} N$$

as $\mathbb{F}_q[V]$-modules. Under the natural map $\mathbb{F}_q[V] \longrightarrow \mathbb{F}_q[W]$ the Dickson polynomials $\mathbf{d}_{n,n-1}, \ldots, \mathbf{d}_{n,n-r}$ in $\mathbb{F}_q[V]$ map to $\mathbf{d}_{r,r-1}^{q^r}, \ldots, \mathbf{d}_{r,0}^{q^r} \in \mathbb{F}_q[W]$: This follows from the Stong–Tamagawa formula (Theorem 6.1.7). These form a regular sequence on $\mathbb{F}_q[W]$ and hence also on $\mathbb{F}_q[W] \otimes N$, since this is just a direct sum of degree-shifted copies of $\mathbb{F}_q[W]$.

We proceed by induction on the dimension of $\mathrm{Tot}(N)$ as an $\mathbb{F}_q$-vector space. If $\dim_{\mathbb{F}_q}(\mathrm{Tot}(N)) = 1$ then N is a trivial $\mathbb{F}_q[V/W]$-module so the preceding calculation applies, and the induction is started. So we suppose that $\dim_{\mathbb{F}_q}(\mathrm{Tot}(N)) > 1$ and the result has been established for all N'' with $\dim_{\mathbb{F}_q}(\mathrm{Tot}(N'')) < \dim_{\mathbb{F}_q}(\mathrm{Tot}(N))$. Since N is nonzero but totally finite, there exists $d \in \mathbb{N}_0$ such that $N_d \neq 0$, but $N_m = 0$ for $m > d$. Let N' be the graded submodule of N defined by

$$N'_m = \begin{cases} N_d & \text{for } m = d, \\ 0 & \text{otherwise.} \end{cases}$$

Then $N' \subseteq N$ is an $\mathbb{F}_q[V] \odot \mathscr{P}^*$-submodule, and there is an exact sequence

$$0 \longrightarrow N' \longrightarrow N \longrightarrow N'' \longrightarrow 0$$

of $\mathbb{F}_q[V] \odot \mathscr{P}^*$-modules. The module N'' is also totally finite with

$$\dim_{\mathbb{F}_q}(\mathrm{Tot}(N'')) < \dim_{\mathbb{F}_q}(\mathrm{Tot}(N)),$$

and N'' is a trivial $\mathbb{F}_q[V] \odot \mathscr{P}^*$-module. Hence by the induction hypothesis we obtain that $\mathbf{d}_{n,n-1}, \ldots, \mathbf{d}_{n,n-r} \in \mathbb{F}_q[V]$ is a regular sequence on both $\mathbb{F}_q[V] \otimes_{\mathbb{F}_q[V/W]} N'$ and $\mathbb{F}_q[V] \otimes_{\mathbb{F}_q[V/W]} N''$. The functor $\mathbb{F}_q[V] \otimes_{\mathbb{F}_q[V/W]} -$ is exact, since $\mathbb{F}_q[V]$ is a free $\mathbb{F}_q[V/W]$-module. Applying this functor to the preceding exact sequence, we see that the end terms are free $\mathbb{F}_q[\mathbf{d}_{n,n-1}, \ldots, \mathbf{d}_{n,n-r}]$-modules, hence so is the middle term. □

Since $\mathsf{E}(V, W, k) = \mathbb{F}_q[V] \otimes_{\mathbb{F}_q[V/W]} \mathsf{J}_{\mathbb{F}_q[V/W]}(k)$ and $\mathsf{J}_{\mathbb{F}_q[V/W]}(k)$ is totally finite, we obtain from the preceding lemma:

PROPOSITION 8.5.7 (D. Bourguiba-S. Zarati): *If V is an n-dimensional $\mathbb{F}_q$-vector space and $W \leq V$ an r-dimensional subspace, then the Dickson polynomials $\mathbf{d}_{n,n-1}, \ldots, \mathbf{d}_{n,n-r} \in \mathbb{F}_q[V]$ form a regular sequence on $\mathsf{E}(V, W, k)$.* □

COROLLARY 8.5.8 (D. Bourguiba-S. Zarati): *If M is a Noetherian unstable $\mathbb{F}_q[V] \odot \mathscr{P}^*$-module, then* $\text{hom-codim}_{\mathbb{F}_q[V]}(\mathsf{E}_{\mathbb{F}_q[V] \odot \mathscr{P}^*}(M))$ *is the minimum integer in the set* $\left\{\dim_{\mathbb{F}_q}(W \subseteq V) \mid a_M(W, k) \neq 0\right\}$. □

We can now prove the main result of D. Bourguiba and S. Zarati.

THEOREM 8.5.9 (D. Bourguiba–S. Zarati): *Let M be a Noetherian unstable $\mathbb{F}_q[V] \odot \mathscr{P}^*$-module with* hom-codim$_{\mathbb{F}_q[V]} \geq r$. *Then the Dickson polynomials $\mathbf{d}_{n,n-1}, \ldots, \mathbf{d}_{n,n-r} \in \mathbb{F}_q[V]$ are a regular sequence on M.*

PROOF: Consider the exact sequence

$$0 \longrightarrow M \xrightarrow{e} \mathsf{E}_{\mathbb{F}_q[V]\odot\mathscr{P}^*}(M) \xrightarrow{f} N \longrightarrow 0$$

arising from the inclusion into the injective hull. By Theorem 8.5.1 the homological codimension hom-codim$_{\mathbb{F}_q[V]}(\mathsf{E}_{\mathbb{F}_q[V]\odot\mathscr{P}^*}(M))$ is at least r. Therefore, one sees by depth chasing[11] that hom-codim$_{\mathbb{F}_q[V]}(N) \geq r-1$.

We proceed by induction on r. If $r = 1$, then

$$1 = \text{hom-codim}_{\mathbb{F}_q[V]}(M) \leq \text{hom-codim}_{\mathbb{F}_q[V]}(\mathsf{E}_{\mathbb{F}_q[V]\odot\mathscr{P}^*}(M)),$$

and hom-codim$_{\mathbb{F}_q[V]}(N) \geq 0$. By Proposition 8.5.7 the bottom Dickson class is a regular element on $\mathsf{E}_{\mathbb{F}_q[V]\odot\mathscr{P}^*}(M)$, i.e.,

$$\mathbf{d}_{n,n-1}\mathsf{E}_{\mathbb{F}_q[V]\odot\mathscr{P}^*}(M) \neq 0.$$

We need to show that $\mathbf{d}_{n,n-1}$ is regular on M. Assume to the contrary that the bottom Dickson class annihilates an element $x \in M$. Then we have that

$$0 = \mathscr{P}^{q^{n-2}}(\mathbf{d}_{n,n-1}x) = -\mathbf{d}_{n,n-2}x,$$

where we made use of the formulae given in Proposition 8.6.1 and Corollary 8.6.2 that describe how the Steenrod algebra acts on the Dickson polynomials. This means that also the next higher Dickson polynomial $\mathbf{d}_{n,n-2}$ annihilates x. Iteratively we obtain from these formulae, loc. cit.,

$$0 = \mathscr{P}^{q^{i-1}}(\mathbf{d}_{n,i}x) = -\mathbf{d}_{n,i-1}x,$$

i.e., all Dickson polynomials annihilate x. In other words there is no element in $\mathbf{D}(n)$ that is regular on M so M has depth zero. This contradicts our hypothesis that $r > 0$.

So by induction we may suppose that $\mathbf{d}_{n,n-1}, \ldots, \mathbf{d}_{n,n-(r-1)}$ is a regular sequence on M and on N. Corollary 8.5.8 and the Lannes–Zarati structure theorem (Theorem 8.5.2) imply that $\mathbf{d}_{n,n-1}, \ldots, \mathbf{d}_{n,n-r}$ is a regular sequence on $\mathsf{E}_{\mathbb{F}_q[V]\odot\mathscr{P}^*}(M)$. Using the same reasoning as in the induction start we conclude from this that $\mathbf{d}_{n,n-1}, \ldots, \mathbf{d}_{n,n-r}$ is a regular sequence on M. □

Since functor $\mathbb{F}_q[V] \otimes_{\mathbf{D}(n)} -$ is exact, and $\mathbf{D}(n) \leq \mathbb{F}_q[V]$ is a finite extension

$$\begin{aligned}\text{hom-codim}_{\mathbf{D}(n)}(M) &= \text{hom-codim}_{\mathbf{D}(n)}(\mathbb{F}_q[V] \otimes_{\mathbf{D}(n)} M)\\ &= \text{hom-codim}_{\mathbb{F}_q[V]}(\mathbb{F}_q[V] \otimes_{\mathbf{D}(n)} M),\end{aligned}$$

[11] E.g., use the characterization of codimension in terms of the functors $\text{Ext}_{\mathbb{F}_q[V]}(-, \mathbb{F}_q)$, [47] Section 1.2 or [372] Section 6.6.

and the sequence $\mathbf{d}_{n,n-1}, \ldots, \mathbf{d}_{n,n-r}$ is a regular sequence on $\mathbb{F}_q[V] \otimes_{\mathbf{D}(n)} M$ if and only if it is a regular sequence on M. Therefore, we obtain the following corollary:

COROLLARY 8.5.10 (D. Bourguiba-S. Zarati): *Let M be a Noetherian unstable $\mathbf{D}(n) \odot \mathscr{P}^*$-module with depth* hom-codim$_{\mathbf{D}(n)}(M) \geq r$. *Then $\mathbf{d}_{n,n-1}, \ldots, \mathbf{d}_{n,n-r} \in \mathbf{D}(n)$ is a regular sequence on M.* □

This corollary includes the depth conjecture as the special case where the module $M = \mathbb{F}_q[V]^G$ is a ring of invariants. However, note that by the imbedding theorem (Theorem 8.4.3), we always find a fractal [12] of the Dickson algebra, i.e., $\mathbf{D}(n)^{q^t}$ for some $t \in \mathbb{N}_0$, in every Noetherian unstable algebra. So we have the following result:

THEOREM 8.5.11 (Generalized Landweber-Stong Conjecture): *Let H be a Noetherian unstable algebra over the Steenrod algebra. Then the homological codimension* hom-codim$(H) \geq r$ *if and only if the elements $\mathbf{d}^{q^t}_{n,n-1}, \ldots, \mathbf{d}^{q^t}_{n,n-r} \in \mathbf{D}(n)^{q^t}$ form a regular sequence on H, for some large $t \in \mathbb{N}$.* □

We close this section with some words about the significance of the *order* of the Dickson polynomials in the Landweber-Stong conjecture. If we reverse the order, that is to say, ask: *Is the depth of a Noetherian unstable algebra over the Steenrod algebra H always at least the smallest $k \in \mathbb{N}$, such that high enough qth powers of the k top* [13] *Dickson classes, $\mathbf{d}^{q^s}_{n,0}, \ldots, \mathbf{d}^{q^s}_{n,k-1} \in H$ form a regular sequence on H?* then this is what one might call the **Reverse Landweber-Stong Conjecture**. As we will see in Chapter 9, when we study the $\mathscr{P}^*$-invariant ideals in $\mathbf{D}(n)$, this would have a number of pleasant features, more so than the original Landweber-Stong conjecture. (Un)fortunately (however you want to view this), this statement is false. We cite the following example from [290], Example 2 in Section 8.4.

EXAMPLE 1: Let $\mathbb{F}_q$ be the field with two elements. We have an integral extension of unstable Noetherian algebras

$$\mathbf{D}(2) = \mathbb{F}_q[xy(x+y), x^2+y^2+xy] \hookrightarrow \mathbb{F}_q[x, y, z]/(yz, z^2) = H.$$

The algebra H has depth 1, but the top Dickson class $\mathbf{d}_{2,0} = xy(x+y) \in H$ is a zero divisor, since

$$\mathbf{d}_{2,0}z = xy(x+y)z = 0.$$

However, the bottom Dickson class $\mathbf{d}_{2,1} = x^2+y^2+xy \in H$ is not a zero divisor, as one can prove in the following way: Let $h \in H$ such that

$$\left(x^2+y^2+xy\right) h = 0 \in (z) = \mathcal{N}il(H) \subset H,$$

[12] See Section 9.2 Equation ❻ and the discussion surrounding it for an explanation of why we call this a fractal property.

[13] The word *top* is used in the sense of counting from the top degree downwards.

where $\mathcal{N}il(H)$ denotes the nilradical of H. Hence $h = z\overline{h} \in (z)$, because (z) is a prime ideal and $\mathbf{d}_{2,1} \notin (z)$. Then

$$0 = \left(x^2 + y^2 + xy\right) h = \left(x^2 + y^2 + xy\right) z\overline{h} = x^2 z\overline{h} + (x + y)yz\overline{h} = x^2 z\overline{h}.$$

Since x^2 is not a zero divisor in H, we conclude that $h = z\overline{h} = 0$, and that's all we wanted.

8.6 The Steenrod Algebra and the Dickson Algebra

The interaction between the Steenrod algebra and the Dickson algebra has been of increasing importance in applications, and we therefore devote this section to showing that certain formulae discovered in the case of the prime field given in Section 3 of [389], Section II of [456] (see also Section 10.6 in [372]) have direct analogues over arbitrary finite fields. This is a good exercise in the use of the Cartan formulae and other properties of Steenrod operations, so we go into some detail in the calculations to illustrate how one can actually use Steenrod operations in concrete cases. We follow the exposition in Appendix A.2 of [290].

As usual, the Dickson algebra of Krull dimension n is denoted by

$$\mathbf{D}(n) = \mathbb{F}_q[\mathbf{d}_{n,0}, \dots, \mathbf{d}_{n,n-1}] \subseteq \mathbb{F}_q[V]^{\mathrm{GL}(n,\mathbb{F}_q)}.$$

Recall that the Stong-Tamagawa formulae, Theorem 6.1.7, describe the Dickson polynomials as

$$\mathbf{d}_{n,i} = (-1)^{n-i} \sum_{\substack{W^* \leq V^* \\ \dim(W^*)=i}} \left(\prod_{v \notin W^*} v \right).$$

They are also the orbit Chern classes of the $\mathrm{GL}(n, \mathbb{F}_q)$-orbit consisting of the nonzero linear forms $V^* \setminus \{0\}$, and therefore occur as the coefficients of the polynomial

$$\varphi(X) = \prod_{v \in V^*} (X + v) = X^{q^n} + \sum_{i=0}^{n-1} (-1)^{n-i} \mathbf{d}_{n,i} X^{q^i} \in \mathbb{F}_q[V][X].$$

It is the interplay of these two descriptions, the additive and the multiplicative, that makes the following work.

To simplify the notation we introduce the conventions

$$\mathbf{d}_{n,n} = 1 \quad \text{and} \quad \mathbf{d}_{n,i} = 0 \quad \forall\, i < 0.$$

If X is given degree one, then the Steenrod algebra acts on $\mathbb{F}_q[V][X]$ since it is just a polynomial algebra in one more generator of degree one than the dimension of V as $\mathbb{F}_q$-vector space. Then using the product form of

$\varphi(X)$, we get by the Cartan formulae,

$$\begin{aligned}
\mathcal{P}^k(\varphi(X)) &= \mathcal{P}^k\Big(\prod_{v\in V^*}(X+v)\Big)\\
&= \sum_{(i_1,\dots,i_k)} \mathcal{P}^1(X+v_{i_1})\cdots\mathcal{P}^1(X+v_{i_k})\Big(\prod_{V^*\setminus\{v_{i_1},\dots,v_{i_k}\}}(X+v)\Big)\\
&= \sum_{(i_1,\dots,i_k)} (X^q+v_{i_1}^q)\cdots(X^q+v_{i_k}^q)\Big(\prod_{V^*\setminus\{v_{i_1},\dots,v_{i_k}\}}(X+v)\Big)\\
&= \varphi(X)\Big(\sum_{(i_1,\dots,i_k)}(X+v_{i_1})^{q-1}\cdots(X+v_{i_k})^{q-1}\Big),
\end{aligned}$$

i.e., $\varphi(X)$ divides every Steenrod power applied to $\varphi(X)$. On the other hand, we can use the sum form of $\varphi(X)$. If we do, we get

$$\mathcal{P}^k(\varphi(X)) = \mathcal{P}^k\Big(X^{q^n}+\sum_{i=0}^{n-1}(-1)^{n-i}\mathbf{d}_{n,i}X^{q^i}\Big)$$

$$= \begin{cases} 0 & k>q^n,\\ X^{q^{n+1}}+\sum\limits_{i=0}^{n-1}(-1)^{n-i}\Big(\mathcal{P}^{q^n}(\mathbf{d}_{n,i})X^{q^i}+\mathcal{P}^{q^n-q^i}(\mathbf{d}_{n,i})\mathcal{P}^{q^i}(X^{q^i})\Big) & k=q^n,\\ \sum\limits_{i=0}^{n-1}(-1)^{n-i}\Big(\mathcal{P}^k(\mathbf{d}_{n,i})X^{q^i}+\mathcal{P}^{k-q^i}(\mathbf{d}_{n,i})\mathcal{P}^{q^i}(X^{q^i})\Big) & 1\le k<q^n,\\ \varphi(X) & k=0,\end{cases}$$

$$= \begin{cases} 0 & k>q^n,\\ X^{q^{n+1}}+\sum\limits_{i=0}^{n-1}(-1)^{n-i}\mathcal{P}^{q^n-q^i}(\mathbf{d}_{n,i})\mathcal{P}^{q^i}(X^{q^i}) & k=q^n,\\ \sum\limits_{i=0}^{n-1}(-1)^{n-i}\Big(\mathcal{P}^k(\mathbf{d}_{n,i})X^{q^i}+\mathcal{P}^{k-q^i}(\mathbf{d}_{n,i})X^{q^{i+1}}\Big) & 1\le k<q^n,\\ \varphi(X) & k=0,\end{cases}$$

$$= \begin{cases} 0 & k>q^n,\\ X^{q^{n+1}}+\sum\limits_{i=0}^{n-1}(-1)^{n-i}(\mathbf{d}_{n,i})^qX^{q^{i+1}} & k=q^n,\\ \sum\limits_{i=0}^{n}(-1)^{n-i}\Big(\mathcal{P}^k(\mathbf{d}_{n,i})+\mathcal{P}^{k-q^{i-1}}(\mathbf{d}_{n,i-1})\Big)X^{q^i} & 1\le k<q^n,\\ \varphi(X) & k=0.\end{cases}$$

Combining the two formulae for $\mathcal{P}^k(\varphi(X))$ and comparing the highest coefficients gives for $1\le k<q^n$

$$\mathcal{P}^k(\varphi(X)) = -\varphi(X)\mathcal{P}^{k-q^{n-1}}(\mathbf{d}_{n,n-1}).$$

For degree reasons we find that $\mathcal{P}^k(\mathbf{d}_{n,i}) = 0$ if $k \geq q^n$; hence we have

$$\mathcal{P}^k(\mathbf{d}_{n,i}) = \begin{cases} 0 & \text{if } k \geq q^n, \\ -\mathcal{P}^{k-q^{n-1}}(\mathbf{d}_{n,n-1})\mathbf{d}_{n,i} + \mathcal{P}^{k-q^{i-1}}(\mathbf{d}_{n,i-1}) & \text{if } 1 \leq k < q^n, \\ \mathbf{d}_{n,i} & \text{if } k = 0. \end{cases}$$

Analogously, we derive a formulae for $\mathcal{P}^{\Delta_k}(\mathbf{d}_{n,i})$. First, we use the product form for $\varphi(X)$ and get

$$\begin{aligned} \mathcal{P}^{\Delta_k}\left(\varphi(X)\right) &= \sum_{v_1 \in V^*} \mathcal{P}^{\Delta_k}(X + v_1) \prod_{v \in V^* \setminus \{v_1\}} (X + v) \\ &= \sum_{v_1 \in V^*} (X^{q^k} + v_1^{q^k}) \prod_{v \in V^* \setminus \{v_1\}} (X + v) \\ &= \varphi(X) \sum_{v_1 \in V^*} (X + v_1)^{q^k - 1}. \end{aligned}$$

Here we made use of the derivation property of the $\mathcal{P}^{\Delta_k}$'s and how they act on linear forms. Again $\varphi(X) \mid \mathcal{P}^{\Delta_k}\left(\varphi(X)\right)$. Using the sum formula for $\varphi(X)$ we get

$$\mathcal{P}^{\Delta_k}\left(\varphi(X)\right) = \mathbf{d}_{n,0}X^{q^k} + \sum_{i=0}^{n-1}(-1)^{n-i}\left(\mathcal{P}^{\Delta_k}\left(\mathbf{d}_{n,i}\right)X^{q^i}\right).$$

Once again we compare the coefficients in the two formulae for $\mathcal{P}^k(\varphi(X))$ and use in addition the defining recursion formulae for the $\mathcal{P}^{\Delta_k}$'s, to obtain

$$\mathcal{P}^{\Delta_k}(\varphi(X)) = \begin{cases} 0 & \text{for } k < n, \\ (-1)^n \mathbf{d}_{n,0}\varphi(X) & \text{for } k = n, \\ \mathcal{P}^{q^{k-1}}\mathcal{P}^{\Delta_{k-1}}\left(\varphi(X)\right) & \text{for } k > n, \end{cases}$$

where we made use of the formulae for $\mathcal{P}^k(\varphi(X))$ already established. Recursively, we get

$$\mathcal{P}^{\Delta_k}(\mathbf{d}_{n,i}) = \begin{cases} 0 & \text{if } 0 \leq k < n \text{ and } k \neq i, \\ (-1)^{i+1}\mathbf{d}_{n,0} & \text{if } 0 \leq k < n \text{ and } k = i, \\ (-1)^n \mathbf{d}_{n,0}\mathbf{d}_{n,i} & \text{if } k = n, \\ \mathcal{P}^{q^{k-1}}\mathcal{P}^{\Delta_{k-1}}\left(\mathbf{d}_{n,i}\right) & \text{if } k > n. \end{cases}$$

We collect these results in a proposition.

PROPOSITION 8.6.1: *With the above notation we have*

$$(1) \qquad \mathcal{P}^k(\mathbf{d}_{n,i}) = \begin{cases} 0 & \text{if } k \geq q^n, \\ -\mathcal{P}^{k-q^{n-1}}(\mathbf{d}_{n,n-1})\mathbf{d}_{n,i} + \mathcal{P}^{k-q^{i-1}}(\mathbf{d}_{n,i-1}) & \text{if } 1 \leq k < q^n, \\ \mathbf{d}_{n,i} & \text{if } k = 0, \end{cases}$$

$$(2) \quad \mathcal{P}^{\Delta_k}(\mathbf{d}_{n,i}) = \begin{cases} 0 & \text{if } 0 \leq k < n \text{ and } k \neq i, \\ (-1)^{i+1}\mathbf{d}_{n,0} & \text{if } 0 \leq k < n \text{ and } k = i, \\ (-1)^n \mathbf{d}_{n,0}\mathbf{d}_{n,i} & \text{if } k = n, \\ \mathcal{P}^{q^{k-1}}\mathcal{P}^{\Delta_{k-1}}(\mathbf{d}_{n,i}) & \text{if } k > n. \quad \square \end{cases}$$

Finally, let us append for handy reference the following useful formulae that explain how the Dickson polynomials are connected with each other via Steenrod operations.

COROLLARY 8.6.2: *With the above notation we have*

$$\mathcal{P}^{q^k}(\mathbf{d}_{n,i}) = \begin{cases} -\mathbf{d}_{n,i-1} & \text{for } k = i-1 \geq 0, \\ -\mathbf{d}_{n,i}\mathbf{d}_{n,n-1} & \text{for } k = n-1 \geq 0, \\ 0 & \text{otherwise,} \end{cases}$$

for all $i = 0, \ldots, n-1$.

PROOF: This formula can be read off directly from part (1) of the preceding proposition. □

Chapter 9
Invariant Ideals

CLASSICAL commutative algebra has its roots in problems arising in algebraic geometry which led to the development of ideal theory. Results from ideal theory have advanced algebra proper, as well as influenced other areas of mathematics. Through what has become known as the algebra-geometry dictionary ideal theory has enriched the older area of algebraic geometry. In short, ideal theory in its current form is significant to many important modern developments. It should not come as a surprise that in an algebra carrying the extra structure of an unstable $\mathscr{P}^*$-algebra those ideals that are closed under the action of the Steenrod algebra, the so-called **$\mathscr{P}^*$-invariant ideals**, play a central role in the study of such an algebra. This chapter is devoted to the investigation of these $\mathscr{P}^*$-invariant ideals.

We denote by Proj(H) the homogeneous prime ideal spectrum of a Noetherian commutative graded algebra H over a Galois field $\mathbb{F}_q$. We construct an idempotent self map, $\mathcal{J}_\infty$, of Proj(H), and the $\mathscr{P}^*$-invariant prime ideals are just the fixed points of this operation. A most remarkable property of an unstable algebra over $\mathscr{P}^*$ is that this is a finite set, i.e., there are only finitely many $\mathscr{P}^*$-invariant prime ideals in an unstable Noetherian algebra over $\mathscr{P}^*$.

The imbedding theorem (Theorem 8.4.3) tells us we can find a fractal of the Dickson algebra in H, and what is more remarkable is that a power of the top Dickson class of this Dickson algebra will be contained in any nonzero $\mathscr{P}^*$-invariant ideal of positive height. These and other properties of $\mathscr{P}^*$-invariant ideals are developed in this chapter. They find applications to invariant theory in Sections 9.3 and 9.4.

9.1 Invariant Ideals and the $\mathcal{J}$-Construction

We start this section with a proper definition of $\mathscr{P}^*$-invariant ideals. Then we introduce a construction, $\mathcal{J}_\infty$, that turns arbitrary ideals into invariant ones, and we study its basic properties. This construction is one of the central technical tools in the investigation of invariant ideals. In this section we use it to prove the $\mathscr{P}^*$-Lasker-Noether theorem, i.e., the existence of a $\mathscr{P}^*$-primary decomposition for a $\mathscr{P}^*$-invariant ideal in a Noetherian unstable algebra over the Steenrod algebra. We are indebted to P. S. Landweber for correspondence that led to this much shorter and intrinsic proof then our original one [300].

DEFINITION: *Let H be an unstable algebra over the Steenrod algebra. An ideal $I \subseteq H$ is called $\mathscr{P}^*$-**invariant** (or just **invariant**) if it is closed under the action of the Steenrod algebra, i.e., $P(f) \in I \;\; \forall f \in I, P \in \mathscr{P}^*$.*

For example, in $\mathbb{F}_q[V]$ an ideal generated by linear forms is $\mathscr{P}^*$-invariant. In addition, for any unstable algebra H over the Steenrod algebra the augmentation ideal $\overline{H}$ is $\mathscr{P}^*$-invariant. But are there any other $\mathscr{P}^*$-invariant ideals, and if so, how can we study them? The following construction will allow us to convert any ideal into a $\mathscr{P}^*$-invariant one.

Let $I \subseteq H$ be an arbitrary (not necessarily invariant) ideal in the unstable algebra H. We set

$$\mathcal{J}(I) = \left\{ x \in I \mid \mathcal{P}^i(x) \in I, \;\; \forall i \geq 0 \right\}.$$

Recall from Section 8.1 the giant Steenrod operation defined by

$$\mathbf{P}(\xi) : H \longrightarrow H[\xi], \quad \mathbf{P}(\xi)(f) = \sum_{i=0}^{\deg(f)} \mathcal{P}^i(f)\xi^i,$$

where ξ is a formal variable of degree $-(q-1)$. For an ideal $I \subseteq H$ we denote by $I[\xi] \subseteq H[\xi]$ the extended ideal in the polynomial ring $H[\xi]$ over H. The following description of $\mathcal{J}(I)$ is taken from [223].

LEMMA 9.1.1 (P. S. Landweber): *With the preceding notation we have that*

$$\mathcal{J}(I) = \mathbf{P}^{-1}(\xi)(I[\xi]).$$

PROOF: Let $f \in \mathcal{J}(I)$. Then $\mathcal{P}^i(f) \in I$ for all $i \in \mathbb{N}_0$. Hence

$$\mathbf{P}(\xi)(f) \in I[\xi],$$

and therefore, $f \in \mathbf{P}^{-1}(\xi)(I[\xi])$. Conversely, suppose that $f \in \mathbf{P}^{-1}(\xi)(I[\xi])$. Then $\mathbf{P}(\xi)(f) \in I[\xi]$. In other words, all coefficients $\mathcal{P}^i(f)$ lie in I for $i = 0, \ldots, \deg(f)$. So $f \in \mathcal{J}(I)$ as desired. □

We proceed with an investigation of the properties of the construction $\mathcal{J}$. The following lemma shows (see also [371] Lemma 2.1 or [372] Lemma

11.2.1) that $\mathcal{J}(I)$ is again an ideal, and moreover, if I was prime, then so is $\mathcal{J}(I)$, [220], and likewise $\mathcal{J}(I)$ is primary if I was, [223]. See also [339] where the same lemma and proof occur in the context of differential ideals, and [427] Theorem 2.4.2 in the context of Hasse-Schmidt differentials.

LEMMA 9.1.2 (S. P. Lam, P. S. Landweber): *Let H be an unstable algebra over the Steenrod algebra and $I \subset H$ an ideal. Then*
(1) *$\mathcal{J}(I) \subset H$ is an ideal,*
(2) *if I is prime so is $\mathcal{J}(I)$, and*
(3) *if I is primary then so is $\mathcal{J}(I)$.*

PROOF: The proof of (1) is immediate from the description of $\mathcal{J}(I)$ given in Lemma 9.1.1. The second and third statements follow by noting that the extended ideal $I[\xi] \subseteq H[\xi]$ is prime whenever $I \subseteq H$ is prime, and primary whenever I is primary. (Do Exercise 7 of Chapter 4 in [20], or see Lemma 2 in [339], or Hilfsatz 5 in [161]). Therefore the contraction $\mathcal{J}(I) = \mathbf{P}^{-1}(\xi)(I[\xi]) \subseteq H$ is also a prime ideal when I is prime and primary when I is primary. □

Next we show that the $\mathcal{J}$-construction commutes with taking radicals and intersections, properties that will turn out to be crucial for a number of applications; see Lemma 1.3 in [300].

LEMMA 9.1.3: *Let H be an unstable algebra over the Steenrod algebra and $I, I', I'' \subset H$ ideals. Then*
(1) *$\mathcal{J}(I' \cap I'') = \mathcal{J}(I') \cap \mathcal{J}(I'')$, and*
(2) *$\mathcal{J}(\sqrt{I}) = \sqrt{\mathcal{J}(I)}$.*

PROOF: The proof of (1) is immediate. By definition, $f \in \mathcal{J}(I' \cap I'')$ is equivalent to $f \in I' \cap I''$ and $\mathcal{P}^k(f) \in I' \cap I''$ for all $k \geq 0$, which is in turn equivalent to $f \in \mathcal{J}(I') \cap \mathcal{J}(I'')$. As for (2), suppose that $f \in \sqrt{\mathcal{J}(I)}$. Then there is an integer m such that $f^m \in \mathcal{J}(I)$. By definition of $\mathcal{J}$ we have

$$\mathcal{P}^k(f^m) \in I, \qquad \forall\, k.$$

In particular, $f^m \in I$, and so $f \in \sqrt{I}$. To show that $f \in \mathcal{J}(\sqrt{I})$, choose $s \in \mathbb{N}$ with $q^s \geq m$, so $f^{q^s} \in \mathcal{J}(I)$, and therefore $\mathcal{P}^k(f^{q^s}) \in I$ for all $k \geq 0$. By the Cartan formulae,

$$(\div) \qquad \mathcal{P}^i(f^{q^s}) = \begin{cases} \mathcal{P}^k(f)^{q^s} & \text{for } i = kq^s, \\ 0 & \text{otherwise.} \end{cases}$$

Therefore, $\mathcal{P}^k(f)^{q^s} \in I$ for all $k \in \mathbb{N}_0$, so $\mathcal{P}^k(f) \in \sqrt{I}$ for all $k \in \mathbb{N}_0$, and hence $f \in \mathcal{J}(\sqrt{I})$ by the definition of $\mathcal{J}$.

To establish the reverse inclusion suppose $f \in \mathcal{J}(\sqrt{I})$. By the definition of $\mathcal{J}$ we have $\mathcal{P}^k(f) \in \sqrt{I}\ k \in \mathbb{N}_0$. Therefore, there are integers $k_0, k_1, \ldots$ such that

$$(\mathcal{P}^i(f))^{k_i} \in I, \quad \forall\, i \in \mathbb{N}_0.$$

Recall that $\mathcal{P}^i(f) = 0$ for $i > \deg(f)$ by unstability. Hence if we choose $s \in \mathbb{N}$ such that $q^s \geq \max\{k_0, k_1, \ldots, k_{\deg(f)}\}$, then

$$(\mathcal{P}^k(f))^{q^s} \in I \quad \forall\, k \in \mathbb{N}_0.$$

Therefore, by equation (⁖) we obtain $\mathcal{P}^i(f^{q^s}) \in I$ for all $i \in \mathbb{N}_0$, and hence $f^{q^s} \in \mathcal{J}(I)$, so $f \in \sqrt{\mathcal{J}(I)}$, as was to be shown. □

We note the following immediate consequence of these lemmas.

Proposition 9.1.4: *Let H be an unstable Noetherian algebra over $\mathcal{P}^*$. Let $I = \mathfrak{q}_1 \cap \cdots \mathfrak{q}_k \subseteq H$ be a primary decomposition of the ideal I with associated prime ideals $\mathfrak{p}_1, \ldots, \mathfrak{p}_k$. Then*

$$\mathcal{J}(I) = \mathcal{J}(\mathfrak{q}_1) \cap \cdots \cap \mathcal{J}(\mathfrak{q}_k)$$

is a primary decomposition of the ideal $\mathcal{J}(I)$ with associated prime ideals $\mathcal{J}(\mathfrak{p}_1), \ldots, \mathcal{J}(\mathfrak{p}_k)$. □

Note that $\mathcal{J}(I) = I$ if and only if I is invariant. If I is not invariant, then $\mathcal{J}(I)$ is closer to being invariant than I was, but may fail to be invariant, since $\mathcal{P}^i(x) \in I$ need not imply $\mathcal{P}^j\mathcal{P}^i(x) \in I$, as the following example taken from [289] shows.

Example 1: Consider the Weyl group of the Lie group $\mathbf{F}_4$; cf. Example 2 in Section 7.1, and its ring of invariants in the defining representation over a field of characteristic 3,

$$\mathbb{F}_q[x_1, x_2, x_3, x_4]^{W(\mathbf{F}_4)} = \mathbb{F}_q[p_1, \overline{p}_2, \overline{p}_5, \overline{p}_9, \overline{p}_{12}]/(r_{15}),$$

where p_i, $i = 1, \ldots, 4$, are the elementary symmetric functions in x_1^2, x_2^2, x_3^2, and x_4^2, and

$$\begin{aligned}
\overline{p}_2 &= p_2 - p_1^2,\\
\overline{p}_5 &= p_4p_1 + p_3\overline{p}_2,\\
\overline{p}_9 &= p_3^3 - p_4p_3p_1^2 + p_3^2\overline{p}_2p_1 - p_4\overline{p}_2p_1^3,\\
\overline{p}_{12} &= p_4^3 + p_4^2p_2^2 + p_4\overline{p}_2^4,\\
r_{15} &= \overline{p}_5^3 + \overline{p}_5^2\overline{p}_2^2p_1 - \overline{p}_{12}p_1^3 - \overline{p}_9\overline{p}_2^3.
\end{aligned}$$

This calculation is due to H. Toda [426].

We claim that the ideal generated by p_1 and $\mathcal{P}^1(p_1)$ is prime of height 2. To see this, pass to the quotient algebra,

$$\begin{aligned}
&\mathbb{F}_q[x_1, x_2, x_3, x_4]^{W(\mathbf{F}_4)}/(p_1, \mathcal{P}^1(p_1))\\
&\quad = \mathbb{F}_q[p_1, \overline{p}_2, \overline{p}_5, \overline{p}_9, \overline{p}_{12}]/(r_{15}, p_1, \mathcal{P}^1(p_1))\\
&\quad = \mathbb{F}_q[p_2, p_3p_2, p_3^3, p_4^3 + p_4^2p_2^2 + p_4p_2^4]\Big/\Big((p_3p_2)^3 - p_3^3p_2^3, p_1, \mathcal{P}^1(p_1)\Big).
\end{aligned}$$

Note that

$$\mathcal{P}^1(p_1) = x_1^4 + x_2^4 + x_3^4 + x_4^4 = p_1^2 - 2p_2 = \overline{p}_2 - p_1^2,$$

hence $(p_1, \mathcal{P}^1(p_1)) = (p_1, \overline{p}_2)$ and

$$\mathbb{F}_q[x_1, x_2, x_3, x_4]^{W(\mathrm{F}_4)} \big/ (p_1, \mathcal{P}^1(p_1)) = \mathbb{F}_q[p_3^3, p_4^3],$$

which is an integral domain of Krull dimension 2.

Next we claim that $(p_1, \mathcal{P}^1(p_1)) \supsetneqq \mathcal{J}((p_1, \mathcal{P}^1(p_1))) = (p_1) \supsetneqq \mathcal{J}((p_1)) = (0)$. Certainly,

$$p_1 \in \mathcal{J}\Big((p_1, \mathcal{P}^1(p_1))\Big),$$

because $\deg(p_1) = 2$ and $\mathcal{P}^1(p_1) \in (p_1, \mathcal{P}^1(p_1))$. On the other hand, consider $\overline{p}_9 \mathcal{P}^1(p_1) \in (p_1, \mathcal{P}^1(p_1))$. Note that $\overline{p}_9$ is not in $(p_1, \mathcal{P}^1(p_1))$. Since

$$\mathcal{P}^i(\mathcal{P}^1(p_1)) = \begin{cases} p_1^3 & \text{for } i = 1, \\ 0 & \text{for } i = 2, \\ -(x_1^{10} + x_2^{10} + x_3^{10} + x_4^{10}) & \text{for } i = 3, \\ \mathcal{P}^1(p_1)^3 & \text{for } i = 4, \\ 0 & \text{otherwise,} \end{cases}$$

it follows after a short computation with these formulae that

$$\mathcal{P}^3(\overline{p}_9 \mathcal{P}^1(p_1)) = \mathcal{P}^3(\overline{p}_9)\mathcal{P}^1(p_1) + \mathcal{P}^2(\overline{p}_9)p_1^3 + \overline{p}_9 \mathcal{P}^3 \mathcal{P}^1(p_1).$$

The first two summands are in $(p_1, \mathcal{P}^1(p_1))$. However, none of the factors of the last summand are in $(p_1, \mathcal{P}^1(p_1))$. Hence, as this ideal is prime, $\overline{p}_9 \mathcal{P}^3 \mathcal{P}^1(p_1) \notin (p_1, \mathcal{P}^1(p_1))$, and therefore

$$\mathcal{P}^3(\overline{p}_9 \mathcal{P}^1(p_1)) \notin (p_1, \mathcal{P}^1(p_1)).$$

A similar calculation shows that the principal ideal (p_1) is *not* $\mathscr{P}^*$-invariant, and in fact, $\mathcal{J}(p_1) = (0)$.

So, in order to get a $\mathscr{P}^*$-invariant ideal from an ideal I it is not enough to form $\mathcal{J}(I)$. The construction must be iterated and we will need the analogue of Lemmas 9.1.2 and 9.1.3 for these iterations. We define inductively

$$\mathcal{J}_n(I) = \mathcal{J}(\mathcal{J}_{n-1}(I)), \quad n \in \mathbb{N},$$

starting from $\mathcal{J}_0(I) = I$, giving a descending chain of ideals

$$I = \mathcal{J}_0(I) \supseteq \mathcal{J}_1(I) \supseteq \cdots.$$

The intersection of the elements in this chain (which might be zero) is then

$$\mathcal{J}_\infty(I) = \bigcap_{i \geq 0} \mathcal{J}_i(I).$$

That $\mathcal{J}_\infty(I)$ is indeed the invariant ideal desired is the content of the next result (see also Lemma 11.2.4 in [372] or Lemma 2.4 in [371]).

LEMMA 9.1.5: *Let H be an unstable algebra over the Steenrod algebra and $I \subset H$ an ideal. Then*

(1) *$\mathcal{I}_\infty(I)$ is the maximal ideal contained in I that is invariant under the action of the Steenrod algebra, and*

(2) *if I is a prime ideal, so is $\mathcal{I}_\infty(I)$.*

PROOF: Let $f \in \mathcal{I}_\infty(I)$. Then $f \in \mathcal{I}_i(I)$ for every $i \in \mathbb{N}_0$ by definition of $\mathcal{I}_\infty$. Hence, by definition of $\mathcal{I}$ we have

$$\mathcal{P}^k(f) \in \mathcal{I}_i(I) \quad \forall\, i \geq 0,$$

which proves that $\mathcal{I}_\infty(I)$ is indeed invariant. The maximality of $\mathcal{I}_\infty(I)$ is clear by construction, so (1) is established. The second statement is true by iterative use of Lemma 9.1.2 and the fact that the intersection of a descending chain of prime ideals is also prime. □

Note that $\mathcal{I}_\infty(I) = I$ if and only if I is $\mathcal{P}^*$-invariant. Lemma 9.1.3 generalizes to the following result (see Lemma 1.4 in [300] and Theorem 5 in [223]).

LEMMA 9.1.6: *Let H be an unstable Noetherian algebra over the Steenrod algebra and $I, I', I'' \subset H$ ideals. Then*

(1) *$\mathcal{I}_\infty(I' \cap I'') = \mathcal{I}_\infty(I') \cap \mathcal{I}_\infty(I'')$, and*

(2) *$\mathcal{I}_\infty(\sqrt{I}) = \sqrt{\mathcal{I}_\infty(I)}$.*

PROOF: By iterating Lemma 9.1.3 we obtain that

$$\mathcal{I}_n(I' \cap I'') = \mathcal{I}_n(I) \cap \mathcal{I}_n(I')$$

for every $n \in \mathbb{N}_0$. Let $f \in \mathcal{I}_\infty(I' \cap I'')$. This is equivalent to $f \in \mathcal{I}_n(I' \cap I'') = \mathcal{I}_n(I) \cap \mathcal{I}_n(I')$ for all $n \in \mathbb{N}_0$, which is in turn equivalent to

$$f \in \mathcal{I}_\infty(I) \cap \mathcal{I}_\infty(I'').$$

We come to (2). As in the proof of (1), iterating Lemma 9.1.3 gives that

$$\mathcal{I}_n(\sqrt{I}) = \sqrt{\mathcal{I}_n(I)} \quad \forall\, n \in \mathbb{N}_0.$$

Hence, we have for all $n \in \mathbb{N}_0$ that

$$\sqrt{\mathcal{I}_\infty(I)} \subseteq \sqrt{\mathcal{I}_n(I)} = \mathcal{I}_n(\sqrt{I}),$$

and therefore,

$$\sqrt{\mathcal{I}_\infty(I)} \subseteq \bigcap_{n \in \mathbb{N}_0} \mathcal{I}_n(\sqrt{I}) = \mathcal{I}_\infty(\sqrt{I}).$$

To show the reverse inclusion, we note that there exists an $s \in \mathbb{N}$ such that

$$\left(\sqrt{I}\right)^{q^s} \subseteq I,$$

because H is Noetherian. Hence, if $f \in \mathcal{I}_\infty(\sqrt{I}) \subseteq \sqrt{I}$, then $f^{q^s} \in I$. By the formulae (⁘) in the proof of Lemma 9.1.3 we have that

$$\mathcal{P}^i(f^{q^s}) = \begin{cases} \mathcal{P}^k(f)^{q^s} & \text{if } i = kq^s, \\ 0 & \text{otherwise.} \end{cases}$$

This implies that $f^{q^s} \in \mathcal{J}_\infty(I)$. Therefore, $f \in \sqrt{\mathcal{J}_\infty(I)}$ as desired. □

From this we can quickly deduce the following result of [300] (see Lemma 2.4 for a direct proof).

COROLLARY 9.1.7: *If H is an unstable algebra over $\mathbb{F}_q$ and $I \subset H$ is a $\mathcal{P}^*$-invariant ideal, then $\sqrt{I}$ is also $\mathcal{P}^*$-invariant.*

PROOF: By the preceding result we have $\mathcal{J}_\infty\left(\sqrt{I}\right) = \sqrt{\mathcal{J}_\infty(I)} = \sqrt{I}$. □

Equipped with the preceding result we are able to prove the essential fact that $\mathcal{J}_\infty$ preserves primary ideals (see Theorem 3.3 in [289]).

PROPOSITION 9.1.8: *Let H be an unstable Noetherian algebra over $\mathcal{P}^*$. Let $\mathfrak{q} \subseteq H$ be a primary ideal with radical $\sqrt{\mathfrak{q}} = \mathfrak{p}$. Then $\mathcal{J}_\infty(\mathfrak{q})$ is a primary ideal with radical $\mathcal{J}_\infty(\mathfrak{p})$.*

PROOF: Suppose that $f', f'' \in H$ satisfy $f'f'' \in \mathcal{J}_\infty(\mathfrak{q})$ and $f' \notin \mathcal{J}_\infty(\mathfrak{q})$. We need to show that $f'' \in \sqrt{\mathcal{J}_\infty(\mathfrak{q})}$. Let $\mathfrak{p} = \sqrt{\mathfrak{q}}$ be the radical of $\mathfrak{q}$, and consider the sequence of prime ideals

$$\mathfrak{p} \supseteq \mathcal{J}(\mathfrak{p}) \supseteq \cdots \supseteq \mathcal{J}_k(\mathfrak{p}) \supseteq \cdots.$$

This sequence must stabilize since H satisfies the descending chain condition on prime ideals. Say $\mathcal{J}_n(\mathfrak{p}) = \mathcal{J}_{n+1}(\mathfrak{p}) = \cdots = \mathcal{J}_\infty(\mathfrak{p})$. Since $\mathcal{J}_k$, $k \in \mathbb{N}_0 \cup \{\infty\}$, preserves radicals we conclude

$$\sqrt{\mathcal{J}_n(\mathfrak{q})} = \sqrt{\mathcal{J}_{n+1}(\mathfrak{q})} = \cdots = \sqrt{\mathcal{J}_\infty(\mathfrak{q})}.$$

As $f' \notin \mathcal{J}_\infty(\mathfrak{q})$ there is an integer $m \in \mathbb{N}_0$ such that $f' \notin \mathcal{J}_m(\mathfrak{q})$. Let $k \geq \max\{m, n\}$. We then have

$$f'f'' \in \mathcal{J}_\infty(\mathfrak{q}) \subseteq \mathcal{J}_k(\mathfrak{q}) \qquad f' \notin \mathcal{J}_k(\mathfrak{q}).$$

Since $\mathcal{J}_k(\mathfrak{q})$ is a primary ideal it follows that $f'' \in \sqrt{\mathcal{J}_k(\mathfrak{q})} = \sqrt{\mathcal{J}_\infty(\mathfrak{q})}$. □

The $\mathcal{P}^*$-invariant version of the Lasker–Noether theorem is a direct corollary of the preceding work. We refer to Theorem 3.5 in [300] for the original proof.

THEOREM 9.1.9 ($\mathcal{P}^*$-invariant Lasker–Noether Theorem): *Let $I \subset H$ be a $\mathcal{P}^*$-invariant ideal in an unstable Noetherian algebra. Then I has a $\mathcal{P}^*$-invariant primary decomposition, i.e., there exist $\mathcal{P}^*$-invariant primary ideals $\mathfrak{q}_1, \ldots, \mathfrak{q}_k \subset H$ such that*

$$I = \mathfrak{q}_1 \cap \cdots \cap \mathfrak{q}_k$$

is a minimal irredundant primary decomposition of I. Moreover, the associated prime ideals $\mathfrak{p}_i = \sqrt{\mathfrak{q}_i}$ are also $\mathscr{P}^*$-invariant.

PROOF: Let $I \subseteq H$ be a $\mathscr{P}^*$-invariant ideal. Since H is Noetherian I has a minimal irredundant primary decomposition, say

$$I = \mathfrak{q}_1 \cap \cdots \cap \mathfrak{q}_k$$

with associated prime ideals $\mathfrak{p}_1, \ldots, \mathfrak{p}_k$. Since I is invariant, we have that $\mathcal{I}_\infty(I) = I$. Since $\mathcal{I}_\infty$ commutes with intersections (see Lemma 9.1.6), we find that

$$I = \mathcal{I}_\infty(\mathfrak{q}_1) \cap \cdots \cap \mathcal{I}_\infty(\mathfrak{q}_k).$$

Since $\mathcal{I}_\infty(\mathfrak{q}_i)$ is $\mathcal{I}_\infty(\mathfrak{p}_i)$-primary by Proposition 9.1.8 this gives a $\mathscr{P}^*$-invariant primary decomposition of I. We make it minimal by combining, and irredundant by omitting superfluous ideals. □

REMARK: The $\mathcal{I}$-construction has a direct generalization to the context of submodules of an $H \odot \mathscr{P}^*$-module with equally nice properties; see Section 1 of [293]. In particular, the $\mathscr{P}^*$-invariant version of the Lasker-Noether theorem for unstable modules is valid (loc.cit.).

9.2 The Invariant Prime Ideal Spectrum

Given the basic tool $\mathcal{I}_\infty$ to produce $\mathscr{P}^*$-invariant ideals from arbitrary ideals, we investigate how the $\mathscr{P}^*$-invariant ideals are situated against the background of all ideals in an unstable Noetherian algebra. As in the classical approaches, we start with the study of the spectrum of $\mathscr{P}^*$-invariant homogeneous prime ideals in an unstable algebra H, which we will denote by $\mathrm{Proj}_{\mathscr{P}^*}(H)$.

Logically enough, we first investigate $\mathrm{Proj}_{\mathscr{P}^*}(\mathbb{F}_q[V])$. This is justified for two reasons. On the one hand, the polynomial algebra $\mathbb{F}_q[V]$ is the standard example of an unstable algebra over the Steenrod algebra; we even used it to define the Steenrod algebra (see Section 8.1). On the other hand, the polynomial algebra $\mathbb{F}_q[V]$ can be viewed as a kind of universal *largest* object,[1] at least when we restrict our attention to Noetherian integral domains H, since the embedding theorem (Theorem 8.4.2) tells us that we can always find an integral embedding $H \hookrightarrow \mathbb{F}_q[V]$ of algebras over $\mathscr{P}^*$. Finally, we mustn't forget invariant theory: $\mathbb{F}_q[V]$ is nothing but the ring of invariants of the smallest possible group, the trivial group.

The following remarkable theorem of J.-P. Serre settles the structure of $\mathrm{Proj}_{\mathscr{P}^*}(\mathbb{F}_q[V])$ completely (see [343]).

THEOREM 9.2.1 (J.-P. Serre): *Let $\mathfrak{p} \subset \mathbb{F}_q[V]$ be a prime ideal. Then $\mathfrak{p}$ is $\mathscr{P}^*$-invariant if and only if it is generated by the linear forms it contains.*

[1] Or in the language of C. W. Wilkerson (see [454]) an algebraically closed object in $\mathcal{K}_{fg}$.

PROOF: If $\mathfrak{p} \subset \mathbb{F}_q[V]$ is generated by linear forms, say $\mathfrak{p} = (\ell_1, \ldots, \ell_k)$, then we obtain by unstability

$$\mathcal{P}^i(\ell_j) = \begin{cases} \ell_j & \text{for } i = 0, \\ \ell_j^q & \text{for } i = 1, \\ 0 & \text{otherwise.} \end{cases}$$

In particular, we have that $\mathcal{P}^i(\ell_j) \in \mathfrak{p}$ for all $i \in \mathbb{N}_0$ and $j = 1, \ldots, k$. Therefore, $\mathfrak{p}$ is $\mathscr{P}^*$-invariant.

To prove the converse we proceed by induction on $\dim(\mathbb{F}_q[V]) = n$. If $n = 1$, then we are looking at the polynomial algebra, $\mathbb{F}_q[x]$, over $\mathbb{F}_q$ in one variable. This is a principal ideal domain and has, in particular, only one nontrivial homogeneous prime ideal at all, namely (x). This happens to be $\mathscr{P}^*$-invariant and also generated by a linear form.

So suppose $n > 1$. We claim that every nontrivial $\mathfrak{p}$ must contain at least one nonzero linear form ℓ. Granted this, we can complete the inductive step as follows: The ideal $(\ell) \subset \mathbb{F}_q[V]$ is itself a $\mathscr{P}^*$-invariant prime ideal. So, if we pass to the quotient algebra, we obtain

$$\mathrm{pr} : \mathbb{F}_q[V] \longrightarrow \mathbb{F}_q[V]/(\ell) = \mathbb{F}_q[W],$$

and $\mathbb{F}_q[W]$ is a polynomial ring of Krull dimension $n-1$. Since $(\ell) \subseteq \mathfrak{p}$ are $\mathscr{P}^*$-invariant ideals, the image, $\mathrm{pr}(\mathfrak{p}) = \overline{\mathfrak{p}} \subset \mathbb{F}_q[W]$, is again a $\mathscr{P}^*$-invariant prime ideal. By induction this is generated by linear forms, or it is zero. If it is zero, $\mathfrak{p} = (\ell)$, and we are done. Otherwise, lifting a basis for the linear forms in $\overline{\mathfrak{p}}$ to $\mathbb{F}_q[V]$, and adjoining ℓ to them, gives the desired set of linear ideal generators for $\mathfrak{p} \subset \mathbb{F}_q[V]$.

So, finally, we are left to show that every nontrivial $\mathscr{P}^*$-invariant prime ideal $\mathfrak{p} \subset \mathbb{F}_q[V]$ contains a nonzero linear form. Assume the contrary, i.e., that $\mathfrak{p}$ is a nontrivial $\mathscr{P}^*$-invariant prime ideal that does not contain any linear form. Then the Euler polynomial $\mathbf{L}$ (see Section 4.4) which is the pre-Euler class of the configuration of all hyperplanes in V, does not belong to $\mathfrak{p}$, since it is a product of linear forms, and $\mathfrak{p}$ is prime. Let $z_1, \ldots, z_n$, be a basis for the dual space V^*. Then the $\mathscr{P}^*$-generalized Jacobian matrix satisfies

$$0 \neq \det\left(\mathcal{P}^{\Delta_i}(z_j)\right)_{\substack{i=0,\ldots,n-1\\ j=1,\ldots,n}} = \det\left(z_j^{q^i}\right)_{\substack{i=0,\ldots,n-1\\ j=1,\ldots,n}} = \mathbf{L} \in \mathbb{F}_q[V]/\mathfrak{p}.$$

Therefore, see Lemma A.4.1, the linear forms $z_1, \ldots, z_n \in \mathbb{F}_q[V]/\mathfrak{p}$ are algebraically independent, so this quotient has Krull dimension at least n. This means that $\mathfrak{p}$ must have height zero, and hence is the trivial ideal (0), contrary to assumption. □

Recall from the preceding chapter that if $V = \mathbb{F}_q^n$ and $W \leq V$ is a linear subspace, then we denote by $\mathfrak{p}_W \subset \mathbb{F}_q[V]$ the ideal generated by all the linear forms $\lambda : V \longrightarrow \mathbb{F}_q$ that vanish on W. Then $\mathfrak{p}_W$ is a $\mathscr{P}^*$-invariant prime ideal in $\mathbb{F}_q[V]$, and every $\mathscr{P}^*$-invariant ideal in $\mathbb{F}_q[V]$ is of this form by Theorem 9.2.1.

Let $I \subset \mathbb{F}_q[V]$ be a $\mathscr{P}^*$-invariant ideal. By the Lasker-Noether theorem for $\mathscr{P}^*$-invariant ideals (Theorem 9.1.9) every associated prime ideal of I is $\mathscr{P}^*$-invariant. Therefore, we conclude that the affine variety $\mathfrak{X}_I$ defined by I in $\overline{\mathbb{F}}_q$, where $\overline{\mathbb{F}}_q$ is the algebraic closure of $\mathbb{F}_q$, is the union of the affine varieties $\mathfrak{X}_\mathfrak{p}$, where $\mathfrak{p}$ ranges over the minimal prime ideals of I. By Theorem 9.2.1 each such $\mathfrak{X}_\mathfrak{p}$ is a linear subspace defined over $\mathbb{F}_q$, so $\mathfrak{X}_I$ is a union of subspaces of $\overline{\mathbb{F}}_q$ defined over $\mathbb{F}_q$: This is Proposition (1) in §2 of [343].

If $\rho : G \hookrightarrow \mathrm{GL}(n, \mathbb{F}_q)$ is a representation of a finite group over the field $\mathbb{F}_q$, then the transfer homomorphism $\mathrm{Tr}^G : \mathbb{F}_q[V] \longrightarrow \mathbb{F}_q[V]^G$ commutes with the action of the Steenrod algebra, and hence $\mathrm{Im}(\mathrm{Tr}^G) \subseteq \mathbb{F}_q[V]^G$, as well as the extended ideal $\mathrm{Im}(\mathrm{Tr}^G)^e \subseteq \mathbb{F}_q[V]$, are $\mathscr{P}^*$-invariant ideals. So Theorem 9.2.1 gives us a decomposition of $\sqrt{\mathrm{Im}(\mathrm{Tr}^G)^e}$ into an intersection of ideals generated by linear forms. M. Feshbach's transfer theorem, Theorem 2.4.5, does the same. Therefore, after striking redundant ideals from the intersections

$$\bigcap_{W \le V} \mathfrak{p}_W = \sqrt{\mathrm{Im}(\mathrm{Tr}^G)^e} = \bigcap_{|g|=p} I_g,$$

we conclude that each of the ideals in the decomposition given by Theorem 9.2.1 corresponds to an element of order p in G. In other words, the ideals $\mathfrak{p}_W$ occurring on the left-and side of the preceding decomposition are all of the form $\mathfrak{p}_{V^g}$, where $g \in G$ is an element of order p, and V^g is maximal in the poset $\left\{V^g \mid g \in G,\ |g| = p\right\}$ ordered under inclusion. For further applications of Steenrod operations to the transfer see Section 9.3.

We proceed to investigate the $\mathscr{P}^*$-invariant homogeneous prime ideal spectrum in an arbitrary Noetherian unstable algebra over the Steenrod algebra. The following lemma is easy, and its proof is left to the reader (see [289]).

LEMMA 9.2.2: *Let $\varphi : H_1 \to H_2$ be a homomorphism between unstable algebras over the Steenrod algebra. Then extensions and contractions of $\mathscr{P}^*$-invariant ideals are again $\mathscr{P}^*$-invariant.* □

So $\mathscr{P}^*$-invariant ideals behave naturally with respect to extensions and contractions. More remarkable is that $\mathscr{P}^*$-invariant prime ideals behave naturally with respect to the classical[2] relations of W. Krull and I. S. Cohen–A. Seidenberg, relating prime ideals under integral extension. A more detailed discussion may be found in the original sources [289], and [371].

---

[2] These are discussed in Section A.2.

THEOREM 9.2.3: *Let $\varphi : H' \hookrightarrow H''$ be an integral extension of unstable Noetherian algebras, and $\mathfrak{p}' \subset H'$ a $\mathscr{P}^*$-invariant prime ideal. Then:*

(i) There exists a $\mathscr{P}^*$-invariant prime ideal $\mathfrak{p}'' \subset H''$ lying over $\mathfrak{p}'$, i.e., $\mathfrak{p}'' \cap H' = \mathfrak{p}'$.

(ii) Every prime ideal $\mathfrak{p}'' \subset H''$ lying over $\mathfrak{p}' \subset H'$ is $\mathscr{P}^*$-invariant.

Finally, if one of the preceding statements is true, then $\mathfrak{p}$ is a $\mathscr{P}^*$-invariant prime ideal.

PROOF: By the usual lying-over lemma, Theorem A.2.3 in Appendix A, there is a prime ideal $\widetilde{\mathfrak{p}} \subset H''$ lying over $\mathfrak{p}'$. The $\mathscr{P}^*$-invariant ideal $\mathcal{I}_\infty(\widetilde{\mathfrak{p}})$ is prime, contained in $\widetilde{\mathfrak{p}}$, and contains $\mathcal{I}_\infty(\mathfrak{p}') = \mathfrak{p}'$, so lies over $\mathfrak{p}'$. Since there are no proper inclusions between the prime ideals lying over a given prime ideal it follows that $\widetilde{\mathfrak{p}} = \mathcal{I}_\infty(\widetilde{\mathfrak{p}})$ proving the second assertion. The last statement is clear by Lemma 9.2.2. □

J.-P. Serre's characterization of the $\mathscr{P}^*$-invariant prime ideals in $\mathbb{F}_q[V]$ can be used to give a particularly elegant, and complete, description of the $\mathscr{P}^*$-invariant prime ideals of a ring of invariants. To explain how this goes, let $\rho : G \hookrightarrow \mathrm{GL}(n, \mathbb{F}_q)$ be a representation of a finite group over the Galois field $\mathbb{F}_q$, and $\mathfrak{p} \in \mathrm{Proj}_{\mathscr{P}^*}(\mathbb{F}_q[V]^G)$ a $\mathscr{P}^*$-invariant prime ideal. The extension $\mathbb{F}_q[V]^G \subseteq \mathbb{F}_q[V]$ is finite, so the group G acts transitively on the set of prime ideals lying over $\mathfrak{p}$; see, e.g., Theorem A.2.4, or [372] Theorem 5.4.5 or do Exercise 13 in Chapter 5 of [20]. By Theorem 9.2.3 these ideals are $\mathscr{P}^*$-invariant. So by J.-P. Serre's theorem, Theorem 9.2.1, each of these prime ideals is generated by linear forms. If $\widetilde{\mathfrak{p}} = (w_1, \ldots, w_d)$ is one such, then $d = ht(\mathfrak{p})$ and $W = \{x \in V \mid w_1(x) = \cdots = w_d(x) = 0\}$ is a linear subspace of V of codimension d. Introduce the **Grassmann variety** $\mathfrak{G}_d(V)$ consisting of all the codimension-d linear subspaces of V. For $d = \dim_{\mathbb{F}_q}(V) - 1$ this is just the projective space $\mathbb{PF}(n)$ of the vector space $V = \mathbb{F}_q^n$, and for $d = 1$ it is just the collection $\mathcal{A}$ of all hyperplanes in $V = \mathbb{F}_q^n$. The action of G on the points of V via ρ extends to an action on the codimension-d subspaces, making $\mathfrak{G}_d(V)$ into a finite G-set. The preceding construction associates to each $\mathscr{P}^*$-invariant prime ideal $\mathfrak{p}$ in $\mathbb{F}_q[V]^G$ of height $ht(\mathfrak{p}) = d$ an orbit of G acting on $\mathfrak{G}_d$. Conversely, given such an orbit, we can associate to it a $\mathscr{P}^*$-invariant ideal of height d in $\mathbb{F}_q[V]^G$ by choosing for the orbit a representative subspace, $W \leq V$ say, and taking the kernel $\mathfrak{p}_W$ of the homomorphism $\mathbb{F}_q[V]^G \hookrightarrow \mathbb{F}_q[V] \xrightarrow{i^*} \mathbb{F}_q[W]$ induced by the inclusion $i : W \hookrightarrow V$. In this way we set up a bijective correspondence between $\mathscr{P}^*$-invariant prime ideals in $\mathbb{F}_q[V]^G$ of height d and orbits of G on $\mathfrak{G}_d(V)$. Thus we have proven the following theorem:

THEOREM 9.2.4: *Let $\rho : G \hookrightarrow \mathrm{GL}(n, \mathbb{F}_q)$ be a representation of the finite group G over the Galois field $\mathbb{F}_q$. Let $\mathfrak{G}(V) = \cup\, \mathfrak{G}_d(V)$ be the Grassmann variety of all subspaces of $V = \mathbb{F}_q^n$. Then there is a bijective corre-*

spondence

$$\mathrm{Proj}_{\mathscr{P}^*}(\mathbb{F}_q[V]^G) \longleftrightarrow \mathfrak{G}(V)/G$$

obtained by associating to the G-orbit $[W]$ of a codimension-d subspace $W \subseteq V$ the prime ideal $\mathfrak{p}_W = \ker(\varphi_W)$, where φ_W is the composition $\mathbb{F}_q[V]^G \hookrightarrow \mathbb{F}_q[V] \xrightarrow{i^*} \mathbb{F}_q[W]$, and $i : W \hookrightarrow V$ is the inclusion. The number of $\mathscr{P}^*$-invariant prime ideals in $\mathbb{F}_q[V]^G$ of height d is finite, and equal to the number of orbits of G on $\mathfrak{G}_d(V)$. □

COROLLARY 9.2.5: *Suppose that $\rho : G \hookrightarrow \mathrm{GL}(n, \mathbb{F}_q)$ is a representation of the finite group G over the Galois field $\mathbb{F}_q$. If G acts transitively on the hyperplanes of V, then $\mathbb{F}_q[V]^G$ contains a unique $\mathscr{P}^*$-invariant prime ideal of height 1. This ideal is minimal over the principal ideal generated by a power of the Euler class $\mathbf{L}_n$, which in turn is indecomposable in $\mathbb{F}_q[V]^G$.*

PROOF: Since G acts transitively on the set of hyperplanes in V, there is a unique $\mathfrak{p} \subset \mathbb{F}_q[V]^G$ that is a $\mathscr{P}^*$-invariant prime ideal of height 1 by Theorem 9.2.4.

If $\mathcal{A}$ denotes the configuration of all hyperplanes in V, then the Euler class $\mathbf{E}_{\mathcal{A}}$ of $\mathcal{A}$ with respect to G is a power of the Euler polynomial $\mathbf{L}_n$, namely, the minimal power, say $\mathbf{L}_n^s$, that belongs to $\mathbb{F}_q[V]^G$ (see Section 4.4). We claim that $\mathbf{E}_{\mathcal{A}} \in \mathbb{F}_q[V]^G$ is indecomposable. To this end, suppose that $f \in \mathbb{F}_q[V]^G$ divides $\mathbf{E}_{\mathcal{A}}$, $\deg(f) \geq 1$. Looking at the relation $f \mid \mathbf{E}_{\mathcal{A}} = \mathbf{L}_n^s \in \mathbb{F}_q[V]$, we see that f is a product of linear forms that is G-invariant. Hence one linear form from every line in V divides f, so $\mathbf{L}_n$ divides f, and by invariance of the quotients $\mathbf{L}_n^s = \mathbf{E}_{\mathcal{A}}$ divides f. Therefore $f = \mathbf{E}_{\mathcal{A}}$ as required. □

COROLLARY 9.2.6 (P. S. Landweber): *Let $\mathfrak{p} \subset \mathbf{D}(n)$ be a homogeneous prime ideal. Then $\mathfrak{p}$ is $\mathscr{P}^*$-invariant if and only if it is one of the ideals in the chain*

$$(0) \subsetneq (\mathbf{d}_{n,0}) \subsetneq (\mathbf{d}_{n,0}, \mathbf{d}_{n,1}) \subsetneq \cdots \subsetneq (\mathbf{d}_{n,0}, \ldots, \mathbf{d}_{n,n-1}) \subseteq \mathbf{D}(n).$$

PROOF: Since $\mathrm{GL}(n, \mathbb{F}_q)$ acts transitively on the Grassmann variety $\mathfrak{G}_d(V)$ of codimension-d subspaces of V, there is a unique $\mathscr{P}^*$-invariant ideal of height d in $\mathbf{D}(n)$. By the computations of Section 8.6 the indicated ideals are $\mathscr{P}^*$-invariant. Since they are also prime, with one of each height, the result follows. □

Having described the two extreme cases $\mathrm{Proj}_{\mathscr{P}^*}(\mathbb{F}_q[V])$ and $\mathrm{Proj}_{\mathscr{P}^*}(\mathbf{D}(n))$ and how they apply to invariant theory, let's move on to other cases. The imbedding theorem (Theorem 8.4.3) tells us that we find a fractal $\mathbf{D}(n)^{q^s}$ of the Dickson algebra in every unstable Noetherian algebra H such that the extension

$$\mathbf{D}(n)^{q^s} \hookrightarrow H$$

is integral. So the Dickson algebra and its fractals constitute the set of universal *minimal* unstable $\mathscr{P}^*$-algebra algebras. For the Dickson algebra itself we have Corollary 9.2.6. For a Dickson algebra fractal, the ingredients needed to determine the spectrum $\mathrm{Proj}_{\mathscr{P}^*}(\mathbf{D}(n)^{q^s})$ are the computations of Section 8.6: the theorem of J.-P. Serre, Theorem 9.2.1, the lying-over property for $\mathscr{P}^*$-invariant prime ideals (Theorem 9.2.3), and a certain fractal-like property of the Dickson algebra. We begin with a short description of the last of these. For a complete discussion see [290] Appendix A.3.

Let V be an n-dimensional vector space over $\mathbb{F}_q$. Choose a subspace $W \subset V$ of codimension 1, and let

$$i^* : \mathbb{F}_q[V] \longrightarrow \mathbb{F}_q[W]$$

be the map induced by the inclusion $i : W \hookrightarrow V$. Recall from the proof of the Stong-Tamagawa formula, Theorem 6.1.7, that

$$i^*(\mathbf{d}_{n,i}) = \begin{cases} \mathbf{d}^q_{n-1,i-1} & \text{for } 1 \leq i \leq n-1, \\ 0 & \text{for } i = 0. \end{cases}$$

Since the Frobenius map $\Phi(f) = f^q,\ \forall\, f \in \mathbb{F}_q[V]$ is an algebra homomorphism, it follows that

$$\mathbf{d}^q_{n-1,i-1}(t_1, \ldots, t_{n-1}) = \mathbf{d}_{n-1,i-1}(t_1^q, \ldots, t_{n-1}^q).$$

In other words,

$$\mathrm{Im}\{\mathbf{D}(n) \xrightarrow{i^*} \mathbb{F}_q[W]\} = \mathbb{F}_q[t_1^q, \ldots, t_{n-1}^q]^{\mathrm{GL}(n-1,\mathbb{F}_q)},$$

while clearly, $\ker(i^*) = (\mathbf{d}_{n,0})$. The usual Frobenius homomorphism

$$\Phi : \mathbb{F}_q[t_1, \ldots, t_{n-1}] \longrightarrow \mathbb{F}_q[t_1^q, \ldots, t_{n-1}^q]$$

is $\mathrm{GL}(n-1, \mathbb{F}_q)$-equivariant, and provides an isomorphism

$$\Phi : \mathbf{D}(n-1) \longrightarrow \mathbb{F}_q[t_1^q, \ldots, t_{n-1}^q]^{\mathrm{GL}(n-1,\mathbb{F}_q)} = \mathbf{D}(n-1)^q$$

multiplying degrees by q. This means that apart from multiplication of the degrees by q, and this is the **fractal property of the Dickson algebra**, there is an isomorphism of unstable algebras over $\mathscr{P}^*$

$$(☯) \qquad \mathbf{D}(n-1) \xrightarrow[\Phi]{\cong} \mathbf{D}(n-1)^q \cong \mathbf{D}(n)/(\mathbf{d}_{n,0}).$$

Moreover, from the Cartan formulae[3] we have

$$\Phi(\mathcal{P}^i(f)) = \mathcal{P}^{iq}(\Phi(f)),$$

and so for any $f \in \mathbb{F}_q[t_1^q, \ldots, t_{n-1}^q]$,

$$\mathcal{P}^j(\Phi(f)) = 0, \qquad j \not\equiv 0 \bmod q.$$

Thus, the isomorphism (☯) sets up a bijective correspondence between $\mathscr{P}^*$-invariant ideals in $\mathbf{D}(n)/(\mathbf{d}_{n,0})$ and $\mathscr{P}^*$-invariant ideals in $\mathbf{D}(n-1)^q$.

---

[3] This leads to the **fractal property of the Steenrod algebra**.

THEOREM 9.2.7: *Let $\mathfrak{p} \subset \mathbf{D}(n)^{q^s}$ be a homogeneous prime ideal. Then $\mathfrak{p}$ is $\mathscr{P}^*$-invariant if and only if it is one of the ideals in the chain*

$$(0) \subsetneq (\mathbf{d}_{n,0}^{q^s}) \subsetneq (\mathbf{d}_{n,0}^{q^s}, \mathbf{d}_{n,1}^{q^s}) \subsetneq \cdots \subsetneq (\mathbf{d}_{n,0}^{q^s}, \ldots, \mathbf{d}_{n,n-1}^{q^s}) \subseteq \mathbf{D}(n)^{q^s}.$$

PROOF: The ideals in the chain are certainly prime, and as follows from the formulae developed in Section 8.6 it is apparent that they are also $\mathscr{P}^*$-invariant.

To prove that any $\mathscr{P}^*$-invariant prime ideal must be one of these we proceed as in J.-P. Serre's theorem, Theorem 9.2.1, by induction on the Krull dimension n. If $n = 1$, there is nothing to show. So consider $n > 1$, and suppose the conclusion has been established for $n - 1$ and all $s \in \mathbb{N}_0$.

By the fractal property of the Dickson algebra it is enough to consider the case $s = 0$ to establish the validity of the conclusion for n. This is, however, just Corollary 9.2.6. □

Combining the preceding results gives us the following intriguing fact generalizing the finiteness statement in Theorem 9.2.4.

COROLLARY 9.2.8: *Let H be an unstable Noetherian algebra over the Steenrod algebra. Then the spectrum of the $\mathscr{P}^*$-invariant homogeneous prime ideals is a finite set.*

PROOF: Let $\mathfrak{p} \subset H$ be a minimal prime ideal, i.e., $ht(\mathfrak{p}) = 0$. Then $\mathfrak{p}$ is $\mathscr{P}^*$-invariant by Theorem 9.1.9. Therefore, taking the quotient with respect to $\mathfrak{p}$ leads to an unstable integral domain $H/\mathfrak{p}$ of Krull dimension, say, n. This quotient, in turn, can be integrally embedded into a polynomial ring $\mathbb{F}_q[V]$ of the same Krull dimension by the embedding theorem, Theorem 8.4.2, say

$$H/\mathfrak{p} \hookrightarrow \mathbb{F}_q[V].$$

By the $\mathscr{P}^*$-invariant version of the Krull relations, Theorem 9.2.3, the set of $\mathscr{P}^*$-invariant prime ideals of $H/\mathfrak{p}$ consists of precisely the contractions of the $\mathscr{P}^*$-invariant prime ideals in $\mathbb{F}_q[V]$. By J.-P. Serre's theorem, Theorem 9.2.1, this is a finite set, since $\mathbb{F}_q$ is a finite field. Since the number of minimal primes in a Noetherian algebra is finite, we are done. □

The preceding proof combines the embedding theorem with J.-P. Serre's result via the Krull relations. Symmetrically, we could use the these relations and combine the imbedding theorem with Theorem 9.2.7.

ALTERNATIVE PROOF OF 9.2.8: By the imbedding theorem (Theorem 8.4.3) we find a fractal of the Dickson algebra in H such that

$$\mathbf{D}(n)^{q^s} \hookrightarrow H$$

is integral. In the fractal of the Dickson algebra there are only finitely many $\mathscr{P}^*$-invariant prime ideals, namely the $n+1$ given by Theorem 9.2.7. Take any of these and call it $\mathfrak{p} \subset \mathbf{D}(n)^{q^s}$. Its extension $\mathfrak{p}^e \subset H$ in H is again $\mathscr{P}^*$-invariant by Lemma 9.2.2. Since H is Noetherian, the ideal $\mathfrak{p}^e$ has a primary decomposition, and in particular, it has only a finite number of isolated primes, all of which are $\mathscr{P}^*$-invariant by Theorem 9.1.9 This set is precisely the set of prime ideals in H lying over $\mathfrak{p}$. So the set of prime ideals in H lying over $\mathscr{P}^*$-invariant prime ideals in $\mathbf{D}(n)^{q^s}$ is finite. However, this is exactly the set of $\mathscr{P}^*$-invariant prime ideals in H, by Theorem 9.2.3. □

This completes our overview of the spectrum of $\mathscr{P}^*$-invariant prime ideals. We round out this discussion of the $\mathscr{P}^*$-invariant prime ideal spectrum of a Noetherian unstable algebra with one of the many applications of the interplay between the Dickson algebra and the Steenrod algebra in rings of invariants over Galois fields (see, e.g., Section 8.5): The Dickson polynomials provide a universal system of parameters and are depth sensitive. They are also sensitive to the height of $\mathscr{P}^*$-invariant ideals as the following proposition shows.[4]

PROPOSITION 9.2.9: *Let* $\varrho : G \hookrightarrow \mathrm{GL}(n, \mathbb{F})$ *be a representation of a finite group* G *over the Galois field* $\mathbb{F}_q$. *If* $I \subset \mathbb{F}_q[V]^G$ *is a* $\mathscr{P}^*$*-invariant ideal of height* h *then*

(i) $\mathbf{d}_{n,0}, \ldots, \mathbf{d}_{n,h-1} \in \sqrt{I}$, *but*

(ii) $\mathbf{d}_{n,h}, \ldots, \mathbf{d}_{n,n-1} \notin \sqrt{I}$.

PROOF: Let $\mathfrak{p}$ be a minimal associated prime ideal of I. Then $\mathfrak{p}$ has height h. By Theorem 9.1.9 the isolated prime ideals of I are $\mathscr{P}^*$-invariant ideals, so $\mathfrak{p}$ is $\mathscr{P}^*$-invariant. Therefore, by P. S. Landweber's result, Corollary 9.2.7, $\mathfrak{p} \cap \mathbf{D}(n)$ contains $\mathbf{d}_{n,0}, \ldots, \mathbf{d}_{n,h-1}$ but does not contain $\mathbf{d}_{n,h}, \ldots, \mathbf{d}_{n,n-1}$. □

Let's look at the $\mathscr{P}^*$-invariant prime ideal spectra of some rings of invariants.

EXAMPLE 1: Let $\varrho : \mathrm{SL}(n, \mathbb{F}_q) \hookrightarrow \mathrm{GL}(n, \mathbb{F}_q)$ be the tautological representation of the special linear group $\mathrm{SL}(n, \mathbb{F}_q)$. We have an integral extension

$$\varphi : \mathbf{D}(n) = \mathbb{F}_q[V]^{\mathrm{GL}(n,\mathbb{F}_q)} \hookrightarrow \mathbb{F}_q[V]^{\mathrm{SL}(n,\mathbb{F}_q)}.$$

By Theorem 9.2.7 the $\mathscr{P}^*$-invariant prime ideal spectrum of the Dickson algebra $\mathbf{D}(n)$ forms a single chain

$$(\bigstar) \qquad (0) \subset (\mathbf{d}_{n,0}) \subset (\mathbf{d}_{n,0}, \mathbf{d}_{n,1}) \subset \cdots \subset (\mathbf{d}_{n,0}, \ldots, \mathbf{d}_{n,n-1}).$$

The invariants of the special linear group are given by

$$\mathbb{F}_q[\mathbf{L}_n, \mathbf{d}_{n,1}, \ldots, \mathbf{d}_{n.n-1}],$$

where $\mathbf{L}_n^{q-1} = \mathbf{d}_{n,0}$ is the Euler class of the $\mathrm{SL}(n, \mathbb{F}_q)$-action on the dual vector space V^*; cf. Theorem 6.1.8. The $\mathscr{P}^*$-invariant prime ideal spectrum

[4] This result extends to any commutative Noetherian graded connected unstable algebra over the Steenrod algebra once one has the imbedding theorem by using Theorem 9.2.7.

of $\mathbb{F}_q[V]^{SL(n,\mathbb{F}_q)}$ also forms just a single chain

$$(0) \subset (\mathbf{L}_n) \subset (\mathbf{L}_n, \mathbf{d}_{n,1}) \subset \cdots \subset (\mathbf{L}_n, \mathbf{d}_{n,1}, \ldots, \mathbf{d}_{n,n-1}),$$

since these are precisely the prime ideals lying over those of the chain (★); cf. Example 1 of Section 2 in [289].

EXAMPLE 2: Let $\mathbb{F}_3$ be the field with 3 elements. If we restrict our attention to $n = 2$, then we find the quaternion group Q_8 in $SL(2, \mathbb{F}_3)$. Its ring of invariants $\mathbb{F}_3[x, y]^{Q_8}$, also has a chain of just two $\mathscr{P}^*$-invariant prime ideals. (This example has already been discussed at various points; see, e.g., Example 10 in Section 7.4.) This can be seen as follows: The representation $\rho : Q_8 \hookrightarrow GL(2, \mathbb{F}_3)$ of the quaternion group Q_8 is afforded by the matrices

$$\mathbf{A} = \begin{pmatrix} 0 & 1 \\ -1 & 0 \end{pmatrix} \text{ and } \mathbf{B} = \begin{pmatrix} -1 & 1 \\ 1 & 1 \end{pmatrix}.$$

The respective rings of invariants we need to discuss are given by

$$\begin{array}{lll} \mathbf{D}(2) = \mathbb{F}_3[x, y]^{GL(2,\mathbb{F}_3)} & \subset \mathbb{F}_3[x, y]^{SL(2,\mathbb{F}_3)} & \subset \mathbb{F}_3[x, y]^{Q_8} \\ \| & \quad\| & \quad\| \\ \mathbb{F}_3[\mathbf{d}_{2,0}, \mathbf{d}_{2,1}] & \subset \mathbb{F}_3[\mathbf{L}_2, \mathbf{d}_{2,1}] & \subset \mathbb{F}_3[\mathbf{L}_2, \mathbf{d}_{2,1}, (x^2+y^2)^2]/(r), \end{array}$$

where $\mathbf{d}_{2,0}$, $\mathbf{d}_{2,1}$ are the Dickson polynomials, $\mathbf{L}_n = xy(x+y)(x-y)$ is the Euler class, and $r = (x^2+y^2)^6 - \mathbf{d}_{2,1}^2 + 2(x^2+y^2)^2\mathbf{L}_n^2$.

In order to find the $\mathscr{P}^*$-invariant prime ideals of height 1 in $\mathbb{F}_3[x, y]^{Q_8}$, there are several possibilities: We could use P. S. Landweber's result, Corollary 9.2.6, that the $\mathscr{P}^*$-invariant prime ideals in the Dickson algebra of degree 2 are the ideals in the chain

$$(0) \subsetneq (\mathbf{d}_{2,0}) \subset (\mathbf{d}_{2,0}, \mathbf{d}_{2,1}) \subset \mathbf{D}(2),$$

and check which prime ideals in $\mathbb{F}_3[x, y]^{Q_8}$ lie over these. Or we could use J.-P. Serre's result, Theorem 9.2.1, that the $\mathscr{P}^*$-invariant prime ideals in $\mathbb{F}_3[x, y]$ are

$$(0) \subset (x), (y), (x+y), (x-y) \subset (x, y) \subset \mathbb{F}_3[x, y],$$

and calculate their contractions in the ring of invariants. Finally, we could use Theorem 9.2.4, because the elements of Q_8 are $\pm\mathbf{I}$, $\pm\mathbf{A}$, $\pm\mathbf{B}$, $\pm\mathbf{AB}$, and inspection of these matrices shows that every nonzero vector in $\mathbb{F}_3^2$ occurs exactly once as a first column. Therefore Q_8 acts transitively on the space of linear forms V^* and hence by Theorem 9.2.4 there is a unique $\mathscr{P}^*$-invariant prime ideal of height one in $\mathbb{F}_3[x, y]^{Q_8}$.

Using one way or the other, the resulting subspectra of $\mathscr{P}^*$-invariant prime ideals of height 1 for the rings of invariants of Q_8, $SL(2, \mathbb{F}_3)$ and $GL(2, \mathbb{F}_3)$, as well as the inclusions between them, are as follows

$$\begin{array}{ccccccc}
\mathbf{D}(2) & \subset & \mathbb{F}_3[x,y]^{\mathrm{SL}(2,\mathbb{F}_3)} & \subset & \mathbb{F}_3[x,y]^{Q_8} & \subset & \mathbb{F}[x,y] \\
\cup & & \cup & & \cup & & \cup \\
(\mathbf{d}_{2,0}) & \subset & (\mathrm{L}_2) & \subset & (\mathrm{L}_2) & \subset & \left\{\begin{array}{l}(x+y)\\(x-y)\\(x)\\(y)\end{array}\right.
\end{array}.$$

Note that for height 2 the only prime ideals are the maximal ideals, while for height 0 the respective prime ideals are (0).

The preceding examples show that the Dickson algebra is *not* characterized by its $\mathscr{P}^*$-invariant prime ideal spectrum.

9.3 Applications to the Transfer

In this section we apply our new knowledge about the $\mathscr{P}^*$-invariant ideal structure of unstable Noetherian algebras to the image of the transfer for rings of invariants over Galois fields. Recall that every ring of invariants $\mathbb{F}_q[V]^G$, over a Galois field is an unstable Noetherian algebra. The transfer homomorphism commutes with the giant Steenrod operation $\boldsymbol{P}(\xi)$ (see Section 8.1), and therefore

$$\mathrm{Tr}^G : \mathbb{F}_q[V] \longrightarrow \mathbb{F}_q[V]^G, \quad \mathrm{Tr}^G(f) = \sum_{g \in G} gf$$

is a $\mathscr{P}^*$-homomorphism, so the image of the transfer is a $\mathscr{P}^*$-invariant ideal. This makes it clear that we may apply the results from the preceding section in order to obtain information about the transfer. Indeed, we have as a special case of Proposition 9.2.9 ([372] Theorem 11.5.3, and [216] Proposition 3.1) the following theorem:

THEOREM 9.3.1 (K. Kuhnigk–L. Smith): *Let $\mathbb{F}_q[V]^G$ be the ring of invariants of a finite group G over a finite field $\mathbb{F}_q$. If $ht(\mathrm{Im}(\mathrm{Tr}^G)) = h$, then $\mathbf{d}_{n,n-1}, \ldots, \mathbf{d}_{n,h} \notin \sqrt{\mathrm{Im}(\mathrm{Tr}^G)}$ and $\mathbf{d}_{n,h-1}, \ldots, \mathbf{d}_{n,0} \in \sqrt{\mathrm{Im}(\mathrm{Tr}^G)}$.* □

So, this hands us a universal element that is in the image of the transfer: Some high enough power of the top Dickson polynomial. Indeed, we can determine a bound for this power. The image of the transfer of the full general linear group was determined in [59] Corollary 9.14: It is the principal ideal generated by $\mathbf{d}_{n,0}^{n-1}$. Next, recall from Lemma 2.4.4 that we have

$$\mathrm{Im}(\mathrm{Tr}^{\mathrm{GL}(n,\mathbb{F}_q)}) \subset \mathrm{Im}(\mathrm{Tr}^G)$$

for every finite group G. Hence

$$\mathbf{d}_{n,0}^{n-1} \in \mathrm{Im}(\mathrm{Tr}^G)$$

for every finite group G.

On the other hand, Theorem 9.3.1 also gives us a sort of universal non-transfer element in the modular case. Namely, if the order of the group is divisible by the characteristic of the ground field, then no power of the bottom Dickson polynomial is in the image of the transfer.

Here is an example to illustrate some of these ideas. It is taken from the unpublished manuscript [216].

EXAMPLE 1 ($\mathbb{Sp}(n, \mathbb{F}_q)$, q odd): By E. Witt's theorem, [457] (see also [17] Theorem 3.9) the finite symplectic group in its defining representations, $\mathbb{Sp}(n, \mathbb{F}_q)$, $q=p^{\nu}$ with p odd, acts transitively on the hyperplanes in $V=\mathbb{F}_q^n$.

Recall that the symplectic group $\mathbb{Sp}(n, \mathbb{F}_q)$ is generated by the pseudoreflections it contains. Let $t \in \mathbb{Sp}(n, \mathbb{F}_q)$ be a transvection. Then $|t| = p$ and $V^t \subset V$ is a codimension-1 subspace. If $g \in \mathbb{SP}(n, \mathbb{F}_q)$ has order p and fixed-point set V^g, then $V^g \subseteq W$ for some hyperplane $W \subset V$. Since $\mathbb{Sp}(n, \mathbb{F}_q)$ acts transitively on the hyperplanes of V, there is an element $g_W \in \mathbb{Sp}(n, \mathbb{F}_q)$ such that $g_W(V^t) = W$. Recall from [372] Lemma 7.1.1 that $g_W(V^t) = V^{g_W t g_W^{-1}}$ and $g_W t g_W^{-1} \in \mathbb{Sp}(n, \mathbb{F}_q)$ is also a transvection. We have $V^g \subseteq V^{g_W t g_W^{-1}}$, so by Corollary 6.4.6 we obtain

$$\mathfrak{X}_{\mathbb{Sp}(n,\mathbb{F}_q)} = \bigcup_{\substack{g\in G\\ |g|=p}} V^g = \bigcup_{\substack{t\in G\\ t \text{ a transvection}}} V^t = \bigcup_{\dim_{\mathbb{F}_q}(W)=n-1} W.$$

If $\overline{\mathbb{F}}_q$ is the algebraic closure of $\mathbb{F}_q$ and $\overline{\mathfrak{X}}_{\mathbb{Sp}(n,\mathbb{F}_q)}$ the transfer variety over $\overline{\mathbb{F}}_q$, then

$$\overline{\mathfrak{X}}_{\mathbb{Sp}(n,\mathbb{F}_q)} = \bigcup_{\substack{W<V\\ \dim_{\bar{\mathbb{F}}_q}(W)=n-1}} W \otimes_{\mathbb{F}_q} \overline{\mathbb{F}}_q$$

is the variety defined by the set of linear forms $\left\{z \otimes_{\mathbb{F}_q} 1 \,|\, z \in V^* \setminus \{0\}\right\}$, and hence

$$\overline{\mathfrak{X}}_{\mathbb{Sp}(n,\mathbb{F}_q)} = \left\{\overline{v} \in \overline{V} = \overline{\mathbb{F}}_q^n \,|\, \mathbf{L}_n(\overline{v}) = 0\right\}.$$

By Hilbert's Nullstellensatz ([25] Corollary 1.1.12) it therefore follows that $\sqrt{\operatorname{Im}(\operatorname{Tr}^{\mathbb{Sp}(n,\mathbb{F}_q)})^{\mathrm{e}}} = (\mathbf{L}_n)$ in $\overline{\mathbb{F}}_q[\overline{V}]$, and, by flat base change, that $\sqrt{\operatorname{Im}(\operatorname{Tr}^{\mathbb{Sp}(n,\mathbb{F}_q)})^{\mathrm{e}}} = (\mathbf{L}_n)$ in $\mathbb{F}_q[V]$, since $\mathbf{L}_n$ is defined over $\mathbb{F}_q$.

Next, we claim that

$$\sqrt{\operatorname{Im}(\operatorname{Tr}^{\mathbb{Sp}(n,\mathbb{F}_q)})} = \mathbb{F}_q[V]^{\mathbb{Sp}(n,\mathbb{F}_q)} \cap \sqrt{(\operatorname{Im}(\operatorname{Tr}^{\mathbb{Sp}(n,\mathbb{F}_q)}))^{\mathrm{e}}}.$$

To see this, note that the inclusion "$\subseteq$" is elementary. To prove the reverse inclusion suppose that

$$F \in \mathbb{F}_q[V]^{\mathbb{S}\mathrm{p}(n,\mathbb{F}_q)} \cap \sqrt{(\mathrm{Im}(\mathrm{Tr}^{\mathbb{S}\mathrm{p}(n,\mathbb{F}_q)}))^e} = (\mathbf{L}_n).$$

We may write F in the form $f \cdot \mathbf{L}_n$ for some $f \in \mathbb{F}_q[V]$. Since in addition F and $\mathbf{L}_n$ are $\mathbb{S}\mathrm{p}(n, \mathbb{F}_q)$-invariant, it follows that $f \in \mathbb{F}_q[V]^{\mathbb{S}\mathrm{p}(n,\mathbb{F}_q)}$. To show that $F \in \sqrt{\mathrm{Im}(\mathrm{Tr}^{\mathbb{S}\mathrm{p}(n,\mathbb{F}_q)})}$ we must show that some power $F^k = f^k \mathbf{L}_n^k$ lies in $\mathrm{Im}(\mathrm{Tr}^{\mathbb{S}\mathrm{p}(n,\mathbb{F}_q)})$. By Theorem 9.3.1 some power of $\mathbf{L}_n$ lies in $\mathrm{Im}(\mathrm{Tr}^{\mathbb{S}\mathrm{p}(n,\mathbb{F}_q)})$, say $\mathbf{L}_n^k \in \mathrm{Im}(\mathrm{Tr}^{\mathbb{S}\mathrm{p}(n,\mathbb{F}_q)})$. Write $\mathbf{L}_n^k = \mathrm{Tr}^{\mathbb{S}\mathrm{p}(n,\mathbb{F}_q)}(L)$. Then

$$\mathrm{Tr}^{\mathbb{S}\mathrm{p}(n,\mathbb{F}_q)}(f^k \cdot L) = f^k \cdot \mathrm{Tr}^{\mathbb{S}\mathrm{p}(n,\mathbb{F}_q)}(L) = f^k \mathbf{L}_n^k = F^k,$$

since the transfer is an $\mathbb{F}_q[V]^{\mathbb{S}\mathrm{p}(n,\mathbb{F}_q)}$-module homomorphism, and $f^k \in \mathbb{F}_q[V]^{\mathbb{S}\mathrm{p}(n,\mathbb{F}_q)}$. This shows that $F^k \in \mathrm{Im}(\mathrm{Tr}^{\mathbb{S}\mathrm{p}(n,\mathbb{F}_q)})$. We therefore have the equalities

$$\begin{aligned}\sqrt{\mathrm{Im}(\mathrm{Tr}^{\mathbb{S}\mathrm{p}(n,\mathbb{F}_q)})} &= \mathbb{F}_q[V]^{\mathbb{S}\mathrm{p}(n,\mathbb{F}_q)} \cap \sqrt{(\mathrm{Im}(\mathrm{Tr}^{\mathbb{S}\mathrm{p}(n,\mathbb{F}_q)}))^e} \\ &= \mathbb{F}_q[V]^{\mathbb{S}\mathrm{p}(n,\mathbb{F}_q)} \cap (\mathbf{L}_n) = (\mathbf{L}_n) \subset \mathbb{F}_q[V]^{\mathbb{S}\mathrm{p}(n,\mathbb{F}_q)}.\end{aligned}$$

The group $\mathbb{S}\mathrm{p}(n, \mathbb{F}_q)$ acts transitively on the set of lines in V^*, and hence $\mathbf{L}_n \in \mathbb{F}_q[V]^{\mathbb{S}\mathrm{p}(n,\mathbb{F}_q)}$ is irreducible. To see that the ideal $(\mathbf{L}_n) \subset \mathbb{F}_q[V]^{\mathbb{S}\mathrm{p}(n,\mathbb{F}_q)}$ is a prime ideal,[5] suppose that $f, h \in \mathbb{F}_q[V]^{\mathbb{S}\mathrm{p}(n,\mathbb{F}_q)}$ and $f \cdot h \in (\mathbf{L}_n)$. Write

$$(\star) \qquad \mathbf{L}_n = \prod_{\substack{0 \neq z_\ell \in \ell \subseteq V^* \\ \dim_{\mathbb{F}_q}(\ell)=1}} z_\ell.$$

Looking at the condition $\mathbf{L}_n \mid f \cdot h$ in $\mathbb{F}_q[V]$, and using the fact that linear forms are prime elements in $\mathbb{F}_q[V]$, we see that each linear form z_ℓ figuring in the product $(\star)$ must divide either f or h. If $z_{\ell'}, z_{\ell''}$ are two such forms, then there is an element $g \in \mathbb{S}\mathrm{p}(n, \mathbb{F}_q)$ such that $g z_{\ell'} = \lambda(\ell', \ell'') z_{\ell''}$, for some $\lambda(\ell', \ell'') \in \mathbb{F}_q^\times$, so if $z_{\ell'}$ divides f, then the computation

$$f = g \cdot f = g \cdot (z_{\ell'} f') = g \cdot z_{\ell'} g \cdot f' = \lambda(\ell', \ell'') z_{\ell''} f''$$

shows that $z_{\ell''}$ also divides f. Since the forms z_ℓ figuring in the product $(\star)$ are relatively prime in $\mathbb{F}_q[V]$, and $\mathbb{F}_q[V]$ is a unique factorization domain, we conclude that $\mathbf{L}_n$ divides f in $\mathbb{F}_q[V]$. Since both f and $\mathbf{L}_n$ are $\mathbb{S}\mathrm{p}(n, \mathbb{F}_q)$-invariant, it follows that the quotient $f/\mathbf{L}_n$ is also, so $\mathbf{L}_n$ divides f in $\mathbb{F}_q[V]^{\mathbb{S}\mathrm{p}(n,\mathbb{F}_q)}$, showing that $(\mathbf{L}_n) = \sqrt{\mathrm{Im}(\mathrm{Tr}^{\mathbb{S}\mathrm{p}(n,\mathbb{F}_q)})}$ is a prime ideal in $\mathbb{F}_q[V]^{\mathbb{S}\mathrm{p}(n,\mathbb{F}_q)}$.

[5] Equivalently, we could just cite that $\mathbb{F}_q[V]^{\mathbb{S}\mathrm{p}(n,\mathbb{F}_q)}$ is a unique factorization domain, [62] or Chapter 14 in [297].

9.4 Applications to Homological Properties

In this section we come back to our result Corollary 9.2.8 that the number of $\mathscr{P}^*$-invariant prime ideals in any unstable Noetherian algebra is finite. So, what does their number count, and how do we compute this number? Not much is known about these questions. However, we broach here what we think is just the tip of an iceberg: The $\mathscr{P}^*$-invariant prime ideal spectrum seems to be related to the homological properties of the algebra in question, [294] and [296].

The results of this section are taken from [294]. Recall from Theorem 8.4.3 that a Noetherian unstable $\mathscr{P}^*$-algebra of Krull dimension n contains a unique fractal of the Dickson algebra $\mathbf{D}(n)^{q^s}$ for s large enough.

Recall that we can characterize a Cohen–Macaulay algebra by the statement that for every ideal its height is equal to its depth (see Theorem 3.3.2 in [25]). We can generalize this to the context of unstable algebras as follows (see [294]).

PROPOSITION 9.4.1: *Let H be an unstable Noetherian algebra over the Steenrod algebra $\mathscr{P}^*$ with maximal ideal $\mathfrak{m}$. The following statements are equivalent:*

(1) *H is Cohen–Macaulay.*
(2) *$dp(\mathfrak{m}) = ht(\mathfrak{m})$.*
(3) *$dp(\mathfrak{p}) = ht(\mathfrak{p})$ for every $\mathscr{P}^*$-invariant prime ideal $\mathfrak{p} \subset H$.*
(4) *$dp(I) = ht(I)$ for every $\mathscr{P}^*$-invariant ideal $I \subset H$.*
(5) *Every $\mathscr{P}^*$-invariant ideal $I \subseteq H$ of height k contains*

$$\mathbf{d}_{n,0}^{q^t}, \dots, \mathbf{d}_{n,k-1}^{q^t} \in I$$

for $t \in \mathbb{N}$ suitably large. Moreover, these elements form a regular sequence of maximal length in I.

PROOF: Since the maximal ideal $\mathfrak{m} \subset H$ is $\mathscr{P}^*$-invariant, the implications (4) $\Rightarrow$ (3) $\Rightarrow$ (2) are obvious. The implication (2) $\Rightarrow$ (1) follows from the corresponding classical result (see Theorem 3.3.2 in [25]). The implication (5) $\Rightarrow$ (4) is clear. So, what is left to show is implication (1) $\Rightarrow$ (5). To this end assume that H is Cohen–Macaulay. By the imbedding theorem (Theorem 8.4.3), since H is Noetherian, it contains a fractal of the Dickson algebra such that

$$\mathbf{D}(n)^{q^s} = \mathbb{F}_q[\mathbf{d}_{n,0}^{q^s}, \dots, \mathbf{d}_{n,n-1}^{q^s}] \hookrightarrow H$$

is an integral extension. Therefore, the powers of the Dickson polynomials

$$\mathbf{d}_{n,0}^{q^s}, \dots, \mathbf{d}_{n,n-1}^{q^s} \in H$$

form a homogeneous system of parameters, and hence a regular sequence of maximal length in H by Macaulay's theorem, Theorem 3.3.5 (vi) in [25]. By Theorem 8.3.4 in [290] every $\mathscr{P}^*$-invariant prime ideal $\mathfrak{p} \subset H$ of height

k contains a $\mathscr{P}^*$-invariant ideal of height k generated by k elements. By the imbedding theorem and the $\mathscr{P}^*$-invariant Krull relations (Theorem 9.2.3) we can choose these k elements to be the fractals of the top k Dickson polynomials

$$\mathbf{d}_{n,0}^{q^s}, \dots, \mathbf{d}_{n,k-1}^{q^s}.$$

A $\mathscr{P}^*$-invariant ideal $I \subset H$ has a $\mathscr{P}^*$-invariant primary decomposition

$$I = \mathfrak{q}_1 \cap \dots \cap \mathfrak{q}_l,$$

by Theorem 9.1.9. By what we have seen above, we know that

$$\mathbf{d}_{n,0}^{q^s}, \dots, \mathbf{d}_{n,k-1}^{q^s} \in \sqrt{I}.$$

Therefore,

$$\mathbf{d}_{n,0}^{q^t}, \dots, \mathbf{d}_{n,k-1}^{q^t} \in I \subset H$$

for some $t \geq s$, and this is a regular sequence, because $\mathbf{d}_{n,0}^{q^s}, \dots, \mathbf{d}_{n,k-1}^{q^s}$ is (see Exercise 12 of Section 3.1 in [25]). □

Let $\varphi : H' \hookrightarrow H''$ be an integral extension of unstable Noetherian algebras over the Steenrod algebra $\mathscr{P}^*$. Consider the induced map φ^* of prime ideal spectra. Then, by Theorem 9.2.3, we have that $(\varphi^*)^{-1}(\mathrm{Proj}_{\mathscr{P}^*}(H')) \subseteq \mathrm{Proj}_{\mathscr{P}^*}(H'')$. Moreover, the restriction map

$$\varphi^* \mid_{\mathscr{P}^*} : \mathrm{Proj}_{\mathscr{P}^*}(H'') \longrightarrow \mathrm{Proj}_{\mathscr{P}^*}(H')$$

is surjective, by Lemma 9.2.2. The following result gives some kind of relative version of what the prime ideal spectra measures, see [294].

THEOREM 9.4.2: *Let $\varphi : H' \hookrightarrow H''$ be an integral extension of unstable Noetherian algebras over the Steenrod algebra $\mathscr{P}^*$. Assume that the induced map $\mathrm{Proj}_{\mathscr{P}^*}(H'') \longrightarrow \mathrm{Proj}_{\mathscr{P}^*}(H')$ is bijective. Then, if H'' is Cohen-Macaulay, so is H'.* □

REMARK: The converse of the preceding theorem is false. Choose a polynomial ring $\mathbb{F}_q[x]$ in one linear variable over a finite field $\mathbb{F}_q$. This is an unstable Cohen-Macaulay algebra. By Theorem 9.2.1 there are exactly two $\mathscr{P}^*$-invariant prime ideals

$$(0) \subsetneq (x) \subsetneq \mathbb{F}_q[x].$$

Let $\mathbb{F}_q[x, y]$ be the polynomial algebra in two linear variables, and consider its quotient by the ideal (xy, y^q). Since this ideal is $\mathscr{P}^*$-invariant, we get an integral extension in the category of unstable algebras as follows:

$$\varphi : \mathbb{F}_q[x] \hookrightarrow \mathbb{F}_q[x, y]/(xy, y^q).$$

Note that the bigger algebra also has a chain of just two $\mathscr{P}^*$-invariant prime ideals

$$(y) \subsetneq (y, x) \subsetneq \mathbb{F}_q[x, y]/(xy, y^q).$$

Hence the restriction map $\varphi^* \mid_{\mathscr{P}^*}$ is bijective. However, the polynomial algebra $\mathbb{F}_q[x]$ is Cohen-Macaulay, but the bigger algebra not: It has depth zero, since every element is a zero divisor.

Let's close with an application to invariant theory.

COROLLARY 9.4.3: *Let* $\rho : G \hookrightarrow \mathrm{GL}(n, \mathbb{F}_q)$ *be a representation of a finite group* G *over the Galois field* $\mathbb{F}_q$. *Suppose that* G *contains a subgroup* H *such that:*

(i) $\mathbb{F}_q[V]^H$ *is Cohen-Macaulay, and*

(ii) G *and* H *have the same orbits on the Grassmannian* $\mathfrak{G}(V)$.

Then $\mathbb{F}_q[V]^G$ *is also Cohen-Macaulay.*

PROOF: By Corollary 9.2.4 the induced map between the $\mathscr{P}^*$-invariant prime ideal spectra $\mathrm{Proj}_{\mathscr{P}^*}(\mathbb{F}_q[V]^H) \longrightarrow \mathrm{Proj}_{\mathscr{P}^*}(\mathbb{F}_q[V]^G)$ is a bijection, so the result follows from Theorem 9.4.2. □

We illustrate this result with an example. This is one of the smallest examples of a group that acts transitively on the Grassmannian. It occurs in the study of the cohomology of the Janko sporadic simple group J_1 with coefficients in $\mathbb{F}_2$ (see [4] Chapter III Example 1.9).

EXAMPLE 1: The general linear group $\mathrm{GL}(3, \mathbb{F}_2)$, which is the simple group of order 168, contains a Frobenius subgroup F of order 21 afforded by the matrices

$$\begin{bmatrix} 0 & 1 & 1 \\ 1 & 0 & 0 \\ 1 & 0 & 1 \end{bmatrix} \quad \text{and} \quad \begin{bmatrix} 0 & 1 & 0 \\ 0 & 0 & 1 \\ 1 & 0 & 0 \end{bmatrix}.$$

The invariants of F form a Cohen-Macaulay algebra. Indeed, they are a complete intersection of embedding dimension 5, to wit,

$$\mathbb{F}_2[x_1, x_2, x_3]^F = \mathbb{F}_2[\mathbf{d}_{3,0}, \mathbf{d}_{3,1}, \mathbf{d}_{3,2}, f_1, f_2]/(r_1, r_2),$$

where (the indices in the following formulae should be read modulo 3)

$$f_1 = \sum_{i=1}^{3} (x_i^3 + x_i x_{i+1}^2) + x_1 x_2 x_3,$$

$$f_2 = \sum_{i=1}^{3} (x_i^5 + x_i x_{i+1}^4 + x_i^2 x_{i+1}^2 x_{i+2}),$$

$$r_1 = f^4 + \mathbf{d}_{3,2}^3 + \mathbf{d}_{3,1}^2 + f_1^2 \mathbf{d}_{3,1} + f_2 \mathbf{d}_{3,0},$$

$$r_2 = f_2^2 + \mathbf{d}_{3,1}\mathbf{d}_{3,2} + f_1^2 \mathbf{d}_{3,2} + f_1 \mathbf{d}_{3,0}$$

(see Appendix B in [30] or Chapter 23 in [297]). Moreover, the group F acts transitively on the Grassmannian $\mathfrak{G}(V)$. This can be seen by direct computation of the orbits, or by computing the $\mathscr{P}^*$-invariant prime ideal spectrum of $\mathbb{F}_2[x, y, z]^F$. Hence, any overgroup of F must have Cohen-Macaulay invariants, even if its order is divisible by 2.

Of course, rings of invariants of degree at most 3 are always Cohen-Macaulay by Proposition 5.6.10. So one might feel this is an accident. The following example taken from [294] shows this is probably not the case.

EXAMPLE 2: Consider the group $\mathrm{Aut}(\mathbb{F}_{2^5})$ as a group of $\mathbb{F}_2$-linear transformations of $\mathbb{F}_{2^5} \cong \mathbb{F}_2^5$. The cyclic permutation π of the basis elements is a cycle of order 5 which normalizes $\mathrm{Aut}(\mathbb{F}_{2^5})$ in $GL(5, \mathbb{F}_2)$ and together with $\mathrm{Aut}(\mathbb{F}_{2^5})$ generates the normalizer of $\mathrm{Aut}(\mathbb{F}_{2^5})$ in $GL(5, \mathbb{F}_2)$, cf. Satz II 7.3 in [179].[6] This group may also be described as the natural embedding of the general semilinear group of the 1-dimensional vector space over the field $\mathbb{F}_{2^5}$, namely $\Gamma(1, \mathbb{F}_{2^5})$ in the notations of Section 2.1 in [206], into $GL(5, \mathbb{F}_2)$. By Proposition 8.4 in [49] the induced action of $\Gamma(1, \mathbb{F}_{2^5})$ on the Grassmannian of subspaces of dimension k in $\mathbb{F}_2^5$ is transitive for $k = 0, \ldots, 5$. Since $\Gamma(1, \mathbb{F}_{2^5})$ has order $31 \cdot 5$ which is not divisible by 2, the ring of invariants $\mathbb{F}_2[V]^{\Gamma(1,\mathbb{F}_{2^5})}$ is Cohen-Macaulay by Theorem 5.5.2. So by Corollary 9.4.3 all groups between $\Gamma(1, \mathbb{F}_{2^5})$ and $GL(5, \mathbb{F}_2)$ have Cohen-Macaulay invariants. Note that the index $|GL(5, \mathbb{F}_2) : \Gamma(\Gamma(1, \mathbb{F}_{2^5}))|$ is $9,999,360/155 = 2^{10} \cdot 3^2 \cdot 7$ and is divisible by the characteristic.

[6] Note that also in the preceding example the group F is the normalizer of a Singer cycle, namely of $\begin{bmatrix} 0 & 1 & 1 \\ 1 & 0 & 0 \\ 1 & 0 & 1 \end{bmatrix}$ of order 7 (see [179] Chapter II Satz 7.3).

Chapter 10
Lannes's T-Functor and Applications

DESPITE being the object of intense study for over half a century new problems arising in connection with the Steenrod algebra continue to yield exciting applications and unexpected links to other parts of mathematics. See the excellent survey article of R. M. W. Wood [458] for several such examples. This chapter is devoted to one such interdependence with invariant theory. In [227] J. Lannes built a remarkable functor $\mathbf{T} : \mathcal{K} \rightsquigarrow \mathcal{K}$, where $\mathcal{K}$ denotes the category of unstable algebras over the Steenrod algebra. He used it and its iterations to study the cohomology of function spaces. Their significance for invariant theory was realized by W. G. Dwyer and C. W. Wilkerson [101] (Proposition 10.1.7), who gave an ingenious new proof of a theorem of H. Nakajima [276]. We will give a variant of their proof in this chapter (see Theorem 10.4.3 and Corollary 10.6.1). Note also that the original proof of the Landweber-Stong conjecture by D. Bourguiba and S. Zariti [41] relies heavily on properties of the $\mathbf{T}$-functor and several related functors.

One of the strengths of the $\mathbf{T}$-functor lies in the fact that it preserves many homological properties, and numerous ring theoretic properties have a homological characterization. For example, a connected commutative graded Noetherian algebra over a field is a polynomial algebra if and only if it has finite homological dimension (see Theorem 5.3.3). This becomes relevant for invariant theory because the ring $\mathbb{F}_q[V]^{G_U}$, where $G_U \leq G$ is the pointwise stabilizer subgroup in G of a subspace $U \leq V$, may be expressed in terms of the $\mathbf{T}$-functor. Combining these two facts often allows one to show that a particular ring theoretic property of $\mathbb{F}[V]^G$ is inherited by $\mathbb{F}_q[V]^{G_U}$. As was the case with the Steenrod algebra we have chosen to present this material in a wider context than actually needed for invariant theory.

10.1 The T-Functor and Invariant Theory

In this section we discuss **Lannes's T-functor** $\mathbf{T}_U$, its basic properties, and explain its relevance for invariant theory.

Recall that $\mathcal{U}$ denotes the category of unstable modules over the Steenrod algebra. The category $\mathcal{U}$ is abelian and has an internal $\otimes$-functor, since the Steenrod algebra is a Hopf algebra. If U is a finite-dimensional $\mathbb{F}_q$-vector space, then the functor

$$\mathcal{U} \rightsquigarrow \mathcal{U}, \qquad M \rightsquigarrow \mathbb{F}_q[U] \otimes M,$$

has a left adjoint denoted by $\mathbf{T}_U$: The functor $\mathbf{T}_U$ is characterized by

$$\mathrm{Hom}_{\mathcal{U}}(M, \mathbb{F}_q[U] \otimes N) \cong \mathrm{Hom}_{\mathcal{U}}(\mathbf{T}_U(M), N).$$

The existence of this functor is a consequence of Freyd's adjoint functor theorem (see page 84 in [125]). If $m = \dim_{\mathbb{F}_q}(U)$ then the functor $\mathbf{T}_U$ is the mth iteration of the functor $\mathbf{T}$ introduced in [227]; see Section 2.4 loc. cit. as well as the book [336]. For the extension from the prime field to the case of a general Galois field see [214] Section 9. For easy reference we cite from [227] the following properties (Théorèmes 2.1.1 and 2.2.1 loc. cit.) of these functors.

THEOREM 10.1.1 (J. Lannes): *The functor* $\mathbf{T}_U$ *is exact and respects tensor products, i.e.,*

$$\mathbf{T}_U(M' \otimes_{\mathbb{F}_q} M'') \cong \mathbf{T}_U(M') \otimes_{\mathbb{F}_q} \mathbf{T}_U(M'')$$

for any objects M' *and* M'' *in* $\mathcal{U}$. □

We denote by $\mathcal{K}$ the category of unstable algebras. As a consequence of Theorem 10.1.1 J. Lannes proves that $\mathbf{T}_U$ maps an object of $\mathcal{K}$ to an object of $\mathcal{K}$ (see Proposition 2.3.1 loc.cit.). In other words, the functor

$$\mathcal{K} \rightsquigarrow \mathcal{K}, \qquad K \rightsquigarrow \mathbb{F}_q[U] \otimes K,$$

also has a left adjoint

$$\mathbf{T}_U : \mathcal{K} \rightsquigarrow \mathcal{K}.$$

If H is an object of $\mathcal{K}$ then it does not matter if we compute $\mathbf{T}_U(H)$ as an object of $\mathcal{K}$ or $\mathcal{U}$, the result is the same. Hence there is no conflict of notation in using $\mathbf{T}_U$ to denote both of these functors. Collectively we refer to the functors $lannesT_U$ for $U = \mathbb{F}_q^m$, $m \in \mathbb{N}$, as **T-functors**.

We follow [101] Section 3 and introduce a slight refinement of these functors arising from the fundamental adjointness relation

$$\mathrm{Hom}_{\mathcal{K}}(H, \mathbb{F}_q[U]) \cong \mathrm{Hom}_{\mathcal{K}}(\mathbf{T}_U(H), \mathbb{F}_q)$$

allowing us to separate $\mathbf{T}_U(H)$ into components,[1] one component for each map $\alpha \in \mathrm{Hom}_{\mathcal{K}}(H, \mathbb{F}_q[U])$. Here is how this goes.

[1] The term **component** arises from the fact that for a nice enough topological space X

A map of algebras $\alpha : H \longrightarrow \mathbb{F}_q[U]$ in $\mathcal{K}$ corresponds by adjointness to a map of algebras $\widetilde{\alpha} : \mathbf{T}_U(H) \longrightarrow \mathbb{F}_q$. The degree-zero component[2] of $\mathbf{T}_U(H)$ is a q-Boolean algebra by unstability. The map $\widetilde{\alpha}$ restricted to the degree-0 component of $\mathbf{T}_U(H)$, viz., $\widetilde{\alpha} : \mathbf{T}_U(H)_0 \longrightarrow \mathbb{F}_q$ is an algebra homomorphism that allows us to regard $\mathbb{F}_q$ as a $\mathbf{T}_U(H)_0$-module. Since[3] $\mathbb{F}_q$ is a flat $\mathbf{T}_U(H)_0$-module the functor $- \rightsquigarrow - \otimes_{\mathbf{T}_U(H)_0} \mathbb{F}_q$ is exact. We define the **component of α in $\mathbf{T}_U(H)$** by

$$\mathbf{T}_{U,\alpha}(H) = \mathbf{T}_U(H) \otimes_{\mathbf{T}_U(H)_0} \mathbb{F}_q.$$

The component functors $\mathbf{T}_{U,\alpha}$ behave better algebraically than $\mathbf{T}_U$ since they preserve connectedness.

An object M that is both an unstable $\mathscr{P}^*$-module and a module over an unstable $\mathscr{P}^*$-algebra H such that the Cartan formulae

$$\mathcal{P}^k(h \cdot x) = \sum_{i+j=k} \mathcal{P}^i(h) \cdot \mathcal{P}^j(x) \qquad \forall\, h \in H,\ x \in M,$$

are satisfied is called an **unstable $H \odot \mathscr{P}^*$-module**, cf. Section 8.5, where we introduced $\mathbf{D}(n) \odot \mathscr{P}^*$-modules. The category whose objects are unstable $H \odot \mathscr{P}^*$-modules and whose morphisms are both H- and $\mathscr{P}^*$-module homomorphisms is denoted by $\mathcal{U}_H$. For an object M in $\mathcal{U}_H$ the object $\mathbf{T}_U(M)$ is a $\mathbf{T}_U(H) \odot \mathscr{P}^*$-module, because $\mathbf{T}_U$ preserves tensor products. We set

$$\mathbf{T}_{U,\alpha}(M) = \mathbf{T}_U(M) \otimes_{\mathbf{T}_U(H)_0} \mathbb{F}_q \cong \mathbf{T}_U(M) \otimes_{\mathbf{T}_U(H)} \mathbf{T}_{U,\alpha}(H).$$

These are the **components** of $\mathbf{T}_U(M)$ for M regarded as an H-module. The functor

$$\mathbf{T}_{U,\alpha} : \mathcal{U}_H \rightsquigarrow \mathcal{U}_{\mathbf{T}_{U,\alpha}(H)}$$

is also exact.

We proceed to some explicit calculations of $\mathbf{T}_U(H)$ and $\mathbf{T}_{U,\alpha}(H)$ that culminate in Proposition 10.1.7 illustrating the importance of the functor $\mathbf{T}_U$ and its components for invariant theory.

LEMMA 10.1.2: *Let U and V be finite-dimensional vector spaces over the Galois field $\mathbb{F}_q$. Then*

$$\mathrm{Hom}_{\mathcal{K}}(\mathbb{F}_q[V], \mathbb{F}_q[U]) \cong \mathrm{Hom}_{\mathbb{F}_q}(U, V).$$

The isomorphism is given by restricting to the linear forms the homomorphisms in $\mathrm{Hom}_{\mathcal{K}}(\mathbb{F}_q[V], \mathbb{F}_q[U])$ and passing to vector space duals.

$\mathbf{T}_U(H^*(X; \mathbb{F}_p))$ computes the cohomology of the function space map(BU, X), where BU denotes the classifying space of the abelian group U. In such cases the cohomology of the components of this mapping space are given by the components of $\mathbf{T}_U(H^*(X; \mathbb{F}_p))$.

[2] The algebra $\mathbf{T}_U(H)$ is in general not connected, even if H was.

[3] Recall that a q-Boolean algebra is **absolutely flat**, i.e., any module over it is flat (Exercise 28 of Chapter 2 in [20]).

PROOF: The statement follows from the universal properties of $\mathbb{F}_q[-]$. □

EXAMPLE 1: The adjoint condition

$$(\because)\qquad \mathrm{Hom}_{\mathcal{K}}(H, \mathbb{F}_q[U] \otimes \mathbb{F}_q) \cong \mathrm{Hom}_{\mathcal{K}}(\mathbf{T}_U(H), \mathbb{F}_q)$$

together with the preceding lemma yields for $H = \mathbb{F}_q[V]$

$$(\maltese)\quad \mathbf{T}_U(\mathbb{F}_q[V]) = \prod_{\alpha \in \mathrm{Hom}_{\mathbb{F}_q}(U,V)} (\mathbb{F}_q[V])_\alpha = \mathrm{map}(\mathrm{Hom}_{\mathbb{F}_q}(U, V), \mathbb{F}_q[V])_*,$$

where for each $\alpha \in \mathrm{Hom}_{\mathbb{F}_q}(U, V)$, $(\mathbb{F}_q[V])_\alpha$ denotes a copy of $\mathbb{F}_q[V]$. The graded algebra $\mathrm{map}(\mathrm{Hom}_{\mathbb{F}_q}(U, V), \mathbb{F}_q[V])_*$ is defined by taking maps of the ungraded vector space $\mathrm{Hom}_{\mathbb{F}_q}(U, V)$ into the graded vector space $\mathbb{F}_q[V]$ degree by degree, i.e.,

$$\mathrm{map}\left(\mathrm{Hom}_{\mathbb{F}_q}(U, V), \mathbb{F}_q[V]\right)_k = \mathrm{map}\left(\mathrm{Hom}_{\mathbb{F}_q}(U, V), \mathbb{F}_q[V]_k\right).$$

The algebra operations are defined pointwise. In particular, in degree zero we have that

$$\mathbf{T}_U(\mathbb{F}_q[V])_0 = \prod_{\alpha \in \mathrm{Hom}_{\mathbb{F}_q}(U,V)} (\mathbb{F}_q)_\alpha = \mathrm{map}(\mathrm{Hom}_{\mathbb{F}_q}(U, V), \mathbb{F}_q).$$

Next, we consider the case where $H = \mathbb{F}_q[V]^G$ for some $G \le \mathrm{GL}(n, \mathbb{F}_q)$. The action of $\mathrm{GL}(n, \mathbb{F}_q)$ on $\mathbb{F}_q[V]$ passes to an action of $\mathrm{GL}(n, \mathbb{F}_q)$ on $\mathbf{T}_U(\mathbb{F}_q[V])$ by functoriality. Although it does not preserve the direct product decomposition ($\maltese$) we do have:

LEMMA 10.1.3: *The functor $\mathbf{T}_U$ commutes with taking invariants, i.e., if $\rho : G \hookrightarrow \mathrm{Aut}(H)$ is a representation of a finite group on the $\mathscr{P}^*$-unstable algebra H by algebra automorphisms commuting with the action of the Steenrod operations, then $\mathbf{T}_U(H^G) = (\mathbf{T}_U(H))^G$.*

PROOF: If H is an algebra with a G-action and $g \in G$ we denote by $H(g)$ a copy of H with index g. Define

$$\Delta_H : H \longrightarrow \prod_{g \in G} H(g)$$

by requiring for $u \in H$ that

$$\Delta_H(u) = \prod_{g \in G} (u - gu).$$

Then $H^G = \ker(\Delta_H)$. If H is in addition an unstable $\mathscr{P}^*$-algebra then

$$\mathbf{T}_U(H^G) = \mathbf{T}_U(\ker(\Delta_H)) = \ker(\mathbf{T}_U(\Delta_H)).$$

Note that $\mathrm{pr}_g \cdot \Delta_H = \mathrm{id}_H - g \cdot \mathrm{id}_H$, where

$$\mathrm{pr}_g : \prod_{h \in G} H(h) \longrightarrow H$$

is the projection onto the factor indexed by $g \in G$ and id_H is the identity map of H. Applying $\mathbf{T}_U$ to this equation gives

$$\mathbf{T}_U(\Delta_H) = \Delta_{\mathbf{T}_U(H)} : \mathbf{T}_U(H) \longrightarrow \prod_{g \in G} \mathbf{T}_U(H)(g),$$

so

$$\mathbf{T}_U(H^G) = \mathbf{T}_U(\ker(\Delta_H)) = \ker(\Delta_{\mathbf{T}_U(H)}) = \mathbf{T}_U(H)^G$$

as claimed. □

To make the action of G on $\mathbf{T}_U(\mathbb{F}_q[V])$ more explicit we use the representation of $\mathbf{T}(\mathbb{F}_q[V])$ from Example 1 and some simple facts about G actions on mapping spaces. This will allow us to describe the components of $\mathbf{T}_U(\mathbb{F}_q[V]^G)$ in invariant theoretic terms (see Proposition 10.1.7).

If G is a group and X and Y are G-sets, then $\mathrm{map}_G(X, Y)$ denotes the set of G-equivariant maps from X to Y. If X is just a set and Y G-set, then the action of the group G on Y induces an action on $\mathrm{map}(X, Y)$ by the requirement $(g \cdot f)(x) = g(f(x))$ for all $x \in X$ and $g \in G$. If in addition X is also a G-set, then there is the intertwining action of G on $\mathrm{map}(X, Y)$ defined by $(g \cdot f)(x) = g(f(g^{-1}x))$ for which one has

$$\mathrm{map}_G(X, Y) \cong \mathrm{map}(X, Y)^G.$$

Identify $\mathbf{T}_U(\mathbb{F}_q[V])$ with the graded algebra $\mathrm{map}(\mathrm{Hom}_{\mathbb{F}_q}(U, V), \mathbb{F}_q[V])_*$ as in Example 1. If $\rho : G \hookrightarrow \mathrm{GL}(n, \mathbb{F}_q)$ is a representation, $V = \mathbb{F}_q^n$, and U is a finite dimensional $\mathbb{F}_q$-vector space, then we let the group G act on $\mathrm{Hom}_{\mathbb{F}_q}(U, V)$ by

$$(g\vartheta)(u) = g \cdot \vartheta(u) \qquad \forall\, \vartheta \in \mathrm{Hom}_{\mathbb{F}_q}(U, V), g \in G.$$

Then the action of G on the graded algebra $\mathrm{map}(\mathrm{Hom}_{\mathbb{F}_q}(U, V), \mathbb{F}_q[V])_*$ via the intertwining action satisfies

$$\begin{aligned}\mathrm{map}_G(\mathrm{Hom}_{\mathbb{F}_q}(U, V), \mathbb{F}_q[V])_* &= \mathrm{map}(\mathrm{Hom}_{\mathbb{F}_q}(U, V), \mathbb{F}_q[V])_*^G \\ &= \mathbf{T}_U(\mathbb{F}_q[V])^G = \mathbf{T}_U(\mathbb{F}_q[V]^G).\end{aligned}$$

If $i : U \leq V = \mathbb{F}_q^n$ is a linear subspace, then the **pointwise stabilizer subgroup** of U in G is denoted by G_U and defined by

$$G_U = \{g \in G \mid g(u) = u \ \forall\, u \in U\}.$$

The inclusion $i : U \hookrightarrow V$ induces a map of $\mathscr{P}^*$-algebras $i^* : \mathbb{F}_q[V] \longrightarrow \mathbb{F}_q[U]$. Let $\alpha : \mathbb{F}_q[V]^G \longrightarrow \mathbb{F}_q[U]$ be the composition $\nu \cdot i^*$ of i^* with the canonical inclusion $\nu : \mathbb{F}_q[V] \hookrightarrow \mathbb{F}_q[V]$. We want to show that the α-component of

$\mathbf{T}_U(\mathbb{F}_q[V]^G)$ is given by the formula $\mathbf{T}_{U,\alpha}(\mathbb{F}_q[V]^G) \cong \mathbb{F}_q[V]^{G_U}$. To do so we need some more elementary facts about G-actions which we collect in the next few lemmas.

LEMMA 10.1.4: *Let G be a finite group and X and Y finite G-sets. If G acts transitively on X then for any $x \in X$ the evaluation map at x induces an identification of $map_G(X, Y)$ with Y^{G_x}.*

PROOF: Since G acts transitively on X a G-map $f: X \longrightarrow Y$ is determined by its value on any single element $x \in X$. Let $x \in X$ be chosen and set $y = f(x)$. If $g \in G_x$ then since f is G-equivariant we have

$$gy = gf(x) = f(gx) = f(x) = y$$

so $y \in Y^{G_x}$. Conversely, if $y \in Y^{G_x}$ then we can define a G-map $f: X \longrightarrow Y$ by the formula $f(gx) = gy$. This is defined on all of X since G acts transitively on X, and is well defined since $g'x = g''x$ implies $g'^{-1}g'' \in G_x$, and y is fixed by G_x. □

LEMMA 10.1.5: *Let G be a finite group, X a finite G-set, and $A \subseteq X$. Then the pointwise stabilizer G_A of A in X is equal to the isotropy group of the inclusion $i: A \hookrightarrow X$ in the G-set* $\mathrm{map}(A, X)$, *where the G-action on* $\mathrm{map}(A, X)$ *is induced from the action of G on X.*

PROOF: This is a routine verification. □

If $\rho: G \longrightarrow \mathrm{GL}(n, \mathbb{F}_q)$ is a representation of a finite group, $V = \mathbb{F}_q^n$, and $U \leq V$ a linear subspace, then the action of G on V induces an action on $\mathrm{Hom}_{\mathbb{F}_q}(U, V)$. The linear analog of the preceding lemma is the following:

LEMMA 10.1.6: *Let $\rho: G \hookrightarrow \mathrm{GL}(n, \mathbb{F}_q)$ be a representation of a finite group, $V = \mathbb{F}_q^n$, and $U \leq V$ a linear subspacc. Then the pointwise stabilizer G_U of U in V is equal to the isotropy group of the inclusion $i \in \mathrm{Hom}_{\mathbb{F}_q}(U, V)$, where the G-action on $\mathrm{Hom}_{\mathbb{F}_q}(U, V)$ is induced from the action of G on V.* □

The following proposition relates the **T**-functors and pointwise stabilizers (see Section 5 in [101] and Section 3.9 in [336]) and explains why the **T**-functor is of relevence to invariant theory. We postpone a discussion of applications of this result to invariant theory until we have developed several additional properties of the **T**-functors in the next few sections.

PROPOSITION 10.1.7: *Let $\rho: G \hookrightarrow \mathrm{GL}(n, \mathbb{F}_q)$ be a representation of the finite group G over the Galois field $\mathbb{F}_q$ and $i: U \hookrightarrow V = \mathbb{F}_q^n$ a linear subspace. Let $\alpha: \mathbb{F}_q[V]^G \overset{\nu}{\hookrightarrow} \mathbb{F}_q[V] \overset{i^*}{\longrightarrow} \mathbb{F}_q[U]$ be the composition of i^* with the canonical inclusion $\nu: \mathbb{F}_q[V]^G \hookrightarrow \mathbb{F}_q[V]$. Then,*

$$\mathbf{T}_{U,\alpha}(\mathbb{F}_q[V]^G) \cong \mathbb{F}_q[V]^{G_U}.$$

PROOF: From the calculations in Example 1 and the discussion following it we have

$$\mathbf{T}_U(\mathbb{F}_q[V]) \cong \operatorname{map}\left(\operatorname{Hom}_{\mathbb{F}_q}(U, V), \mathbb{F}_q[V]\right)_*.$$

The action of the group G on V induces the action of G on $\operatorname{Hom}_{\mathbb{F}_q}(U, V)$, and the action on $\operatorname{map}(\operatorname{Hom}_{\mathbb{F}_q}(U, V), \mathbb{F}_q[V])_*$ is the intertwining action. Hence we have

$$\begin{aligned}\mathbf{T}_U(\mathbb{F}_q[V]^G) = \mathbf{T}_U(\mathbb{F}_q[V])^G &\cong \operatorname{map}\left(\operatorname{Hom}_{\mathbb{F}_q}(U, V), \mathbb{F}_q[V]\right)_*^G \\ &\cong \operatorname{map}_G\left(\operatorname{Hom}_{\mathbb{F}_q}(U, V), \mathbb{F}_q[V]\right)_*.\end{aligned}$$

Under this identification the component of $\mathbf{T}_U(\mathbb{F}_q[V]^G)$ corresponding to the map $\alpha : \mathbb{F}_q[V]^G \subseteq \mathbb{F}_q[V] \longrightarrow \mathbb{F}_q[U]$ is $\operatorname{map}_G([i], \mathbb{F}_q[V])_*$, where $[i]$ denotes the G-orbit of the inclusion $i : U \hookrightarrow V$ in $\operatorname{Hom}_{\mathbb{F}_q}(U, V)$. Since G acts transitively on $[i]$ Lemma 10.1.4 gives

$$\operatorname{map}_G([i], \mathbb{F}_q[V]) = \mathbb{F}_q[V]^{G_i},$$

where $G_i \leq G$ is the isotropy group of $i \in \operatorname{Hom}_{\mathbb{F}_q}(U, V)$. By Lemma 10.1.6 $G_i = G_U$ and the result follows. □

Let's have a look at an example.

EXAMPLE 2: Let $i : U \leq V$ be a linear subspace of dimension m and $i^* : \mathbb{F}_q[V] \longrightarrow \mathbb{F}_q[U]$ the induced map. Denote the composition of i^* with the canonical inclusion $\nu : \mathbb{F}_q[V]^{\mathrm{GL}(n, \mathbb{F}_q)} \hookrightarrow \mathbb{F}_q[V]$ of the invariants by $\alpha : \mathbb{F}_q[V]^{\mathrm{GL}(n, \mathbb{F}_q)} \overset{\nu}{\hookrightarrow} \mathbb{F}_q[V] \xrightarrow{i^*} \mathbb{F}_q[U]$. Let $\mathrm{GL}(n, \mathbb{F}_q)_U$ be the pointwise stabilizer of U in $\mathrm{GL}(n, \mathbb{F}_q)$. Write $\mathbf{D}(n)_U$ for $\mathbf{T}_{U,\alpha}(\mathbf{D}(n))$. Then by Proposition 10.1.7 and Theorem 6.1.6

$$\mathbf{D}(n)_U = \mathbb{F}_q[V]^{\mathrm{GL}(n, \mathbb{F}_q)_U} = \mathbb{F}_q[h_1, \dots, h_n]$$

for suitable polynomials $h_1, \dots, h_n$. Here is why: By Proposition 10.1.7 $\mathbf{D}(n)_U = \mathbb{F}_q[V]^{\mathrm{GL}(n, \mathbb{F}_q)_U}$. Then, by the definition of $\mathrm{GL}(n, \mathbb{F}_q)_U$, we may choose a basis for V such that the matrices in $\mathrm{GL}(n, \mathbb{F}_q)_U$ have the block form

$$\begin{bmatrix} \mathbf{A} & \mathbf{0} \\ \mathbf{B} & \mathbf{I} \end{bmatrix},$$

where $\mathbf{I}$ is the identity matrix of size $m \times m$ and $\mathbf{A} \in \mathrm{GL}(n - m, \mathbb{F}_q)$. The result then follows from Theorem 6.1.6.

Let Φ be the functor that assigns to an algebra H the algebra $\Phi(H)$ defined by

$$\Phi(H)_k = \begin{cases} 0 & \text{if } k \text{ is not divisible by } q, \\ H_m & \text{if } k = q \cdot m. \end{cases}$$

For $u \in H_m$ we denote by $\Phi(u)$ the corresponding element of $\Phi(H)_{qm}$. If H is an unstable $\mathscr{P}^*$-algebra, then the Cartan formulae show that $\Phi(H)$ becomes an unstable $\mathscr{P}^*$-algebra if we define

$$\mathscr{P}^i(\Phi(u)) = \begin{cases} 0 & \text{if } i \text{ is not divisible by } q, \\ \Phi(\mathscr{P}^j(u)) & \text{if } i = qj. \end{cases}$$

The functor Φ commutes with the functor $\mathbf{T}_U$, i.e., $\mathbf{T}_U(\Phi(H)) \cong \Phi(\mathbf{T}_U(H))$; see Section 3.4 in [336]. There is a map $\varphi_H : \Phi(H) \longrightarrow H$ defined by $\varphi_H(\Phi(u)) = u^q$. This is a map of algebras over $\mathscr{P}^*$ whose image is the subalgebra H generated by the q-th powers.

From Example 2 we then obtain

$$\mathbf{T}_{U,\alpha}(\mathbf{D}(n)^{q^s}) = \mathbf{D}(n)_U^{q^s},$$

since the $\mathbf{T}$-functor commutes with Φ and

$$\Phi^s(\mathbf{D}(n)_U) - \mathbb{F}_q[h_1^{q^s}, \dots, h_n^{q^s}] = \mathbf{D}(n)_U^{q^s}.$$

10.2 The T-Functor and Noetherian Finiteness

The purpose of this section is to show that the $\mathbf{T}$-functors preserves Noetherian finiteness properties. Although this is a bit of a digression from our main concern, the invariant theory of finite groups, it is justified by two facts. First, we can make use of invariant theory, in particular the imbedding and embedding theorems of [290] (see Theorems 8.4.3 and 8.4.4), to do much of the work. Second, the original proofs [100] (see also Theorem VI in [160]) are quite lengthy and somewhat inaccessible to nonspecialists. The following lemma starts things off: It treats the case of rings of invariants.

LEMMA 10.2.1: *Let $\rho : G \hookrightarrow \mathrm{GL}(n, \mathbb{F}_q)$ be a representation of a finite group G. Then $\mathbf{T}_U(\mathbb{F}_q[V]^G)$ is Noetherian and $\mathbf{T}_U(\mathbb{F}_q[V])$ is a finitely generated $\mathbf{T}_U(\mathbb{F}_q[V]^G)$-module.*

PROOF: By Example 1 of Section 10.1 we have that

$$\mathbf{T}_U(\mathbb{F}_q[V]) = \prod_{\alpha \in \mathrm{Hom}_{\mathbb{F}_q}(U,V)} (\mathbb{F}_q[V])_\alpha.$$

So $\mathbf{T}_U(\mathbb{F}_q[V])$ is a finite product of copies of $\mathbb{F}_q[V]$, and hence Noetherian. Since

$$\mathbf{T}_U(\mathbb{F}_q[V]^G) = (\mathbf{T}_U(\mathbb{F}_q[V]))^G$$

(see Lemma 10.1.3) is the subring of invariants of a finite group the extension

$$(\mathbf{T}_U(\mathbb{F}_q[V]))^G \hookrightarrow \mathbf{T}_U(\mathbb{F}_q[V])$$

is integral. Since $\mathbf{T}_U(\mathbb{F}_q[V])$ is finitely generated as an algebra, the extension is also finite and $\mathbf{T}_U(\mathbb{F}_q[V]^G)$ is Noetherian. □

The following finiteness result is adequate for most purposes related to invariant theory.

THEOREM 10.2.2 (W. G. Dwyer - C. W. Wilkerson): *Let H be a connected reduced Noetherian unstable algebra over $\mathscr{P}^*$. Then $\mathbf{T}_U(H)$ is also Noetherian.*

PROOF: We begin with the special case that H is an integral domain. By the imbedding theorem (Theorem 8.4.3) and embedding theorem (Theorem 8.4.2) we have integral extensions

$$\mathbf{D}(n)^{q^t} \hookrightarrow H \hookrightarrow \mathbb{F}_q[V]$$

for t suitably large. Applying the functor $\mathbf{T}_U$ yields extensions

$$\mathbf{T}_U(\mathbf{D}(n)^{q^t}) \hookrightarrow \mathbf{T}_U(H) \hookrightarrow \mathbf{T}_U(\mathbb{F}_q[V])$$

by exactness (Theorem 10.1.1). Since $\mathbf{T}_U$ commutes with Φ and $\mathbf{D}(n)^{q^t} = \Phi^t(\mathbf{D}(n))$ it follows from Lemma 10.2.1 that

$$\mathbf{T}_U(\mathbf{D}(n)^{q^t}) = \Phi^t(\mathbf{T}_U(\mathbf{D}(n)))$$

is a Noetherian ring and that $\mathbf{T}_U(\mathbb{F}_q[V])$ is a finitely generated module over $\mathbf{T}_U(\mathbf{D}(n))$. Since $\mathbf{T}_U(\mathbf{D}(n))$ in turn is Noetherian and so finitely generated over $\Phi^t(\mathbf{T}_U(\mathbf{D}(n)))$ we have by transitivity that $\mathbf{T}_U(\mathbb{F}_q[V])$ is finitely generated as a module over $\mathbf{T}_U(\mathbf{D}(n)^{q^t})$. So, finally we conclude that the $\Phi^t(\mathbf{T}_U(\mathbf{D}(n)))$-submodule $\mathbf{T}_U(H)$ is finitely generated, and hence a Noetherian algebra.

Next we consider a reduced algebra H. It can be embedded into the finite direct product

$$H \hookrightarrow \prod_{\mathfrak{p}} H/\mathfrak{p},$$

where the direct sum runs over the minimal prime ideals of H. Each of the rings $H/\mathfrak{p}$ is a domain, so that $\mathbf{T}_U(H/\mathfrak{p})$ is Noetherian by what was just shown. If we apply $\mathbf{T}_U$ to the inclusion into the direct sum we get

$$\mathbf{T}_U(H) \hookrightarrow \prod_{\mathfrak{p}} \mathbf{T}_U(H/\mathfrak{p}),$$

since $\mathbf{T}_U$ commutes with direct products (Theorem 10.1.1) and is exact. By the imbedding theorem (Theorem 8.4.3) H contains a fractal of the Dickson algebra. The composition

$$\mathbf{D}(n)^{q^t} \hookrightarrow H \longrightarrow H/\mathfrak{p}$$

is a monomorphism. To see this note the kernel must be a prime ideal since $H/\mathfrak{p}$ is an integral domain. Furthermore since $\mathbf{D}(n)$ and $H/\mathfrak{p}$ both have Krull dimension n it is a minimal prime. But $\mathbf{D}(n)$ is itself an integral domain, so the only minimal prime is the zero ideal. Hence $\mathbf{D}(n) \hookrightarrow H/\mathfrak{p}$ is an integral extension for each minimal prime ideal $\mathfrak{p}$ of H. By the preceding discussion we have that $\mathbf{T}_U(H/\mathfrak{p})$ is a finitely generated $\mathbf{T}_U(\mathbf{D}(n)^{q^t})$-module. Therefore it follows that $\prod_{\mathfrak{p}} \mathbf{T}_U(H/\mathfrak{p})$ is a finitely generated $\mathbf{T}_U(\mathbf{D}(n)^{q^t})$-module. Since $\mathbf{T}_U(\mathbf{D}(n)^{q^t})$ is Noetherian it follows that $\mathbf{T}_U(H)$ is also Noetherian. □

If we attempt to extend the preceding argument to the case where the nil-radical $\mathcal{N}il(H)$ of H is nontrivial we encounter a problem. The direct approach would be to proceed as follows: Since H is Noetherian, $\mathcal{N}il(H)$ is finitely generated and there exists a $t \in \mathbb{N}$ such that

$$\mathcal{N}il(H)^{q^t} = 0.$$

In other words, the subalgebra K of H consisting of q^tth powers is reduced and H is a finite extension of it. By what we have proved above we know that $\mathbf{T}_U(K)$ is Noetherian. If we knew the extension

$$\mathbf{T}_U(K) \hookrightarrow \mathbf{T}_U(H)$$

were also finite, we would have that $\mathbf{T}_U(H)$ is Noetherian as desired. This is indeed the case but would take us a bit too far a field to prove here. What is needed is a result for modules: to wit, that $\mathbf{T}$ preserves Noetherian modules over Noetherian rings. See instead the original sources or [246] where this is deduced by relating degree bounds for generators of H and $\mathbf{T}_{U,\alpha}(H)$.

From the preceding theorem we can extract the following additional information.

COROLLARY 10.2.3: *If $H' \hookrightarrow H''$ is an integral extension of reduced Noetherian unstable algebras over the Steenrod algebra, then the extension $\mathbf{T}_U(H') \hookrightarrow \mathbf{T}_U(H'')$ is also integral.*

PROOF: As in the proof of Theorem 10.2.2 we may construct a finite extension

$$\mathbf{D} \hookrightarrow H'' \hookrightarrow H' \hookrightarrow \mathbb{F}_q[V],$$

where $\mathbf{D}$ is a fractal of the Dickson algebra and $\dim(H') = \dim_{\mathbb{F}_q}(V) = \dim(H'')$. The argument used in the proof of Theorem 10.2.2 shows that the extension

$$\mathbf{T}_U(\mathbf{D}) \hookrightarrow \mathbf{T}_U(H') \hookrightarrow \mathbf{T}_U(H'') \hookrightarrow \mathbf{T}_U(\mathbb{F}_q[V])$$

is finite. So in particular $\mathbf{T}_U(H') \hookrightarrow \mathbf{T}_U(H'')$ remains finite since $\mathbf{T}_U(\mathbf{D})$ is a Noetherian ring. By Theorem 10.2.2 $\mathbf{T}_U(H')$ and $\mathbf{T}_U(H'')$ are Noetherian, so this extension is also integral. □

COROLLARY 10.2.4: *Let H be a reduced Noetherian unstable algebra over the Steenrod algebra. Then $\dim(\mathbf{T}_U(H)) = \dim(H)$.*

PROOF: As before we employ the imbedding theorem (Theorem 8.4.3) and get a finite (and therefore integral) extension

$$\mathbf{T}_U(\mathbf{D}(n)^{q^t}) \hookrightarrow \mathbf{T}_U(H),$$

where $n = \dim(H)$. It therefore suffices to show that $n = \dim(\mathbf{T}_U(\mathbf{D}(n)^{q^t}))$. Since $\mathbf{T}_U$ commutes with Φ and $\mathbf{D}(n)^{q^t} = \Phi^t(\mathbf{D}(n))$ this means we must show that $n = \dim(\mathbf{T}_U(\mathbf{D}(n))$. By Corollary 10.2.3 $\mathbf{T}(\mathbf{D}(n)) \hookrightarrow \mathbf{T}(\mathbb{F}_q[V])$ is integral so we are reduced to showing that

$$\mathbf{T}(\mathbb{F}_q[V]) = \prod_{\alpha \in \mathrm{Hom}(U,V)} (\mathbb{F}[V])_\alpha$$

has Krull dimension n. This graded algebra has only finitely many minimal prime ideals, viz.,

$$\mathfrak{p}(\alpha) = \mathbb{F}_q[V]_\beta \times \cdots \times \mathbb{F}_q[V]_\lambda \times (0) \times \mathbb{F}_q[V]_\mu \times \cdots \times \mathbb{F}_q[V]_\gamma,$$

where the (0) occurs in the α-component. This means that

$$\dim\Big(\prod_{\alpha \in \mathrm{Hom}_{\mathbb{F}_q}(U,V)} (\mathbb{F}_q[V])_\alpha\Big) \le \dim(\mathbb{F}_q[V]) = \dim\left(\frac{\mathbf{T}_U(\mathbb{F}_q[V])}{\mathfrak{p}(\alpha)}\right).$$

In $\mathbf{T}(\mathbb{F}_q[V])$ the zero ideal can be written as

$$(0) = \bigcap_{\alpha \in \mathrm{Hom}_{\mathbb{F}_q}(U,V)} \mathfrak{p}(\alpha).$$

Hence (0) is a radical ideal and the associated prime ideals $\mathfrak{p}(\alpha)$ are all isolated of height zero. Hence

$$\dim\Big(\prod_{\alpha \in \mathrm{Hom}_{\mathbb{F}_q}(U,V)} (\mathbb{F}_q[V])_\alpha\Big) = \dim(\mathbb{F}_q[V])$$

as desired. □

The fact that $\mathbf{T}$ commutes with Φ which we used in the proof of Theorem 10.2.2 leads to a short proof that $\mathbf{T}_U(H)$ is reduced whenever H is reduced: this result occurs for example in [100] as Proposition 3.3.

PROPOSITION 10.2.5: *Let H be an unstable algebra over the Steenrod algebra that is reduced and Noetherian, then $\mathbf{T}_U(H)$ is also reduced.*

PROOF: A Noetherian algebra H is reduced if and only if for all large $k \in \mathbb{N}$ the q^kth power map in H is a monomorphism. Consider the map $\varphi_H : \Phi(H) \longrightarrow H$. By iteration we obtain a map $\varphi_H^k : \Phi^k(H) \longrightarrow H$ whose image is the subalgebra of H consisting of the q^kth powers. The map φ_H^k is a monomorphism if and only if the q^kth power map in H is a monomorphism. By hypothesis H is reduced so φ_H^k is a monomorphism for all large k. Since $\mathbf{T}_U$ preserves monomorphisms $\mathbf{T}_U(\varphi_H^k)$ is also a monomorphism. Since $\mathbf{T}_U$ and Φ commute we conclude that raising to the q^kth power is a monomorphism in $\mathbf{T}_U(H)$ for all large k, so $\mathbf{T}_U(H)$ is reduced. □

10.3 Change of Rings for Components

We have already seen in Proposition 10.1.7 that it is the components of the $\mathbf{T}$-functors that are relevant for invariant theory. If H is an unstable algebra over the Steenrod algebra, then $\mathbf{T}_U(H)_0 = \mathrm{Hom}_{\mathcal{K}}(H, \mathbb{F}_q[U])$. For each map of $\mathscr{P}^*$-algebras $\alpha : H \longrightarrow \mathbb{F}_q[U]$ the component of $\mathbf{T}_U(H)$ corresponding to α is

$$\mathbf{T}_{U,\alpha}(H) = \mathbb{F}_q \otimes_{\mathbf{T}_U(H)_{(0)}} \mathbf{T}_U(H).$$

In other words, a component is just a quotient by an ideal, and hence is Noetherian if $\mathbf{T}_U(H)$ is Noetherian. Together with Theorem 10.2.2 this proves the following result.

THEOREM 10.3.1 (W. G. Dwyer - C. W. Wilkerson): *Let H be a connected reduced unstable algebra over $\mathscr{P}^*$. Suppose that H is Noetherian and that $\alpha : H \longrightarrow \mathbb{F}_q[U]$ is a morphism in $\mathcal{K}$, where $U = \mathbb{F}_q^m$ is a finite-dimensional vector space over $\mathbb{F}_q$. Then $\mathbf{T}_{U,\alpha}(H)$ is also Noetherian.* □

REMARK: The ideal occuring in the preceding can be made a bit more explicit. A map of algebras $\alpha : H \longrightarrow \mathbb{F}_q[U]$ in $\mathcal{K}$ corresponds by adjointness to a map of algebras $\widetilde{\alpha} : \mathbf{T}_U(H) \longrightarrow \mathbb{F}_q$. The kernel of the map $\widetilde{\alpha}$ restricted to the degree-0 component of $\mathbf{T}_U(H)$, viz.,

$$\mathfrak{m}_\alpha = \ker\{\widetilde{\alpha} : \mathbf{T}_U(H)_0 \longrightarrow \mathbb{F}_q\},$$

is a maximal ideal. If we extend this ideal up to $\mathbf{T}_U(H)$ then $\mathbf{T}_{U,\alpha}(H) = \mathbf{T}_U(H)/\mathfrak{m}_\alpha^e$.

Alternatively, one can associate to the ideal $\mathfrak{m}_\alpha$ an idempotent $e_{\mathfrak{m}_\alpha} \in \mathbf{T}_U(H)$ with the properties listed in Figure 10.3.1 (see e.g., [20] Exercise 28 of Chapter 2).

$$e_{\mathfrak{m}_\alpha}^2 = e_{\mathfrak{m}_\alpha}$$
(*so $e_{\mathfrak{m}_\alpha}$ is an idempotent*)
$$e_{\mathfrak{m}_\alpha} \equiv 1 \bmod \mathfrak{m}_\alpha$$
$$e_{\mathfrak{m}_\alpha} \in \mathfrak{m} \quad \forall\, \mathfrak{m} \in \mathit{Spec}(\mathbf{T}_U(H)_0) \setminus \mathfrak{m}_\alpha$$

The element $e_{\mathfrak{m}_\alpha}$ does not belong to $\mathfrak{m}_\alpha$, but does belong to every other maximal ideal in $\mathbf{T}_U(H)_0$. In a q-Boolean algebra prime ideals are maximal, and the existence of such idempotent elements is the first step toward showing that any q-Boolean algebra B is isomorphic to the algebra $\mathrm{map}_C(\mathit{Spec}(B), \mathbb{F}_q)$ of continuous maps from $\mathit{Spec}(B)$ to $\mathbb{F}_q$.

FIGURE 10.3.1: The idempotent $e_{\mathfrak{m}_\alpha}$

Choosing such an idempotent element $e_{\mathfrak{m}_\alpha}$ gives $\mathbf{T}_{U,\alpha}(H) \cong e_{\mathfrak{m}_\alpha} \cdot \mathbf{T}_U(H)$.

This leads to our first change of rings result for components. Suppose $\varphi : H' \longrightarrow H''$ is map of unstable $\mathscr{P}^*$-algebras and U a finite dimensional $\mathbb{F}_q$ vector space. The algebra $\mathbf{T}_U(H')$ decomposes into components

$$\mathbf{T}_U(H') = \prod_{\alpha \in \mathrm{Hom}_{\mathcal{K}}(H', \mathbb{F}_q[U])} \mathbf{T}_{U,\alpha}(H'),$$

and similarly for $\mathbf{T}_U(H'')$. The map φ allows us to regard H' as a module over H'' and so we also have the decomposition of $\mathbf{T}_U(H')$ into components regarded as an $\mathbf{T}_U(H'')$-module, say

$$\mathbf{T}_U(H'') = \prod_{\gamma \in \mathrm{Hom}_{\mathcal{K}}(H'', \mathbb{F}_q[U])} \mathbf{T}_{U,\gamma}(H').$$

Recall from Section 10.1, that for $\gamma \in \mathrm{Hom}_{\mathcal{K}}(H'', \mathbb{F}_q[U])$

$$\mathbf{T}_{U,\gamma}(H') = \mathbb{F}_q \otimes_{\mathbf{T}_U(H'')_0} \mathbf{T}_U(H'),$$

where $\mathbb{F}_q$ is regarded as an $\mathbf{T}_U(H'')_0$-module via the map $\widetilde{\gamma} : \mathbf{T}_U(H'') \longrightarrow \mathbb{F}_q$ adjoint to $\mathbf{T}_U(\gamma)$ restricted to the degree zero component of $\mathbf{T}_U(H'')$. The following Lemma explains how these two decompositions are related.

LEMMA 10.3.2: *Let $\varphi : H'' \longrightarrow H'$ be a map of $\mathcal{P}^*$-algebras and U a finite dimensional $\mathbb{F}_q$ vector space. For each map $\alpha : H' \longrightarrow \mathbb{F}_q[U]$ of $\mathcal{P}^*$-algebras introduce the set $\Theta(\alpha)$ of all maps of $\mathcal{P}^*$-algebras $\gamma : H'' \longrightarrow \mathbb{F}_q[U]$ such that the diagram*

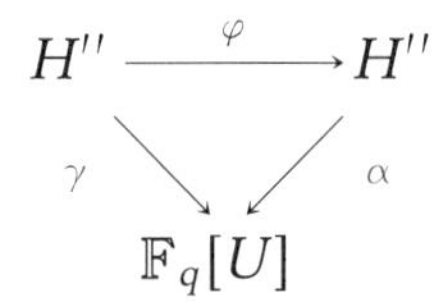

commutes. Then

$$\mathbf{T}_{U,\alpha}(H') \cong \prod_{\gamma \in \Theta(\alpha)} \mathbf{T}_{U,\gamma}(H').$$

PROOF: The elements $\gamma \in \Theta(\alpha)$ are precisely those elements of $\mathrm{Hom}_{\mathcal{K}}(H'', \mathbb{F}_q[U])$ with the property that the corresponding idempotent $e_{\mathfrak{m}_\gamma}$ when applied to $\mathbf{T}_{U,\alpha}(H')$ is nonzero. Since

$$1 = \sum_{\gamma \in \mathrm{Hom}_{\mathcal{K}}(H'', \mathbb{F}_q[U])} e_{\mathfrak{m}_\gamma} \in \mathbf{T}_U(H''))$$

the result follows. □

For the applications to invariant theory we will need to show that the components of the $\mathbf{T}_U$-functor have many of the same properties as the full functor $\mathbf{T}_U$. Theorem 10.2.2 is one such example and preservation of freeness another (see Section 10.4). To deal with this and other matters we will need a number of additional change of rings lemmas for the components of the functor $\mathbf{T}_U$. For this we need to discuss some properties of q-Boolean algebras.

LEMMA 10.3.3: *If B is a q-Boolean ring, then B contains no nonzero nilpotent elements, and hence the nilradical $\mathcal{N}il(B)$ is zero.*

PROOF: Suppose that $b \in B$ is nilpotent. Then we can find a $k \in \mathbb{N}_0$ such that $b^{q^k} = 0$, and hence

$$0 = b^{q^k} = b$$

as desired. □

LEMMA 10.3.4: *Let $\varphi : B' \longrightarrow B''$ be a map of finite q-Boolean rings. Let A be an algebra over B''. Suppose we have a commutative square*

$$\begin{array}{ccc} B' & \xrightarrow{\varphi} & B'' \\ \gamma \downarrow & & \downarrow \alpha \\ \mathbb{F}_q & = & \mathbb{F}_q \end{array}$$

of algebra homomorphisms. Then the induced map

$$\varphi_* : \mathbb{F}_q \otimes_{B'} A \longrightarrow \mathbb{F}_q \otimes_{B''} A$$

is a split epimorphism of A-modules.

PROOF: First note that

$$\mathbb{F}_q \otimes_{B'} A \cong (\mathbb{F}_q \otimes_{B'} B'') \otimes_{B''} A$$

as A-modules. We introduce the map of B''-modules

$$f : \mathbb{F}_q \otimes_{B'} B'' \longrightarrow \mathbb{F}_q$$

given by $f(\lambda \otimes b'') = \lambda\alpha(b'')$. Since B'' is a finite q-Boolean algebra its nilradical is trivial by Lemma 10.3.3. Moreover, it satisfies the descending chain condition on ideals, since it is a finite dimensional $\mathbb{F}_q$-vector space. Therefore B'' is a semisimple ring with minimum condition. By the structure theorem for such rings (see, e.g., Chapter 2 in [184]) every finitely generated module is projective. In particular, $\mathbb{F}_q$ is a projective B''-module, so the map f is a split epimorphism of B''-modules. Hence

$$f \otimes \mathrm{id} : (\mathbb{F}_q \otimes_{B'} B'') \otimes_{B''} A \longrightarrow \mathbb{F}_q \otimes_{B''} A$$

is a split epimorphism of A-modules. Since we have the factorization

$$\varphi_* : \mathbb{F}_q \otimes_{B'} A \cong (\mathbb{F}_q \otimes_{B'} B'') \otimes_{B''} A \xrightarrow{f \otimes \mathrm{id}} \mathbb{F}_q \otimes_{B''} A$$

the result follows. □

This has the following application to the $\mathbf{T}$-functor.

LEMMA 10.3.5: *Let U be a finite dimensional vector space over $\mathbb{F}_q$. Let $\delta : H' \longrightarrow H''$ be an map of Noetherian unstable algebras and let $\alpha : H'' \longrightarrow \mathbb{F}_q[U]$ be a map in the category $\mathcal{K}$. Denote the composition $\alpha \circ \delta$ by $\gamma : H' \longrightarrow \mathbb{F}_q[U]$. Then δ induces a split epimorphism*

$$\delta_* : \mathbf{T}_{U,\gamma}(H'') \longrightarrow \mathbf{T}_{U,\alpha}(H'')$$

of unstable $\mathbf{T}_U(H'')$-modules.

PROOF: The map $\mathbf{T}_U(\delta)$ restricted to the degree zero component yields a commutative diagram

$$\begin{array}{ccc} \mathbf{T}_U(H')_0 & \xrightarrow{\mathbf{T}_U(\delta)_0} & \mathbf{T}_U(H'')_0 \\ \gamma \downarrow & & \downarrow \alpha \\ \mathbb{F}_q & = & \mathbb{F}_q . \end{array}$$

We set

$$\mathbf{T}_{U,\gamma}(H'') = \mathbb{F}_q \otimes_{\mathbf{T}_U(H')_0} \mathbf{T}_U(H''),$$
$$\mathbf{T}_{U,\alpha}(H'') = \mathbb{F}_q \otimes_{\mathbf{T}_U(H'')_0} \mathbf{T}_U(H'').$$

Since H' and H'' are Noetherian, the degree zero components $\mathbf{T}_U(H')_0$ and $\mathbf{T}_U(H'')_0$ are finite q-Boolean rings (see, e.g., Section 3.8 of [336]). Therefore, the map

$$\delta_* : \mathbf{T}_{U,\gamma}(H'') \longrightarrow \mathbf{T}_{U,\alpha}(H'')$$

is a split epimorphism of $\mathbf{T}_U(H'')$-modules by Lemma 10.3.4. □

Equipped with this we can prove the following result about the Krull dimension of the components.

PROPOSITION 10.3.6: *Let H be a connected unstable integral domain over $\mathscr{P}^*$. Suppose that H is Noetherian and that $\alpha : H \longrightarrow \mathbb{F}_q[U]$ is a morphism in $\mathcal{K}$, where $U = \mathbb{F}_q^m$ is a finite-dimensional vector space over $\mathbb{F}_q$. Then* $\dim(\mathbf{T}_{U,\alpha}(H)) = \dim(H)$.

PROOF: Since a component $\mathbf{T}_{U,\alpha}(H)$ is a quotient of $\mathbf{T}_U(H)$ and the latter algebra has dimension $\dim(H)$, we have that

$$\dim(\mathbf{T}_{U,\alpha}(H)) \leq \dim(H).$$

To prove the converse inequality we employ the embedding theorem (Theorem 8.4.2). Let $\alpha : H \longrightarrow \mathbb{F}_q[U]$ be a map in $\mathcal{K}$, and $\psi : H \hookrightarrow \mathbb{F}_q[V]$ given by the embedding theorem. Since $\mathbb{F}_q[U]$ is injective (see, e.g., Chapter 3 of [336]) the map α factors through $\mathbb{F}_q[V]$ and we get

$$\alpha : H \overset{\psi}{\hookrightarrow} \mathbb{F}_q[V] \xrightarrow{\eta} \mathbb{F}_q[U].$$

Since the degree zero component $\mathbf{T}_U(H)_{(0)}$ is a q-Boolean algebra, every module over it is flat. Therefore we obtain an inclusion

$$\mathbf{T}_{U,\alpha}(H) \hookrightarrow \mathbb{F}_q(\alpha) \otimes_{\mathbf{T}_U(H)_{(0)}} \mathbf{T}_U(\mathbb{F}_q[V]),$$

which is a finite extension of algebras. The (α) emphasizes that we take the component with respect to the map α. The map

$$\mathbb{F}_q(\alpha) \otimes_{\mathbf{T}_U(H)_{(0)}} \mathbf{T}_U(\mathbb{F}_q[V]) \longrightarrow \mathbb{F}_q(\eta) \otimes_{\mathbf{T}_U(\mathbb{F}_q[V])_{(0)}} \mathbf{T}_U(\mathbb{F}_q[V]) = \mathbf{T}_{U,\eta}(\mathbb{F}_q[V])$$

is a split epimorphism by Lemma 10.3.5. Composing the two maps yields that $\mathbf{T}_{U,\eta}(\mathbb{F}_q[V])$ is a finitely generated $\mathbf{T}_{U,\alpha}(H)$-module, and we obtain the converse inequality

$$\dim(\mathbf{T}_{U,\alpha}(H)) \geq \dim(\mathbf{T}_{U,\eta}(\mathbb{F}_q[V]) = \dim(\mathbb{F}_q[V]) = \dim(H)$$

as desired. □

REMARK: The first part of the proof of Proposition 10.3.6 works even in the presence of zero divisors. Hence whenever H is a Noetherian unstable $\mathscr{P}^*$-algebra, U a finite dimensional $\mathbb{F}_q$-vector space, and $\alpha : H \longrightarrow \mathbb{F}_q[U]$ a map of $\mathscr{P}^*$-algebras we have $\dim(\mathbf{T}_{U,\alpha}(H)) \leq \dim(H)$.

In a similar manner one establishes the following:

PROPOSITION 10.3.7: *Let H be an unstable algebra over the Steenrod algebra that is reduced and Noetherian. If $\varphi : H \longrightarrow \mathbb{F}_q[U]$ a map of algebras over the Steenrod algebra then $\mathbf{T}_{U,\varphi}(H)$ is also reduced.* □

10.4 The T-Functor and Freeness

In this section we will develop some homological properties of the **T**-functor allowing us to show that **T** preserves a number of basic ring-theoretic properties of interest in invariant theory (see e.g., the discussion in Section 5.7), such as being a polynomial algebra, or a Cohen-Macaulay algebra. In Section 10.5 we will also show that **T** preserves complete intersections. These results depend on exploiting the exactness of **T**, as well as the appropriate homological characterization of the property in question. This, and the next section correct a number of unfortunate mistakes in [380]. We are indebted to D. M. Meyer for considerable help, and for allowing us to use some of her arguments here.

The central property we need is that the **T**-functor preserves freeness in several senses that will shortly be made precise. These results are due to W. G. Dwyer and C. W. Wilkerson [101]. To do this we need some preliminary manœuvres. We begin with the simple fact that $\mathbf{T}_U$-preserves the bar construction since it is exact and commutes with tensor products. This is at the heart of many useful relationships with homological algebra. A result of this type appears in Proposition 6.4.2 of [336].

LEMMA 10.4.1: *If H is an object of $\mathcal{K}$ and M, N are objects in $\mathcal{U}_H$, then $\mathrm{Tor}_k^H(M, N)$, $k \in \mathbb{N}$ is an unstable $\mathscr{P}^*$-module and*

$$\mathbf{T}_U(\mathrm{Tor}_k^H(M, N)) \cong \mathrm{Tor}_k^{\mathbf{T}_U(H)}(\mathbf{T}_U(M), \mathbf{T}_U(N)).$$

If $\alpha : H \longrightarrow \mathbb{F}_q[U]$ is a map of $\mathscr{P}^*$-algebras then

$$\mathbf{T}_{U,\alpha}(\mathrm{Tor}_k^H(M, N)) \cong \mathrm{Tor}_k^{\mathbf{T}_{U,\alpha}(H)}(\mathbf{T}_{U,\alpha}(M), \mathbf{T}_{U,\alpha}(N)).$$

PROOF: Since the $\mathbf{T}$-functor is exact and commutes with taking tensor products it sends the bar construction $M \otimes \mathscr{B}(H) \otimes N$ into the bar construction $\mathbf{T}_U(M) \otimes \mathscr{B}(\mathbf{T}_U(H)) \otimes \mathbf{T}_U(N)$. Again, since $\mathbf{T}_U(-)$ is exact it commutes with taking homology yielding the first conclusion. Let $e_{\mathfrak{m}_\alpha} \in \mathbf{T}_U(H)_0$ be an idempotent element not belonging to $\mathfrak{m}_\alpha = \ker(\alpha : \mathbf{T}(H)_0 \longrightarrow \mathbb{F}_q)$, but belonging to every other maximal ideal in $\mathbf{T}_U(H)_0$ (See the discussion following Theorem 10.3.1). The second commutation rule follows likewise from exactness and the fact that $\mathbf{T}_{U,\alpha}(\mathrm{Tor}_k^H(M, N))$ may be defined (see the discussion of the components of the functor $\mathbf{T}_U$ in Section 10.3) by applying the idempotent to $\mathrm{Tor}^{e_{\mathfrak{m}_\alpha}}(e_{\mathfrak{m}_\alpha}, e_{\mathfrak{m}_\alpha})$ to $\mathbf{T}_U(\mathrm{Tor}_k^H(M, N))$. □

In [227] J. Lannes shows in Proposition 4.5 that for any finite dimensional $\mathbb{F}_q$-vector space and any unstable $\mathbb{F}_q[U] \odot \mathscr{P}^*$-module M that $\mathbf{T}_{U,\iota}(M)$ is free as an $\mathbf{T}_{U,\iota}(\mathbb{F}_q[U])$-module, where $\iota : \mathbb{F}_q[U] \longrightarrow \mathbb{F}_q[U]$ is the identity map. We need the following generalization of this result.

LEMMA 10.4.2: *Let H be an unstable $\mathscr{P}^*$-algebra, $\varphi : H \longrightarrow \mathbb{F}_q[U]$ a map of $\mathscr{P}^*$-algebras, and L an unstable $\mathbb{F}_q[U] \odot \mathscr{P}^*$-module. Then $\mathbf{T}_{U,\varphi}(L)$ is a free $\mathbf{T}_{U,\varphi}(\mathbb{F}_q[U])$-module.*

PROOF: Let Ω be the set of all maps $\omega : \mathbb{F}_q[U] \circlearrowleft$ of $\mathscr{P}^*$-algebras making the following triangle

$$\begin{array}{ccc} & H & \\ {}^{\varphi}\swarrow & & \searrow{}^{\varphi} \\ \mathbb{F}_q[U] & \xrightarrow{\ \omega\ } & \mathbb{F}_q[U] \end{array}$$

commute. Note that the identity map ι belongs to Ω. We may regard $\mathbf{T}_U(L)$ as both $\mathbf{T}_U(\mathbb{F}_q[U])$-module as well as a $\mathbf{T}_U(H)$-module along the map $\mathbf{T}_U(\varphi) : \mathbf{T}_U(H) \longrightarrow \mathbf{T}_U(\mathbb{F}_q[U])$. Note that by Lemma 10.3.2

$$\mathbf{T}_{U,\varphi}(L) = \prod_{\omega \in \Omega} \mathbf{T}_{U,\omega}(L)$$

where

$$\mathbf{T}_{U,\varphi}(L) = \mathbb{F}_q(\varphi) \otimes_{\mathbf{T}_U(H)_0} \mathbf{T}_U(L)$$

and

$$\mathbf{T}_{U,\omega}(L) = \mathbb{F}_q(\omega) \otimes_{\mathbf{T}_U(\mathbb{F}_q[U])_0} \mathbf{T}_U(L).$$

In other words, the component $\mathbf{T}_{U,\varphi}(L)$ of $\mathbf{T}_U(L)$ regarded as a $\mathbf{T}_U(H)$-module is the direct sum of the components $\mathbf{T}_{U,\omega}(L)$ of $\mathbf{T}_U(L)$ regarded as an $\mathbf{T}_U(\mathbb{F}_q[U])$-module where ω runs through Ω. If we apply this to $L = \mathbb{F}_q[U]$ we obtain

$$\mathbf{T}_{U,\varphi}(\mathbb{F}_q[U]) = \prod_{\omega \in \Omega} \mathbf{T}_{U,\omega}(\mathbb{F}_q[U]).$$

Let

$$\pi : \mathbf{T}_{U,\varphi}(\mathbb{F}_q[U]) \longrightarrow \mathbf{T}_{U,\iota}(\mathbb{F}_q[U])$$

be the projection onto the component corresponding to $\iota \in \Omega$. Regarded as a $\mathbf{T}_{U,\varphi}(\mathbb{F}_q[U])$-module along π the module $\mathbf{T}_{U,\iota}(\mathbb{F}_q[U])$ is flat. Hence by flat base change ([68] Chapter VI Section 4) we obtain an isomorphism

$$(\text{✣}) \quad \begin{aligned} &\mathrm{Tor}_s^{\mathbf{T}_{U,\varphi}(\mathbb{F}_q[U])}(\mathbb{F}_q, \mathbf{T}_{U,\varphi}(L)) \\ &\quad \cong \mathrm{Tor}_s^{\mathbf{T}_{U,\iota}(\mathbb{F}_q[U])}(\mathbb{F}_q, \mathbf{T}_{U,\iota}(\mathbb{F}_q[U]) \otimes_{\mathbf{T}_{U,\varphi}(\mathbb{F}_q[U])} \mathbf{T}_{U,\varphi}(L)). \end{aligned}$$

Since

$$\mathbf{T}_{U,\iota}(\mathbb{F}_q[U]) \otimes_{\mathbf{T}_{U,\varphi}(\mathbb{F}_q[U])} \mathbf{T}_{U,\varphi}(L) \cong \mathbf{T}_{U,\iota}(L)$$

we conclude

$$\mathrm{Tor}_s^{\mathbf{T}_{U,\varphi}(\mathbb{F}_q[U])}(\mathbb{F}_q, \mathbf{T}_{U,\varphi}(L)) \cong \mathrm{Tor}_s^{\mathbf{T}_{U,\iota}(\mathbb{F}_q[U]}(\mathbb{F}_q, \mathbf{T}_{U,\iota}(L)).$$

By [227] Proposition 4.5 $\mathbf{T}_{U,\iota}(L)$ is a free $\mathbf{T}_{U,\iota}(\mathbb{F}_q[U])$-module so

$$\mathrm{Tor}_s^{\mathbf{T}_{U,\iota}(\mathbb{F}_q[U]}(\mathbb{F}_q, \mathbf{T}_{U,\iota}(L)) = 0$$

for all $s > 0$. The isomorphism (✣) then yields

$$\mathrm{Tor}_s^{\mathbf{T}_{U,\varphi}(\mathbb{F}_q[U])}(\mathbb{F}_q, \mathbf{T}_{U,\varphi}(L)) = 0$$

so $\mathbf{T}_{U,\varphi}(L)$ is a free $\mathbf{T}_{U,\varphi}(\mathbb{F}_q[U])$-module by a standard result (see e.g., [372] Proposition 6.1.1) of homological algebra. □

PROPOSITION 10.4.3 (W. G. Dwyer - C. W. Wilkerson): *Let H be an unstable $\mathscr{P}^*$-algebra, $\varphi : H \longrightarrow \mathbb{F}_q[U]$ a map of $\mathscr{P}^*$-algebras, and M an $H \odot \mathscr{P}^*$-module that is free as an H-module. Then $\mathbf{T}_{U,\varphi}(M)$ is free as a $\mathbf{T}_{U,\varphi}(H)$-module.*

PROOF: The map φ induces a map $\mathbf{T}_{U,\varphi}(H) \longrightarrow \mathbf{T}_{U,\varphi}(\mathbb{F}_q[U])$. Consider the change of rings spectral sequence along this map ([68] Chapter XVI Section 5), viz.,

$$\begin{aligned} &E^r \Rightarrow \mathrm{Tor}^{\mathbf{T}_{U,\varphi}(H)}(\mathbb{F}_q, \mathbf{T}_{U,\varphi}(M)) \\ &E^2_{s,t} = \mathrm{Tor}_s^{\mathbf{T}_{U,\varphi}(\mathbb{F}_q[U])}(\mathbb{F}_q, \mathrm{Tor}_t^{\mathbf{T}_{U,\varphi}(H)}(\mathbf{T}_{U,\varphi}(\mathbb{F}_q[U]), \mathbf{T}_{U,\varphi}(M))). \end{aligned}$$

By Lemma 10.4.1 we find

$$E^2_{s,t} = \mathrm{Tor}_s^{\mathbf{T}_{U,\varphi}(\mathbb{F}_q[U])}(\mathbb{F}_q, \mathbf{T}_{U,\varphi}(\mathrm{Tor}_t^H(\mathbb{F}_q[U], M)).$$

By hypothesis M is a free H-module, so

$$\mathrm{Tor}_t^H(\mathbb{F}_q[U], M) = 0 \quad \forall\, t > 0$$

and the spectral sequence collapses to give an isomorphism

$$\mathrm{Tor}_s^{\mathbf{T}_{U,\varphi}(H)}(\mathbb{F}_q, \mathbf{T}_{U,\varphi}(M)) \cong \mathrm{Tor}_s^{\mathbf{T}_{U,\varphi}(\mathbb{F}_q[U])}(\mathbb{F}_q, \mathbf{T}_{U,\varphi}(\mathbb{F}_q[U]) \otimes_H \mathbf{T}_{U,\varphi}(M)).$$

If we apply Lemma 10.4.2 to the $\mathbb{F}_q[U] \odot \mathscr{P}^*$-module $L = \mathbb{F}_q[U] \otimes_H M$ we conclude that $\mathbf{T}_{U,\varphi}(\mathbb{F}_q[U] \otimes_H M)$ is free as $\mathbf{T}_{U,\varphi}(\mathbb{F}_q[U])$-module. Hence the right hand side of this equation is zero for all $s > 0$. Since flat and free agree in the case of graded connected algebras over a field (see e.g., [372] Proposition 6.1.1) this completes the proof. □

COROLLARY 10.4.4: *Suppose that H is an unstable $\mathscr{P}^*$-algebra and M an unstable $H \odot \mathscr{P}^*$-module. Let $\varphi : H \longrightarrow \mathbb{F}_q[U]$ be a map of $\mathscr{P}^*$-algebras. If* $\text{hom-dim}_H(M) \leq d$ *then* $\text{hom-dim}_{\mathbf{T}_{U,\varphi}(H)}(\mathbf{T}_{U,\varphi}(M)) \leq d$ *also.*

PROOF: Choose a resolution

$$\cdots \longrightarrow F_k \xrightarrow{\partial_k} F_{k-1} \xrightarrow{\partial_{k-1}} \cdots \xrightarrow{\partial_1} F_0 \longrightarrow M \longrightarrow 0$$

of M in which each F_i is an unstable $H \odot \mathscr{P}^*$-module that is free as an H-module. This is possible for example by using the functor $U_H(\text{-})$ introduced in [240] as is shown in [241]. Form the truncated sequence

$$0 \longrightarrow K_d \longrightarrow F_{d-1} \xrightarrow{\partial_{d-1}} \cdots \xrightarrow{\partial_1} F_0 \longrightarrow M \longrightarrow 0$$

where K_d is the kernel of ∂_{d-1}. Then K_d is an unstable $H \odot \mathscr{P}^*$-module that is free as an H-module (see e.g., [372] Proposition 6.3.3). Applying $\mathbf{T}_{U,\varphi}(\text{-})$ to this sequence yields the exact sequence

$$0 \longrightarrow \mathbf{T}_{U,\varphi}(K_d) \longrightarrow \mathbf{T}_{U,\varphi}(F_{d-1}) \xrightarrow{\partial_{d-1}} \cdots \xrightarrow{\partial_1} \mathbf{T}_{U,\varphi}(F_0) \longrightarrow \mathbf{T}_{U,\varphi}(M) \longrightarrow 0$$

and Proposition 10.4.3 tells us that all the terms apart from perhaps $\mathbf{T}_{U,\varphi}(M)$ itself are free $\mathbf{T}_{U,\varphi}(H)$-modules, so the result follows. □

COROLLARY 10.4.5: *Suppose that H is an unstable $\mathscr{P}^*$-algebra and $\varphi : H \longrightarrow \mathbb{F}_q[U]$ is a map of $\mathscr{P}^*$-algebras. If H is reduced and Cohen-Macaulay then so is $\mathbf{T}_{U,\varphi}(H)$.*

PROOF: By the imbedding theorem, Theorem 8.4.3, we may find a fractal of the Dickson algebra $\mathbf{D}(n)$ in H where $n = \dim(H)$, say $\mathbf{D}$, such that $\vartheta : \mathbf{D} \subseteq H$ is a finite ring extension. Then H is a free $\mathbf{D}$-module because it is Cohen-Macaulay. Form the diagram

$$\begin{array}{ccc} \mathbf{D} & \xrightarrow{\vartheta} & H \\ & \searrow^{\gamma} & \downarrow \varphi \\ & & \mathbb{F}_q[U] \end{array}$$

defining the map γ. In Proposition 10.2.3 we saw that for reduced Noetherian algebras the functor $\mathbf{T}_{U,\gamma}$ preserves integral extensions. Hence $\mathbf{T}_{U,\gamma}(\mathbf{D}) \subseteq \mathbf{T}_{U,\gamma}(H)$ is a finite ring extension also. In Example 2 of Section 10.1 we saw that $\mathbf{T}_{U,\gamma}(\mathbf{D})$ is a polynomial algebra on n generators, and by Theorem 10.4.3 $\mathbf{T}_{U,\gamma}(H)$ is a free $\mathbf{T}_{U,\gamma}(\mathbf{D})$-module. Lemma 10.3.5 tells us that $\mathbf{T}_{U,\varphi}(H)$ is a direct summand in $\mathbf{T}_{U,\gamma}(H)$, so it too is free as a $\mathbf{T}_{U,\gamma}(\mathbf{D})$-module. Therefore the map $\mathbf{T}_{U,\gamma}(\mathbf{D}) \longrightarrow \mathbf{T}_{U,\varphi}(H)$ is a monomorphism making $\mathbf{T}_{U,\varphi}(H)$ into a finitely generated free module over $\mathbf{T}_{U,\gamma}(\mathbf{D})$, and hence $\mathbf{T}_{U,\varphi}(H)$ is Cohen–Macaulay. □

We next show (see [101]) that a component of the functor $\mathbf{T}_U$ applied to a polynomial algebra is again a polynomial algebra. To do so we make use of the criterion [372] Theorem 6.4.4. To wit: If $H \subset \mathbb{F}[V]$ is a finite extension of commutative graded connected algebras then H is a polynomial algebra if and only if $\mathbb{F}[V]$ is a free H-module.

THEOREM 10.4.6 (W. G. Dwyer and C. W. Wilkerson): *Suppose that the object P in $\mathcal{K}$ is a connected finitely generated polynomial algebra over the ground field $\mathbb{F}_q$. If $U = \mathbb{F}_q^m$ and $\alpha : P \longrightarrow \mathbb{F}_q[U]$ is a map in $\mathcal{K}$, then $\mathbf{T}_{U,\alpha}(P)$ is also a finitely generated polynomial algebra over $\mathbb{F}_q$ with the same number of generators as P.*

PROOF: By the embedding theorem, Theorem 8.4.4, we can find an embedding of $\mathscr{P}^*$-algebras $\psi : H \hookrightarrow \mathbb{F}_q[z_1, \ldots, z_n]$, where $n = \dim(H)$ and $\mathbb{F}_q[z_1, \ldots, z_n]$ is a polynomial algebra on generators of degree one which is finitely generated as an H-module. By [372] Theorem 6.4.4 $\mathbb{F}_q[z_1, \ldots, z_n]$ is a free H-module. Consider the diagram

$$\begin{array}{ccc} H & \xrightarrow{\psi} & \mathbb{F}_q[z_1, \ldots, z_n] \\ {\scriptstyle\alpha}\downarrow & \swarrow{\scriptstyle\varphi} & \\ \mathbb{F}_q[U] & & \end{array}$$

and note that as ψ is monic and $\mathbb{F}_q[U]$ is an injective in the category $\mathcal{K}$ there is a map φ of $\mathscr{P}^*$-algebras making the diagram commute. By Proposition 10.4.3 $\mathbf{T}_{U,\alpha}(\mathbb{F}_q[z_1, \ldots, z_n])$ is a free $\mathbf{T}_{U,\alpha}(H)$-module. By Lemma 10.3.5 the map

$$\begin{array}{c} \mathbb{F}_q(\alpha) \otimes_{\mathbf{T}_U(H)_0} \mathbf{T}_{U,\varphi}(\mathbb{F}_q[z_1, \ldots, z_n]) = \mathbf{T}_{U,\alpha}(\mathbb{F}_q[z_1, \ldots, z_n]) \\ \downarrow \\ \mathbb{F}_q(\varphi) \otimes_{\mathbf{T}_U(\mathbb{F}_q[z_1, \ldots, z_n])_0} \mathbf{T}_{U,\psi}(\mathbb{F}_q[z_1, \ldots, z_n]) = \mathbf{T}_{U,\varphi}(\mathbb{F}_q[z_1, \ldots, z_n]) \end{array}$$

is a split epimorphism of $\mathbf{T}_{U,\alpha}(H)$-modules. Hence $\mathbf{T}_{U,\varphi}(\mathbb{F}_q[z_1, \ldots, z_n])$ is also free as a $\mathbf{T}_{U,\alpha}(H)$-module. By Example 1 in Section 10.1

$$\mathbf{T}_{U,\varphi}(\mathbb{F}_q[z_1, \ldots, z_n]) \cong \mathbb{F}_q[z_1, \ldots, z_n].$$

Thus, viewed as a module over $\mathbf{T}_{U,\alpha}(H)$ along the map

$$\mathbf{T}_{U,\alpha}(H) \longrightarrow \mathbf{T}_{U,\alpha}(\mathbb{F}_q[z_1, \ldots, z_n]) \longrightarrow \mathbf{T}_{U,\varphi}(\mathbb{F}_q[z_1, \ldots, z_n]) = \mathbb{F}[z_1, \ldots, z_n]$$

$\mathbb{F}_q[z_1, \ldots, z_n]$ becomes a free $\mathbf{T}_{U,\alpha}(H)$-module. Hence this map is an embedding

$$\mathbf{T}_{U,\alpha}(H) \hookrightarrow \mathbb{F}_q[z_1, \dots, z_n].$$

It follows from [372] Theorem 6.4.4 that $\mathbf{T}_{U,\alpha}(H)$ is a polynomial algebra, and by Corollary 10.3.6 has $n = \dim(H)$ generators . □

10.5 The T-Functor and Complete Intersections

In this section we discuss the behavior of the **T**-functor and complete intersections. For this we need an appropriate characterization of graded complete intersections along the lines of [113]. The following recognition principle as well as its proof was communicated to us by S. Iyengar [181]. We are indebted to him for permission to make use of it here.[4]

The characterization of complete intersections we employ is based on the module of **Kähler differentials** for which there are many equivalent definitions. See e.g., [243] Chapter 9 or [104] Chapter 16. We choose to work with the following definition because it makes it easy to relate Kähler differentials to the **T**-functors.

DEFINITION: *Let A be a commutative graded connected algebra over the field $\mathbb{F}$. The module $\Omega_{A\,|\,\mathbb{F}}$ of Kähler differentials is defined to be*

$$\Omega_{A\,|\,\mathbb{F}} = \mathrm{Tor}_1^{A \otimes A}(A, A)$$

where $A \otimes A$ acts on A via the multiplication map $A \otimes A \xrightarrow{\mu} A$.

The module of Kähler differentials is an A-module. The A-module structure is obtained by letting A act on either the left or right hand variable of $\mathrm{Tor}^{A\otimes A}(-, -)$ via left multiplication or right multiplication as the case may be.

Let K be the kernel of the multiplication map $\mu : A \otimes A \longrightarrow A$ so that

$$0 \longrightarrow K \longrightarrow A \otimes A \xrightarrow{\mu} A \longrightarrow 0$$

is a short exact sequence of $A \otimes A$-modules. This gives a long exact sequence

$$\cdots \longrightarrow \mathrm{Tor}_1^{A\otimes A}(A \otimes A, A) \xrightarrow{\mathrm{Tor}_1^{A\otimes A}(\mu, 1)} \mathrm{Tor}_1^{A\otimes A}(A, A) \xrightarrow{\partial} K \otimes_{A\otimes A} A \longrightarrow (A \otimes A) \otimes_{A\otimes A} A \longrightarrow A \otimes_{A\otimes A} A \longrightarrow 0.$$

Using that $A \otimes A$ is free over itself this yields an exact sequence

$$0 \longrightarrow \Omega_{A\,|\,\mathbb{F}} \xrightarrow{\partial} K/K^2 \longrightarrow A \xrightarrow{\mathrm{id}} A \longrightarrow 0$$

[4] This result is no doubt well known to the experts, but we were unable to locate any published proof for it.

So $\Omega_{A|\mathbb{F}} \cong K/K^2$, and from this one sees that $\Omega_{\mathbb{F}[x_1,\ldots,x_m]|\mathbb{F}}$ is a free module over $\mathbb{F}[x_1,\ldots,x_m]$ of rank m (see e.g., [431] Theorem 1.1 or [285] Proposition 4.1).

If $\varphi : B \longrightarrow A$ is a map of commutative graded connected algebras over the field $\mathbb{F}$ then φ induces a map $\Omega(\varphi) : \Omega_{B|\mathbb{F}} \longrightarrow \Omega_{A|\mathbb{F}}$ from which we may construct the map of A-modules

$$\Omega(\varphi) \otimes_B A : \Omega_{B|\mathbb{F}} \otimes_B A \longrightarrow \Omega_{A|\mathbb{F}}.$$

It is not hard to see that for φ an epimorphism the map $\Omega(\varphi)$ is also an epimorphism; see [243] 25.1 or [104] 16.3. If J denotes the kernel of φ then J/J^2 is in a natural way an A-module and one has the exact sequence

$$J/J^2 \longrightarrow \Omega_{B|\mathbb{F}} \otimes_B A \longrightarrow \Omega_{A|\mathbb{F}} \longrightarrow 0$$

which is called the **conormal sequence**.

PROPOSITION 10.5.1: *Let* $B = \mathbb{F}[x_1,\ldots,x_m] \xrightarrow{\varphi} A$ *be an epimorphism of commutative graded connected algebras over the perfect field* $\mathbb{F}$, *and suppose that the kernel* J *of* φ *is generated by a regular sequence, so* A *is a graded complete intersection. Assume in addition that* A *is reduced. Then the conormal sequence is short exact and yields a free resolution*

$$0 \longrightarrow J/J^2 \longrightarrow \Omega_{B|\mathbb{F}} \otimes_B A \longrightarrow \Omega_{A|\mathbb{F}} \longrightarrow 0$$

of $\Omega_{A|\mathbb{F}}$ *as* A*-module. Hence* $\text{hom-dim}_A(\Omega_{A|\mathbb{F}}) \leq 1$.

PROOF: We saw above that $\Omega_{B|\mathbb{F}}$ is a free B-module of rank m, so $\Omega_{B|\mathbb{F}} \otimes_B A$ is a free A-module also of rank m.. Since J is generated by a regular sequence J/J^2 is a free A-module by [431] Theorem 1. The rank of this module is the length c of the regular sequence. What remains to be proved is that the map

$$J/J^2 \longrightarrow \Omega_{B|\mathbb{F}} \otimes_B A$$

is a monomorphism. Let C be the kernel of this map, so we have the exact sequence

$$0 \longrightarrow C \longrightarrow \bigoplus_c A \longrightarrow \bigoplus_m A \longrightarrow \Omega_{A|\mathbb{F}} \longrightarrow 0.$$

Let $\mathfrak{p} \subset A$ be a minimal prime ideal. Since A is reduced the localization of A at $\mathfrak{p}$ is a field $\mathbb{K}$, the residue field of A at $\mathfrak{p}$. (See for example the discussion following Corollary 3.13 in [20].) Localization commutes with the formation of Kähler differentials and $\mathbb{K}$ is a flat A-module so we obtain an exact sequence

$$(\star) \qquad 0 \longrightarrow C \otimes_A \mathbb{K} \longrightarrow \bigoplus_c \mathbb{K} \longrightarrow \bigoplus_m \mathbb{K} \longrightarrow \Omega_{A|\mathbb{F}} \otimes_A \mathbb{K} \longrightarrow 0.$$

The field $\mathbb{K}$ contains $\mathbb{F}$, and the field $\mathbb{F}$ is perfect. It follows from a Theorem of S. Mac Lane [231], or [104] Theorem A.1.3 and Corollary A.1.7, that the field extension $\mathbb{K} \mid \mathbb{F}$ is separable. Hence by [104] Corollary 16.17

$$\dim_{\mathbb{K}}(\Omega_{A \mid \mathbb{F}} \otimes_A \mathbb{K}) = \dim_{\mathbb{K}}(\Omega_{\mathbb{K} \mid \mathbb{F}}) = \operatorname{tr}-\deg(\mathbb{K} \mid \mathbb{F}) = \dim(A) = m - c,$$

where $\operatorname{tr}-\deg(\mathbb{K} \mid \mathbb{F})$ denotes the transcendence degree of the field extension $\mathbb{K} \mid \mathbb{F}$. Putting this into the exact sequence ($\bigstar$) and taking Euler characteristics shows that $C \otimes_A \mathbb{K} = 0$. Therefore $A \setminus \mathfrak{p}$ annihilates C for any minimal prime ideal $\mathfrak{p}$ of A. Since A is reduced

$$\bigcup_{\mathfrak{p} \in \operatorname{Ass}(0)} (A \setminus \mathfrak{p}) = A \setminus \bigcap_{\mathfrak{p} \in \operatorname{Ass}(0)} \mathfrak{p} = A,$$

so A annhilates C and hence C is zero as was claimed. □

PROPOSITION 10.5.2: *Let $\mathbb{F}$ be a perfect field and A a commutative graded connected algebra over $\mathbb{F}$ which is reduced and Noetherian. If* $\operatorname{hom-dim}_A(\Omega_{A \mid \mathbb{F}}) \leq 1$ *then A is a complete intersection.*

PROOF: Choose an epimorphism $B = \mathbb{F}[x_1, \ldots, x_m] \xrightarrow{\varphi} A$ of commutative graded connected algebras over $\mathbb{F}$ and let

$$J/J^2 \longrightarrow \Omega_{B \mid \mathbb{F}} \otimes_B A \longrightarrow \Omega_{A \mid \mathbb{F}} \longrightarrow 0$$

be the associated conormal sequence. If $L = \operatorname{Im}(J/J^2 \longrightarrow \Omega_{B \mid \mathbb{F}} \otimes_B A)$ then

$$0 \longrightarrow L \longrightarrow \Omega_{B \mid \mathbb{F}} \otimes_B A \longrightarrow \Omega_{A \mid \mathbb{F}} \longrightarrow 0$$

is exact. Since $\Omega_{A \mid \mathbb{F}}$ has projective dimension at most 1 the A-module L is a free A-module. We need to estimate its rank.

Let $\mathfrak{p} \subset A$ be a minimal prime ideal and $\mathbb{K}$ the residue field of A at $\mathfrak{p}$. The functor $- \otimes_A \mathbb{K}$ is exact since $\mathbb{K}$ is a flat A module, hence we obtain an exact sequence

$$(\substack{\bigstar \\ \bigstar}) \qquad 0 \longrightarrow L \otimes_A \mathbb{K} \longrightarrow (\Omega_{B \mid \mathbb{F}} \otimes_B A) \otimes_A \mathbb{K} \longrightarrow \Omega_{\mathbb{K} \mid \mathbb{F}} \longrightarrow 0,$$

where we have identified $\Omega_{A \mid \mathbb{F}} \otimes_A \mathbb{K}$ with $\Omega_{\mathbb{K} \mid \mathbb{F}}$: This is valid because localization commutes with the formation of Kähler differentials. Since $\mathbb{F}$ is perfect the field extension $\mathbb{K} \mid \mathbb{F}$ is separable by Mac Lane's theorem ([231] or [104] Theorem A.1.3 and Corollary A.1.7). So by [104] Corollary 16.17 the dimension of $\Omega_{\mathbb{K} \mid \mathbb{F}}$ as a $\mathbb{K}$-vector space is equal to the Krull dimension of A. If $\dim(A) = n$ then taking Euler characteristics in ($\substack{\bigstar \\ \bigstar}$) gives that $L \otimes_A \mathbb{K}$ has dimension $m - n = c$ as a vector space over $\mathbb{K}$. Hence the rank of the free A-module L is c, which is the height of the ideal J.

The map $J/J^2 \longrightarrow L$ is an epimorphism of A-modules, and since L is a free A-module, it splits. Hence J/J^2 contains a free summand of rank equal to the height of the ideal J. By Theorem 1.1 of [431] this implies that J is generated by a regular sequence of length c so A is a complete intersection. □

Combining Propositions 10.5.1 and 10.5.2 gives us the following homological characterization of graded complete intersections over perfect fields.

COROLLARY 10.5.3: *Let $\mathbb{F}$ be a perfect field and A a commutative graded connected algebra over $\mathbb{F}$ which is reduced and Noetherian. Then A is a graded complete intersection if and only if* $\text{hom-dim}_A(\Omega_{A\,|\,\mathbb{F}}) \leq 1$.

PROOF: Combine 10.5.1 and 10.5.2. □

We next show that the functor $\mathbf{T}_U$ commutes with the formation of Kähler differentials ([101] Proposition 3.1).

PROPOSITION 10.5.4 (W. G. Dwyer - C. W. Wilkerson): *Let H be an unstable-$\mathscr{P}^*$ algebra over $\mathbb{F}_q$ and $\varphi : H \longrightarrow \mathbb{F}_q[U]$ a map of $\mathscr{P}^*$-unstable algebras. Then $\mathbf{T}_{U,\varphi}(\Omega_{H\,|\,\mathbb{F}_q}) \cong \Omega_{\mathbf{T}_{U,\varphi}(H)\,|\,F_q}$ as $\mathbf{T}_{U,\varphi}(H)$-modules.*

PROOF: Let $\overline{\varphi} : H \otimes H \xrightarrow{\mu} H \xrightarrow{\varphi} \mathbb{F}_q[U]$. Note that Lannes' tensor product theorem, Theorem 10.1.1, implies for any $H \odot \mathscr{P}^*$-module M that

$$\begin{aligned} \mathbf{T}_{U,\overline{\varphi}}(M) &= \mathbb{F}_q(\overline{\varphi}) \otimes_{\mathbf{T}_U(H\otimes H)_0} \mathbf{T}(M) \cong \mathbb{F}_q(\overline{\varphi}) \otimes_{\mathbf{T}_U(H\otimes H)_0} \mathbf{T}(M) \\ &\cong \mathbb{F}_q(\varphi) \otimes_{\mathbf{T}_U(H)_0} \mathbf{T}(M) \end{aligned}$$

since the map $\mathbf{T}_U(\mu)$ is an epimorphism. By the Dwyer-Wilkerson tensor product theorem [100] Proposition 2.2 $\mathbf{T}_{U,\overline{\varphi}}(H \otimes H) \cong \mathbf{T}_{U,\varphi}(H) \otimes \mathbf{T}_{U,\varphi}(H)$. From these two facts, the definitions, and Lemma 10.4.1 we have

$$\begin{aligned} \mathbf{T}_{U,\varphi}(\Omega_{H\,|\,\mathbb{F}_q}) &= \mathbf{T}_{U,\overline{\varphi}}(\Omega_{H\,|\,\mathbb{F}_q}) = \mathbf{T}_{U,\overline{\varphi}}(\mathrm{Tor}_1^{H\otimes H}(H, H)) \\ &\cong \mathrm{Tor}_1^{\mathbf{T}_{U,\bar{\varphi}}(H\otimes H)}(\mathbf{T}_{U,\overline{\varphi}}(H), \mathbf{T}_{U,\overline{\varphi}}(H)) \\ &\cong \mathrm{Tor}_1^{\mathbf{T}_{U,\varphi}(H)\otimes \mathbf{T}_{U,\varphi}(H)}(\mathbf{T}_{U,\overline{\varphi}}(H), \mathbf{T}_{U,\overline{\varphi}}(H)) \\ &\cong \mathrm{Tor}_1^{\mathbf{T}_{U,\varphi}(H)\otimes \mathbf{T}_{U,\varphi}(H)}(\mathbf{T}_{U,\varphi}(H), \mathbf{T}_{U,\varphi}(H)) \\ &\cong \mathrm{Tor}_1^{\mathbf{T}_{U,\varphi}(H)\otimes \mathbf{T}_{U,\varphi}(H)}(\mathbf{T}_{U,\varphi}(H), \mathbf{T}_{U,\varphi}(H)) \\ &\cong \Omega_{\mathbf{T}_{U,\varphi}(H)\,|\,\mathbb{F}_q} \end{aligned}$$

and the result is established. □

Since finite fields are perfect we may avail ourselves of Corollary 10.5.3 to show that the functor $\mathbf{T}_U$ preserves reduced complete intersections.

THEOREM 10.5.5: *Let H be a $\mathscr{P}^*$-unstable algebra that is a Noetherian reduced graded complete intersection over $\mathbb{F}_q$. If $\varphi : H \longrightarrow \mathbb{F}_q[U]$ is a map of unstable algebras then $\mathbf{T}_{U,\varphi}(H)$ is also a Noetherian reduced graded complete intersection over $\mathbb{F}_q$.*

PROOF: That $\mathbf{T}_{U,\varphi}(H)$ is Noetherian and reduced follows from Theorem 10.3.1 and Proposition 10.3.7. To verify it is a complete intersection, note that Proposition 10.5.1 applies to tell us that $\text{hom-dim}_A(\Omega_{A\,|\,\mathbb{F}}) \leq 1$.

By Proposition 10.5.4 $\mathbf{T}_{U,\varphi}(\Omega_{H|\mathbb{F}_q}) \cong \Omega_{\mathbf{T}_{U,\varphi}(H)|F_q}$ as $\mathbf{T}_{U,\varphi}(H)$-modules and by Corollary 10.4.4 $\text{proj-dim}_{\mathbf{T}_{U,\varphi}(H)}(\mathbf{T}_{U,\varphi}(\Omega_{H|\mathbb{F}_q})) \le 1$. Since $\mathbb{F}_q$ is perfect Proposition 10.5.2 yields the desired conclusion. □

10.6 Invariants of Stabilizer Subgroups

In this section we collect some applications of the **T**-functors to invariant theory as corollaries to the more general theorems of the preceding sections. The key result is Proposition 10.1.7 relating the **T**-functor with stabilizer subgroups.

R. Steinberg, [407], and H. Nakajima, [277], have shown that the ring of invariants $\mathbb{F}[V]^{G_U}$ is a polynomial algebra whenever $\mathbb{F}[V]^G$ is polynomial, where $G_U \le G$ is the pointwise stabilizer subgroup of a subspace $U \le V$. For a finite field $\mathbb{F}$ this follows from the theorem of W. G. Dwyer and C. W. Wilkerson, Theorem 10.4.6.

COROLLARY 10.6.1 (H. Nakajima): *Let $\varrho : G \hookrightarrow \mathrm{GL}(n, \mathbb{F}_q)$ be a representation of a finite group G over the Galois field $\mathbb{F}_q$. Let $U \le V = \mathbb{F}_q^n$ be a linear subspace. If $\mathbb{F}_q[V]^G$ is a polynomial algebra, then so is $\mathbb{F}_q[V]^{G_U}$, where G_U is the pointwise stabilizer of U in G.*

PROOF: Let $\alpha : \mathbb{F}_q[V]^G \longrightarrow \mathbb{F}_q[U]$ be the composition of i^* with the canonical inclusion $\mathbb{F}_q[V]^G \hookrightarrow \mathbb{F}_q[V]$. By Theorem 10.4.6, applying a component of the **T**-functor to a polynomial algebra leads to a polynomial algebra. By Proposition 10.1.7 $\mathbf{T}_{U,\alpha}(\mathbb{F}_q[V]^G) \cong \mathbb{F}_q[V]^{G_U}$. □

Similarly, we can combine Proposition 10.1.7 with Theorem 10.4.4 and Auslander - Buchsbaum equality to get our next result:

COROLLARY 10.6.2: *Let $\varrho : G \hookrightarrow \mathrm{GL}(n, \mathbb{F}_q)$ be a representation of a finite group G over the Galois field $\mathbb{F}_q$. Let $U \le V = \mathbb{F}_q^n$. Then*

$$\text{hom-codim}(\mathbb{F}_q[V]^{G_U}) \ge \text{hom-codim}(\mathbb{F}_q[V]^G).$$

In particular, if $\mathbb{F}_q[V]^G$ is Cohen-Macaulay, then so is $\mathbb{F}_q[V]^{G_U}$. □

It can be easily seen that the inequality in the preceding Corollary need not be an equality. Take, e.g., the regular representation of $\mathbb{Z}/4$ over a field of characteristic two (Example 3 of Section 1.1). The ring of invariants has depth 3, but taking a stabilizer subgroup of, say, $U = \mathrm{Span}_{\mathbb{F}}\{x_1\}$ gives us the trivial group with polynomial invariants, which of course has depth 4.

As a further result of this nature we obtain the analog for complete intersections of H. Nakajima's theorem (Corollary 10.6.1). As with polynomial algebras, the question of when a ring of invariants is a complete intersection has been much studied. In this connection note that in Theorem A of

[188] V. Kac and K. Watanabe show that a group G, such that its ring of invariants $\mathbb{F}[V]^G$ is a complete intersection, is generated by **bireflections**, i.e., elements $g \in G$ such that the rank of the linear transformation $g - \mathbf{I}$ is at most two. Moreover, they show in Theorem C, loc. cit., that in this case a pointwise stabilizer subgroup G_U, $U \subset V$, is also generated by bireflections. This was also shown by H. Nakajima in [279], and both results have been obtained independently by N. L. Gordeev, [147]. In [282] H. Nakajima and K. Watanabe present examples to illustrate that the converse of Theorem A is not true: Not every bireflection group has a ring of invariants that is a complete intersection. Indeed, the converse of Theorem A in [188] is only true in small dimensions, namely for $n \le 3$. In this sense, the following corollary sharpens things. It also improves on Theorem 2.3 of [380] as well as correcting its proof.

COROLLARY 10.6.3: *Let $\varrho : G \hookrightarrow \mathrm{GL}(n, \mathbb{F}_q)$ be a representation of a finite group over the Galois field $\mathbb{F}_q$ and $V = \mathbb{F}_q^n$. Suppose that $\mathbb{F}_q[V]^G$ is a complete intersection. If $U \le V$ is a subspace of V and G_U the pointwise stabilizer of U in G, then $\mathbb{F}_q[V]^{G_U}$ is also a complete intersection.*

PROOF: This follows from Proposition 10.1.7 and Theorem 10.5.5. □

We turn to some examples. Note that the classical groups obtained as isotropy groups of certain bilinear forms, such as the orthogonal, symplectic, and unitary groups, can be considered as forming infinite chains of groups, where each is a stabilizer subgroup of the next. We illustrate how this can be used with the (special) orthogonal groups; cf. Sections 13.4 and 13.5 in [297].

EXAMPLE 1: Let p be an odd prime and $\mathbb{F}_q$ the finite field with $q = p^\nu$ elements. Let Q be a nondegenerate quadratic form in n variables over $\mathbb{F}_q$. Up to change of basis Q can be brought into one of the following canonical forms [87]

$$Q = \begin{cases} x_1x_2 + \cdots + x_{n-2}x_{n-1} + x_n^2 & \text{if } n \text{ is odd,} \\ x_1x_2 + \cdots + x_{n-1}x_n & \text{if } n \text{ is even,} \\ x_1x_2 + \cdots + x_{n-3}x_{n-2} + x_{n-1}^2 - \lambda x_n^2 & \text{if } n \text{ is even, } \lambda \in \mathbb{F}_q^\times \text{ not a square.} \end{cases}$$

In addition, if n is even and λ', λ'' are nonsquares in $\mathbb{F}_q$ then the quadratic forms

$$x_1x_2 + \cdots + x_{n-3}x_{n-2} + x_{n-1}^2 - \lambda' x_n^2$$
$$x_1x_2 + \cdots + x_{n-3}x_{n-2} + x_{n-1}^2 - \lambda'' x_n^2$$

can be transformed one into the other by a change of coordinates. The **orthogonal group** of Q is, by definition, the isotropy subgroup of the quadratic form Q in the general linear group $\mathrm{GL}(n, \mathbb{F}_q)$. Hence we have, up to conjugation in $\mathrm{GL}(n, \mathbb{F}_q)$, according to the three possible normal forms of Q, three orthogonal groups:[5]

$$\begin{array}{ll} \mathbb{O}(n, \mathbb{F}_q) & \text{for odd } n, \\ \mathbb{O}_+(n, \mathbb{F}_q) & \text{for even } n, \\ \mathbb{O}_-(n, \mathbb{F}_q) & \text{for even } n. \end{array}$$

We use the notation $\mathbb{O}_\pm(n, \mathbb{F}_q)$ to mean either of the orthogonal groups when n is even, and similarly for the special orthogonal groups.

The defining representation of the orthogonal groups yield inclusions

$$\begin{bmatrix} \begin{matrix} 1 & 0 \\ 0 & 1 \end{matrix} & 0 \\ 0 & \mathbb{O}_\pm(2m, \mathbb{F}_q) \end{bmatrix} \leq \mathbb{O}_\pm(2m+2, \mathbb{F}_q),$$

$$\begin{bmatrix} \mathbb{O}_+(2m, \mathbb{F}_q) & 0 \\ 0 \cdots 0 & 1 \end{bmatrix} \leq \mathbb{O}(2m+1, \mathbb{F}_q).$$

So, for suitably chosen vector subspaces W and U we obtain

$$\mathbf{T}_{U,i^*}(\mathbb{F}_q[V]^{\mathbb{O}_\pm(2m+2,\mathbb{F}_q)}) = \mathbb{F}_q[W]^{\mathbb{O}_\pm(2m,\mathbb{F}_q)} \otimes \mathbb{F}_q[U], \text{ and}$$
$$\mathbf{T}_{U,i^*}(\mathbb{F}_q[V]^{\mathbb{O}(2m+1,\mathbb{F}_q)}) = \mathbb{F}_q[W]^{\mathbb{O}_+(2m,\mathbb{F}_q)} \otimes \mathbb{F}_q[U].$$

If we restrict our attention to the special orthogonal groups, i.e., orthogonal matrices of determinant 1, then in the same way as above we obtain

$$\begin{bmatrix} \begin{matrix} 1 & 0 \\ 0 & 1 \end{matrix} & 0 \\ 0 & \mathbb{SO}_\pm(2m, \mathbb{F}_q) \end{bmatrix} \leq \mathbb{SO}_\pm(2m+2, \mathbb{F}_q), \text{ and}$$

$$\begin{bmatrix} \mathbb{SO}_+(2m, \mathbb{F}_q) & 0 \\ 0 \cdots 0 & 1 \end{bmatrix} \leq \mathbb{SO}(2m+1, \mathbb{F}_q).$$

So, for suitably chosen vector subspaces W and U we get

$$\mathbf{T}_{U,i^*}(\mathbb{F}_q[V]^{\mathbb{SO}_\pm(2m+2,\mathbb{F}_q)}) = \mathbb{F}_q[W]^{\mathbb{SO}_\pm(2m,\mathbb{F}_q)} \otimes \mathbb{F}_q[U], \text{ and}$$
$$\mathbf{T}_{U,i^*}(\mathbb{F}_q[V]^{\mathbb{SO}(2m+1,\mathbb{F}_q)}) = \mathbb{F}_q[W]^{\mathbb{SO}_+(2m,\mathbb{F}_q)} \otimes \mathbb{F}_q[U].$$

The point of all this is that the rings of invariants of the special orthogonal groups in dimension 2 are not polynomial, but hypersurfaces:

$$\mathbb{F}_q[x, y]^{\mathbb{SO}_+(2,\mathbb{F}_q)} = \mathbb{F}_q[xy, x^{q-1}, y^{q-1}] / \left((xy)^{q-1} - x^{q-1}y^{q-1}\right)$$

and

$$\mathbb{F}_q[x, y]^{\mathbb{SO}_-(2,\mathbb{F}_q)} = \mathbb{F}_q[x^2 + y^2, x^{q+1} + y^{q+1}, x^q y - xy^q]/(r),$$

[5] These groups, as all of the classical groups over finite fields, were investigated by L. E. Dickson in [87]: He calculates the order of these groups, finds generators, and shows many of them are simple groups, or have simple subgroups of small index.

where the relation r is given by

$$r = (x^{q+1} + y^{q+1})^2 - (x^2 + y^2)^{q+1} + (x^q y - xy^q)^2.$$

This generalizes Example 6 of Section 7.4 to all Galois fields. By applying $\mathbf{T}_{U,i^*}(-)$, with $U < V$ of codimension 2, to the rings of invariants of a special orthogonal group of higher dimension $2n$, we receive

$$\mathbf{T}_{U,i^*}(\mathbb{F}_q[z_1, \ldots, z_n]^{\mathrm{SO}_\pm(n,\mathbb{F}_q)}) \cong \mathbb{F}_q[z_1, z_2]^{\mathrm{SO}_\pm(2,\mathbb{F}_q)} \otimes \mathbb{F}_q[z_3, \ldots, z_n].$$

If U has codimension 1 and n is odd then

$$\mathbf{T}_{U,i^*}(\mathbb{F}_q[z_1, \ldots, z_n]^{\mathrm{SO}(n,\mathbb{F}_q)}) \cong \mathbb{F}_q[z_1, \ldots, z_{n-1}]^{\mathrm{SO}_+(n-1,\mathbb{F}_q)} \otimes \mathbb{F}_q[z_n].$$

Therefore none of these can be a polynomial ring. So put optimistically, the rings of invariants of the finite special orthogonal groups are at best hypersurfaces.

10.7 A Last Look at the Transfer

In this section we apply **T**-technology to study the transfer homomorphism. See Sections 2.2 and 2.4 for the basic notation and results on the transfer homomorphism.

Let $\rho : G \hookrightarrow \mathrm{GL}(n, \mathbb{F}_q)$ be a representation of a finite group G over the Galois field $\mathbb{F}_q$, and $U \leq V$ a linear subspace of $V = \mathbb{F}_q^n$, with $i : U \hookrightarrow V$ the inclusion. We are going to show how the functor $\mathbf{T}_U$ allows us to relate the image of the transfer homomorphism $\mathrm{Im}(\mathrm{Tr}^G) \subseteq \mathbb{F}_q[V]^G$ for G to that of the transfer homomorphism $\mathrm{Im}(\mathrm{Tr}^{G_U}) \subseteq \mathbb{F}_q[V]^{G_U}$ of the pointwise stabilizer subgroup G_U.

To this end fix $\rho : G \hookrightarrow \mathrm{GL}(n, \mathbb{F}_q)$ and $U \leq V = \mathbb{F}_q^n$. Introduce the notation $\nu : \mathbb{F}_q[V]^G \hookrightarrow \mathbb{F}_q[V]$ for the inclusion and $\alpha = i^* \cdot \nu : \mathbb{F}_q[V]^G \longrightarrow \mathbb{F}_q[U]$ for the composite map of $\mathscr{P}^*$-algebras. From Proposition 10.1.7 we know that $\mathbf{T}_{U,\alpha}(\mathbb{F}_q[V]^G)$ is isomorphic with $\mathbb{F}[V]^{G_U}$, the ring of invariants of the pointwise stabilizer $G_U \leq G$ of U in V.

If we regard $\mathbb{F}_q[V]$ as an $\mathbb{F}_q[V]^G$-module via ν, then the transfer homomorphism $\mathrm{Tr}^G : \mathbb{F}_q[V] \longrightarrow \mathbb{F}_q[V]$ is a map of $\mathbb{F}_q[V]^G$-modules. Therefore it induces a map $\mathbf{T}_{U,\alpha}(\mathrm{Tr}^G_U) : \mathbf{T}_{U,\alpha}(\mathbb{F}_q[V]) \longrightarrow \mathbf{T}_{U,\alpha}(\mathbb{F}_q[V])$. The group G acts on $\mathbf{T}_{U,\alpha}(\mathbb{F}[V])$ and $\mathbf{T}_{U,\alpha}(\mathrm{Tr}^G)$ coincides with the transfer homomorphism associated to this action. We begin our analysis of this map with a lemma.

LEMMA 10.7.1: *Let $\rho : G \hookrightarrow \mathrm{GL}(n, \mathbb{F}_q)$ be a representation of a finite group G over the Galois field $\mathbb{F}_q$, and $U \leq V$ a linear subspace of $V = \mathbb{F}_q^n$, with $i : U \hookrightarrow V$ the inclusion. Let $\nu : \mathbb{F}_q[V]^G \hookrightarrow \mathbb{F}_q[V]$ be the canonical inclusion and $\alpha = i^* \cdot \nu : \mathbb{F}_q[V]^G \longrightarrow \mathbb{F}_q[U]$ the composite map of $\mathscr{P}^*$-algebras.*

Then $\mathbf{T}_{U,\alpha}(\mathbb{F}_q[V])$ *is a direct product of components of* $\mathbf{T}_U(\mathbb{F}_q[V])$, *viz.,*

$$\mathbf{T}_{U,\alpha}(\mathbb{F}_q[V]) = \prod_{\vartheta\in\Theta} \mathbf{T}_{U,\vartheta}(\mathbb{F}_q[V]),$$

where the product is over the set Θ *of all the* $\vartheta \in \mathrm{Hom}_{\mathbb{F}_q}(U, V)$ *such that the diagram*

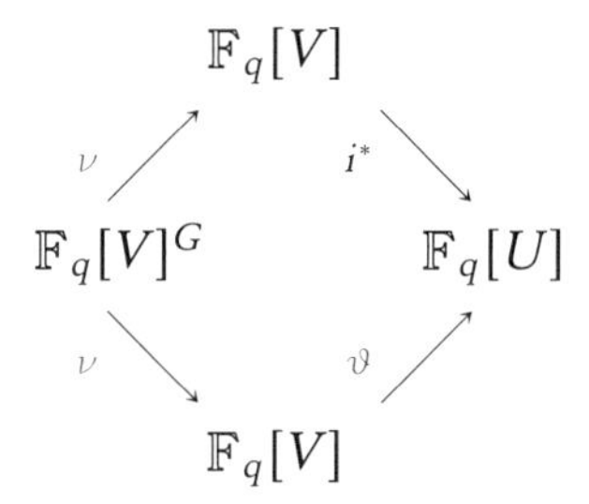

commutes. The elements in Θ *are monomorphisms and the subset* $\Theta \subseteq \mathrm{Hom}_{\mathbb{F}_q}(U, V)$ *is* G*-invariant.*

PROOF: The commutatitivity of the diagram is a special case of Lemma 10.3.2. Since $\mathrm{Im}(\vartheta^* \cdot \nu) = \mathrm{Im}(i^* \cdot \nu)$ is a subalgebra of $\mathbb{F}_q[U]$ of Krull dimension $\dim_{\mathbb{F}_q}(U)$ the map $\vartheta : U \longrightarrow V$ must be a monomorphism. To see that $\Theta \subseteq \mathrm{Hom}_{\mathbb{F}_q}(U, V)$ is G-invariant note that for $\vartheta \in \Theta$, $g \in G$, and $f \in \mathbb{F}_q[V]$ that $(g\vartheta)^*(f) = \vartheta^*(gf)$. If $f \in \mathbb{F}_q[V]^G$ then $gf = f$ and the result follows. □

For $\vartheta \in \Theta$ we introduce the notation $\mathrm{pr}^{\alpha}_{\vartheta} : \mathbf{T}_{U,\alpha}(\mathbb{F}_q[V]) \longrightarrow \mathbf{T}_{U,\vartheta}(\mathbb{F}_q[V])$ for the projection of $\mathbf{T}_{U,\alpha}$ onto the direct factor $\mathbf{T}_{U,\vartheta}(\mathbb{F}_q[V])$ in the decomposition of Lemma 10.7.1. Note that the split epimorphism δ^* of Lemma 10.3.5 is just $\mathrm{pr}^{\alpha}_{i^*}$. To summarize: there is a commutative diagram

$$\begin{array}{ccc}
\mathbf{T}_{U,\alpha}(\mathbb{F}_q[V]) & \xrightarrow{\mathbf{T}_{U,\alpha}(\mathrm{Tr}^G)} & \mathbf{T}_{U,\alpha}(\mathbb{F}_q[V]) \\
\cong\big\downarrow & & \big\downarrow\cong \\
\prod_{\vartheta\in\Theta} \mathbf{T}_{U,\vartheta}(\mathbb{F}_q[V]) & \xrightarrow{\mathrm{Tr}^G} & \prod_{\vartheta\in\Theta} \mathbf{T}_{U,\vartheta}(\mathbb{F}_q[V]).
\end{array}$$

DIAGRAM 10.7.1: Product Representation for $\mathbf{T}_{U,\alpha}(\mathrm{Tr}^G)$

If $f \in \mathbb{F}_q[V]$ and $\vartheta \in \Theta$ we write (f, ϑ) for the element of $\mathbf{T}_{U,\alpha}(\mathbb{F}_q[V])$ whose ϑ-component in the decomposition of Lemma 10.7.1 is f and all other components are zero. The action of G on $\prod_{\vartheta\in\Theta} \mathbf{T}_{U,\vartheta}(\mathbb{F}_q[V])$ is given by $g(f, \vartheta) = (gf, g\vartheta)$ for $g \in G$. It follows from the decomposition displayed in Diagram 10.7.1 that to describe $\mathbf{T}_{U,\alpha}(\mathrm{Tr}^G) : \mathbf{T}_{U,\alpha}(\mathbb{F}_q[V]) \longrightarrow \mathbf{T}_{U,\alpha}(\mathbb{F}_q[V])$ it is sufficient to describe for each $f \in \mathbb{F}_q[V]$ and $\vartheta \in \Theta$ the components of $\mathbf{T}_{U,\alpha}(\mathrm{Tr}^G)(f, \vartheta)$. The following lemma provides us with what we need.

LEMMA 10.7.2: *Let $\rho : G \hookrightarrow \mathrm{GL}(n, \mathbb{F}_q)$ be a representation of a finite group G over the Galois field $\mathbb{F}_q$. Let $U \leq V = \mathbb{F}_q^n$ be a linear subspace and Θ as in Lemma 10.7.1. Let $\omega \in \Theta$ and $G_{\omega(U)}$ be the pointwise stabilizer of the subspace $\omega(U) \leq V = \mathbb{F}_q^n$. Choose a right transversal $\mathfrak{T}(\omega) = \{t_1(\omega), \ldots, t_k(\omega)\}$ for $G_{\omega(U)}$ in G. Then*

$$\mathrm{pr}_{\omega}^{\alpha}(\mathbf{T}_{U,\alpha}(\mathrm{Tr}^G)(f, \vartheta)) = \mathrm{Tr}^{G_{\omega(U)}}\Big(\sum_{t(\omega) \in \mathfrak{T}(\omega) \mid t(\omega)\vartheta = \omega} t(\omega) f \Big).$$

PROOF: If we write

$$\widehat{\mathrm{Tr}^{G}_{G_{\omega(U)}}} = \sum_{t(\omega) \in \mathfrak{T}(\omega)} t(\omega)$$

then we have the factorization $\mathrm{Tr}^G = \mathrm{Tr}^{G_{\omega}(U)} \cdot \widehat{\mathrm{Tr}^{G}_{G_{\omega(U)}}}$ (compare with the proof of Lemma 2.4.4). Therefore we obtain

$$\begin{aligned}
\mathbf{T}_{U,\alpha}(\mathrm{Tr}^G)(f, \vartheta) &= \mathrm{Tr}^G(f, \vartheta) = \mathrm{Tr}^{G_{\omega(U)}} \cdot \widehat{\mathrm{Tr}^{G}_{G_{\omega(U)}}}(f, \vartheta) \\
&= \mathrm{Tr}^{G_{\omega(U)}}\Big(\sum_{t(\omega) \in \mathfrak{T}(\omega)} t(\omega)(f, \vartheta) \Big) = \mathrm{Tr}^{G_{\omega(U)}}\Big(\sum_{t(\omega) \in \mathfrak{T}(\omega)} (t(\omega)f, t(\omega)\vartheta) \Big)
\end{aligned}$$

and applying $\mathrm{pr}_{\omega}^{\alpha}$ to this yields the stated formula. □

If the transfer homomorphism is regarded as a map $\mathrm{Tr}^G : \mathbb{F}_q[V] \longrightarrow \mathbb{F}_q[V]^G$ of $\mathbb{F}_q[V]^G$-modules, it induces

$$\mathbf{T}_{U,\alpha}(\mathrm{Tr}^G) : \mathbf{T}_{U,\alpha}(\mathbb{F}_q[V]) \longrightarrow \mathbf{T}_{U,\alpha}(\mathbb{F}_q[V]^G).$$

Moreover $\mathbf{T}_{U,\alpha}(\mathbb{F}_q[V]^G) \cong \mathbb{F}_q[V]^{G_U}$ and the composite $\mathrm{pr}_{i^*}^{\alpha} \cdot \mathbf{T}_{U,\alpha}(\nu)$ coincides with the natural inclusion $\mathbb{F}_q[V]^{G_U} \hookrightarrow \mathbb{F}_q[V]$. Making these identification, one has the following commutative diagram.

$$\begin{array}{ccc}
\mathbf{T}_{U,\alpha}(\mathbb{F}_q[V]) & \xrightarrow{\mathbf{T}_{U,\alpha}(\mathrm{Tr}^G)} & \mathbf{T}_{U,\alpha}(\mathbb{F}_q[V]) \\
\Big\downarrow {\scriptstyle \mathbf{T}_{U,\alpha}(\mathrm{Tr}^G)} & & \Big\downarrow {\scriptstyle \mathrm{pr}_{i^*}^{\alpha}} \\
\mathbf{T}_{U,\alpha}(\mathbb{F}_q[V]^G) & \xrightarrow{\mathrm{pr}_{i^*}^{\alpha} \cdot \mathbf{T}_{U,\alpha}(\nu)} & \mathbf{T}_{U,i^*}(\mathbb{F}_q[V])
\end{array}$$

Recall from Proposition 10.1.7 that $\mathbb{F}_q[V]^{G_U} \cong \mathbf{T}_{U,\alpha}(\mathbb{F}_q[V]^G)$. From Lemma 10.7.2 we therefore obtain the following formula.

PROPOSITION 10.7.3: *Let $\rho : G \hookrightarrow \mathrm{GL}(n, \mathbb{F}_q)$ be a representation of a finite group G over the Galois field $\mathbb{F}_q$ and $U \leq V = \mathbb{F}_q^n$ a linear subspace. Let Θ be as in Lemma 10.7.1. Choose a right transversal for G_U in G, $\mathfrak{T} = \{t_1, \ldots, t_k\}$. If we identify $\mathbf{T}_{U,\alpha}(\mathbb{F}_q[V]^G)$ with $\mathbb{F}_q[V]^{G_U}$ then the map $\mathbf{T}_{U,\alpha}(\mathrm{Tr}^G) : \mathbf{T}_{U,\alpha}(\mathbb{F}_q[V]) \longrightarrow \mathbf{T}_{U,\alpha}(\mathbb{F}_q[V]^G)$ is given by the following for-*

mula

$$\mathbf{T}_{U,\alpha}(\mathrm{Tr}^G)(f, \vartheta)) = \mathrm{Tr}^{G_U}\Big(\sum_{t \in \mathfrak{T} \mid t\vartheta = i^*} tf \Big),$$

for any $f \in \mathbb{F}_q[V]$ *and* $\vartheta \in \Theta$. □

REMARK: Although it would seem in the formula of Proposition 10.7.3 that the inclusion $i : U \hookrightarrow V$ plays a special role among all the elements of Θ this is not the case. Since the elements of Θ are monomorphisms, and $\mathbf{T}_{U,\vartheta}(\mathbb{F}_q[V]^G) \cong \mathbb{F}_q[V]^{G_{\vartheta(U)}}$ an analagous formula applies to the map $\mathbf{T}_{U,\gamma}(\mathrm{Tr}^G)$ where γ is the composite $\mathbb{F}_q[V]^{G_{\omega(U)}} \hookrightarrow \mathbb{F}_q[V] \xrightarrow{\vartheta^*} \mathbb{F}_q[U]$.

PROPOSITION 10.7.4: *Let* $\rho : G \hookrightarrow \mathrm{GL}(n, \mathbb{F}_q)$ *be a representation of a finite group* G *over the Galois field* $\mathbb{F}_q$, *and* $U \leq V$ *a linear subspace of* $V = \mathbb{F}_q^n$, *with* $i : U \hookrightarrow V$ *the inclusion. Let* $\nu : \mathbb{F}_q[V]^G \hookrightarrow \mathbb{F}_q[V]$ *be the canonical inclusion and* $\alpha = i^* \cdot \nu : \mathbb{F}_q[V]^G \longrightarrow \mathbb{F}_q[U]$ *the composite map of* $\mathscr{P}^*$*-algebras. Then there is a split epimorphism of* $\mathbb{F}_q[V]^{G_U}$*-modules* $\mathbf{T}_{U,\alpha}(\mathrm{Im}(\mathrm{Tr}^G)) \longrightarrow \mathrm{Im}(\mathrm{Tr}^{G_U})$.

PROOF: By Proposition 10.1.7 $\mathbf{T}_{U,\alpha}(\mathbb{F}_q[V]^G) \cong \mathbb{F}_q[V]^{G_U}$ and by Example 1 in Section 10.1 $\mathbf{T}_{U,i^*}(\mathbb{F}_q[V]) \cong \mathbb{F}_q[V]$. Under these identifications the composition

$$\mathbf{T}_{U,\alpha}(\mathbb{F}_q[V]^G) \xrightarrow{\mathbf{T}_{U,\alpha}(\nu)} \mathbf{T}_{U,\alpha}(\mathbb{F}_q[V]) \xrightarrow{\mathrm{pr}^{\alpha}_{i^*}} \mathbf{T}_{U,i^*}(\mathbb{F}_q[V])$$

corresponds with the inclusion $\mathbb{F}_q[V]^{G_U} \hookrightarrow \mathbb{F}_q[V]$. From Proposition 10.7.4 it follows that $\mathrm{pr}^{\alpha}_{i^*}(\mathrm{Im}(\mathbf{T}_{U,\alpha}(\mathrm{Tr}^G)) \subseteq \mathrm{Im}(\mathrm{Tr}^{G_U})$. On the other hand, from Lemma 10.7.2 we obtain the commutative diagram

$$\begin{array}{ccc}
\mathbf{T}_{U,\alpha}(\mathbb{F}_q[V]) & \xrightarrow{\mathbf{T}_{U,\alpha}(\mathrm{Tr}^G)} & \mathbf{T}_{U,\alpha}(\mathbb{F}_q[V]) \\
{\scriptstyle \mathrm{i}^{\alpha}_{i^*}} \uparrow & & \downarrow {\scriptstyle \mathrm{pr}^{\alpha}_{i^*}} \\
\mathbf{T}_{U,i^*}(\mathbb{F}_q[V]) & \xrightarrow{\mathbf{T}_{U,i^*}(\mathrm{Tr}^G)} & \mathbf{T}_{U,i^*}(\mathbb{F}_q[V]),
\end{array}$$

where $\mathrm{i}^{\alpha}_{i^*}$ is the inclusion of the i^* component, so $\mathrm{pr}^{\alpha}_{i^*}(\mathrm{Im}(\mathbf{T}_{U,\alpha}(\mathrm{Tr}^G)) \supseteq \mathrm{Im}(\mathrm{Tr}^{G_U})$. Therefore $\mathrm{pr}^{\alpha}_{i^*}$ maps $\mathbf{T}_{U,\alpha}(\mathrm{Im}(\mathrm{Tr}^G))$ onto $\mathrm{Im}(\mathrm{Tr}^{G_U})$ with $\mathrm{i}^{\alpha}_{i^*}$ as splitting map. □

THEOREM 10.7.5: *Let* $\rho : G \hookrightarrow \mathrm{GL}(n, \mathbb{F}_q)$ *be a representation of a finite group* G *over the Galois field* $\mathbb{F}_q$, *and* $U \leq V$ *a linear subspace of* $V = \mathbb{F}_q^n$, *with* $i : U \hookrightarrow V$ *the inclusion. If* $\mathrm{Im}(\mathrm{Tr}^G) \subseteq \mathbb{F}_q[V]^G$ *is a principal ideal then so is* $\mathrm{Im}(\mathrm{Tr}^{G_U}) \subseteq \mathbb{F}_q[V]^{G_U}$.

PROOF: If $\mathrm{Im}(\mathrm{Tr}^G)$ is a principal ideal then as $\mathbb{F}[V]^G$-module it is free. Therefore by Propositions 10.4.3 and 10.1.7 $\mathbf{T}_{U,\alpha}(\mathrm{Im}(\mathrm{Tr}^G))$ is a free $\mathbb{F}_q[V]^{G_U}$-module. By Proposition 10.7.4 there is a split epimorphism of $\mathbb{F}_q[V]^{G_U}$-modules $\mathbf{T}_{U,\alpha}(\mathrm{Im}(\mathrm{Tr}^G)) \longrightarrow \mathrm{Im}(\mathrm{Tr}^{G_U})$. Therefore $\mathrm{Im}(\mathrm{Tr}^{G_U})$ is also a free $\mathbb{F}_q[V]^{G_U}$-module, and since it is a submodule of $\mathbb{F}_q[V]^{G_U}$ it must have rank one. So as an ideal it is principal. □

Appendix A
Review of Commutative Algebra

WE COLLECT here as much of the terminology and notation from commutative algebra used in this book that we feel is necessary to orient the reader further. The reader should consult [24], [25], [104], [243], or [372] for missing proofs, examples, and details. In several places we supply proofs because contrary to a widespread myth, it is simply not true that the graded and local cases have the same theorems: There are often subtle differences, and moreover, it is sometimes possible to find proofs in the graded case that are distinctly different from the local analogues, when they exist; see, e.g., Propositions A.1.1 and A.3.2 for examples of what we mean. Caveat! Having said this: If we do quote a result in the text from the literature, which is stated and proven in the cited source for the ungraded case, and we use it in the graded case, then the cited proof needs little or no change to obtain the graded result. We illustrate this in the proof of prime avoidance (see Lemma A.2.1).

A.1 Gradings

Our attitude toward gradings has been strongly influenced by J. C. Moore, and so we use this section to make this viewpoint clear. [1]

DEFINITION: *A* **graded vector space** *over* $\mathbb{F}$ *is a family of vector spaces* $M = \{M_i \mid i \in \mathbb{Z}\}$. *The vector space* M_i *is called the* **component** *of degree* i. *If* $M_i = 0$ *for all* $i < 0$, *we call* M *a* **positively graded** *vector space over* $\mathbb{F}$. *If* $\dim_{\mathbb{F}}(M_i)$ *is finite for each* i, *we say that* M *is of* **finite type**. *If*

[1] There is a long tradition, particularly among algebraic topologists, for this viewpoint, and first used in a text book by S. Mac Lane ([232] page 177). If the reader feels uneasy with this, then the reader can apply what we call the totalization functor (see below) with the proviso that when we speak of elements of a graded object we mean ***homogeneous*** elements, unless explicitly indicated otherwise.

M and N are graded vector spaces over $\mathbb{F}$, a **morphism** $f : M \longrightarrow N$ **of graded vector spaces of degree** d *is a sequence of linear transformations* $\{f_i : M_i \longrightarrow N_{i+d} \mid i \in \mathbb{Z}\}$. *If $d = 0$, we speak simply of a morphism of graded vector spaces.*

NOTATION: The set of all morphisms of degree d between two graded vector spaces M and N over the field $\mathbb{F}$ will be denoted by $\mathrm{Hom}_{\mathbb{F}}(M, N)_d$, and $\mathrm{Hom}_{\mathbb{F}}(M, N)_*$ for the graded object whose components are the $\mathrm{Hom}_{\mathbb{F}}(M, N)_d$ for $d \in \mathbb{Z}$.

Graded vector spaces with $\mathrm{Hom}_{\mathbb{F}}(M, N)_*$ as the morphisms between M and N do not form an abelian category: There are difficulties defining kernels, images, cokernels, and coimages caused by the graded maps of nonzero degrees. For example, if $f : M \longrightarrow N$ has degree $d \neq 0$ how should we grade $\mathrm{Im}(f)$? With the degree coming from M or from N? If we use $\mathrm{Hom}_{\mathbb{F}}(M, N) = \mathrm{Hom}_{\mathbb{F}}(M, N)_0$ as morphism set between M and N, then all is well and we get an abelian category.

REMARK: Graded objects and morphisms over any category are defined analogously. If $\mathcal{C}$ is a category, then the graded objects over $\mathcal{C}$ are the families $C = \{C_i \mid i \in \mathbb{Z}\}$. We often write C_* to indicate that C is a graded object, and $\mathrm{Hom}_{\mathcal{C}}(-, -)_*$ for the graded morphisms of any degree, and $\mathrm{Hom}_{\mathcal{C}}(-, -)$ for those of degree zero. N.b., if no degree is explicitly mentioned for a morphism $f : C' \longrightarrow C''$ between graded objects, then it is understood that the morphism has degree zero.

It would be more precise to call what we have defined $\mathbb{Z}$-graded objects, but we will not do so unless we are forced to for the sake of precision. In this context, the k**-fold suspension** of an object C for $k \in \mathbb{Z}$ is defined by $\Sigma^k(C)_i = C_{i+k}$. Some text books call the suspension the k-fold shift of C and use the notation $C[k]$ for what we denote by $\Sigma^k C$.

An oft used convention for moving grading indices from subscripts to superscripts is $C^k = C_{-k}$. This convention was introduced by S. Eilenberg to turn chain complexes into cochain complexes, thereby obviating the need for twice as many proofs (see [103]).

REMARK: The use of the group $\mathbb{Z}$ to index the grading is not essential. It should be clear how to define **bigraded**, **trigraded**, etc., objects, where the grading indices come from $\mathbb{Z} \times \mathbb{Z}$, $\mathbb{Z} \times \mathbb{Z} \times \mathbb{Z}$, etc. But be warned: Contrary to popular belief, all is not well when the grading indices do not form a well-ordered set. See, for example, Section A.3, Example 1.

DEFINITION: *A* **graded algebra** A *over the field $\mathbb{F}$ is a graded vector space together with morphisms* $\eta : \mathbb{F} \longrightarrow A$ *and* $\mu : A \otimes A \longrightarrow A$ *such that*[2]

[2] So a graded algebra is not an algebra in the nongraded sense.

the diagrams

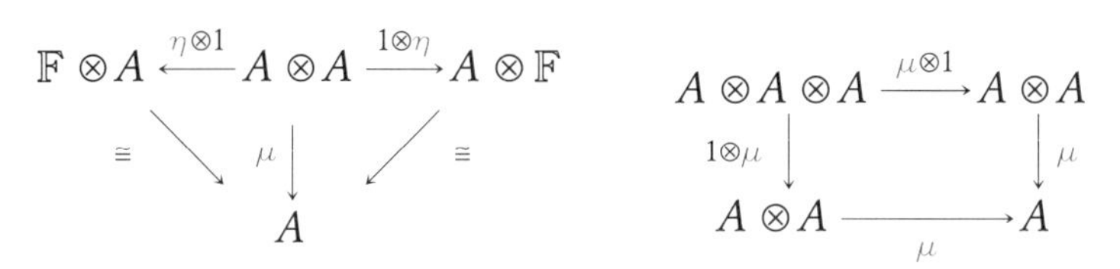

are commutative.

Here we regard $\mathbb{F}$ as a graded vector space over itself concentrated in degree 0, i.e., all components apart from the zero component are the zero vector space, and the zero component is $\mathbb{F}$. The first diagram expresses the fact that $\eta(1)\in A$ is a two-sided unit, and the second diagram that the associative law holds. The distributive laws are a consequence of μ being a linear map. If A is positively graded and η is an isomorphism on the component of degree 0, we say that A is **connected**. If A is connected, the map $\varepsilon : A \longrightarrow \mathbb{F}$ defined by η^{-1} in degree 0 and 0 in positive degrees is called the **augmentation homomorphism**; its kernel is called the **augmentation ideal**, and is denoted by $\overline{A}$.

The usual notions of ring theory, such as ideals, modules, finite generation, etc. may be carried over to graded algebras. For example if

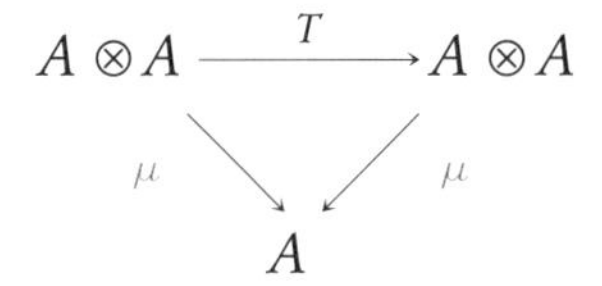

commutes, where $T(a'\otimes a'') = a''\otimes a'$ is the (unsigned) twisting map, then we say that A is **commutative**. N.b. this is ***not*** the graded commutativity prevalent in algebraic topology.

Likewise, one can introduce graded modules as follows:

DEFINITION: *Let A be a graded algebra over the field $\mathbb{F}$. A* **(left) graded module** *over A consists of a graded vector space M together with a map $\alpha : A\otimes M \longrightarrow M$ such that the diagrams*

$$\begin{array}{ccc} A\otimes A\otimes M & \xrightarrow{\mu\otimes 1} & A\otimes M \\ {\scriptstyle 1\otimes\mu}\downarrow & & \downarrow{\scriptstyle\alpha} \\ A\otimes M & \xrightarrow[\alpha]{} & M \end{array} \qquad\qquad \begin{array}{ccc} \mathbb{F}\otimes M & \xrightarrow{\alpha\cdot\eta\otimes 1} & M \\ & {\scriptstyle\cong}\searrow & \downarrow{\scriptstyle 1} \\ & & M \end{array}$$

are commutative.

In Section 4.5 we use the graded field of fractions of a graded integral domain. More generally a graded field is defined as follows:

DEFINITION: *A* **graded field** *is a commutative graded algebra with* *unit, which has the property that every nonzero (homogeneous) element is invertible.*

A graded field $\mathbb{K}$ is not a field in the usual sense, but its degree-zero part $\mathbb{K}_0$ is. Moreover, the set $S = \{s \mid \mathbb{K}_s \neq 0\} \subseteq \mathbb{Z}$ is an additive subgroup. If $S \neq \{0\}$ and we choose a nonzero element $k \in \mathbb{K}$ of minimal positive degree m then one sees $\mathbb{K}_0[k,\ k^{-1}]$ (see, e.g., [242] Lemma 1.2). For if $n \in \mathbb{Z}$ and $x \in K_{nm}$ then $x/k^n \in \mathbb{K}_0$ so $x = \ell \cdot k^n$ with $\ell \in \mathbb{K}_0$. The element k is clearly transcendental over $\mathbb{K}_0$ so we find $\mathbb{K}$ is isomorphic to the ring of Laurent polynomials $\mathbb{K}_0[k,\ k^{-1}]$; see Chapter 2 in [290] for more information on graded field theory.

There are several standard ways to pass back and forth between graded and ungraded modules, rings, etc. In one direction we can replace a graded object A by its **totalization** defined by $\mathrm{Tot}(A) = \oplus A_i$, thereby throwing away the grading. In the other direction we can regard a nongraded object B as graded by **concentrating** it in degree 0, i.e., by declaring $B_0 = B$ and $B_i = 0$ for $i \neq 0$. We often use the same symbol for a graded object and its ungraded totalization, as well as for an ungraded object and that object graded by being concentrated in degree 0: Be warned.

DEFINITION: *If A is a graded connected algebra over a field $\mathbb{F}$ and M is a graded A-module, then $Q_A(M) = M/(\overline{A} \cdot M)$ is the module of* **indecomposable** *elements of M. The indecomposable elements of the algebra A, denoted by QA, are by definition $Q\overline{A}$, where $\overline{A}$ is regarded as a module over A.*

If A is a connected algebra over $\mathbb{F}$, then $A/\overline{A} \cong \mathbb{F}$, and it is easy to see that $Q_A(M) \cong \mathbb{F} \otimes_A M$. When A is clear from context it is customary to write QM for $Q_A(M)$: $Q_A(M)$ is just a graded vector space over $\mathbb{F}$.

PROPOSITION A.1.1 (Graded Nakayama Lemma): *If A is a connected commutative algebra over a field $\mathbb{F}$ and M is a positively graded A-module, then $M = 0$ if and only if $Q_A(M) = 0$.*

PROOF: If $M = 0$, then $Q_A(M) = 0$. Suppose, on the other hand, that $Q_A(M) = 0$. Since M is positively graded, $M_i = 0$ for $i < 0$, so we may suppose inductively that we have shown that $M_i = 0$ for $i < s$. If $x \in M_s$, then since $Q_A(M) = M/(\overline{A} \cdot M) = 0$, it follows that $x = \sum a_r \cdot x_r$ with $a_j \in \overline{A}$. Hence $\deg(a_j) > 0$, and therefore $\deg(x_j) < s$, so $x_j = 0$, and hence $x = 0$. □

COROLLARY A.1.2: *If A is a commutative graded connected algebra over a field and M', M'' are positively graded A-modules, then an A-module morphism $f : M' \longrightarrow M''$ is surjective if and only if the map induced on the module of indecomposables $Qf : Q_A(M)' \longrightarrow Q_A(M)''$ is surjective.*

PROOF: This follows immediately from the fact that the functor $\mathbb{F} \otimes_A -$ sends epimorphisms to epimorphisms and Proposition A.1.1. □

DEFINITION: *If X is a graded set and A a graded algebra, then a free A-module generated by X consists of an A-module F and a map $X \longrightarrow F$ of graded sets, such that for any graded A-module M and any map $X \longrightarrow M$ of graded sets, there exists a unique morphism of graded A-modules $F \longrightarrow M$ making the triangle*

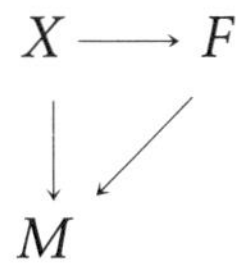

commute.

A free module on any graded set always exists, and is unique up to isomorphism: It it is isomorphic to the direct sum of copies of A indexed by the elements of X and shifted by their grading.

COROLLARY A.1.3: *Let A be a commutative graded connected algebra over a field and M an A-module. Then $x_1, \ldots, x_n, \ldots \in M$ generate M if and only if they project in $Q_A(M)$ to a spanning set.*

PROOF: Form a free A-module F generated by elements z_i of degree $\deg(x_i)$, $i = 1, \ldots, n$, and define an A-module map $\zeta : F \longrightarrow M$ by requiring that $z_i \longmapsto x_i$ for $i = 1, \ldots, n$. Then apply the preceding corollary. □

Thus a basis for $Q_A(M)$ lifts to an A-module generating set for M. Likewise a basis for $Q(A)$ lifts to a minimal generating set for A as an algebra. If M is finitely generated as an A-module then M cannot be generated by fewer than $\dim_{\mathbb{F}}(\mathrm{Tot}(Q_A(M)))$ elements, and the integers

$$\sigma_A^+(M) = \max\{k \mid (Q_A(M))_k \neq 0\}$$
$$\sigma_A^-(M) = \min\{k \mid (Q_A(M))_k \neq 0\}$$

are the maximum and minimum degrees of generators in a minimal generating set. These integers have another interpretation in terms of graded A-module homomorphisms as shown in the following lemma.

LEMMA A.1.4: *Let A be a graded connected algebra over the field $\mathbb{F}$. If M and N are finitely generated A-modules then the minimal possible degree of a nonzero homomorphism of graded A-modules $M \longrightarrow N$ is $\sigma_A^-(N) - \sigma_A^+(M)$. If M is a free A module than this minimum is achieved.*

PROOF: Let $L = A \otimes Q_A(M)$ be the free A-module generated by $Q_A(M)$. Define an epimorphism $\vartheta : L \longrightarrow M$ in the usual way, i.e., lift the elements of a basis $x_1, \ldots, x_s$ for $Q_A(M)$ to elements $\widehat{x}_1, \ldots, \widehat{x}_s$ in M and demand that $\vartheta(x_i) = \widehat{x}_i$ for $i = 1, \ldots, s$. If $\varphi : M \longrightarrow N$ is nonzero then so is $\varphi \cdot \vartheta : L \longrightarrow N$. We have

$$\mathrm{Hom}_A(L, N)_* \cong \mathrm{Hom}_{\mathbb{F}}(Q_A(M), N)_*$$

since L is free with basis $Q_A(M)$. The module N is bounded below by

$\sigma_A^-(N)$ because A is nonnegatively graded. The vector space $Q_A(M)$ is bounded above by $\sigma_A^+(M)$ so $\deg(\varphi \cdot \vartheta) = \deg(\varphi) \geq \sigma_A^-(N) - \sigma_A^+(M)$. If we choose a nonzero $\mathbb{F}$ linear map $L_{\sigma_A^+(M)} \longrightarrow N_{\sigma_A^-(N)}$ then it extends to a nonzero homomorphism of A-modules $L \longrightarrow N$ of degree $\sigma_A^-(N) - \sigma_A^+(M)$. □

Let A be a graded connected commutative algebra over a field $\mathbb{F}$ and M an A-module. M is called **flat** if the functor $- \otimes_A M$ is exact. M is called **projective** if every exact sequence of the form

$$0 \longrightarrow L \longrightarrow N \longrightarrow M \longrightarrow 0$$

splits or, equivalently, the functor $\mathrm{Hom}_A(M, -)$ is exact. Over a general ring (i.e., not graded and connected) flat, projective, and free are different concepts, but in our case we have the following result:

PROPOSITION A.1.5: *Let A be a commutative graded connected algebra over a field $\mathbb{F}$ and M a positively graded A-module. Then the following are equivalent:*

(i) M is a free A-module.
(ii) M is a projective A-module.
(iii) M is a flat A-module.
(iv) $\mathrm{Tor}_A^1(\mathbb{F}, M) = 0$.

PROOF: Clearly, *(i)* ⇒ *(ii)* ⇒ *(iii)* ⇒ *(iv)*, so it will suffice to show *(iv)* ⇒ *(i)*. Let $q : M \longrightarrow Q_A(M) \cong \mathbb{F} \otimes_A M$ denote the canonical projection. Choose a splitting $\sigma : Q_A(M) \longrightarrow M$ as graded vector spaces and introduce the map of A-modules $\varphi : A \otimes_{\mathbb{F}} Q_A(M) \longrightarrow M$ defined by $\varphi(a \otimes v) = a \cdot \sigma(v)$. It is elementary to see that $Q(A \otimes_{\mathbb{F}} Q_A(M)) \cong Q_A(M)$ and $Q\varphi$ is an isomorphism. By the corollary to Nakayama's lemma (Corollary A.1.2), φ is an epimorphism, and there is the exact sequence

$$0 \longrightarrow K \longrightarrow A \otimes_{\mathbb{F}} Q_A(M) \longrightarrow M \longrightarrow 0.$$

Applying the functor $\mathbb{F} \otimes_A -$ to this sequence yields by virtue of *(iv)* the short exact sequence

$$0 \longrightarrow QK \longrightarrow A \otimes_{\mathbb{F}} Q_A(M) \xrightarrow{\cong} Q_A(M) \longrightarrow 0,$$

so $QK = 0$, and hence $K = 0$ by Nakayama's lemma (Proposition A.1.1). □

A.2 Primary Decompositions and Integral Extensions

In this section we list some of the classical results in the ideal theory of commutative Noetherian rings in their graded versions; in particular, we review the Lasker-Noether theorem and the Krull and Cohen-Seidenberg relations. The following fact is basic to all that follows.

LEMMA A.2.1 (Prime Avoidance Lemma): *Let A be a commutative graded connected algebra over the field $\mathbb{F}$.*

(i) If an ideal I of A is contained in a finite union of prime ideals $\mathfrak{p}_1, \ldots, \mathfrak{p}_k$, then I is contained in some $\mathfrak{p}_i$.

(ii) If a prime ideal $\mathfrak{p}$ of A contains a finite intersection of ideals $I_1, \ldots, I_k$, then $\mathfrak{p}$ contains I_j for some $j \in \{1, \ldots, k\}$.

PROOF: If we reexamine the classical proof of the first statement, e.g., [20] Proposition 1.11, we observe that it involves the construction of an explicit element, which might not be homogeneous, i.e., might not exist in A regarded as a graded algebra. So we need to revise the proof a bit.

We proceed by induction. If $k = 1$, there is nothing to prove, so we may suppose that $k > 1$ and that the result has been established for $k - 1$. Suppose that I is an ideal of A that is not contained in any of the $\mathfrak{p}_1, \ldots, \mathfrak{p}_k$. We can, of course, then choose for each $i = 1, \ldots, k$ an element $a_i \in I$ such that $a_i \notin \mathfrak{p}_j$ whenever $j \neq i$. If for some i we have $a_i \notin \mathfrak{p}_i$ also, then $a_i \notin \cup \mathfrak{p}_i$, so $I \not\subseteq \cup \mathfrak{p}_i$, and we are done. Otherwise, let $d_i = \deg(a_i)$ and d be the least common multiple of $d_1, \ldots, d_k$. Since $a_i \notin \mathfrak{p}_j$ for $j \neq i$ and $\mathfrak{p}_j$ is a prime ideal, $\bar{a}_i = a_i^{d/d_i}$ also does not belong to $\mathfrak{p}_j$ for $j \neq i$. The elements $\bar{a}_i$ all have degree d, so we can form

$$b = \sum_{i=1}^{k} \bar{a}_1 \cdots \widehat{\bar{a}_i} \cdots \bar{a}_k,$$

which is an element of A of degree $(k-1)d$. It belongs to I but not to any $\mathfrak{p}_1, \ldots, \mathfrak{p}_k$, so $I \not\subseteq \cup \mathfrak{p}_i$ as was to be shown.

The proof of the second statement is straightforward: If $\mathfrak{p}$ does not contain any of the $I_1, \ldots, I_k$, then we may choose elements $a_i \in I_i$ such that $a_i \notin \mathfrak{p}$. The product of these elements lies in each I_i and hence also in their intersection, but not in $\mathfrak{p}$, since it is prime. This is a contradiction. □

REMARK: Note that we used the primality of all the ideals $\mathfrak{p}_i$. So, the usual remark after proving *(i), we used only the fact that $\mathfrak{p}_n$ was prime, and even this is necessary only if $n > 2$. So in fact, it suffices to assume that all except possibly two of the $\mathfrak{p}_i$ are prime* does not apply to our graded proof. In fact, as the following example shows, this strengthened form of the statement of prime avoidance is false in the graded case.

EXAMPLE 1: Consider the algebra $A = \mathbb{F}[x, y]/(xy, y^2)$, where x and y have degree one. The only elements in this ring are scalar multiples of $1, y, x, x^2, \ldots$ and the ideal (y) is prime in A since $A/(y) \cong \mathbb{F}[x]$. Moreover, $(x, y) = (x) \cup (y)$, but (x, y) is not a subset of either (x) or (y).

The prime avoidance lemma is one of the key ingredients that is needed for much of classical ideal theory, e.g., the Lasker-Noether theorem.

Let A be a commutative graded connected algebra over a field. Classically, an ideal $\mathfrak{m}$ is called **maximal** if it is not contained in any proper ideal of A. Since A is connected and positively graded, it has a unique maximal ideal consisting of all the elements of A of strictly positive degree. This is the augmentation ideal $\overline{A}$. Recall that an ideal I of A is a **primary** ideal if whenever $a, b \in A$ with $a \cdot b \in I$, and a not in I, then some positive power of b belongs to I. The **radical** of an ideal I, denoted by $\sqrt{I}$, is the ideal of all $a \in A$ such that some positive power of a lies in I. If $\mathfrak{q}$ is primary, then $\sqrt{\mathfrak{q}}$ is prime.

These properties of ideals are equivalent to certain properties of the quotient algebra obtained from A by dividing by the ideal. For example:

- An ideal $\mathfrak{m} \subset A$ is maximal if and only if $A/\mathfrak{m}$ is a field.
- An ideal $\mathfrak{p} \subset A$ is prime if and only if $A/\mathfrak{p}$ is an integral domain.
- An ideal $\mathfrak{q} \subset A$ is a primary ideal if and only if every zero divisor in $A/\mathfrak{q}$ is nilpotent.

The following theorem provides us with a description of an arbitrary ideal in terms of primary ones.

THEOREM A.2.2 (E. Lasker, E. Noether): *If I is any ideal in a commutative graded connected Noetherian algebra over a field, then there are a finite number of primary ideals $\mathfrak{q}_1, \ldots, \mathfrak{q}_n$ such that*

(i) $I = \mathfrak{q}_1 \cap \cdots \cap \mathfrak{q}_n$,

(ii) no $\mathfrak{q}_i$ contains $\bigcap_{i \neq j} \mathfrak{q}_j$, and

(iii) if $j \neq i$, then $\sqrt{\mathfrak{q}_i} \neq \sqrt{\mathfrak{q}_j}$. □

A representation of I as the intersection of primary ideals as in Theorem A.2.2 is called an **irredundant primary decomposition** of I. If

$$\mathfrak{q}_1' \cap \cdots \cap \mathfrak{q}_{n'}' = I = \mathfrak{q}_1'' \cap \cdots \cap \mathfrak{q}_{n''}''$$

are two irredundant primary decompositions of I, then $n' = n''$, and after possibly reordering, $\sqrt{\mathfrak{q}_i'} = \sqrt{\mathfrak{q}_i''}$. The ideals $\sqrt{\mathfrak{q}_1}, \ldots, \sqrt{\mathfrak{q}_n}$ in an irredundant primary decomposition of I are called the **associated prime ideals of** I, and denoted by Ass(I). A minimal associated prime[3] of I is called an **isolated prime** of I. The isolated primes of I are the minimal prime ideals among those that include I, and $\sqrt{I}$ is the intersection of the isolated primes of I.

Finally, we recall the Theorems of W. Krull and I. S. Cohen–A. Seidenberg.[4] Although the usual proofs of these theorems use localization techniques,

[3] I.e., an associated prime of I containing no other associated prime of I.

[4] One of us, having a German mathematical education, learned to call the statements of Theorem A.2.3 the Krull relations. However, the other author, with an American background, refers to them as the Cohen–Seidenberg theorems. Indeed, both names are somehow justified: W. Krull proved these results first in the context of integral domains, [212], while I. S. Cohen and A. Seidenberg generalized them to rings with zero divisors [75]. Moreover, concerning

these are compatible with the grading, and so can be considered as adequate to prove the following graded analogues; cf. Section 5.4 in [372], in particular Theorems 5.4.2 and 5.4.5.

THEOREM A.2.3: *Suppose that $A' \supseteq A''$ is an integral extension of commutative graded connected algebras over the field $\mathbb{F}$.*

(i) **(Lying-Over):** *If $\mathfrak{p}''$ is a prime ideal of A'', then there is a prime ideal $\mathfrak{p}'$ of A' with $\mathfrak{p}' \cap A'' = \mathfrak{p}''$. There are no strict inclusions between such prime ideals $\mathfrak{p}'$. In this situation we say that $\mathfrak{p}'$* **lies over** *$\mathfrak{p}''$.*

(ii) **(Going-Up):** *If $\mathfrak{p}_1'' \supseteq \mathfrak{p}_0''$ are prime ideals in A'' and $\mathfrak{p}_0'$ is a prime ideal in A' lying over $\mathfrak{p}_0''$, then there is a prime ideal $\mathfrak{p}_1'$ in A' lying over $\mathfrak{p}_1''$ with $\mathfrak{p}_1' \supseteq \mathfrak{p}_0'$.*

(iii) **(Preserving Maximality):** *If $\mathfrak{p}' \subset A'$ lies over $\mathfrak{p}'' \subset A''$, then one is maximal if and only if the other is.*

(iv) **(Incomparability):** *If $\mathfrak{p}_0', \mathfrak{p}_1' \subset A'$ lie over the same prime ideal $\mathfrak{p}'' \subset A''$, then they are incomparable, i.e., neither is contained in the other.* □

THEOREM A.2.4: *Suppose that $A' \supseteq A''$ is an integral extension of commutative graded integral domains over the field $\mathbb{F}$, A'' is integrally closed, and the corresponding extension $\mathbb{FF}(A') \supseteq \mathbb{FF}(A'')$ of graded fields of fractions is normal.*[5] *Then:*

(1) **(Transitivity):** *The Galois group $G = \mathrm{Gal}(\mathbb{FF}(A') \mid \mathbb{FF}(A''))$ acts transitively on the prime ideals $\mathfrak{p}'$ of A' lying over a given prime ideal $\mathfrak{p}''$ of A''.*

(ii) **(Going-Down):** *If $\mathfrak{p}_0'' \subseteq \mathfrak{p}_1''$ are prime ideals in A'' and $\mathfrak{p}_1'$ is a prime ideal of A' lying over $\mathfrak{p}_1''$, then there is a prime ideal $\mathfrak{p}_0'$ in A' lying over $\mathfrak{p}_0''$, with $\mathfrak{p}_0' \subseteq \mathfrak{p}_1'$.* □

A.3 Noetherian Algebras

A **system of parameters** for an algebra A over the field $\mathbb{F}$ is a finite set of elements $h_1, \ldots, h_n$ in A such that the ring extension $\mathbb{F}[h_1, \ldots, h_n] \subseteq A$ is finite. For the proof of the following theorem we refer to Theorem 5.3.3 in [372].

Theorem A.2.4, in particular the going-down statement, we find that W. Krull indeed observed that going-down does not hold in as great a generality as going-up, Satz 4 in [212]. He calls this "einen unerwarteten und sehr schmerzlichen Schönheitsfehler," page 751 loc. cit. Again, he proves the result in the absence of zero divisors, Satz 6 loc. cit., while I. S. Cohen and A. Seidenberg give a proof in full generality, Theorem 5 in [75]. They also give examples that show that none of the assumptions can be removed, Section 3 loc. cit.

[5] The field extension $\mathbb{FF}(A') \subseteq \mathbb{FF}(A'')$ is called **normal** when every irreducible polynomial over $\mathbb{FF}(A'')$ with a root in $\mathbb{FF}(A')$ splits completely in $\mathbb{FF}(A'')$. It is not necessary for the extension to be separable.

THEOREM A.3.1 (Noether Normalization Theorem): *Let A be a finitely generated commutative graded connected algebra over a field $\mathbb{F}$. Then A has a system of parameters. Any system of parameters for A consists of algebraically independent elements. Moreover, the following integers are equal:*

(i) the number of elements in a system of parameters,

(ii) the smallest integer r such that there exist r elements $a_1, \ldots, a_r \in A$ with $A/(a_1, \ldots, a_r)$ totally finite,

(iii) the largest integer s such that there exist s algebraically independent elements in A,

(iv) the length t of the longest strictly increasing chain

$$\mathfrak{p}_0 \subset \mathfrak{p}_1 \subset \cdots \subset \mathfrak{p}_t \subset A$$

of prime ideals in A.

The common value of the integer in (i), (ii), (iii), and (iv) is the ***Krull dimension*** *of A, and is denoted by* $\dim(A)$. □

This result allows us to consider any finitely generated commutative graded connected algebra A over a field as a finite extension of a polynomial algebra of the same Krull dimension, i.e., as a finitely generated module over a polynomial algebra. The choice of such a polynomial subalgebra $\mathbf{N} \subseteq A$, i.e., of a system of parameters $a_1, \ldots, a_n \in A$ we call a **Noether normalization** for A. A system of parameters always consists of algebraically independent elements, but the converse can be false: For example, $xy, y^2 \in \mathbb{F}[x, y]$ are algebraically independent, but not a system of parameters. Thus, while the numbers in *(i)* and *(ii)* of Theorem A.3.1 are always equal, this does not imply that any algebraically independent set of elements of length $\dim(A)$ form a system of parameters.

EXAMPLE 1: The Noether normalization theorem is ***not*** correct for multigraded algebras. Consider the bigraded polynomial algebra $\mathbb{F}[x, y]$ where $\deg(x) = (1, 0)$ and $\deg(y) = (0, 1)$. Let A be the quotient algebra $\mathbb{F}[x, y]/(xy)$. Then A has no bigraded Noether normalization. The usual choice, $\mathbb{F}[x + y]$, is no longer a bigraded subalgebra of A because the element $x + y$ does not belong to A: The elements x and y lie in different bidegrees, so cannot be added together.

PROPOSITION A.3.2: *Let A be a commutative graded connected algebra over the field $\mathbb{F}$. Then A is Noetherian if and only if A is finitely generated.*

PROOF: If A is finitely generated, then there is an epimorphism $\varphi : \mathbb{F}[X_1, \ldots, X_k] \longrightarrow A$ obtained by choosing algebra generators $a_1, \ldots, a_k \in A$ and demanding that $\varphi(X_i) = a_i$, for $i = 1, \ldots, k$. As a quotient of the Noetherian algebra $\mathbb{F}[X_1, \ldots, X_k]$, A is also Noetherian. Conversely, if A is Noetherian, let $\mathfrak{m} \subset A$ be the maximal ideal. By D. Hilbert's basis theorem the ideal $\mathfrak{m}$ is finitely generated by, say, $m_1, \ldots, m_k$. Let $B \subseteq A$ be the subalgebra generated by $m_1, \ldots, m_k$. Then A is finite over B because

B contains $m_1, \ldots, m_k$. Hence B contain a system of parameters, say $b_1, \ldots, b_n$ for A. If $a_1, \ldots, a_l$ generate A as a B-module, then

$$b_1, \ldots, b_n, a_1, \ldots, a_l$$

generate A as an algebra over $\mathbb{F}$. □

The nonconnected version of Theorem A.3.2 is false: See the ingenious example of Nagata, [269], Appendix, Example 1. Here is another example of a result that does not remain valid in the nongraded and unconnected case: the wallflower theorem (Theorem 3.1) in [291]:

THEOREM A.3.3 (Wallflower): *Let H be a reduced graded connected algebra over a field. Then H is Noetherian if and only if its integral closure $\overline{H}$ is Noetherian.* □

REMARK: Let H be a reduced graded connected algebra over a field $\mathbb{F}$ as in the preceding theorem. If $\mathbb{F}$ is not algebraically closed then the integral closure $\overline{H}$ may have a different ground field, i.e., degree zero component. Even worse, it might not be connected, [7]:

(i) If H is an integral domain, then the ground field of $\overline{H}$ becomes the algebraic closure $\overline{\mathbb{F}}$ of $\mathbb{F}$.

(ii) If H contains zerodivisors, then the degree-zero part $\overline{H}_0$ of the integral closure consists of the algebraic closure of $\mathbb{F}$ and, possibly, some zerodivisors, i.e., $\overline{H}$ is possibly no longer connected.

Conversely, if $\overline{H}$ is connected with ground field $\mathbb{K}$, then H is connected with ground field $\mathbb{F}$, where $\mathbb{F} \hookrightarrow \mathbb{K}$ is an algebraic field extension.

For example, consider the polynomial algebra $\mathbb{F}[x]$ in one linear generator over a field $\mathbb{F}$ that is not algebraically closed. Choose an element $f \in \overline{\mathbb{F}} \setminus \mathbb{F}$, where $\overline{\mathbb{F}}$ is the algebraic closure of $\mathbb{F}$, of degree zero. Consider the $\mathbb{F}$-subalgebra of $\overline{\mathbb{F}}[x]$ generated by x and fx. Then $H = \mathbb{F}[x, fx]$ is connected with ground field $\mathbb{F}$. However, its integral closure

$$\overline{H} = \overline{\mathbb{F}}[x]$$

has changed its ground field to $\overline{\mathbb{F}} = H_0$, while remaining connected in the sense that $H_0 = \overline{\mathbb{F}}$ is a field and H_i a vector space over $\overline{\mathbb{F}}$ for $i \in \mathbb{N}_0$.

Finally, in the category of unstable algebras over the Steenrod algebra the above phenomena cannot occur. In other words, if H is a connected unstable $\mathbb{F}_q$-algebra then $\overline{H}$ is a connected unstable $\mathbb{F}_q$-algebra. This is because the degree zero component of an unstable algebra is a q-Boolean algebra; see [291].

The Krull dimension of $\mathbb{F}[u_1, \ldots, u_n]$ is obviously n, since $u_1, \ldots, u_n \in \mathbb{F}[u_1, \ldots, u_n]$ are of course a system of parameters. If A is a graded connected algebra over $\mathbb{F}$ and $\varphi : \mathbb{F}[u_1, \ldots, u_n] \longrightarrow A$ is an epimorphism, then any proper chain of prime ideals

$$\mathfrak{p}_0 \subset \mathfrak{p}_1 \subset \cdots \subset \mathfrak{p}_t \subset A$$

pulls back via φ to a proper chain of prime ideals

$$\varphi^{-1}(\mathfrak{p}_0) \subset \varphi^{-1}(\mathfrak{p}_1) \subset \cdots \subset \varphi^{-1}(\mathfrak{p}_t) \subset \mathbb{F}[u_1, \ldots, u_n],$$

and hence $t \leq n$. From this we draw the following conclusion:

COROLLARY A.3.4: *Let A be a finitely generated commutative graded algebra over the field $\mathbb{F}$, Then* $\dim(A) \leq \dim_{\mathbb{F}}(QA)$.

PROOF: By Proposition A.1.2 we can lift a vector space basis for QA to a minimal generating set $a_1, \ldots, a_n \in A$ for A as an algebra. So the map $\varphi : \mathbb{F}[u_1, \ldots, u_n] \longrightarrow A$ defined by $\varphi(u_i) = a_i$ for $i = 1, \ldots, n$ is an epimorphism, and the result follows from the preceding discussion. □

If there is a system of parameters $a_1, \ldots, a_n \in A$ such that A is a free $\mathbb{F}[a_1, \ldots, a_n]$-module, then A is called **Cohen–Macaulay**. Luckily, Macaulay's theorem, [25] Theorem 3.3.5 or [372] Corollary 6.7.7, assures us that if one system of parameters has this property, then all systems of parameters do. To wit:

THEOREM A.3.5 (F. S. Macaulay): *Let A be a commutative graded connected algebra over the field $\mathbb{F}$. If A is Cohen-Macaulay then every system of parameters is a regular sequence.* □

Notice that when $\mathsf{N} \subseteq A$ is a Noether normalization and A is Cohen-Macaulay, then, since A is a finite extension of N, the Poincaré series of $A//\mathsf{N} = \mathbb{F} \otimes_{\mathsf{N}} A$ is a polynomial with nonnegative integral coefficients. Since, moreover, $A \cong \mathsf{N} \otimes_{\mathbb{F}} A//\mathsf{N}$ as N-modules, and hence also as graded vector spaces, we have an equality of Poincaré series

$$P(A, t) = P(\mathsf{N}, t) \cdot P(A//\mathsf{N}, t) = \frac{P(A//\mathsf{N}, t)}{(1 - t^{|a_1|})(1 - t^{|a_2|}) \cdots (1 - t^{|a_n|})},$$

where $\mathsf{N} = \mathbb{F}[a_1, \ldots, a_n]$ and $a_1, \ldots, a_n \in A$ is a system of parameters. $P(A//\mathsf{N}, t)$ is a polynomial with nonnegative integral coefficients, and being able to write the Poincaré series in this form is an additional requirement for an algebra to be Cohen-Macaulay; see, e.g., [400]. In fact, phrased properly, this condition is necessary and sufficient [127].

We will often make use of a consequence of Hilbert's Nullstellensatz ([25] Corollary 1.1.12), which provides a criterion for when $h_1, \ldots, h_n \in \mathbb{F}[V]$, $\dim_{\mathbb{F}}(V) = n$, are a system of parameters; see, e.g., [372] Proposition 5.3.7.

PROPOSITION A.3.6: *$h_1, \ldots, h_n \in \mathbb{F}[z_1, \ldots, z_n]$ are a system of parameters if and only if for every field extension $\overline{\mathbb{F}} \supseteq \mathbb{F}$ the variety of common zeros $\mathfrak{V}(h_1, \ldots, h_n; \overline{\mathbb{F}})$ of $h_1, \ldots, h_n$ in $\overline{\mathbb{F}}^n$, namely*

$$\{(x_1, \ldots, x_n) \in \overline{\mathbb{F}}^n \mid h_i(x_1, \ldots, x_n) = 0 \text{ for } i = 1, \ldots, n\}$$

consists of the point $(0, \ldots, 0)$ *alone.* □

A.4 Graded Algebras and Modules

It is often necessary to verify that elements $h_1, \ldots, h_n$ in a commutative graded algebra H over the field $\mathbb{F}$ are algebraically independent over $\mathbb{F}$. The following lemma, while elementary in character, seems not to be as well known as it ought to be.

LEMMA A.4.1 (Derivation Lemma): *Let* H *be a commutative connected graded algebra over the field* $\mathbb{F}$, $h_1, \ldots, h_n \in H$, *and* $\partial_1, \ldots, \partial_n : H \circlearrowleft$ *derivations. Suppose that*

$$\det \begin{bmatrix} \partial_1(h_1) & \ldots & \partial_1(h_n) \\ \vdots & \vdots & \vdots \\ \partial_n(h_1) & \ldots & \partial_n(h_n) \end{bmatrix} \neq 0$$

and is not a zero divisor. Then $h_1, \ldots, h_n$ *are algebraically independent.*

PROOF: Without loss of generality we can assume that $\mathbb{F}$ is perfect by passing to an algebraic closure.

Introduce the polynomial algebra $\mathbb{F}[X_1, \ldots, X_n]$, where $\deg(X_i) = \deg(h_i)$ for $i = 1, \ldots, n$, and define the map $\varphi : \mathbb{F}[X_1, \ldots, X_n] \longrightarrow H$ by the requirement that $\varphi(X_i) = h_i$ for $i = 1, \ldots, n$. We need to show that φ is a monomorphism. So, suppose it is not, and choose $r(X_1, \ldots, X_n) \in \ker(\varphi)$ to be a nonzero element of minimal degree. Then

$$r(h_1, \ldots, h_n) = 0 \in H.$$

If we apply ∂_i to this equation, we receive by the chain rule

$$0 = \partial_i r(h_1, \ldots, h_n) = \sum_{j=1}^{n} \frac{\partial h_j}{\partial X_j} \Big|_{(X_1, \ldots, X_n) = (h_1, \ldots, h_n)} \partial_i r,$$

for $i = 1, \ldots, n$. If we rearrange the terms and write this as one matrix equation instead of n separate equations, we obtain

$$0 = \begin{bmatrix} \partial_1 h_1 & \cdots & \partial_1 h_n \\ \vdots & \vdots & \vdots \\ \partial_n h_1 & \cdots & \partial_n h_n \end{bmatrix} \cdot \begin{bmatrix} \frac{\partial r}{\partial X_1} \\ \vdots \\ \frac{\partial r}{\partial X_n} \end{bmatrix} \Bigg|_{(X_1, \ldots, X_n) = (h_1, \ldots, h_n)}.$$

Since $\det[\partial_i h_j] \neq 0$, and also is not a zero divisor, we conclude from Cramer's rule that

$$0 = \frac{\partial r}{\partial X_1} \Big|_{(X_1, \ldots, X_n) = (h_1, \ldots, h_n)} = \cdots = \frac{\partial r}{\partial X_n} \Big|_{(X_1, \ldots, X_n) = (h_1, \ldots, h_n)}.$$

If for some $1 \leq i \leq n$, $\frac{\partial r}{\partial X_i} \neq 0 \in \mathbb{F}[X_1, \ldots, X_n]$, then, since $\frac{\partial r}{\partial X_i} \in \ker(\varphi)$ and $\deg(\frac{\partial r}{\partial X_i}) \leq \deg(r)$, we would have a contradiction to the choice of $r \in \ker(\varphi)$ as a nonzero element of minimal degree. Therefore,

$$0 = \frac{\partial r}{\partial X_1} = \cdots = \frac{\partial r}{\partial X_n} \in \mathbb{F}[X_1, \ldots, X_n].$$

If $\mathbb{F}$ has characteristic zero, this says that $r = 0$, whereas if $\mathbb{F}$ has characteristic $p \neq 0$, it says that $r = s^p$. In either case, we again have a contradiction to the choice of $r \in \ker(\varphi)$ as a nonzero element of minimal degree. □

REMARK: The converse of this lemma can fail in characteristic $p \neq 0$. For example, if $\mathbb{F}_p$ is the Galois field with p elements, then $x^p, y^p \in \mathbb{F}_p[x, y]$ are algebraically independent but their Jacobian matrix is identically zero.

We next review some of the theory of associated primes for graded modules over Noetherian algebras over a field. This material is all standard and, in ungraded form, may be found in [24] and [25]. We adopt the definition that a prime ideal $\mathfrak{p} \subset A$ is an **associated prime of an A-module M** if there exists a nonzero element $x_{\mathfrak{p}} \in M$ whose annihilator ideal is exactly $\mathfrak{p}$.

LEMMA A.4.2: *Let A be a commutative graded connected Noetherian algebra over the field $\mathbb{F}$ and M a nonzero A-module. Then the set $\mathrm{Ass}_A(M)$ of associated prime ideals of M is nonempty.*

PROOF: Consider the set $\mathrm{Ann}_A(M) = \{\mathrm{Ann}_A(x) \mid x \in M\}$ and order it by inclusion. Since A is Noetherian, there exist maximal elements in $\mathrm{Ann}_A(M)$. Let I be one such. We claim that $I \subset A$ is a prime ideal. To see this let $a, b \in A$ satisfy $a \notin I$ but $ab \in I$. We must show that $b \in I$.

Choose $x \in M$ with $I = \mathrm{Ann}_A(x)$. Since $a \notin I$, the element ax is not zero. If $c \in I$, then

$$c(ax) = a(cx) = 0,$$

so $I \subseteq \mathrm{Ann}_A(ax)$. Since $I \in \mathrm{Ass}_A(x)$ was chosen to be maximal, it follows that $I = \mathrm{Ann}_A(x)$. Next note that

$$b(ax) = (ab)x = 0,$$

since $ab \in I$. Hence $b \in \mathrm{Ann}_A(ax) = I$ as required. □

NOTATION: The set of associated prime ideals of the A-module M is denoted by $\mathrm{Ass}_A(M)$. This leads to the conflict of notation and usage that for an ideal $I \subset A$, $\mathrm{Ass}(I)$ does ***not*** denote the associated prime ideals of I as an A-module. The associated prime ideals of I as an A-module are usually uninteresting. The associated prime ideals of I as an ideal as explained following Theorem A.2.2 are the associated primes of $(0) \subset A/I$ regarded as an A- module. The context in which the phrase *associated primes* is used usually clarifies the meaning.

The next goal is to show that $\mathrm{Ass}_A(M)$ is finite when M is a Noetherian module over a Noetherian ring.

LEMMA A.4.3: *Let A be a commutative graded connected algebra over the field $\mathbb{F}$ and*

$$0 \longrightarrow M' \xrightarrow{\varphi'} M \xrightarrow{\varphi''} M'' \longrightarrow 0$$

a short exact sequence of A-modules. Then

$$\mathrm{Ass}_A(M) \subseteq \mathrm{Ass}_A(M') \cup \mathrm{Ass}_A(M'').$$

PROOF: Let $\mathfrak{p} \in \mathrm{Ass}_A(M)$ and choose $x \in M$ with $\mathfrak{p} = \mathrm{Ann}_A(x)$. Let N be the submodule of M generated by $x \in M$. Then $A/\mathfrak{p} \xrightarrow{\cong} N$ under the map[6] $a + \mathfrak{p} \longmapsto ax$. If $N \cap M' \neq \{0\}$, choose $y \in N \cap M'$ and note that the annihilator ideal of y is $\mathfrak{p}$, for $y = ax \neq 0$ for some $a \in A$ with $a \notin \mathfrak{p}$, and $b(ax) = 0$ if and only if $ba \in \mathfrak{p}$, which is equivalent to $b \in \mathfrak{p}$. Therefore $\mathfrak{p} \in \mathrm{Ass}_A(M')$. On the other hand, if $N \cap M' = \{0\}$ then the map φ'' induces a monomorphism $N \cong A/\mathfrak{p} \hookrightarrow M''$, so setting $z = \varphi''(x)$ we see that $\mathfrak{p} = \mathrm{Ann}_A(z) \in \mathrm{Ann}_A(M'')$. This means that $\mathfrak{p} \in \mathrm{Ass}(M')$ or $\mathfrak{p} \in \mathrm{Ass}(M'')$. □

LEMMA A.4.4: *Let A be a commutative graded connected Noetherian algebra over the field $\mathbb{F}$ and M a finitely generated A-module. Then the set $\mathrm{Ass}_A(M)$ is finite.*

PROOF: If $M = \{0\}$, there is nothing to prove. If $M \neq \{0\}$, then by Lemma A.4.2 $\mathrm{Ass}_A(M) \neq \emptyset$, so we may choose $\mathfrak{p}_1 = \mathrm{Ann}_A(x_1) \in \mathrm{Ass}_A(M)$. Let M_1 denote the A-submodule of M generated by x_1, so $M_1 \cong A/\mathfrak{p}_1$. Assume inductively that we have constructed a chain of submodules

$$\{0\} = M_0 \subset M_1 \subset \cdots \subset M_k \subset M$$

with the property that

$$M_i/M_{i-1} \cong A/\mathfrak{p}_i, \qquad i = 1, \ldots, k,$$

where $\mathfrak{p}_1, \ldots, \mathfrak{p}_k \subset A$ are prime ideals. If $M_k \neq M$, then we form M/M_k and choose an element $\overline{x}_{k+1} \in M/M_k$ with $\mathrm{Ann}_A(\overline{x}_{k+1}) = \mathfrak{p}_{k+1} \subset A$ a prime ideal. Since M is Noetherian, this process comes to an end after finitely many, say s, steps with a chain of submodules

$$\{0\} = M_0 \subset M_1 \subset \cdots \subset M_s = M$$

having the property that

$$M_i/M_{i-1} \cong A/\mathfrak{p}_i, \qquad i = 1, \ldots, s,$$

where $\mathfrak{p}_1, \ldots, \mathfrak{p}_s \subset A$ are prime ideals. Successive application of Lemma

[6] This map may have a nonzero degree, but this is of no consequence for the argument here.

A.4.3 to the short exact sequences

$$0 \longrightarrow M_{i-1} \longrightarrow M_i \longrightarrow M_i/M_{i-1} \longrightarrow 0$$

for $i = 1, \ldots, s$ then yields

$$\mathrm{Ass}_A(M) \subseteq \{\mathfrak{p}_1, \ldots, \mathfrak{p}_s\}$$

as desired. □

An important step in the proof of Lemma A.4.4 is worthy of being singled out as a result for future reference.

LEMMA A.4.5 (Prime Filtration Lemma): *Let A be a connected, commutative, graded Noetherian algebra over the field $\mathbb{F}$ and M a finitely generated A-module. Then there exists a filtration*

$$\{0\} = M_0 \subset M_1 \subset \cdots \subset M_s = M$$

with the property that

$$M_i/M_{i-1} \cong A/\mathfrak{p}_i, \qquad i = 1, \ldots, s,$$

where $\mathfrak{p}_1, \ldots, \mathfrak{p}_s \subset A$ are prime ideals and $\mathrm{Ass}(M) \subseteq \{\mathfrak{p}_1, \ldots, \mathfrak{p}_s\}$. □

References

[1] J. F. Adams and C. W. Wilkerson, *Finite H-spaces and Algebras over the Steenrod Algebra,* Annals of Math. 111 (1980), 95-143.

[2] J. F. Adams and C. W. Wilkerson, *Finite H-spaces and Algebras over the Steenrod Algebra: a correction,* Annals of Math. 113 (1981), 621-622.

[3] A. Adem, J. Maginnis, and R. J. Milgram, *Symmetric Invariants and Cohomology of Groups,* Math. Annalen 287 (1990), 391-411.

[4] A. Adem and R. J. Milgram, *Cohomology of Finite Groups,* Springer-Verlag, Heidelberg, Berlin, New York 1994.

[5] J. Aguadé and L. Smith, *Modular Cohomology Algebras,* Amer. J. of Math. 107 (1985), 507-530.

[6] C. Albrecht, *Elemente der Kommutativen Graduierten Algebra,* Diplomarbeit, University of Göttingen 2000.

[7] C. Albrecht and M. D. Neusel, *private communication,* Göttingen-New Haven CT 2000.

[8] G. Almkvist, *The Number of Nonfree Components in the Decomposition of Symmetric Powers in Charactersitic p,* Pacific J. of Mathematics 77 (1978), 293-301.

[9] G. Almkvist, *Invariants, Mostly Old,* Pacific J. of Mathematics 86 (1980), 1-13.

[10] G. Almkvist, *Representations of* $\mathbb{Z}/p\mathbb{Z}$ *in Characteristic p and Reciprocity Theorems,* J. of Algebra 68 (1981), 1-27.

[11] G. Almkvist, *Rings of Invariants,* preprint, Lund 1981.

[12] G. Almkvist, *Some Formulas in Invariant Theory,* J. of Algebra 77 (1982), 338-359.

[13] G. Almkvist, *Invariants of* $\mathbf{Z}/p\mathbf{Z}$ *in Characteristic p,* in: *Invariant Theory (Proceedings of the 1982 Montecatini Conference),* pp. 109-117. Lecture Notes in Math. 996, Springer-Verlag, Heidelberg, Berlin 1983.

[14] G. Almkvist, *Commutative and Noncommutative Invariant Theory,* in: *Topics in Algebra,* pp. 259-268. Banach Center Publications 26, Part 2, PWN-Polish Scientific Publishers, Warsaw 1990.

[15] G. Almkvist and R. Fossum, *Decompositions of Exterior and Symmetric Powers of Indecomposable* $\mathbf{Z}/p\mathbf{Z}$*-Modules in Characteristic p and Relations to Invariants,* in: *Séminaire d'Algèbre Paul Dubreil, Paris 1976/7,* pp. 1-111. Lecture Notes in Math. 641, Springer-Verlag, Heidelberg, Berlin 1978.

[16] J. L. Alperin, *Local representation Theory,* Cambridge Studies in Advanced Mathematics 11, Camb. Univ. Press, Cambridge 1986.

[17] E. Artin, *Geometric Algebra,* Interscience Tracts in Pure Math. 3, Interscience Publishers, New York, 1957.

[18] E. Artin and J. Tate, *A Note on Finite Ring Extensions,* J. of the Japanese Mathematical Society 3 (1951), 74-77.

[19] E. F. Assmus, *On the Homology of Local Rings,* Ill. J. of Math. 3 (1959), 187-199.

[20] M. F. Atiyah and I. G. Macdonald, *Introduction to Commutative Algebra,* Addison-Wesley Publ., Menlo Park, CA, 1969.

[21] M. Auslander and D. A. Buchsbaum, *Codimension and Multiplicity,* Annals of Math. 68 (1958), 625-657.

[22] L. L. Avramov, *Pseudoreflection Group Actions on Local Rings,* Nagoya Math. J. 88 (1982), 161-180.

[23] H. F. Baker, *Note Introductory to the Study of Klein's Group of Order 168,* Math. Proc. Camb. Phil. Soc. 31 (1935), 468-481.

[24] S. Balcerzyk and T. Józefiak, *Commutative Noetherian and Krull Rings,* Polish Scientific Publishers, Warsaw, 1989.

[25] S. Balcerzyk and T. Józefiak, *Commutative Rings: Dimension, Multiplicity and Homological Methods,* Polish Scientific Publishers, Warsaw, 1989.

[26] G. Barbançon and M. Raïs, *Sur le théorème de Hilbert différentiable pour les groupes linéaires finis,* Ann. Scient. Éc. Norm. Sup. 4e série t. 16 (1983), 355-373.

[27] H. Bass, *The Ubiquity of Gorenstein Rings,* Math. Zeit. 82 (1963), 8-28.

[28] P. F. Baum and L. Smith, *Real Cohomology of Differentiable Fibre Bundles,* Comment. Math. Helv. 42 (1967), 171-179.

[29] D. J. Benson, *Representations and Cohomology,* 2 vols., Cambridge Studies in Advanced Mathematics 30 & 31, Camb. Univ. Press, Cambridge 1991.

[30] D. J. Benson, *Polynomial Invariants of Finite Groups,* London Math. Soc. Lecture Series 190, Camb. Univ. Press, Cambridge 1993.

[31] D. J. Benson and W. W. Crawley-Boevey, *A Ramification Formula for Poincaré Series, and a Hyperplane Formula for Modular Invariants,* Bull. London Math. Soc. 27 (1995), 435-440.

[32] D. J. Benson and M. D. Neusel, *private communication,* Athens GA-Notre Dame IN 2000.

[33] E. R. Berlekamp, *An Analog of the Discriminant over Fields of Characteristic Two,* J. of Algebra 38 (1976), 315-317.

[34] M. Bernard, *Schur Indices and Splitting Fields of Unitary Reflection Groups,* J. of Algebra 38 (1976), 318-342.

[35] M. -J. Bertin, *Anneau des invariants du groupe alterné en caractéristique 2,* Bull. Sci. Math. de France 94 (1970), 65-72.

[36] M. -J. Bertin, *Anneaux d'invariants d'anneaux de polynômes en caractéristique p,* C. R. Acad. Sci. Paris 277 (Série A) (1973), 691-694.

[37] T. P. Bisson and A. Joyal, *Q-Rings and the Homology of the Symmetric Group,* Contemp. Math. 202 (1997), 235-286.

[38] I. Böcker, *Invarianten endlicher Gruppen,* Examensarbeit, University of Göttingen, 1993.

[39] N. Bourbaki, *Groupes et Algèbres de Lie, Ch.* 4, 5 *et* 6, Masson, Paris 1981.

[40] D. Bourguiba, *Profondeur et algèbre de Steenrod,* Thèse, Université de Tunis II, 1997.

[41] D. Bourguiba and S. Zarati, *Depth and Steenrod Operations,* Inventiones Math. 128 (1997), 589-602.

[42] R. Bringhurst, *The Elements of Typographic Style,* second edition, Hartley and Marks, Point Roberts, WA, 1999.

[43] A. Broer, *Remarks on the Invariant Theory of Finite Groups,* Preprint, Université de Montréal, 1997.

[44] C. Broto, *Àlgebres d'Invariants i Àlgebres sobre l'Àlgebra de Steenrod,* Bull. Soc. Cat. Ciėn. VIII (1) (1986), 117-145.

[45] C. Broto, L. Smith, and R. E. Stong, *Thom Modules,* J. of Pure and Appl. Algebra 60 (1989), 1-20.

[46] K. S. Brown, *Cohomology of Groups,* Graduate Texts in Math. 87, Springer-Verlag, New York, 1982.

[47] W. Bruns and J. Herzog, *Cohen-Macaulay Rings,* Cambridge Studies in Advanced Math, 39, Camb. Univ. Press, Cambridge 1993.

[48] S. R. Bullett and I. G. Macdonald, *On the Adem Relations,* Topology 21 (1982), 329-332.

[49] P. J. Cameron and W. M. Kantor, *2-Transitive and Antiflag Transitive Collineation Groups of Finite Projective Spaces,* J. of Algebra 60 (1979), 384-422.

[50] H. E. A. Campbell, *Upper Triangular Invariants,* Canad. Math. Bull. 28 (1985), 243-248.

[51] H. E. A. Campbell, A. V. Geramita, I. P. Hughes, R.J. Shank, and D.L. Wehlau, *Non-Cohen-Macaulay Vector Invariants and a Noether Bound for a Gorenstein Ring of Invariants,* Canad. Math. Bull. 42 (1999), 155-161.

[52] H. E. A. Campbell, J. C. Harris, and D. L. Wehlau, *On Rings of Invariants of Non-modular Abelian Groups,* Bull. Austral. Math. Soc. 60 (1999), 509-520.

[53] H. E. A. Campbell and I. P. Hughes, *2-Dimensional Invariants of $GL_2(\mathbb{F}_p)$ and some of its Subgroups over the Field $\mathbb{F}_p$,* J. of Pure and Appl. Algebra 112 (1996), 1-12.

[54] H. E. A. Campbell and I. P. Hughes, *Vector Invariants of $U_2(\mathbb{F}_p)$: A Proof of a Conjecture of David Richman,* Adv. in Math. 126 (1997), 1-20.

[55] H. E. A. Campbell, I. P. Hughes, G. Kemper, R. J. Shank, and D. L. Wehlau, *Depth of Modular Invariant Rings,* Transformation Groups 5 (2000), 21-34.

[56] H. E. A. Campbell, I. Hughes, F. Pappalardi, and P. S. Selik, *On the Ring of Invariants of $\mathbb{F}^*_{2^n}$,* Comment. Math. Helv. 66 (1991), 322-331.

[57] H. E. A. Campbell, I. P. Hughes, and R. D. Pollack, *Vector Invariants of Symmetric Groups,* Canad. Math. Bull. 33 (1990), 391-397.

[58] H. E. A. Campbell, I. P. Hughes, and R. D. Pollack, *Rings of Invariants and p-Sylow Subgroups,* Canad. Math. Bull. 34 (1991), 42-47.

[59] H. E. A. Campbell, I. P. Hughes, R. J. Shank, and D. L. Wehlau, *Bases for Rings of Coinvariants,* Transformation Groups 1 (1996), 307-336.

[60] D. P. Carlisle, P. Eccles, S. Hilditch, N. Ray, L. Schwartz, G. Walker, and R. M. W. Wood, *Modular Representations of GL(n, p), splitting $\Sigma(CP^\infty\times\cdots\times CP^\infty)$ and the β-family as framed hypersurfaces,* Math. Zeit. 189 (1985), 239-261.

[61] D. P. Carlislc and P. Kropholler, *Rational Invariants of Certain Orthogonal and Unitary Groups,* Proc. London Math. Soc. 24 (1992), 57-60.

[62] D. P. Carlisle and P. Kropholler, *Modular Invariants of Finite Symplectic Groups,* Preprint, Manchester 1992.

[63] D. P. Carlisle and P. Kropholler, *Invariants of some Finite Classical Groups over $GF(2)$,* Preprint, Manchester 1993.

[64] G. Carlsson, *G. B. Segal's Burnside Ring Conjecture for $(\mathbb{Z}/2)^k$,* Topology 21 (1982), 329-332.

[65] H. Cartan, *Sur l'Iteration des Opérations de Steenrod,* Comment. Math. Helv. 29 (1955), 40-58.

[66] H. Cartan, *Algèbres d'Eilenberg-Mac Lane et Homotopie,* Séminaire Henri Cartan, Ecole Normale Supérieure Paris, 7e année 1954/55, Secrétariat Mathématique, Paris 1956, W. A. Benjamin, New York 1967.

[67] H. Cartan, *Quotient d'un Éspace Analytique par un Groupe d'Automorphismes,* in: *Algebraic Geometry and Topology, A Symposium in Honor of S. Lefschetz,* ed. by R. H. Fox, D. C. Spencer and A. W. Tucker, pp 90-102. Princeton Univ. Press, Princeton 1957.

[68] H. Cartan and S. Eilenberg, *Homological Algebra,* Princeton Univ. Press, Princeton 1956.

[69] R. W. Carter, *Finite Groups of Lie Type, Conjugacy Classes and Complex Characters,* John Wiley & Sons, Ltd., Chichester 1985.

[70] C. Chevalley, *Invariants of Finite Groups Generated by Reflections,* Amer. J. of Math. 67 (1955), 778-782.

[71] L. Chiang and Y.-C. Hung, *The Invariants of Orthogonal Group Actions,* Bull. Austral. Math. Soc. 48 (1993), 313-319.

[72] A. Chin, *The Cohomology Rings of Some p-Groups,* Publ. Research Inst. for Math. Sci. Kyoto University 31 (1995), 1031-1044.

[73] A. Clark and J. Ewing, *The Realization of Polynomial Algebras as Cohomology Rings,* Pacific J. of Mathematics 50 (1974), 425-434.

[74] A. M. Cohen, *Finite Complex Reflection Groups,* Ann. Scient. Éc. Norm. Sup.4e série t. 9 (1976), 379-436.

[75] I. S. Cohen and A. Seidenberg, *Prime Ideals and Integral Dependence,* Bull. of the Amer. Math. Soc. 52 (1946), 252-261.

[76] S. D. Cohen, *Rational Functions Invariant under an Orthogonal Group,* Bull. London Math. Soc. 22 (1990), 217-221.

[77] J. H. Conway, R. T. Curtis, S. P. Norton, R. A. Parker, and R. A. Wilson, *Atlas of Finite Groups,* Clarendon Press, Oxford 1985.

[78] J. H. Conway and N. J. A. Sloane, *Sphere Packings, Lattices, and Groups,* second edition, Springer-Verlag, Heidelberg, Berlin, 1993.

[79] H. S. M. Coxeter, *Discrete Groups Generated by Reflections,* Annals of Math. 35 (1934), 588-621.

[80] H. S. M. Coxeter and W. O. J. Moser, *Generators and Relations for Discrete Groups,* Springer-Verlag, Heidelberg, Berlin 1957.

[81] C. W. Curtis and I. Reiner, *Representation Theory of Finite Groups and Associative Algebras,* Interscience Publishers, New York 1962.

[82] M. Demazure, *Invariants Symétriques Entiers des Groupes de Weyl et Torsion,* Inventiones Math. 21 (1973), 287-301.

[83] H. Derksen, *Computation of Invariants for Reductive Groups,* Adv. in Math. 141 (1999), 366-384.

[84] L. E. Dickson, *On Finite Algebras,* Nachr. Akad. Wiss. Göttingen (1905), 358-393.

[85] L. E. Dickson, *A Fundamental System of Invariants of the General Modular Linear Group with a Solution of the Form Problem,* Trans. of the Amer. Math. Soc. 12 (1911), 75-98.

[86] L. E. Dickson, *Binary Modular Groups and Their Invariants,* Amer. J. of Math. 33 (1911), 175-192.

[87] L. E. Dickson, *Linear Groups,* Dover Publications Inc., New York 1958.

[88] L. E. Dickson, *The Collected Mathematical Papers of Leonard Eugene Dickson,* 6 vols, ed. by A. A. Albert, Chelsea Pub. Co., The Bronx, New York 1975.

[89] J. Dieudonné and J. B. Carrell, *Invariant Theory, Old and New,* Academic Press, New York 1971.

[90] J. Dixmier, *Sur les Invariants du Groupe Symétrique dans Certain Représentations,* J. of Algebra 103 (1986), 184-192.

[91] J. Dixmier, *Sur les Invariants du Groupe Symétrique dans Certaines Représentations II,* in: *Topics in Invariant Theory,* ed. by Marie-Paule Malliavin, pp. 1-34. Lecture Notes in Math. 1478, Springer-Verlag, Heidelberg, Berlin 1991.

[92] M. Domokos and P. Hegedús, *Noether's Bound for Polynomial Invariants of Finite Groups,* Archiv der Math. 74 (2000), 161-167.

[93] S. Doty and G. Walker, *The Composition Factors of* $\mathbb{F}_p[x_1, x_2, x_3]$ *as* $GL(3, \mathbb{F}_p)$*-module,* J. of Algebra 147, (1992), 411-441.

[94] A. Dress, *On Finite Groups Generated by Pseudoreflections,* J. of Algebra 11 (1969), 1-5.

[95] M.-L. Dubriel-Jacotin, *Etude Algébrique des Transformations de Reynolds,* in: *Colloque d'Algèbre Supérieure, Bruxelles 1956,* pp. 9-27. CBRM Louvain 1957.

[96] J. Duflot, P. S. Landweber, and R. E. Stong, *On a Problem of Adams on* $H^*(BG;\mathbb{Z}/p)$, in: *Algebraic Topology, Göttingen 1984,* ed. by L. Smith, Lecture Notes in Math. 1172, pp. 73-79. Springer-Verlag, Heidelberg, Berlin 1985.

[97] M.-J. Dumas (the later M.-J. Bertin), *Sous-Anneaux d'Invariants d'Anneaux de Polynomes,* C. R. Acad. Sci. Paris 260 (31 mai 1965), 5655-5658.

[98] W. G. Dwyer, H. Miller, and C. W. Wilkerson, *Homotopy Uniqueness of* BG, private communication, Chicago 1986.

[99] W. G. Dwyer, H. Miller, and C. W. Wilkerson, *Homotopy Uniqueness of Classifying Spaces,* Topology 31 (1992), 29-45.

[100] W. G. Dwyer and C. W. Wilkerson, *Smith Theory and the Functor* $\mathbf{T}$, Comment. Math. Helv. 66 (1991), 1-17.

[101] W. G. Dwyer and C. W. Wilkerson, *Kähler Differentials, the* **T**-*Functor, and a Theorem of Steinberg,* Trans. of the Amer. Math. Soc. 350 (1998), 4919-4930.

[102] W. L. Edge, *The Klein Group in Three Dimensions,* Acta Mathematica 79 (1947), 153-223.

[103] S. Eilenberg and N. Steenrod, *Foundations of Algebraic Topology,* Princeton. Univ. Press, Princeton, N.J. 1952.

[104] D. Eisenbud, *Commutative Algebra,* Graduate Texts in Math. 150, Springer-Verlag, Heidelberg, Berlin 1995.

[105] G. Ellingsrud and T. Skjelbred, *Profondeur d'Anneaux d'Invariants en Caractéristique p,* Comp. Math. 41 (1980), 233-244.

[106] S. Endo and T. Miyata, *Invariants of Finite Abelian Groups,* J. Math. Soc. Japan 25-1 (1973), 7-26.

[107] D. Engelmann, *Polynominvarianten einer Darstellung der symmetrischen Gruppe,* Diplomarbeit, University of Göttingen 1995.

[108] D. Engelmann, *Rings of Invariants,* Preprint, Humboldt University, Berlin 1996.

[109] L. Evens, *The Cohomology of Groups,* Clarendon Press, Oxford 1991.

[110] W. Feit, *Characters of Finite Groups,* Benjamin, New York 1967.

[111] W. Feit, *The Representation Theory of Finite Groups,* North-Holland Mathematical Library 25, North-Holland Pub. Co., Amsterdam, New York, Oxford 1982.

[112] M. M. Feldstein, *Invariants of the Linear Group Modulo p^k,* Trans. of the Amer. Math. Soc. 25 (1923), 223-238.

[113] D. Ferrand, *Suites regulière et Intersections Complet,* C.R.A.S Paris 264 (1967), 427-428.

[114] M. Feshbach, *The Image of the Trace in the Ring of Invariants,* Preprint, University of Minnesota, Minneapolis 1981.

[115] M. Feshbach, *The Mod 2 Cohomology Rings of the Symmetric Groups,* Topology, to appear.

[116] C. S. Fisher, *The Death of a Mathematical Theory, a Study in the Sociology of Knowledge,* Archive for History of Exact Sciences 3 (1966/7), 137-159.

[117] L. Flatto, *Invariants of Finite Reflection Groups,* L'Enseign. de Math. 24 (1978), 235-292.

[118] P. Fleischmann, *On the Ring of Vector Invariants for the Symmetric Group,* Preprint, Institute for Experimental Mathematics, Essen 1996.

[119] P. Fleischmann, *Relative Trace Ideals and Cohen-Macaulay Quotients of Modular Invariant Rings,* in: *Proceedings of the Euroconference on Computational Methods for Representations of Groups and Algebras,* ed. by P. Dräxler, G. O. Michler and C. M. Ringel, pp. 211-233. Progress in Math. 173, Birkhäuser, Basel 1999.

[120] P. Fleischmann, *The Noether Bound in Invariant Theory of Finite Groups,* Adv. in Math. 156 (2000), 23-32.

[121] P. Fleischmann and W. Lempken, *On Generators of Modular Invariant Rings of Finite Groups,* Bull. London Math. Soc. 29 (1997), 585-591.

[122] J. Fogarty, *On Noether's Bound for Polynomial Invariants of Finite Groups,* Electronic Research Announcements of the AMS 7 (2001), 5-7.

[123] R. M. Fossum, *The Divisor Class Group of a Krull Domain,* Ergebnisse der Math. und ihrer Grenzgebiete 74, Springer-Verlag, Heidelberg, Berlin 1973.

[124] R. M. Fossum and P. A. Griffith, *Complete Local Factorial Rings which are not Cohen-Macaulay in Characteristic p,* Ann. Scient. Éc. Norm. Sup. 4e série, t. 8 (1975), 189-200.

[125] P. Freyd, *Abelian Categories: An Introduction to the Theory of Functors,* Harper and Row, New York 1964.

[126] P. Gabriel, *Des Catégories Abéliennes,* Bull. Soc. Math. de France 90 (1962), 323-428.

[127] A. M. Garsia, *Combinatorial Methods in the Theory of Cohen-Macaulay Rings,* Adv. in Math. 38 (1980), 229-266.

[128] A. M. Garsia and M. Haiman, *Orbit Harmonics and Graded Representations,* UCSD Lecture Notes 1991-1992.

[129] A. M. Garsia and D. Stanton, *Group Actions on Stanley-Reisner Rings and Invariants of Permutation Groups,* Adv. in Math. 51 (1984), 107-201.

[130] I. M. Gessel, *Generating Functions and Generalized Dedekind Sums,* Electron. J. Combin 4 N o. 2 (1997).

[131] S. Glaz, *Fixed Rings of Coherent Regular Rings,* Comm. in Algebra 20 (1992), 2635-2651.

[132] O. E. Glenn, *Modular Invariant Processes,* Bull. of the Amer. Math. Soc. 21 (1914-15), 167-173.

[133] D. J. Glover, *A Study of Certain Modular Representations,* J. of Algebra 51 (1978), 425-475.

[134] M. Göbel, *Computing Bases for Permutation Invariant Polynomials,* J. of Symbolic Computation 19 (1995), 285-291.

[135] M. Göbel, *Computing Bases for Permutation-Invariant Polynomials,* Doktorarbeit, University of Tübingen, Shaker Verlag, Aachen 1996.

[136] M. Göbel, *On the Number of Special Permutation-Invariant Orbits and Terms,* Appl. Algebra Engrg. Comm. Comput. 8 (1997), 505-509.

[137] M. Göbel, *Fast Rewriting of Symmetric Polynomials,* in: *Rewriting Techniques and Applications, 10th Intl. Conf.,* ed. by P. Narendran and M. Rusinowitch, pp. 371-381. RTA'99, Vol. 1631 of LNCS, Trento, Springer-Verlag, Heidelberg, Berlin 1999.

[138] M. Göbel, *Rewriting Techniques and Degree Bounds for Higher Order Symmetric Polynomials,* Appl. Algebra Engrg. Comm. Comput. 9 (1999), 559-573.

[139] M. Göbel and H. Kredel, *Reduction of G-Invariant Polynomials for Arbitrary Permutation Groups G - The Non-Commutative Case,* preprint, Berkeley 1997.

[140] P. Goerss, L. Smith, and S. Zarati, *Sur les A-Algèbres Instabiles,* in: *Algebraic Topology, Barcelona 1986,* ed. J. Aguadé and R. Kane, pp. 148-161. Lecture Notes in Math. 1296, Springer-Verlag, Heidelberg, Berlin 1987.

[141] R. Goodman and N. R. Wallach, *Representations and Invariants of the Classical Groups,* Encyclopedia of Mathematics and its Applications, Camb. Univ. Press, Cambridge 1998.

[142] P. Gordan, *Beweis, dass jede Covariante und Invariante einer binären Form eine ganze Function mit numerischen Coefficienten einer endlichen Anzahl solcher Formen ist,* Journal für die Reine und Angewandte Mathematik 69 (1868), 323-354.

[143] P. Gordan, *Ueber endliche Gruppen linearer Transformationen einer Veränderlichen,* Math. Annalen 12 (1877), 23-46.

[144] P. Gordan, *Ueber die typische Darstellung der ternären biquadratischen Form* $f = x_1^3x_2 + x_2^3x_3 + x_3^3x_1$, Math. Annalen 17 (1880), 359-378.

[145] P. Gordan, *Dr. Paul Gordan's Vorlesungen über Invariantentheorie, Erster Band, Determinanten,* ed. by G. Kerschensteiner, B. G. Teubner Verlag, Leipzig 1885.

[146] P. Gordan, *Dr. Paul Gordan's Vorlesungen über Invariantentheorie, Zweiter Band, Binäre Formen,* ed. by G. Kerschensteiner, B. G. Teubner Verlag, Leipzig 1887.

[147] N. L. Gordeev, *Finite Linear Groups Whose Algebras of Invariants Are Complete Intersections,* Math. USSR Izvestiya 28 (1987), 335-379.

[148] N. L. Gordeev, *Coranks of Elements of Linear Groups and the Complexity of Algebraic Invariants,* Leningrad Math. J. 2 (1991), 245-267.

[149] D. Gorenstein, *Finite Groups,* Harper & Row, Publishers, New York, Evanston, London 1968.

[150] J. H. Grace and A. Young, *The Algebra of Invariants,* Camb. Univ. Press, Cambridge 1903.

[151] A. Grothendieck, *Sur quelques Pointes d'Algèbre Homologique,* Tohoku Math. J. 9 (1957), 119-221.

[152] A. Grothendieck, *Séminaire de Geometrie Algébrique,* SGA 2, I.H.E.S., Bois de Marie, 1964.

[153] L. C. Grove and C. T. Benson, *Finite Reflection Groups,* second edition, Graduate Texts in Math. 99, Springer-Verlag, Heidelberg, Berlin 1985.

[154] L. C. Grove and J. M. McShane, *Polynomial Invariants of Finite Groups,* Algebras, Groups and Geometries 10 (1993), 1-12.

[155] W. J. Haboush, *Reductive Groups Are Geometrically Reductive,* Annals of Math. 102 (1975), 67-83.

[156] M. Hall Jr., *The Theory of Groups,* The Macmillan Co., New York NY 1959.

[157] J. Hartmann, *Polynomial Tensor Exterior Invariants of Finite Groups,* Diplomarbeit, University of Göttingen 1999.

[158] J. Hartmann, *Transvection Free Groups and Invariants of Polynomial Tensor Exterior Algebras,* Transformation Groups 6 (2001), 157-164.

[159] B. A. Hedman, *An Earlier Date for "Cramer's Rule",* Historia Mathematica 26 (1999), 365-368.

[160] H.-W. Henn, *Finiteness Propeties of Injective Resolutions of Certain Unstable Modules over the Steenrod Algebra and Applications,* Math. Ann. 291 (1991), 191-203.

[161] G. Hermann, *Der Frage der endliche vielen Schritte in der Theorie der Polynomideale,* Math. Ann 95 (1926), 736-738.

[162] A. E. Heydtmann, *Generating Invariant Rings of Finite Groups,* Diplomarbeit, University of Saarbrücken 1996.

[163] D. Hilbert, *Über die Theorie der Algebraischen Formen,* Math. Annalen 36 (1890), 473-534.

[164] D. Hilbert, *Über die vollen Invariantensysteme,* Math. Annalen 42 (1893), 313-373.

[165] D. Hilbert, *Gesammelte Abhandlungen,* 3 vols., Springer-Verlag, Heidelberg, Berlin 1932/1933/1935.

[166] D. Hilbert, *Hilbert's Invariant Theory Papers,* translated by M. Ackerman, commented by R. Hermann, *Lie Groups: History, Frontiers and Applications, Volume VIII*, Math. Sci. Press, Brookline, Mass. 1978.

[167] D. Hilbert, *Theory of Algebraic Invariants,* translated by Reinhard C. Laubenbacher, Camb. Univ. Press, Cambridge 1994.

[168] H. Hiller, *Geometry of Coxeter Groups,* Pitman Books Ltd., London 1982.

[169] H. Hiller and L. Smith, *On the Realization and Classification of Cyclic Extensions of Polynomial Algebras over the Steenrod Algebra,* Proc. of the Amer. Math. Soc. 100 (1987), 731–738.

[170] J. W. P. Hirschfeld, *Projective Geometries over Finite Fields,* Clarendon Press, Oxford 1979.

[171] F. Hirzebruch and D. Zagier, *The Atiyah-Singer Theorem and Elementary Number Theory,* Mathematics Lecture Series, Publish or Perish Inc., Boston 1974.

[172] M. Hochster and J. A. Eagon, *Cohen-Macaulay Rings, Invariant Theory, and the Generic Perfection of Determinantal Loci,* Amer. J. of Math. 93 (1971), 1020–1058.

[173] R. Howe, *The Classical Groups and Invariants of Binary Forms,* in: *The mathematical heritage of Hermann Weyl (Durham NC 1987),* pp. 133-166. Proc. Symp. Pure Math. 48 (1988), Amer. Math. Soc., Providence RI 1988.

[174] S.-J. Hu and M.-C. Kang, *Efficient Generation of Rings of Invariants,* J. of Algebra 180 (1996), 341–363.

[175] W. C. Huffman, *Polynomial Invariants of Finite Linear Groups of Degree Two,* Canad. J. Math. 32-2 (1980), 317–330.

[176] W. C. Huffman and N. J. A. Sloane, *Most Primitive Groups Have Messy Invariants,* Adv. in Math. 32 (1979), 118-127.

[177] I. Hughes and G. Kemper, *Symmetric Powers of Modular Representations, Hilbert Series and Degree Bounds,* Comm. in Algebra 28 (2000), 2059–2088.

[178] J. E. Humphreys, *Reflection Groups and Coxeter Groups,* Camb. Univ. Press, Cambridge 1990.

[179] B. Huppert, *Endliche Gruppen I,* Die Grundlehren der mathematischen Wissenschaften Bd 134, Springer-Verlag, Heidelberg, Berlin 1967.

[180] H. C. Hutchins, *Examples of Commutative Rings,* Polygonal Publishing House, Passaic, NJ 1981.

[181] S. Iyengar, *Complete Intersections and the Jacobian Criterion, private communication,* University of Sheffield, and Göttingen University, 2001.

[182] N. Jacobson, *Lectures in Abstract Algebra III, Theory of Fields and Galois Theory,* (third corrected printing), GTM 32, Springer-Verlag, Heidelberg, Berlin, New York, 1980.

[183] G. James and M. Liebeck, *Representations and Characters of Groups,* Cambridge Mathematical Textbooks, Camb. Univ. Press, Cambridge 1993.

[184] J. P. Jans, *Rings and Homology,* Holt, Reinhart and Winston, New York, 1964.

[185] V. G. Kac, *Root Systems, Representations of Quivers, and Invariant Theory,* in: *Invariant Theory (Montecatini 1982),* pp. 74–108 Lecture Notes in Math. 996, Springer-Verlag, Heidelberg, Berlin 1983.

[186] V. G. Kac, *Torsion in Cohomology of Compact Lie Groups and Chow Rings of Reductive Algebraic Groups,* Inventiones Math. 80 (1985), 69–79.

[187] V. G. Kac and D. H. Peterson, *Generalized Invariants of Groups Generated by Reflections,* in: *Geometry Today, Roma 1984,* pp. 231-249. Progress in Mathematics 60, Birkhäuser Verlag, Boston 1985.

[188] V. G. Kac and K. Watanabe, *Finite Linear Groups Whose Ring of Invariants is a Complete Intersection,* Bull. of the Amer. Math. Soc. 6 (1982), 221-223.

[189] R. Kane, *Poincaré Duality and the Ring of Coinvariants,* Canad. Math. Bull 37 (1994), 82-88.

[190] M.-C. Kang, *Picard Groups of Some Rings of Invariants,* J. of Algebra 58 (1979), 455-461.

[191] W. Kantor, *Subgroups of Classical Groups Generated by Long Root Elements,* Trans. of the Amer. Math. Soc. 248 (1979), 347-379.

[192] D. K. Karaguezian and P. Symonds, *The Module Structure of a Group Action on a Polynomial Ring,* J. of Algebra 218 (1999), 672-392.

[193] N. E. Kechagias (ed.), *Interactions Between Algebraic Topology and Invariant Theory,* Proceedings of the Summer School held at Ioannina, Greece, 26 - 30 June 2000, Dept. of Math. Univ. of Ioannina 2000.

[194] G. Kemper, *Calculating Invariant Rings of Finite Groups over Arbitrary Fields,* J. of Symbolic Computation 21 (1996), 351-356.

[195] G. Kemper, *Lower Degree Bounds for Modular Invariants and a Question of I. Hughes,* Transformation Groups 3 (1998), 135-144.

[196] G. Kemper, *On the Cohen-Macaulay Property of Modular Invariant Rings,* J. of Algebra 215 (1999), 330-351.

[197] G. Kemper, *An Algorithm to Calculate Optimal Homogeneous Systems of Parameters,* J. of Symb. Comp. 27 (1999), 171-184.

[198] G. Kemper, *The Depth of Invariant Rings and Cohomology,* with an appendix by K. Magaard, J. of Algebra (to appear).

[199] G. Kemper and G. Malle, *The Finite Irreducible Linear Groups with Polynomial Ring of Invariants,* Transformation Groups 2 (1997), 57-89.

[200] G. Kempf, *The Hochster-Roberts Theorem of Invariant Theory,* Michigan Math. J. 26 (1979), 19-32.

[201] A. Kerber, *Algebraic Combinatorics via Finite Group Actions,* Wissenschaftsverlag, Mannheim 1991.

[202] M. Kervaire, *Fractions Rationnelles Invariantes,* in: *Séminaire Bourbaki 26ᵉ année (1973/4),* pp. 170-189. Lecture Notes in Math. 431, Springer-Verlag, Heidelberg, Berlin 1975.

[203] M. Kervaire and T. Vust, *Fractions Rationnelles Invariantes par un Groupe Fini: Quelques Exemples,* in: *Algebraische Transformationsgruppen und Invariantentheorie,* ed. by H. Kraft, P. Slodowy, and T. A. Springer, pp. 157-179. DMV Seminar 13, Birkhäuser Verlag, Basel 1989.

[204] N. Killius, *Göbel's Bound for Permutation Groups,* unpublished notes, Evanston 1995.

[205] N. Killius, *Some Modular Invariant Theory of Finite Groups with Particular Emphasis on the Cyclic Group,* Diplomarbeit, University of Göttingen 1996.

[206] P. Kleidman and M.Liebeck, *The Subgroup Structure of the Finite Classical Groups,* London Math. Soc. Lecture Series 129, Camb. Univ. Press, Cambridge, Cambridge 1990.

[207] F. Klein, *Vorlesung über das Ikosaeder und die Auflösung der Gleichung vom fünften Grad,* B. G. Teubner Verlag, Leipzig 1884.

[208] F. Klein, *Vorlesungen über das Ikosaeder,* Birkhäuser, Basel,Boston,Berlin, B.G. Teubner Verlag, Stuttgart, Leipzig 1993.

[209] B. Kostant, *The McKay Correspondence, the Coxeter Element and Representation Theory,* in: *The mathematical heritage of Élie Cartan (Lyon 1984),* pp. 209-255. Astérique, Numéro Hors Série, Soc. Math. de France, Paris 1985.

[210] H. Kraft, *Geometrische Methoden in der Invariantentheorie,* Aspects of Math., Vieweg Verlag, Braunschweig 1984.

[211] D. Krause, *Die Noethersche Gradgrenze in der Invariantentheorie,* Diplomarbeit, University of Göttingen 1999.

[212] W. Krull, *Beiträge zur Arithmetik kommutativer Integritätsbereiche III,* Mathematische Zeitschrift 42 (1937), 745-766.

[213] N. J. Kuhn, *Generic Representations of the Finite General Linear Groups and the Steenrod Algebra I,* Amer. J. of Math. 116 (1994), 327-360.

[214] N. J. Kuhn, *Generic Representation Theory of the Finite General Linear Groups and the Steenrod Algebra III,* K-Theory 9 (1995), 273-303.

[215] K. Kuhnigk, *Der Transferhomomorphismus in der modularen Invariantentheorie,* Diplomarbeit, University of Göttingen 1998.

[216] K. Kuhnigk and L. Smith, *Feshbach's Transfer Theorem and Applications,* Preprint, AG-Invariantentheorie 1998.

[217] S. Kühnlein, *Torsion Classes in Cohomology and Galois Representations,* J. of Number Theory 70 (1998), 184-190.

[218] S. Kühnlein, *Some Families of Finite Groups and Their Rings of Invariants,* Acta Arithmetica 91 (1999), 133-146.

[219] J. P. S. Kung and G.-C. Rota, *The Invariant Theory of Binary Forms,* Bull. of the Amer. Math. Soc. (New Series) 10-1 (1984), 27-85.

[220] S. P. Lam, *Unstable Algebras over the Steenrod Algebra,* in: *Algebraic Topology, Aarhus 1982,* pp. 374-392 Lecture Notes in Math. 1051, Springer-Verlag, Berlin, New York 1984.

[221] P. S. Landweber, *Dickson Invariants and Prime Ideals Invariant under Steenrod Operations,* seminar talk, Princeton 1984.

[222] P. S. Landweber, *Dickson Invariants and Steenrod Operations on Cohomology Rings,* talk at the AMS Summer Conference on Algebraic Topology held at Minneapolis, 1984.

[223] P. S. Landweber, *Primary Decomposition for $\mathscr{P}^*$-Invariant Ideals,* Rutgers University, March 2001.

[224] P. S. Landweber and R. E. Stong, *The Depth of Rings of Invariants over Finite Fields,* in: *Number Theory, NY 1984/5,* pp. 259-274. Lecture Notes in Math. 1240, Springer-Verlag, Berlin, New York 1987.

[225] P. S. Landweber and R. E. Stong, *Invariants of Finite Groups,* unpublished correspondence, 1981-1993.

[226] S. Lang, *Cyclotomic Fields,* Graduate Texts in Math. 59, Springer-Verlag, Heidelberg, Berlin, New York 1978.

[227] J. Lannes, *Sur les espaces fonctionnels dont la source est le classifiant d'un p-groupe abélien élementaire,* Publ. Math. de l'I.H.E.S. 75 (1992), 135-244.

[228] J. Lannes and S. Zarati, *Foncteurs dérivées de la déstabilitazion,* Mathematische Zeitschrift 194 (1987), 25-59.

[229] J. Lannes and S. Zarati, *Théorie de Smith Algébrique et Classification des $H^*V - \mathcal{U}$-Injectifs,* Bull. Soc. Math. de France 123 (1995), 189-224.

[230] I. G. Macdonald, *Symmetric Functions and Hall Polynomials,* Clarendon Press, Oxford 1995.

[231] S. Mac Lane, *Modular Fields (I),* Duke J. of Math 5 (1939), 372 -393.

[232] S. Mac Lane, *Homology,* Springer-Verlag, Heidelberg, Berlin 1963.

[233] F. J. Macwilliams, *Orthogonal Matrices over Finite Fields,* Amer. Math. Monthly 76 (1969), 152-164.

[234] Z. Mahmud and L. Smith, *Maps Between Rings of Invariants,* Math. Proc. Camb. Phil. Soc. 120 (1996), 103-116.

[235] C. L. Mallows and N. J. A. Sloane, *On the Invariants of a Linear Group of Order 336,* Math. Proc. Camb. Phil. Soc. 74 (1973), 435-440.

[236] H. Maschke, *Aufstellung des vollen Formensystems einer quaternären Gruppe von 51840 linearen Substitutionen,* Math. Annalen 33 (1889), 317-344.

[237] H. Maschke, *The Invariants of a Group of $2 \cdot 168$ Linear Quaternary Substitutions,* in: *Mathematical Papers read at the International Math. Congress Chicago 1893,* pp. 173-186. Macmillan, New York 1896.

[238] H. Maschke, *Über den arithmetischen Charakter der Coeffizienten der Substitutionen endlicher linearer Substitutionsgruppen,* Math. Annalen 50 (1898), 482-498.

[239] W. S. Massey, *Unstable Modules over the Steenrod Algebra Are Noetherian,* Dittoed Notes, Yale University September 1965.

[240] W. S. Massey and F. P. Peterson, *The Cohomology Structure of Certain Fibre Spaces I,* Topology 4 (1965), 47-65.

[241] W. S. Massey and F. P. Peterson, *The mod 2 Cohomology Structure of Certain Fibre Spaces,* Memoirs of the Amer. Math. Soc. 74, Amer. Math. Soc., Providence RI 1967.

[242] J. Matijevic, *Three Local Conditions on a Graded Ring,* Trans. of the Amer. Math. Soc. 205 (1975), 275-284.

[243] H. Matsumura, *Commutative Ring Theory,* translated by M. Ried, Cambridge Studies in Advanced Mathematics 8, Camb. Univ. Press, Cambridge 1986.

[244] J. M. McShane, *Computation of Polynomial Invariants of Finite Groups,* Ph.D. Thesis, University of Arizona 1992.

[245] D. M. Meyer, *Injective Objects in Categories of Unstable K-modules,* Bonner Math. Schriften. 316 (1999).

[246] D. M. Meyer and L. Smith, *Lannes T-Functor and Noetherean Finiteness,* Preprint, AG-Invariantentheorie, 2001.

[247] F. Meyer, *Bericht über den gegenwärtigen Stand der Invariantentheorie,* Jahresbericht der DMV 1 (1892), 79-292.

[248] T. Meyer, *Untersuchungen der Poincaré Reihe im modularen Fall,* Diplomarbeit, University of Göttingen 1998.

[249] R. J. Milgram and S. B. Priddy, *Invariant Theory and* $H^*(\mathrm{GL}_n(\mathbf{F}_p);\mathbf{F}_p)$, J. of Pure and Appl. Algebra 44 (1987), 291-302.

[250] H. R. Miller, *The Sullivan Conjecture on Maps from Classifying Spaces,* Annals of Math. 120 (1984), 39-87.

[251] H. R. Miller, *The Sullivan Conjecture on Maps from Classifying Spaces. Corrigendum,* Annals of Math. 121 (1985), 605-609.

[252] J. W. Milnor, *The Steenrod Algebra and Its Dual,* Annals of Math. (2) 67 (1958), 150-171.

[253] J. W. Milnor and J. C. Moore, *The Structure of Hopf Algebras,* Ann. of Math. 81 (1965), 211-265.

[254] H. Minkowski, *Über positive quadratische Formen,* Journal für die Reine und Angewandte Mathematik 99 (1886), 1-9.

[255] H. Minkowski, *Zur Theorie der positiven quadratischen Formen,* Journal für die Reine und Angewandte Mathematik 101 (1887), 196-202.

[256] H. Minkowski, *Gesammelte Abhandlungen,* 2 vols., ed. by D. Hilbert, B. G. Teubner Verlag, Leipzig, Berlin 1911.

[257] S. A. Mitchell, *Finite Complexes with* $\mathcal{A}(n)$ *Free Cohomology,* Topology 24 (1985), 227-248.

[258] T. Miyata, *Invariants of Certain Groups I,* Nagoya Math. J. 41 (1971), 69-73.

[259] T. Molien, *Über die Invarianten der linearen Substitutionsgruppen,* Sitzungsber. König. Preuss. Akad. Wiss. (1897), 1152-1156.

[260] K. G. Monks, *Nilpotence in the Steenrod Algebra,* Bol. Soc. Math. Mex. (2) 37 (1992) 401-416.

[261] K. G. Monks, *Nilpotence and Torsion in the Steenrod Algebra,* Trans. of the Amer. Math. Soc. 333 (1992) 903-912.

[262] K. G. Monks, *The Nilpotence Height of* P^t_s, Proc. of the Amer. Math. Soc. 124 (1996), 1297-1303.

[263] H. Morikawa, *On the Invariants of Finite Nilpotent Groups,* Osaka Math. J. 10 (1958), 53-56.

[264] H. Mùi, *Modular Invariant Theory and Cohomology Algebras of Symmetric Groups,* J. of Fac. Sci. Univ. Tokyo Sect. 1A Math. 22 (1975), 319-369.

[265] H. Mùi, *Cohomology Operations Derived from Modular Invariants,* Mathematische Zeitschrift 193 (1986), 151-163.

[266] D. Mumford, *Geometric Invariant Theory,* Ergebnisse der Math. und ihrer Grenzgebiete, Neue Folge 34, Academic Press, Inc., New York, Springer-Verlag, Berlin, New York 1965.

[267] D. Mumford, *Hilbert's Fourteenth Problem - The Finite Generation of Subrings such as Rings of Invariants,* in: *Mathematical Developments Arising from Hilbert Problems, Proceedings of the Symposium held at Nothern Illinois University, Dekalb IL 1974,* pp. 431-444. ed. by F. E. Browder, Proc. of Symp. Pure Math 28, Amer. Math. Soc., Providence, RI 1976.

[268] M. Nagata, *On the* 14*-th problem of Hilbert,* Amer. J. of Math. 81 (1959), 766-772.

[269] M. Nagata, *Local Rings,* Interscience, John Wiley & Sons, New York 1962.

[270] H. Nakajima, *Invariants of Reflection Groups in Positive Characteristics,* Proc. Japan Acad. Ser. A 55 (1979), 219-221.

[271] H. Nakajima, *Invariants of Finite Groups Generated by Pseudoreflections in Positive Characteristic,* Tsukuba J. of Math. 3 (1979), 109-122.

[272] H. Nakajima, *On some Invariant Subrings of Polynomial Rings in Positive Characteristics,* Proc. 13th Symp. on Ring Theory, Okayama (1980), 91-107.

[273] H. Nakajima, *Invariants of Finite Abelian Groups Generated by Transvections,* Tokyo J. Math. 3 (1980), 201-214.

[274] H. Nakajima, *Modular Representations of p-Groups with Regular Rings of Invariants,* Proc. Japan Acad. Ser. A 56 (1980), 469-473.

[275] H. Nakajima, *Relative Invariants of Finite Groups,* J. of Algebra 79 (1982), 218-234.

[276] H. Nakajima, *Modular Representations of Abelian Groups with Regular Rings of Invariants,* Nagoya Math. J. 86 (1982), 229-248.

[277] H. Nakajima, *Regular Rings of Invariants of Unipotent Groups,* J. of Algebra 85 (1983), 253-286.

[278] H. Nakajima, *Rings of Invariants of Finite Groups Which are Hypersurfaces,* J. of Algebra 80 (1983), 279-294.

[279] H. Nakajima, *Quotient Singularities Which are Complete Intersections,* Manuscripta Math. 48 (1984), 163-187.

[280] H. Nakajima, *Quotient Complete Intersections of Affine Spaces by Finite Linear Groups,* Nagoya Math. J. 98 (1985), 1-36.

[281] H. Nakajima, *Rings of Invariants of Finite Groups Which are Hypersurfaces II,* Adv. in Math. 65 (1987), 39-64.

[282] H. Nakajima and K. Watanabe, *The Classification of Quotient Singularities Which are Complete Intersections,* in: *Complete Intersections, Acireale 1983,* ed. by S. Greco and R. Strano, pp. 102-120. Lecture Notes in Math. 1092, Springer-Verlag, Berlin, New York 1984.

[283] A. Neeman, *The Connection between a Conjecture of Carlisle and Kropholler, now a Theorem of Benson and Crawley-Boevey, and Grothendieck's Riemann-Roch and Duality Theorems,* Comment. Math. Helv. 70 (1995), 339-349.

[284] E. Netto, *Vorlesung über Algebra in zwei Bände,* Teubner Verlag, Leipzig, 1900.

[285] F. Neumann, M. D. Neusel, and L. Smith, *Rings of Generalized and Stable Invariants of Finite Groups,* J. of Algebra 182 (1996), 85-122.

[286] F. Neumann, M. D. Neusel, and L. Smith, *Rings of Generalized Invariants and Classifying Spaces of Compact Lie groups,* in: *Higher Homotopy Structures in Topology and Mathematical Physics,* ed. by J. McCleary, pp. 267-285. Contemp. Math. 227, Amer. Math. Soc., Providence 1999.

[287] M. D. Neusel, *Cubic Invariants of $GL(2, \mathbb{Z})$,* Comm. in Algebra 24 (1996), 247-257.

[288] M. D. Neusel, *Invariants of some abelian p-Groups in Characteristic p,* Proc. of the Amer. Math. Soc. 125 (1997), 1921-1931.

[289] M. D. Neusel, *Integral Extensions of Unstable Algebras over the Steenrod Algebra,* Forum Mathematicum 12 (2000), 155-166.

[290] M. D. Neusel, *Inverse Invariant Theory and Steenrod Operations,* Memoirs of the Amer. Math. Soc. N o. 692 Vol.146, AMS, Providence RI 2000.

[291] M. D. Neusel, *Localizations over the Steenrod Algebra. The lost Chapter,* Math. Zeit. 235 (2000), 353-378.

[292] M. D. Neusel, *The Transfer in the Invariant Theory of Modular Permutation Representations,* Pacific J. of Mathematics 199 (2001), 121-136.

[293] M. D. Neusel, *The Lasker-Noether Theorem in the Category $\mathcal{U}(H^*)$,* J. of Pure and Appl. Algebra 163 (2001), 221-233.

[294] M. D. Neusel, *Unstable Cohen-Macaulay Algebras,* Math. Research Letters 8 (2001), 347-360.

[295] M. D. Neusel, *The Transfer in the Invariant Theory of Modular Permutation Representations II,* Can. Math. Bull. (to appear).

[296] M. D. Neusel, *The Lasker-Noether Theorem in the Category $\mathcal{U}(H^*)$. Applications,* preprint, AG-Invariantentheorie and Yale University, New Haven 2000.

[297] M. D. Neusel, *A Selection from the 1001 Examples : The Atlas of Polynomial Invariants of Finite Groups,* in preparation.

[298] M. D. Neusel, *$\mathscr{P}^*$-Commutative Algebra,* in preparation.

[299] M. D. Neusel and L. Smith, *Polynomial Invariants of Groups associated to Configurations of Hyperplanes over Finite Fields,* J. of Pure and Appl. Algebra 122 (1997), 87-106.

[300] M. D. Neusel and L. Smith, *The Lasker-Noether Theorem for $\mathscr{P}^*$-invariant Ideals,* Forum Mathematicum 10 (1998), 1-18.

[301] E. Noether, *Der Endlichkeitssatz der Invarianten endlicher Gruppen,* Math. Annalen 77 (1916), 89-92.

[302] E. Noether, *Idealtheorie in Ringenbereichen,* Math. Annalen 83 (1921), 24-66.

[303] E. Noether, *Der Endlichkeitssatz der Invarianten endlicher linearer Gruppen der Characteristik p,* Nachr. v. d. Ges. d. Wiss. zu Göttingen (1926), 28-35.

[304] E. Noether, *Gesammelte Abhandlungen,* ed. by N. Jacobson, Springer-Verlag, Heidelberg, Berlin 1983.

[305] M. Noether, *Paul Gordan (Nachruf),* Math. Annalen 75 (1914), 1-45.

[306] D. G. Northcott, *An Introduction to Homological Algebra,* Camb. Univ. Press, Cambridge 1960.

[307] J. E. Olson, *A Combinatorial Problem on Finite Abelian Groups I,* J. of Number Theory 1 (1969), 8-10.

[308] J. E. Olson, *A Combinatorial Problem on Finite Abelian Groups II,* J. of Number Theory 1 (1969), 195-199.

[309] P. J. Olver, *Classical Invariant Theory,* London Math. Soc. Student Texts 44, Camb. Univ. Press, Cambridge 1999.

[310] O. Ore, *On a Special Class of Polynomials,* Trans. of the Amer. Math. Soc. 35 (1933), 559-584.

[311] P. Orlik and L. Solomon, *The Hessian Map in the Invariant Theory of Reflection Groups,* Nagoya Math. J. 109 (1988), 1-21.

[312] P. Orlik and L. Solomon, *Discriminants in the Invariant Theory of Reflection Groups,* Nagoya Math. J. 109 (1988), 23-45.

[313] P. Orlik and H. Terao, *Arrangements of Hyperplanes,* Springer-Verlag, Heidelberg, Berlin 1992.

[314] M. Oura, *Molien Series Related to Certain Finite Unitary Reflection Groups,* Kyushu Journal of Mathematics 50 (1996), 297-310.

[315] S. Papadima, *Rigidity Properties of Compact Lie Groups Modulo Maximal Tori,* Math. Annalen 275 (1986), 637-652.

[316] K. V. H. Parshall, *The One-Hundredth Anniversay of the Death of Invariant Theory?,* The Mathematical Intelligencer 12, N o. 4 (1990), 10-16.

[317] K. Pommerening, *Invariants of Unipotent Groups, a Survey,* in: *Invariant Theory,* ed. by S. S. Koh, pp. 8-17. Lecture Notes in Math. 1278, Springer-Verlag, Heidelberg, Berlin 1987.

[318] V. L. Popov, *On Hilbert's Theorem on Invariants,* Soviet Math. Dokl. 20 (1979), 1318-1322.

[319] V. L. Popov, *Syzygies in the Theory of Invariants,* Math. USSR Izvestiya 47 (1983), 544-622.

[320] D. L. Rector, *Noetherian Cohomology Rings and Finite Loop Spaces with Torsion,* J. of Pure and Appl. Algebra 32 (1984), 191-217.

[321] V. Reiner, *Free Modules of Relative Invariants of Finite Groups,* Studies in Applied Math. 81 (1989), 181-184.

[322] V. Reiner, *On Göbel's Bound for Invariants of Permutation Groups,* Arch. der Math. 65 (1995), 475-480.

[323] V. Reiner and L. Smith, *Systems of Parameters for Rings of Invariants,* Preprint, Göttingen, 1996.

[324] P. Revoy, *Anneau des invariants du groupe alterné,* Bull. Sci. Math. de France 106 (1982), 427-431.

[325] D. R. Richman, *On Vector Invariants over Finite Fields,* Adv. in Math. 81 (1990), 30-65.

[326] D. R. Richman, *Invariants of Finite Groups over Fields of Characteristic p,* Adv. in Math. 124 (1996), 25-48.

[327] D. R. Richman, *Explicit Generators of the Invariants of Finite Groups,* Adv. in Math. 124 (1996), 49-76.

[328] O. Riemenschneider, *Die Invarianten der endlichen Untergruppen von* GL(2, $\mathbb{C}$), Mathematische Zeitschrift 153 (1977), 37-50.

[329] J. Riordan, *An Introduction to Combinatorial Analysis,* John Wiley & Sons, New York 1958.

[330] D. Rotillon and K. Watanabe, *Invariant Subrings of* $\mathbb{C}[x, y, z]$ *Which Are Complete Intersections,* Manuscripta Math. 39 (1982), 339-357.

[331] J. J. Rotman, *An Introduction to the Theory of Groups,* Fourth Edition, Springer-Verlag, Heidelberg, Berlin, New York 1995.

[332] B. J. Schmid, *Generating Invariants of Finite Groups,* C. R. Acad. Sci. Paris t. 308, Série I (1989), 1-6.

[333] B. J. Schmid, *Finite Groups and Invariant Theory,* in: *Topics in Invariant Theory,* Séminaire d'Algèbre P. Dubriel et M.-P. Malliavin, Paris 1989/1990, pp. 35-66. Lecture Notes in Math. 1478, Springer-Verlag, Berlin, New York 1991.

[334] C. Schulte, *G-Dimension und G-Multiplizität für Graduierte Algebren,* Diplomarbeit, University of Göttingen 2000.

[335] I. Schur, *Vorlesung über Invariantentheorie,* Bearbeitet und herausgegeben von Helmut Grunsky, Springer-Verlag, Heidelberg, Berlin 1968.

[336] L. Schwartz, *Unstable Modules over the Steenrod Algebra and Sullivan's Fixed Point Set Conjecture,* University of Chicago Press, Chicago 1994.

[337] G. W. Schwarz and D. L. Wehlau, *Invariants of Four Subspaces,* Ann. Inst. Fourier, Grenoble 48 (1998), 101-131.

[338] J. Segal, *Pointwise Conjugacy and Invariant Theory,* Doktorarbeit, University of Göttingen 1999.

[339] A. Seidenberg, *Differential Ideals in Rings of Finitely Generated Type,* Amer. J. of Math. 89 (1967), 22-42.

[340] H. Seifert and W. Threlfall, *Lehrbuch der Topologie,* Chelsea Publ. Co, New York NY 1947.

[341] J.-P. Serre, *Cohomologie modulo 2 des complexes d'Eilenberg-Mac Lane,* Comment. Math. Helv. 27 (1953), 198-232.

[342] J.-P. Serre, Corps Loceaux, Herman, Paris 1962.

[343] J.-P. Serre, *Sur la dimension cohomologique des groupes profinis,* Topology 3 (1965), 413-420.

[344] J.-P. Serre, *Algèbre locale–multiplicités,* Lecture Notes in Math. 11, Springer-Verlag, Heidelberg, Berlin 1965.

[345] J.-P. Serre, *Représentations linéaires des groupes finis,* Hermann, Paris 1967.

[346] J.-P. Serre, *Groupes finis d'automorphismes d'anneaux locaux réguliers,* Colloq. d'Alg. Éc. Norm. Sup. de Jeunes Filles, Paris, 8-01—8-11, 1967.

[347] R. J. Shank, *S.A.G.B.I. Bases for Rings of Formal Modular Seminvariants,* Comment. Math. Helv. 73 (1998), 548-565.

[348] R. J. Shank and D. L. Wehlau, *On the Depth of the Invariants of the Symmetric Power Representations of* $SL_2(\mathbb{F}_p)$, J. of Algebra 218 (1999), 642-653.

[349] R. J. Shank and D. L. Wehlau, *The Transfer in Modular Invariant Theory,* J. of Pure and Appl. Algebra 142 (1999), 63-77.

[350] G. C. Shephard, *Unitary Groups Generated by Reflections,* Canad. J. of Math. 5 (1953), 364-383.

[351] G. C. Shephard, *Abstract Definitions for Reflection Groups,* Canad. J. of Math. 9 (1957), 373-376.

[352] G. C. Shephard and J. A. Todd, *Finite Unitary Reflection Groups,* Canad. J. of Math. 6 (1954), 274-304.

[353] K. Shoda, *Über die Invarianten endlicher Gruppen linearer Substitutionen im Körper der Charakteristik p,* Jap. J. of Math. 17 (1940), 109-115.

[354] W. M. Singer, *The Iterated Transfer in Homological Algebra,* Math. Zeit. 202 (1989), 493-523.

[355] B. Singh, *Invariants of Finite Groups Acting on Local Unique Factorization Domains,* J. of the Indian Math. Soc. 34 (1970), 31-38.

[356] N. J. A. Sloane, *Error Correcting Codes and Invariant Theory,* Amer. Math. Monthly 84 (1977), 82-107.

[357] L. Smith, *Homological Algebra and the Eilenberg-Moore Spectral Sequence,* Trans. of the Amer. Math. Soc. 129 (1967), 58-93.

[358] L. Smith, *On the Finite Generation of* $MU^*(X)$, J. of Math. and Mechanics 18 (1969), 1017-1024.

[359] L. Smith, *The e-Invariant and Finite Coverings,* Indiana Math. Journal 24 (1975), 659-675.

[360] L. Smith, *The Nonrealizibility of Modular Rings of Polynomial Invariants by the Cohomology of a Topological Space,* Proc. of the Amer. Math. Soc. 86 (1982), 339-340.

[361] L. Smith, *A Note on the Realization of Complete Intersection Algebras by the Cohomology of a Space,* Quartely. J. of Math. Oxford 33 (1982), 379-384.

[362] L. Smith, *Polynomial and Related Algebras as Cohomology Rings,* Pub. Mat. Uni. Autonome Barcelona 26 (1982), 161-197.

[363] L. Smith, *On the Realization and Classification of Symmetric Algebras as Cohomology Rings,* Proc. of the Amer. Math. Soc. 87 (1983), 144-148.

[364] L. Smith, *A Remark on Realizing Dickson Covariants as Cohomology Rings,* Quartely. J. of Math. Oxford (2) 36 (1984), 113-115.

[365] L. Smith, *On the Invariant Theory of Finite Pseudo Reflection Groups,* Archiv der Math. 44 (1985), 225–228.

[366] L. Smith, *Realizing Nonmodular Polynomial Algebras as the Cohomology of Spaces of Finite Type Fibered over* $\times B\mathbb{U}(d)$, Pacific J. of Mathematics 127 (1987), 361–387.

[367] L. Smith, *Finite Loop Spaces with Maximal Tori Have Finite Weyl Groups,* Proc. of the Amer. Math. Soc. 119 (1993), 299–302.

[368] L. Smith, *The e-Invariant and Finite Coverings II,* Trans. of the Amer. Math. Soc. 347 (1995), 5009–5021.

[369] L. Smith, *Some Rings of Invariants that are Cohen-Macaulay,* Canad. Math. Bull. 39 (1996), 238–240.

[370] L. Smith, *Noether's Bound in the Invariant Theory of Finite Groups,* Arch. der Math. 66 (1996), 89–92.

[371] L. Smith, $\mathcal{P}^*$*-Invariant Ideals in Rings of Invariants,* Forum Mathematicum 8 (1996), 319–342.

[372] L. Smith, *Polynomial Invariants of Finite Groups,* A. K. Peters, Ltd., Wellesley, MA, 1995, second printing 1997.

[373] L. Smith, *Polynomial Invariants of Finite Groups: A Survey of Recent Developments,* Bull. of the Amer. Math. Soc. 34 (1997), 211–250.

[374] L. Smith, *Linear Algebra,* third edition, Springer-Verlag, Heidelberg, Berlin, New York 1998.

[375] L. Smith, *Homological Codimension of Modular Rings of Invariants and the Koszul Complex,* J. of Math of Kyoto Univ. 38 (1998), 727–747.

[376] L. Smith, *Putting the Squeeze on the Noether Gap: The Case of the Alternating Groups* A_n, Math. Annalen 315 (1999), 503–510.

[377] L. Smith, *The Ring of Invariants of* $O(3, \mathbb{F}_q)$, Finite Fields and Their Applications 5 (1999), 96–101.

[378] L. Smith, *Modular Vector Invariants of Cyclic Permutation Groups,* Canad. Math. Bull. 42 (1999) 125–128.

[379] L. Smith, *Noether's Bound in the Invariant Theory of Finite Groups and Iterated Wreath Products of Symmetric Groups,* Quartely. J. of Math. Oxford 51 (2000), 93–105.

[380] L. Smith, *Lannes's* $\mathbf{T}$*-Functor and the Invariants of Pointwise Stabilizers,* Forum Math. 12 (2000), 461–476.

[381] L. Smith, *On a Theorem of Barbara Schmid,* Proc. of the Amer. Math. Soc. 128 (2000), 2199–2201.

[382] L. Smith, *Cohomology Automorphisms over Galois Fields,* AG-Invariantentheorie, Preprint 2000.

[383] L. Smith, *An Algebraic Introduction to the Steenrod Algebra,* Course Notes from the Summer School in: *Interactions Between Invariant Theory and Algebraic Topology,* ed. by N. E. Kechagias, pp 49–64. Ionnina, Greece, 26 - 30 June 2000.

[384] L. Smith, *Invariants and Coinvariants of Pseudoreflection Groups, Jacobian Determinants and Steenrod Operations,* Edinb. J. of Math (to appear).

[385] L. Smith, *Invariants of 2×2 Matrices over Finite Fields,* J. of Finite Fields (to appear).

[386] L. Smith, *Invariant Theory and the Koszul Complex: Representations of $\mathbb{Z}/p$ in Characteristic p,* J. of Math of Kyoto Univ. (to appear).

[387] L. Smith and R. E. Stong, *On the Invariant Theory of Finite Groups: Orbit Polynomials, Chern Classes and Splitting Principles,* J. of Algebra 110 (1987), 134-157.

[388] L. Smith and R. E. Stong, *Invariants of Finite Groups,* unpublished correspondence, 1979-1993.

[389] L. Smith and R. M. Switzer, *Realizability and Nonrealizability of Dickson Algebras as Cohomology Rings,* Proc. of the Amer. Math. Soc. 89 (1983), 303-313.

[390] L. Smith and R. M. Switzer, *Polynomial Algebras over the Steenrod Algebra,* Proc. Edinburgh Math. Soc. 27 (1984), 11-19.

[391] W. Smoke, *Dimension and Multiplicity for Graded Algebras,* J. of Algebra 21 (1972), 149-173.

[392] L. Solomon, *Invariants of Finite Reflection Groups,* Nagoya Math. J. 22 (1963), 57-64.

[393] L. Solomon, *Invariants of Euclidean Reflection Groups,* Trans. of the Amer. Math. Soc. 113 (1964) 274-286.

[394] L. Solomon, *Partition Identities and Invariants of Finite Groups,* J. of Combinatorial Theory (A) 23 (1977), 148-175.

[395] T. Sperlich, *Automorphisms of $P[V]_G$,* Proc. of the Amer. Math. Soc. 120 (1994) 5-11.

[396] T. A. Springer, *Invariant Theory,* Lecture Notes in Math. 585, Springer-Verlag, Heidelberg, Berlin 1977.

[397] T. A. Springer, *On the Invariant Theory of SU_2,* Indag. Math. 42 (1980), 339-345.

[398] R. P. Stanley, *Linear Homogeneous Diophantine Equations and Magic Labelings of Graphs,* Duke J. of Math. 40 (1973), 607-632.

[399] R. P. Stanley, *Relative Invariants of Finite Groups generated by Pseudoreflections,* J. of Algebra 49 (1977), 134-148.

[400] R. P. Stanley, *Hilbert Functions of Graded Algebras,* Adv. in Math. 28 (1978), 57-83.

[401] R. P. Stanley, *Invariants of Finite Groups and Their Applications to Combinatorics,* Bull. of the Amer. Math. Soc. (3) 1 (1979), 475-511.

[402] R. P. Stanley, *Enumerative Combinatorics,* 2 vols., Cambridge Studies in Advanced Mathematics 49 and 62, Camb. Univ. Press, Cambridge, 1997/1999.

[403] R. P. Stanley, *Combinatorics and Commutative Algebra,* (second edition), Prog. in Math 41, Birkhauser Verlag, Basel, 1996.

[404] N. E. Steenrod, *Polynomial Algebras over the Algebra of Cohomology Operations,* in: *H-Spaces, Actes réunion Neuchâtel (Suisse), Auot 1970,* pp. 85–99. Lecture Notes in Math. 196, Springer-Verlag, Berlin, New York 1971.

[405] N. E. Steenrod and D. B. A. Epstein, *Cohomology Operations,* Annals of Math. Studies 50, Princeton University Press, Princeton 1962.

[406] R. Steinberg, *Invariants of Finite Reflection Groups,* Canad. J. of Math. 12 (1960), 616–618.

[407] R. Steinberg, *Differential Equations Invariant under Finite Reflection Groups,* Trans. of the Amer. Math. Soc. 112 (1964), 392–400.

[408] R. Steinberg, *On Dickson's Theorem on Invariants,* J. of Fac. Sci. Univ. Tokyo Sect. 1A Math. 34 (1987), 699–707.

[409] R. Steinberg, *Collected Papers,* Amer. Math. Soc., Providence RI 1997.

[410] E. Stiefel, *Kristallographische Bestimmung der Charaktere der geschlossenen Lie'schen Gruppen,* Comment. Math. Helv. 17 (1944-45), 160–200.

[411] E. Stiefel, *Über eine Beziehung zwischen geschlossenen Lie'schen Gruppen und diskontinuierlichen Bewegungsgruppen euklidischer Räume und ihre Anwendung auf die Aufzählung der einfachen Lie'schen Gruppen,* Comment. Math. Helv. 14 (1941-42), 350–380.

[412] M. Stillman and H. Tsai, *Using SAGBI Bases to Compute Invariants,* J. of Pure and Appl. Algebra 139 (1999), 285–302.

[413] R. E. Stong, *Polynomial Invariants of Finite Groups and Algebraic Topology,* unpublished correspondence, 1982–1993.

[414] C. W. Strom, *On Complete Systems under Certain Finite Groups,* Bull. of the Amer. Math. Soc. 37 (1931), 570–574.

[415] C. W. Strom, *A Complete System for the Simple Group* G_{60}^{6}, Bull. of the Amer. Math. Soc. 43 (1937), 438–440.

[416] C. W. Strom, *Complete Systems of Invariants of the Cyclic Group of Equal Order and Degree,* Proc. of the Iowa Academy of Science 55 (1948), 287–290.

[417] B. Sturmfels, *Gröbner Bases and Invariant Theory,* Advances in Math. 76 (1989), 245–259.

[418] B. Sturmfels, *Algorithms in Invariant Theory,* Springer-Verlag, Heidelberg, Berlin, Vienna 1993.

[419] R. G. Swan, *Noether's Problem in Galois Theory,* in: *Emmy Noether in Bryn Mawr,* ed. by J. Sally and B. Srinivasan, pp. 21–40. Springer-Verlag, Heidelberg, Berlin 1983.

[420] P. Symonds, *Group Actions on Polynomial and Power Series Rings,* Pac. J. of Math. 195 (2000), 225–230.

[421] K.-I. Tahara, *On the Finite Subgroups of* $\mathrm{GL}(3, \mathbb{Z})$, Nagoya Math. J. 41 (1971), 169–209.

[422] T. Tamagawa, *Dickson's Theorem on Invariants of Finite General Linear Groups,* private communication, Yale University, New Haven 1990.

[423] K. Tanabe, *On Molien Series of the Invariant Ring Associated to the Unitary Reflection Group $G(m,p,n)$,* Kyushu Journal of Mathematics 50 (1996), 437–458.

[424] J. Tate, *Homology of Noetherian Rings and Local Rings,* Ill. J. of Math. 1 (1957), 14–27.

[425] J. G. Thompson, *Invariants of Finite Groups,* J. of Algebra 69 (1981), 143–145.

[426] H. Toda, *Cohomology mod 3 of the Classifying Space BF_4 of the Exceptional Group F_4,* J. of Math. Kyoto Univ. 13 (1973), 97–115.

[427] W. N. Traves, *Differential Operators and Nakai's Conjecture,* Ph.D. Thesis, University of Toronto, 1998.

[428] J. Tschichold, *Die Neue Typographie,* Maro Verlag, Augsburg 1996.

[429] J. Tschichold, *Erfreuliche Drucksachen durch gute Typographie,* Maro Verlag, Augsburg 1996.

[430] J. S. Turner, *A Fundamental System of Invariants of a Modular Group of Transformations,* Trans. of the Amer. Math. Soc. 24 (1922), 129–134.

[431] W. V. Vasconcelos, *Ideals Generated by R-Sequences,* J. of Algebra 6 (1967), 309–316.

[432] V. E. Voskresenskii, *On Two-Dimensional Algebraic Tori,* in: AMS Translations 73, pp. 190-195. Amer. Math. Soc. , Providence RI 1968.

[433] B. L. van der Waerden, *Modern Algebra I, II,* translated by F. Blum, F. Ungar Pub., New York NY 1949.

[434] A. Wagner, *Determination of the Finite Primitive Reflection Groups over an Arbitrary Field of Characteristic not 2, I,* Geom. Dedicata 9 (1980), 239–253.

[435] A. Wagner, *Determination of the Finite Primitive Reflection Groups over an Arbitrary Field of Characteristic not 2, II,* Geom. Dedicata 10 (1980), 183–189.

[436] A. Wagner, *Determination of the Finite Primitive Reflection Groups over an Arbitrary Field of Characteristic not 2, III,* Geom. Dedicata 10 (1980), 475–523.

[437] G. Walker and R. M. H. Wood, *The Nilpotence Height of* Sq^{2^n}, Proc. of the Amer. Math. Soc. 124 (1996), 1291–1295.

[438] G. Walker and R. M. H. Wood, *The Nilpotence Height of* $\mathcal{P}^{p^n}$, Math. Proc. Camb. Phil. Soc. 123 (1998), 85–93.

[439] G. Walker and R. M. W. Wood, *The Hit Problem for the Steinberg Module over the Steenrod Algebra,* Preprint, Manchester, 1999.

[440] G. Walker and R. M. W. Wood, *First Occurrence Polynomials over $\mathbb{F}_2$ by Steenrod Operations,* Preprint 2000/21, Manchester 2000.

[441] C. T. C. Wall, *Generators and Relations in the Steenrod Algebra,* Annals of Math 72 (1960), 429–444.

[442] N. R. Wallach, *Invariant Differential Operators on a Reductive Lie Algebra and Weyl Group Representations,* J. of the Amer. Math. Soc. 6 (1993), 779–816.

[443] K. Watanabe, *Certain Invariant Subrings are Gorenstein I,* Osaka J. of Math. 11 (1974), 1–8.

[444] K. Watanabe, *Certain Invariant Subrings are Gorenstein II,* Osaka J. of Math. 11 (1974), 379–388.

[445] K. Watanabe, *Invariant Subrings which are Complete Intersections I (Invariant Subrings of Finite Abelian Groups),* Nagoya Math. J. 77 (1980), 89–98.

[446] H. Weber, *Lehrbuch der Algebra,* 2 vols., 2[te] Auflage, Vieweg Verlag, Braunschweig 1899.

[447] D. L. Wehlau, *A Proof of the Popov Conjecture for Tori,* Proc. of the Amer. Math. Soc. 114 (2) (1992), 839–845.

[448] C. Weibel, *Introduction to Homological Algebra,* Camb. Univ. Press, Cambridge 1994.

[449] M. Weisfeld, *On Derivations in Division Rings,* Pacific J. of Mathematics 10 (1960), 335–343.

[450] R. Weitzenböck, *Invariantcnthcoric,* P. Noordhoff, Groningen 1923.

[451] H. Weyl, *The Classical Groups,* second edition, Princeton Univ. Press, Princeton 1946.

[452] H. Wielandt, *Finite Permutation Groups,* Academic Press, New York 1964.

[453] C. W. Wilkerson, *Some Polynomial Algebras over the Steenrod Algebra,* Bull. of the Amer. Math. Soc. 79 (1973), 1274–1276.

[454] C. W. Wilkerson, *Classifying Spaces, Steenrod Operations and Algebraic Closure,* Topology 16 (1977), 227–237.

[455] C. W. Wilkerson, *Integral Closure of Unstable Steenrod Algebra Actions,* J. of Pure and Appl. Algebra 13 (1978), 49–55.

[456] C. W. Wilkerson, *A Primer on the Dickson Invariants,* Proc. of the Northwestern Homotopy Theory Conference, Contemp. Math. 19, Amer. Math. Soc. 1983, 421–434.

[457] E. Witt, *Theorie der quadratischen Formen in beliebigen Körpern,* Journal für die Reine und Angewandte Mathematik 176 (1937), 31–44.

[458] R. M. W. Wood, *Differential Operators and the Steenrod Algebra,* Proc. London Math. Soc. (3) 75 (1997), 194–220.

[459] R. M. W. Wood, *Problems in the Steenrod Algebra,* Bull. London Math. Soc. 30 (1998), 449–517.

[460] R. M. W. Wood, *Hit Problems and the Steenrod Algebra,* Course Notes from the Summer School *Interactions Between Invariant Theory and Algebraic Topology,* ed. by N. E. Kechagias, pp 65–103. Ionnina, Greece, 26 - 30 June 2000.

[461] Wu Wen-Tsün, *Sur les puissances de Steenrod,* Colloque de Topologie de Strasbourg, 1951, N o. IX, La Bibliothèque Nationale et Universitaire de Strasbourg 1952.

[462] D. B. Zagier, *Equivariant Pontrjagin Classes and Applications to Orbit Spaces,* Lecture Notes in Math. 290, Springer-Verlag, Berlin, New York 1972.

[463] A. E. Zalesskiĭ and V. N. Serežkin, *Linear Groups Generated by Transvections,* Izv. Akad. Nauk, SSSR, Ser. Mat. 40 (1976), 26-49, ibid. 44 (1980), 1279-1307, (English Translation: Math USSR Izv. 10 (1976), 25-46, ibid. 17 (1981), 477-503).

[464] S. Zarati, *Quelques propriétés du Foncteur* $\mathrm{Hom}_{\mathcal{U}_p}(-, H^*V)$, in: *Algebraic Topology, Göttingen 1984,* ed. by L. Smith, pp. 203-209. Lecture Notes in Math. 1172, Springer-Verlag, Heidelberg, Berlin 1985.

[465] O. Zariski and P. Samuel, *Commutative Algebra,* 2 vols., Graduate Texts in Math. 28, 29, Springer-Verlag, Heidelberg, Berlin 1975.

[466] H. J. Zassenhaus, *Theory of Groups,* second edition, Chelsea, NY 1958.

Typography

BLACK ART is what many have named the magic worked by a master printer when they have seen the results of their own feeble efforts transformed into print. Designing a layout and choosing type faces can indeed be mystifying subjects for one not versed in the intricacies of the printing profession. Yet, *"If typography is to make any sense at all then it must make visual and historical sense"* [42]. TEX and the computer have not changed this, they have only moved the burden of choices from the printer to us, the authors.

Invariant theory is a classical mathematical theme: It deserves a classical type face. In this case the choice fell to Garamond, originally attributed to Claude Garamond, who died in 1561: At present it is generally accepted to have been designed by Jean Jannon in 1615. This is a type face that has seen little use in mathemtical publications in recent years. The actual text face used in this book is URW Design Studio's URWGaramondT-Regular, which is a digital rendition of the classic Garamond typeface. The companion fonts of the Garamond family,[1] viz., *italic*, **bold**, and ***bold italic*** make their appearance where appropriate. The nominal 10 point type of this implementation is very small on the sort and has been set here at 11.5 point. The *slant* font used for the statements of theorems, propositions, etc., and the text of definitions, problems, etc., is produced from the roman by PostScript trickery. A companion sansserif font was not easy to find, and in this instance the face is URWImperialT in its four variations; regular, **bold**, *italic*, and ***bold italic***.

[1] It should be noted that the notion of a *type family* post dates the design of Garamond by many hundred years: bold face types became part ot the printers' repertoire only in the 19-th century, and it was not always the case that a type face had a *companion* italic face. In the past, type faces for the roman and italic of distinct provenance would be mixed when they complemented each other properly.

As an aid to the reader we have tried to be consistent in the usage of certain fonts for specific purposes.

We have used the bold face companion font **URWGaramond-RegularBold** to indicate a term that is being defined, whether in a definition or the running text. The italic font *URWGaramond-RegularOblique* has been used for emphasis. On a few occasions where nothing else seemed to work we resorted to ***URWGaramond-BoldOblique*** to emphasize our point.

The font **URWImperialTBold** has been used for matrices and linear transformations, as in, e.g., $\mathbf{T} : V \longrightarrow W$, as well as Noether normalizations, as in $\mathbf{N} = \mathbb{F}[h_1, \ldots, h_n] \subseteq A$, the Dickson algebra $\mathbf{D}(n)$, and the Dickson polynomials $\mathbf{d}_{n,i}$.

The font URWAltSchwabacherD has been used for special kinds of ideals, such as prime ideals, e.g., $\mathfrak{p}$, as well as for varieties, e.g., $\mathfrak{X}$, whether projective or affine, and now and again some older concepts, such as transvections, $\mathfrak{t}$, orbit sums, $\mathfrak{S}(x^K)$, rings of integers, $\mathfrak{D}$, in number fields, and the Hilbert ideal, $\mathfrak{h}$.

Particularly in connection with the Steenrod algebra we make heavy use of various script fonts of unknown origins, e.g., $\mathscr{P}^*$, $\mathscr{P}^i$, and $\boldsymbol{P}$, as well as the font *Zapf Chancery Medium Italic*, designed by Hermann Zapf in 1979 for the International Typographical Corporation.

The lowercase Greek letters are based on the data from the Greek character set created by Dr. A. V. Hershey while working at the U. S. National Bureau of Standards. The $\mathbb{B}$lack$\mathbb{B}$oard$\mathbb{B}$old$\mathbb{F}$ace font is derived from a digitalization of the font Caslon, designed by William Caslon in 1725, and manipulation of the PostScript source to turn it into an outline font. Making it available to TeX required considerable kerning.

As for the layout, the second author is responsible for that. It is based on what he learned as an apprentice printer many decades ago.

Notation

□ indicates the end of a proof, or that what has been stated will not be proved.

map(X, Y) the set of maps from the set X to the set Y
map$(X, Y)_*$ the set of graded maps from the graded set X to the graded set Y
$|-|$ number of element of $-$
$a \mid b$ a divides b
$a \cdots \widehat{b} \cdots c$ the symbol under the $\widehat{}$ is omitted

map(X, Y) is the set of all maps from X to Y
map$(X, Y)_*$ is the graded set of graded maps from the graded set X to the graded ste Y

$\mathbb{N} = \{1, 2, 3, \ldots, \}$, the natural numbers
$\mathbb{N}_0 = \{0, 1, 2, \ldots, \}$, the nonnegative integers
$\mathbb{Z} = \{, \ldots, -2, -1, 0, 1, 2, \ldots, \}$, the ring of integers
$\mathbb{Z}/m$ the integers modulo m, cyclic group of order m
$\mathbb{Z}_p$ denotes $\mathbb{Z}$ localized at the prime ideal (p)
$\mathbb{Z}_{\widehat{p}}$ denotes $\mathbb{Z}$ completed at the prime ideal (p)
$\mathbb{Q}$ field of rational numbers
$\mathbb{Q}_p$ field of fractions of $\mathbb{Z}_p$
$\mathbb{Q}_{\widehat{p}}$ field of fractions of $\mathbb{Z}_{\widehat{p}}$
$\mathbb{R}$ field of real numbers
$\mathbb{C}$ field of complex numbers
$\mathbb{F}_q$ field with q elements ($q = p^\nu$, $p \in \mathbb{N}$ a prime)
$\mathbb{F}^\times$ invertible elements in field $\mathbb{F}$
$\overline{\mathbb{F}}$ the algebraic closure of field $\mathbb{F}$
$\mathbb{PF}(n)$ projective space of $\mathbb{F}^n$, i.e., the set of 1-dimensional subspaces of $\mathbb{F}^n$
$\mathfrak{G}_d(V)$ the Grassmann variety of codimension d linear subspaces of V
$\mathfrak{G}(V)$ the Grassmann variety of all linear subspaces of V, i.e., $\bigcup \mathfrak{G}_d(V)$

$GL(n, \mathbb{F})$ the group of $n \times n$ invertible matrices over $\mathbb{F}$
$SL(n, \mathbb{F})$ subgroup of $GL(n, \mathbb{F})$ of matrices $\mathbf{A}$ with $\det \mathbf{A} = 1$
Σ_n the symmetric group of the n-element set

A_n the alternating subgroup of Σ_n
D_{2k} dihedral group of order $2k$
Q_8 quaternion group, of order 8
Q_{4k} generalized quaternion group of order $4k$
$|G : H|$ the index of the subgroup H in G
$\mathbb{F}(G)$ the group algebra of the group G over the field $\mathbb{F}$

$\mathbb{F}(G)$ the group ring of the group G over field $\mathbb{F}$
$s(G)$ the set of pseudorefelctions in G
$s_\Delta(G)$ the set of diagonalizable pseudoreflections in G
$s_{\not\Delta}(G)$ the set of nondiagonalizable pseudoreflections in G
$\mathcal{T}(H)$ the set of all transvections with hyperplane $H \subsetneq V$

V^* dual vector space of V, if $\mathbb{F}$ is the groundfield then
$V^* = \mathrm{Hom}_{\mathbb{F}}(V, \mathbb{F})$
$\mathbb{F}[V]$ algebra of homogeneous polynomial functions on the vector space V
$S(V)$ symmetric algebra on the vector space V
$E(V)$ exterior algebra on the vector space V
$T(V)$ tensor algebra on the vector space V
$H(V) = E(V) \otimes_{\mathbb{F}} \mathbb{F}[V]$ (the cohomology of V with coefficients in $\mathbb{F}$ when suitably graded)

$z^E = z_1^{e_1} \cdot z_2^{e_2} \cdots z_n^{e_n}$ a monomial in $z_1, \ldots, z_n$, E is called the exponent sequence
e_i ith elementary symmetric polynomial
$e(I)$ Ith polarized elementary symmetric polynomial

I^e if $A \subseteq B$ is an inclusion of rings and $I \subseteq A$ an ideal, then I^e is the extended ideal in B
J^c if $A \subseteq B$ is an inclusion of rings and $J \subseteq B$ an ideal, then $J^c = J \cap A$ is the contracted ideal in B
$\sqrt{I}$ the radical of the ideal I

$A//B$ if $\varphi : B \longrightarrow A$ then $\mathbb{F} \otimes_B A$
$\mathbb{FF}(A)$ field of fractions of A
$\mathrm{Aut}(A)$ the automorphisms of A
$\mathrm{Hom}_R(M, N)$ the R-homomorphisms from M to N

$\mathbf{I}$ identity matrix
$\det \mathbf{A}$ the determinant of the matrix $\mathbf{A}$
$\mathrm{t}(\varphi, x)$ the transvection defined by $\varphi \in V^*$ and $x \in V$
$\mathrm{tr}(\mathbf{A})$ the trace of the matrix $\mathbf{A}$
$\mathrm{Mat}_{m,n}(\mathbb{F})$ vector space of $m \times n$ matrices over $\mathbb{F}$
$\mathrm{Mat}^{\mathrm{sym}}_{n,n}(\mathbb{F})$ vector space of $n \times n$ symmetric matrices over $\mathbb{F}$
$\odot$ semitensor product bifunctor

$ht(I)$ the height of an ideal I
$dp(A)$ the depth of a ring A
$\mathrm{Ass}_A(M)$ the set of associated prime ideals of the A-module M
$Zero_A(M)$ the zero divisors in A on the A-module M
$Reg_A(M)$ the regular elements in A on the A-module M
$Spec(A)$ prime ideal spectrum of the algebra A
hom-codim(A) homological codimension of a ring A (an historically older term for the depth of A)
hom-dim(A) homological dimension of a ring A, i.e., an upper bound for the lengths of projective resolutions of A-modules

$\mathbb{F}[V]^G$ ring of invariants of G
$\mathfrak{h}(G)$ Hilbert ideal of G

$\mathbb{F}[V]^G_\chi$ module of χ-relative invariants of G

$\mathcal{P}^i$ ith Steenrod reduced power operation over a Galois field

Sq^i ith Steenrod squaring operation over the field $\mathbb{F}_2$

$\mathscr{P}^*$ Steenrod algebra over a Galois field

$\mathcal{U}$ category of unstable modules over $\mathscr{P}^*$

$\mathcal{U}_{\mathbb{F}_q[V]}$ category of unstable $\mathbb{F}_q[V] \otimes \mathscr{P}^*$-modules

$\mathcal{K}$ category of unstable algebras over $\mathscr{P}^*$

$\mathcal{K}_{fg}$ category of unstable finitely generated algebras over $\mathscr{P}^*$

$\mathrm{Proj}_{\mathscr{P}^*}(H)$ spectrum of homogeneous unstable prime ideals of the unstable algebra H

$\mathcal{J}$ Lam's $\mathcal{J}$-functor on ideals

$\mathsf{E}_{\mathbb{F}_q[V]\odot\mathscr{P}^*}(H)$ injective hull of H in $\mathcal{U}_{\mathbb{F}_q[V]}$

$\mathsf{J}(m)$ mth dual Brown-Gitler module in $\mathcal{U}$

$\mathsf{J}_{\mathbb{F}_q[V]\odot\mathscr{P}^*}(m)$ mth dual Brown-Gitler module in the category $\mathcal{U}_{\mathbb{F}_q[V]}$

$\sqrt[\mathscr{P}^*]{H}$ $\mathscr{P}^*$-inseparable closure of the unstable algebra H

Index

Selected Titles in This Series

(Continued from the front of this publication)